VOLUME FIVE HUNDRED AND SIX

METHODS IN ENZYMOLOGY

Imaging and Spectroscopic Analysis of Living Cells

Imaging Live Cells in Health and Disease

METHODS IN ENZYMOLOGY

VOLUME FIVE HUNDRED AND SIX

METHODS IN ENZYMOLOGY

Imaging and Spectroscopic Analysis of Living Cells

Imaging Live Cells in Health and Disease

EDITED BY

P. MICHAEL CONN
Divisions of Reproductive Sciences and Neuroscience (ONPRC)
Departments of Pharmacology and Physiology
Cell and Developmental Biology, and
Obstetrics and Gynecology (OHSU)
Beaverton, OR, USA

ELSEVIER

AMSTERDAM • BOSTON • HEIDELBERG • LONDON
NEW YORK • OXFORD • PARIS • SAN DIEGO
SAN FRANCISCO • SINGAPORE • SYDNEY • TOKYO
Academic Press is an imprint of Elsevier

Academic Press is an imprint of Elsevier
525 B Street, Suite 1900, San Diego, CA 92101-4495, USA
225 Wyman Street, Waltham, MA 02451, USA
32 Jamestown Road, London NW1 7BY, UK

First edition 2012

Notice
No responsibility is assumed by the publisher for any injury and/or damage to persons or property as a matter of products liability, negligence or otherwise, or from any use or operation of any methods, products, instructions or ideas contained in the material herein. Because of rapid advances in the medical sciences, in particular, independent verification of diagnoses and drug dosages should be made

For information on all Academic Press publications
visit our website at elsevierdirect.com

ISBN: 978-0-12-391856-7
ISSN: 0076-6879

Printed and bound in United States of America
12 13 14 15 10 9 8 7 6 5 4 3 2 1

CONTENTS

Contributors *xiii*
Preface *xxi*
Volumes in Series *xxiii*

Section I. Viruses, Microbes, and Parasites **1**

1. Microbial Cells Analysis by Atomic Force Microscopy **3**

David Alsteens

1. Introduction 4
2. Immobilization of Microbial Cells 6
3. Applications of AFM to Nanomicrobiology 11
4. Future Challenges 14

Acknowledgments 14
References 15

2. Understanding Parasite Transmission Through Imaging Approaches **19**

Mirko Singer and Freddy Frischknecht

1. Why Imaging Parasites? 20
2. *In Vivo* Models: "How Much Wrong Can Be Right?" 20
3. The Choice of the Appropriate Microscope 24
4. New Materials Provide the Basis for New Approaches to Dissect Cell Migration 25
5. Visualizing Cellular Functions in Motile Cells 27
6. Applying Imaging for Drug Discovery 29

Acknowledgments 31
References 31

3. Multiphoton Microscopy Applied for Real-Time Intravital Imaging of Bacterial Infections *In Vivo* **35**

Ferdinand X. Choong, Ruben M. Sandoval, Bruce A. Molitoris, and Agneta Richter-Dahlfors

1. Introduction 36
2. Selection of Reporters for a Complex Site of Infection 37
3. Setup of a Live Imaging Platform of Infection 41

4. Studying Tissue Responses in Real-Time During Infection 50
5. Postimaging Analysis 57
6. Extending Real-Time Imaging into Molecular Analysis 58
Acknowledgments 60
References 60

4. Analysis of Virus Entry and Cellular Membrane Dynamics by Single Particle Tracking 63

Helge Ewers and Mario Schelhaas

1. Introduction 64
2. Fluorescent Labeling of Viruses 66
3. Data Acquisition and Analysis 73
4. Conclusions and Outlook 78
Acknowledgments 79
References 79

5. Imaging of Live Malaria Blood Stage Parasites 81

Christof Grüring and Tobias Spielmann

1. Introduction 82
2. Imaging Malaria Blood Stages at Ambient Temperature 83
3. Long-Term 4D Imaging of *P. falciparum* Blood Stages 86
Acknowledgments 91
References 91

6. *Escherichia coli* K1 Invasion of Human Brain Microvascular Endothelial Cells 93

Lip Nam Loh and Theresa H. Ward

1. Introduction 94
2. HBMEC Cell Culture and Transfection 95
3. Making Fluorescent *E. coli* K1 that Retain Virulence 101
4. Live Imaging of *E. coli* K1-Infected HBMEC 107
Acknowledgments 110
References 111

Section II. Disease States 115

7. Real-Time Imaging After Cerebral Ischemia: Model Systems for Visualization of Inflammation and Neuronal Repair 117

Pierre Cordeau and Jasna Kriz

1. Introduction 118

2. Design and Generation of Dual Reporter Mouse Models for *In Vivo* Imaging of Brain Response to Ischemic Injury 119
3. Induction of Cerebral Ischemia 124
4. *In Vivo* Bioluminescence Imaging 125
5. Additional Tips for Successful *In Vivo* Bioluminescence Imaging 130
6. Summary 132
Acknowledgments 132
References 132

8. *In Vivo* Real-Time Visualization of Leukocytes and Intracellular Hydrogen Peroxide Levels During a Zebrafish Acute Inflammation Assay 135

Luke Pase, Cameron J. Nowell, and Graham J. Lieschke

1. Introduction 136
2. Visualizing Leukocyte Behavior During Acute Inflammation in Zebrafish 137
3. Things to Consider When Planning *In Vivo* Visualization of the Inflammatory Response 143
Acknowledgments 154
References 154

9. Imaging Protein Oligomerization in Neurodegeneration Using Bimolecular Fluorescence Complementation 157

Federico Herrera, Susana Gonçalves, and Tiago Fleming Outeiro

1. Introduction 158
2. Required Equipment and Materials 161
3. Protocol 162
Acknowledgements 172
References 172

10. 3D/4D Functional Imaging of Tumor-Associated Proteolysis: Impact of Microenvironment 175

Kamiar Moin, Mansoureh Sameni, Bernadette C. Victor, Jennifer M. Rothberg, Raymond R. Mattingly, and Bonnie F. Sloane

1. Introduction 176
2. Assays for Functional Imaging of Proteolysis 178
3. 3D/4D Models for Analysis of Biological Processes Linked to Proteolysis 184
4. Live-Cell Imaging of MAME Models: A Screening Tool for Drug Discovery 186
Acknowledgments 189
References 189

Section III. Other Techniques 195

11. Live Cell Imaging in Live Animals with Fluorescent Proteins 197

Robert M. Hoffman

1. Introduction 199
2. Imaging Angiogenesis *In Vivo* 201
3. Imaging Cell Trafficking 203
4. Methods 209
5. Technical Details 213
6. Imaging 216
7. Chamber Imaging Systems 218
8. Histological Techniques 219
9. Conclusions 220

References 221

12. Protein Activation Dynamics in Cells and Tumor Micro Arrays Assessed by Time Resolved Förster Resonance Energy Transfer 225

Véronique Calleja, Pierre Leboucher, and Banafshé Larijani

1. Introduction 226
2. Protein–Protein Interactions and Conformation Dynamics in Fixed and Live Cells 227
3. Automated High-Throughput Analysis of Protein Activation in Tumor Micro Arrays 240

Acknowledgments 245

References 245

13. Imaging Stem Cell Differentiation for Cell-Based Tissue Repair 247

Zhenghong Lee, James Dennis, Jean Welter, and Arnold Caplan

1. Introduction 248
2. Identifying Marker Genes Associated with Stem Cell Differentiation 250
3. Lenti-Viral Constructs for Event-Specific Reporter Gene Expression 251
4. Applications of the Reporter System in Tissue Repair and Regeneration 254
5. Concluding Remarks 261

Acknowledgments 261

References 261

14. Understanding the Initiation of B Cell Signaling Through Live Cell Imaging 265

Angel M. Davey, Wanli Liu, Hae Won Sohn, Joseph Brzostowski, and Susan K. Pierce

1. Introduction 266
2. Experiment Preparation 268

3. Image Acquisition and Analysis 281
Acknowledgments 288
References 289

15. A Quantitative Method for Measuring Phototoxicity of a Live Cell Imaging Microscope 291

Jean-Yves Tinevez, Joe Dragavon, Lamya Baba-Aissa, Pascal Roux, Emmanuelle Perret, Astrid Canivet, Vincent Galy, and Spencer Shorte

1. Introduction 292
2. Measuring Phototoxicity 293
3. Discussion 304
4. Conclusion 308
Acknowledgments 308
References 308

16. High Content Screening of Defined Chemical Libraries Using Normal and Glioma-Derived Neural Stem Cell Lines 311

Davide Danovi, Amos A. Folarin, Bart Baranowski, and Steven M. Pollard

1. Introduction 312
2. Protocol: Chemical Screening Using Human NS and GNS Cells 315
3. Protocol: Automated Quantitation of Cellular Responses 320
Acknowledgments 327
References 327

17. High-Throughput Screening in Primary Neurons 331

Punita Sharma, D.Michael Ando, Aaron Daub, Julia A. Kaye, and Steven Finkbeiner

1. Introduction 332
2. Sample Preparation 333
3. Image Acquisition 341
4. Data Analysis 346
5. Statistical Approaches to HCS and Multivariate Data 352
6. Future Directions 356
7. Concluding Remarks 357
Acknowledgments 358
References 358

18. Live Imaging Fluorescent Proteins in Early Mouse Embryos 361

Panagiotis Xenopoulos, Sonja Nowotschin, and Anna-Katerina Hadjantonakis

1. Introduction 362
2. Genetically Encoded FPs for Live Imaging Morphogenetic Events in the Early Mouse Embryo 363

3. Tools for Live Cell Imaging 372
4. Methodology 378
5. Conclusions 384
Acknowledgments 385
References 385

19. Methods for Three-Dimensional Analysis of Dendritic Spine Dynamics 391

Enni Bertling, Anastasia Ludwig, Mikko Koskinen, and Pirta Hotulainen

1. Introduction 392
2. Methods 393
3. Imaging Conditions and Procedures 395
4. Illustrative Experiments 397
5. Quantitative Methods for Analyzing Dendritic Spines 398
Acknowledgments 404
References 404

20. Imaging Cell Competition in *Drosophila* Imaginal Discs 407

Shizue Ohsawa, Kaoru Sugimura, Kyoko Takino, and Tatsushi Igaki

1. Introduction 407
2. Live Imaging of Cell Competition 409
References 412

21. Live Cell Imaging of the Oviduct 415

Sabine Kölle

1. Introduction 416
2. Preparation of the Oviduct 417
3. Live Cell Imaging: Qualitative Analysis 419
4. Live Cell Imaging: Quantitative Analysis 420
Acknowledgments 423
References 423

22. Computational Quantification of Fluorescent Leukocyte Numbers in Zebrafish Embryos 425

Felix Ellett and Graham J. Lieschke

1. Introduction 426
2. Computing LUs 427
3. Example of Applicability: Enumerating Leukocyte Populations Following Perturbation of Macrophage/Neutrophil Specification by *irf8* Misexpression 433

Acknowledgments 433
References 434

23. Four-Dimensional Tracking of Lymphocyte Migration and Interactions in Lymph Nodes by Two-Photon Microscopy 437
Masahiro Kitano and Takaharu Okada

1. Introduction 438
2. Sample Preparation 439
3. Imaging Preparations and Data Acquisition 444
4. Data Analysis 447

Acknowledgments 452
References 453

Author Index *455*
Subject Index *479*

Contributors

David Alsteens
Institute of Condensed Matter & Nanosciences, Université catholique de Louvain, Louvain-la-Neuve, Belgium

D.Michael Ando
Gladstone Institute of Neurological Disease, and Biomedical Sciences Graduate Program, University of California, San Francisco, California, USA

Lamya Baba-Aissa
Institut Pasteur, Imagopole, Plateforme d'imagerie dynamique, Paris, France

Bart Baranowski
Samantha Dickson Brain Cancer Unit, Department of Cancer Biology, UCL Cancer Institute, University College London, London, United Kingdom

Enni Bertling
Neuroscience Center, University of Helsinki, Helsinki, Finland

Joseph Brzostowski
Laboratory of Immunogenetics, National Institute of Allergy and Infectious Diseases, NIH, Rockville, Maryland, USA

Véronique Calleja
Cell Biophysics Laboratory, Cancer Research UK, London Research Institute, London, United Kingdom

Astrid Canivet
Institut Pasteur, Imagopole, Plateforme d'imagerie dynamique, Paris, France

Arnold Caplan
Department of Biology, Case Western Reserve University, Cleveland, Ohio, USA

Ferdinand X. Choong
Department of Neuroscience, Swedish Medical Nanoscience Center, Karolinska Institutet, Stockholm, Sweden

Pierre Cordeau
Department of Psychiatry and Neuroscience, Laval University, Centre de Recherche du Centre Hospitalier de l'Université Laval, 2705 boul. Laurier, Québec, Québec City, Canada

Davide Danovi
Samantha Dickson Brain Cancer Unit, Department of Cancer Biology, UCL Cancer Institute, University College London, London, United Kingdom

Aaron Daub
Gladstone Institute of Neurological Disease; Medical Scientist Training Program and Program in Bioengineering, University of California, and Taube-Koret Center for Huntington's Disease Research, The Hellman Family Foundation Program in Alzheimer's Disease Research and the Consortium for Frontotemporal Dementia Research San Francisco, San Francisco, California, USA

Angel M. Davey
Laboratory of Immunogenetics, National Institute of Allergy and Infectious Diseases, NIH, Rockville, Maryland, USA

James Dennis
Hope Heart Matrix Biology Program, Benaroya Research Institute, Seattle, Washington, USA

Joe Dragavon
Institut Pasteur, Imagopole, Plateforme d'imagerie dynamique, Paris, France

Felix Ellett
Australian Regenerative Medicine Institute, Monash University, Clayton; Cancer and Haematology Division, The Walter and Eliza Hall Institute of Medical Biology, and Department of Medical Biology, University of Melbourne, Parkville, Victoria, Australia

Helge Ewers
Institute of Biochemistry, ETH Zurich, Schafmattstrasse, Zurich, Switzerland

Steven Finkbeiner
Gladstone Institute of Neurological Disease; Taube-Koret Center for Huntington's Disease Research, The Hellman Family Foundation Program in Alzheimer's Disease Research and the Consortium for Frontotemporal Dementia Research San Francisco, and Departments of Neurology and Physiology, Graduate Programs in Neuroscience, Biomedical Sciences, and Cell Biology, University of California, San Francisco, California, USA

Amos A. Folarin
Samantha Dickson Brain Cancer Unit, Department of Cancer Biology, UCL Cancer Institute, University College London, London, United Kingdom

Freddy Frischknecht
Parasitology, Department of Infectious Diseases, University of Heidelberg Medical School, Heidelberg, Germany

Vincent Galy
UMR7622, CNRS-UPMC, 9 quai St Bernard, Paris, France

Susana Gonçalves
Cell and Molecular Neuroscience Unit, Instituto de Medicina Molecular, Lisboa, Portugal

Christof Grüring
Bernhard Nocht Institute for Tropical Medicine, Parasitology Section, Hamburg, Germany

Anna-Katerina Hadjantonakis
Developmental Biology Program, Sloan-Kettering Institute, New York, New York, USA

Federico Herrera
Cell and Molecular Neuroscience Unit, Instituto de Medicina Molecular, Lisboa, Portugal

Robert M. Hoffman
AntiCancer, Inc., and Department of Surgery, University of California San Diego, San Diego, California, USA

Pirta Hotulainen
Neuroscience Center, University of Helsinki, Helsinki, Finland

Tatsushi Igaki
Department of Cell Biology, G-COE, Kobe University Graduate School of Medicine, Kobe, and PRESTO, Japan Science and Technology Agency (JST), Saitama, Japan

Julia A. Kaye
Gladstone Institute of Neurological Disease, and Taube-Koret Center for Huntington's Disease Research, The Hellman Family Foundation Program in Alzheimer's Disease Research and the Consortium for Frontotemporal Dementia Research San Francisco, San Francisco, California, USA

Masahiro Kitano
Research Unit for Immunodynamics, RIKEN, Research Center for Allergy and Immunology, Yokohama, Japan

Sabine Kölle
Department of Urology, Ludwig Maximilians University of Munich, Munich, Germany

Mikko Koskinen
Neuroscience Center, University of Helsinki, Helsinki, Finland

Jasna Kriz
Department of Psychiatry and Neuroscience, Laval University, Centre de Recherche du Centre Hospitalier de l'Université Laval, 2705 boul. Laurier, Québec, Québec City, Canada

Banafshé Larijani
Cell Biophysics Laboratory, Cancer Research UK, London Research Institute, London, United Kingdom

Pierre Leboucher
Cell Biophysics Laboratory, Cancer Research UK, London Research Institute, London, United Kingdom, and Centre Emotion, Hôpital de la Pitié Salpêtrière, Paris, France

Zhenghong Lee
Department of Radiology, Case Western Reserve University, Cleveland, Ohio, USA

Graham J. Lieschke
Australian Regenerative Medicine Institute, Monash University, Clayton; Cancer and Haematology Division, Walter and Eliza Hall Institute of Medical, Research, and Department of Medical Biology, University of Melbourne, Parkville, Victoria, Australia

Wanli Liu
Laboratory of Immunogenetics, National Institute of Allergy and Infectious Diseases, NIH, Rockville, Maryland, USA

Lip Nam Loh
Department of Immunology and Infection, Faculty of Infectious and Tropical Diseases, London School of Hygiene and Tropical Medicine, London, United Kingdom

Anastasia Ludwig
Neuroscience Center, University of Helsinki, Helsinki, Finland

Raymond R. Mattingly
Department of Pharmacology, and Barbara Ann Karmanos Cancer Institute, Wayne State University School of Medicine, Detroit, Michigan, USA

Kamiar Moin
Department of Pharmacology, and Barbara Ann Karmanos Cancer Institute, Wayne State University School of Medicine, Detroit, Michigan, USA

Bruce A. Molitoris
Indiana University School of Medicine, Indianapolis, Indiana, USA

Cameron J. Nowell
Centre for Advanced Microscopy, Ludwig Institute for Cancer Research Melbourne-Parkville Branch, Parkville, Victoria, Australia

Sonja Nowotschin
Developmental Biology Program, Sloan-Kettering Institute, New York, New York, USA

Shizue Ohsawa
Department of Cell Biology, G-COE, Kobe University Graduate School of Medicine, Kobe, Japan

Takaharu Okada
Research Unit for Immunodynamics, RIKEN, Research Center for Allergy and Immunology, Yokohama, Japan

Tiago Fleming Outeiro
Cell and Molecular Neuroscience Unit, Instituto de Medicina Molecular; Instituto de Fisiologia, Faculdade de Medicina da Universidade de Lisboa, Av. Professor Egas Moniz, Lisboa, Portugal, and Department of Neurodegeneration and Restaurative Research, Center for Molecular Physiology of the Brain, University of Göttingen, Waldweg 33, Göttingen, Germany

Luke Pase
Australian Regenerative Medicine Institute, Monash University, Clayton; Cancer and Haematology Division, Walter and Eliza Hall Institute of Medical Research, and Department of Medical Biology, University of Melbourne, Parkville, Victoria, Australia

Emmanuelle Perret
Institut Pasteur, Imagopole, Plateforme d'imagerie dynamique, Paris, France

Susan K. Pierce
Laboratory of Immunogenetics, National Institute of Allergy and Infectious Diseases, NIH, Rockville, Maryland, USA

Steven M. Pollard
Samantha Dickson Brain Cancer Unit, Department of Cancer Biology, UCL Cancer Institute, University College London, London, United Kingdom

Agneta Richter-Dahlfors
Department of Neuroscience, Swedish Medical Nanoscience Center, Karolinska Institutet, Stockholm, Sweden

Jennifer M. Rothberg
Barbara Ann Karmanos Cancer Institute, Wayne State University School of Medicine, Detroit, Michigan, USA

Pascal Roux
Institut Pasteur, Imagopole, Plateforme d'imagerie dynamique, Paris, France

Mansoureh Sameni
Department of Pharmacology, Wayne State University School of Medicine, Detroit, Michigan, USA

Ruben M. Sandoval
Indiana University School of Medicine, Indianapolis, Indiana, USA

Mario Schelhaas
Emmy-Noether Group, Virus Endocytosis', Institutes of Molecular Virology and Medical Biochemistry, ZMBE, University of Münster, Münster, Germany

Punita Sharma
Gladstone Institute of Neurological Disease, San Francisco, California, USA

Spencer Shorte
Institut Pasteur, Imagopole, Plateforme d'imagerie dynamique, Paris, France

Mirko Singer
Parasitology, Department of Infectious Diseases, University of Heidelberg Medical School, Heidelberg, Germany

Bonnie F. Sloane
Department of Pharmacology, and Barbara Ann Karmanos Cancer Institute, Wayne State University School of Medicine, Detroit, Michigan, USA

Hae Won Sohn
Laboratory of Immunogenetics, National Institute of Allergy and Infectious Diseases, NIH, Rockville, Maryland, USA

Tobias Spielmann
Bernhard Nocht Institute for Tropical Medicine, Parasitology Section, Hamburg, Germany

Kaoru Sugimura
Institute for Integrated Cell-Material Sicences (iCeMS), Kyoto University iCeMS Complex 2, Sakyo-ku, Kyoto, Japan

Kyoko Takino
Department of Cell Biology, G-COE, Kobe University Graduate School of Medicine, Kobe, Japan

Jean-Yves Tinevez
Institut Pasteur, Imagopole, Plateforme d'imagerie dynamique, Paris, France

Bernadette C. Victor
Barbara Ann Karmanos Cancer Institute, Wayne State University School of Medicine, Detroit, Michigan, USA

Theresa H. Ward
Department of Immunology and Infection, Faculty of Infectious and Tropical Diseases, London School of Hygiene and Tropical Medicine, London, United Kingdom

Jean Welter
Department of Biology, Case Western Reserve University, Cleveland, Ohio, USA

Panagiotis Xenopoulos
Developmental Biology Program, Sloan-Kettering Institute, New York, New York, USA

PREFACE

Going back to the dawn of light microscopy, imaging techniques have provided the opportunity for developing models of cellular function. Over the past 40 years, the availability of technology for high-resolution imaging and for evaluation of those images in live cells has extended our reach and, accordingly, our ability to understand cell function. Given the large number of choices in equipment and approaches, sorting out the best approach can be challenging, even to seasoned investigators.

The present volumes provide descriptions of methods used to image living cells, with particular reference to the technical approaches and reagents needed and approaches to selecting the best techniques. The authors explain how these methods are able to provide important biological insights in normal and pathological cells.

Authors were selected based on research contributions in the area about which they have written and based on their ability to describe their methodological contribution in a clear and reproducible way. They have been encouraged to make use of graphics, comparisons to other methods, and to provide tricks and approaches not revealed in prior publications that make it possible to adapt methods to other systems.

The editor wants to express appreciation to the contributors for providing their contributions in a timely fashion, to the senior editors for guidance, and to the staff at Academic Press for helpful input.

P. MICHAEL CONN

Methods in Enzymology

Volume I. Preparation and Assay of Enzymes
Edited by Sidney P. Colowick and Nathan O. Kaplan

Volume II. Preparation and Assay of Enzymes
Edited by Sidney P. Colowick and Nathan O. Kaplan

Volume III. Preparation and Assay of Substrates
Edited by Sidney P. Colowick and Nathan O. Kaplan

Volume IV. Special Techniques for the Enzymologist
Edited by Sidney P. Colowick and Nathan O. Kaplan

Volume V. Preparation and Assay of Enzymes
Edited by Sidney P. Colowick and Nathan O. Kaplan

Volume VI. Preparation and Assay of Enzymes *(Continued)*
Preparation and Assay of Substrates
Special Techniques
Edited by Sidney P. Colowick and Nathan O. Kaplan

Volume VII. Cumulative Subject Index
Edited by Sidney P. Colowick and Nathan O. Kaplan

Volume VIII. Complex Carbohydrates
Edited by Elizabeth F. Neufeld and Victor Ginsburg

Volume IX. Carbohydrate Metabolism
Edited by Willis A. Wood

Volume X. Oxidation and Phosphorylation
Edited by Ronald W. Estabrook and Maynard E. Pullman

Volume XI. Enzyme Structure
Edited by C. H. W. Hirs

Volume XII. Nucleic Acids (Parts A and B)
Edited by Lawrence Grossman and Kivie Moldave

Volume XIII. Citric Acid Cycle
Edited by J. M. Lowenstein

Volume XIV. Lipids
Edited by J. M. Lowenstein

Volume XV. Steroids and Terpenoids
Edited by Raymond B. Clayton

VOLUME XVI. Fast Reactions
Edited by KENNETH KUSTIN

VOLUME XVII. Metabolism of Amino Acids and Amines (Parts A and B)
Edited by HERBERT TABOR AND CELIA WHITE TABOR

VOLUME XVIII. Vitamins and Coenzymes (Parts A, B, and C)
Edited by DONALD B. MCCORMICK AND LEMUEL D. WRIGHT

VOLUME XIX. Proteolytic Enzymes
Edited by GERTRUDE E. PERLMANN AND LASZLO LORAND

VOLUME XX. Nucleic Acids and Protein Synthesis (Part C)
Edited by KIVIE MOLDAVE AND LAWRENCE GROSSMAN

VOLUME XXI. Nucleic Acids (Part D)
Edited by LAWRENCE GROSSMAN AND KIVIE MOLDAVE

VOLUME XXII. Enzyme Purification and Related Techniques
Edited by WILLIAM B. JAKOBY

VOLUME XXIII. Photosynthesis (Part A)
Edited by ANTHONY SAN PIETRO

VOLUME XXIV. Photosynthesis and Nitrogen Fixation (Part B)
Edited by ANTHONY SAN PIETRO

VOLUME XXV. Enzyme Structure (Part B)
Edited by C. H. W. HIRS AND SERGE N. TIMASHEFF

VOLUME XXVI. Enzyme Structure (Part C)
Edited by C. H. W. HIRS AND SERGE N. TIMASHEFF

VOLUME XXVII. Enzyme Structure (Part D)
Edited by C. H. W. HIRS AND SERGE N. TIMASHEFF

VOLUME XXVIII. Complex Carbohydrates (Part B)
Edited by VICTOR GINSBURG

VOLUME XXIX. Nucleic Acids and Protein Synthesis (Part E)
Edited by LAWRENCE GROSSMAN AND KIVIE MOLDAVE

VOLUME XXX. Nucleic Acids and Protein Synthesis (Part F)
Edited by KIVIE MOLDAVE AND LAWRENCE GROSSMAN

VOLUME XXXI. Biomembranes (Part A)
Edited by SIDNEY FLEISCHER AND LESTER PACKER

VOLUME XXXII. Biomembranes (Part B)
Edited by SIDNEY FLEISCHER AND LESTER PACKER

VOLUME XXXIII. Cumulative Subject Index Volumes I-XXX
Edited by MARTHA G. DENNIS AND EDWARD A. DENNIS

VOLUME XXXIV. Affinity Techniques (Enzyme Purification: Part B)
Edited by WILLIAM B. JAKOBY AND MEIR WILCHEK

Volume LIII. Biomembranes (Part D: Biological Oxidations)
Edited by Sidney Fleischer and Lester Packer

Volume LIV. Biomembranes (Part E: Biological Oxidations)
Edited by Sidney Fleischer and Lester Packer

Volume LV. Biomembranes (Part F: Bioenergetics)
Edited by Sidney Fleischer and Lester Packer

Volume LVI. Biomembranes (Part G: Bioenergetics)
Edited by Sidney Fleischer and Lester Packer

Volume LVII. Bioluminescence and Chemiluminescence
Edited by Marlene A. DeLuca

Volume LVIII. Cell Culture
Edited by William B. Jakoby and Ira Pastan

Volume LIX. Nucleic Acids and Protein Synthesis (Part G)
Edited by Kivie Moldave and Lawrence Grossman

Volume LX. Nucleic Acids and Protein Synthesis (Part H)
Edited by Kivie Moldave and Lawrence Grossman

Volume 61. Enzyme Structure (Part H)
Edited by C. H. W. Hirs and Serge N. Timasheff

Volume 62. Vitamins and Coenzymes (Part D)
Edited by Donald B. McCormick and Lemuel D. Wright

Volume 63. Enzyme Kinetics and Mechanism (Part A: Initial Rate and Inhibitor Methods)
Edited by Daniel L. Purich

Volume 64. Enzyme Kinetics and Mechanism
(Part B: Isotopic Probes and Complex Enzyme Systems)
Edited by Daniel L. Purich

Volume 65. Nucleic Acids (Part I)
Edited by Lawrence Grossman and Kivie Moldave

Volume 66. Vitamins and Coenzymes (Part E)
Edited by Donald B. McCormick and Lemuel D. Wright

Volume 67. Vitamins and Coenzymes (Part F)
Edited by Donald B. McCormick and Lemuel D. Wright

Volume 68. Recombinant DNA
Edited by Ray Wu

Volume 69. Photosynthesis and Nitrogen Fixation (Part C)
Edited by Anthony San Pietro

Volume 70. Immunochemical Techniques (Part A)
Edited by Helen Van Vunakis and John J. Langone

Volume 89. Carbohydrate Metabolism (Part D)
Edited by Willis A. Wood

Volume 90. Carbohydrate Metabolism (Part E)
Edited by Willis A. Wood

Volume 91. Enzyme Structure (Part I)
Edited by C. H. W. Hirs and Serge N. Timasheff

Volume 92. Immunochemical Techniques (Part E: Monoclonal Antibodies and General Immunoassay Methods)
Edited by John J. Langone and Helen Van Vunakis

Volume 93. Immunochemical Techniques (Part F: Conventional Antibodies, Fc Receptors, and Cytotoxicity)
Edited by John J. Langone and Helen Van Vunakis

Volume 94. Polyamines
Edited by Herbert Tabor and Celia White Tabor

Volume 95. Cumulative Subject Index Volumes 61–74, 76–80
Edited by Edward A. Dennis and Martha G. Dennis

Volume 96. Biomembranes [Part J: Membrane Biogenesis: Assembly and Targeting (General Methods; Eukaryotes)]
Edited by Sidney Fleischer and Becca Fleischer

Volume 97. Biomembranes [Part K: Membrane Biogenesis: Assembly and Targeting (Prokaryotes, Mitochondria, and Chloroplasts)]
Edited by Sidney Fleischer and Becca Fleischer

Volume 98. Biomembranes (Part L: Membrane Biogenesis: Processing and Recycling)
Edited by Sidney Fleischer and Becca Fleischer

Volume 99. Hormone Action (Part F: Protein Kinases)
Edited by Jackie D. Corbin and Joel G. Hardman

Volume 100. Recombinant DNA (Part B)
Edited by Ray Wu, Lawrence Grossman, and Kivie Moldave

Volume 101. Recombinant DNA (Part C)
Edited by Ray Wu, Lawrence Grossman, and Kivie Moldave

Volume 102. Hormone Action (Part G: Calmodulin and Calcium-Binding Proteins)
Edited by Anthony R. Means and Bert W. O'Malley

Volume 103. Hormone Action (Part H: Neuroendocrine Peptides)
Edited by P. Michael Conn

Volume 104. Enzyme Purification and Related Techniques (Part C)
Edited by William B. Jakoby

Volume 105. Oxygen Radicals in Biological Systems
Edited by Lester Packer

Volume 106. Posttranslational Modifications (Part A)
Edited by Finn Wold and Kivie Moldave

Volume 107. Posttranslational Modifications (Part B)
Edited by Finn Wold and Kivie Moldave

Volume 108. Immunochemical Techniques (Part G: Separation and Characterization of Lymphoid Cells)
Edited by Giovanni Di Sabato, John J. Langone, and Helen Van Vunakis

Volume 109. Hormone Action (Part I: Peptide Hormones)
Edited by Lutz Birnbaumer and Bert W. O'Malley

Volume 110. Steroids and Isoprenoids (Part A)
Edited by John H. Law and Hans C. Rilling

Volume 111. Steroids and Isoprenoids (Part B)
Edited by John H. Law and Hans C. Rilling

Volume 112. Drug and Enzyme Targeting (Part A)
Edited by Kenneth J. Widder and Ralph Green

Volume 113. Glutamate, Glutamine, Glutathione, and Related Compounds
Edited by Alton Meister

Volume 114. Diffraction Methods for Biological Macromolecules (Part A)
Edited by Harold W. Wyckoff, C. H. W. Hirs, and Serge N. Timasheff

Volume 115. Diffraction Methods for Biological Macromolecules (Part B)
Edited by Harold W. Wyckoff, C. H. W. Hirs, and Serge N. Timasheff

Volume 116. Immunochemical Techniques
(Part H: Effectors and Mediators of Lymphoid Cell Functions)
Edited by Giovanni Di Sabato, John J. Langone, and Helen Van Vunakis

Volume 117. Enzyme Structure (Part J)
Edited by C. H. W. Hirs and Serge N. Timasheff

Volume 118. Plant Molecular Biology
Edited by Arthur Weissbach and Herbert Weissbach

Volume 119. Interferons (Part C)
Edited by Sidney Pestka

Volume 120. Cumulative Subject Index Volumes 81–94, 96–101

Volume 121. Immunochemical Techniques (Part I: Hybridoma Technology and Monoclonal Antibodies)
Edited by John J. Langone and Helen Van Vunakis

Volume 122. Vitamins and Coenzymes (Part G)
Edited by Frank Chytil and Donald B. McCormick

VOLUME 123. Vitamins and Coenzymes (Part H)
Edited by FRANK CHYTIL AND DONALD B. MCCORMICK

VOLUME 124. Hormone Action (Part J: Neuroendocrine Peptides)
Edited by P. MICHAEL CONN

VOLUME 125. Biomembranes (Part M: Transport in Bacteria, Mitochondria, and Chloroplasts: General Approaches and Transport Systems)
Edited by SIDNEY FLEISCHER AND BECCA FLEISCHER

VOLUME 126. Biomembranes (Part N: Transport in Bacteria, Mitochondria, and Chloroplasts: Protonmotive Force)
Edited by SIDNEY FLEISCHER AND BECCA FLEISCHER

VOLUME 127. Biomembranes (Part O: Protons and Water: Structure and Translocation)
Edited by LESTER PACKER

VOLUME 128. Plasma Lipoproteins (Part A: Preparation, Structure, and Molecular Biology)
Edited by JERE P. SEGREST AND JOHN J. ALBERS

VOLUME 129. Plasma Lipoproteins (Part B: Characterization, Cell Biology, and Metabolism)
Edited by JOHN J. ALBERS AND JERE P. SEGREST

VOLUME 130. Enzyme Structure (Part K)
Edited by C. H. W. HIRS AND SERGE N. TIMASHEFF

VOLUME 131. Enzyme Structure (Part L)
Edited by C. H. W. HIRS AND SERGE N. TIMASHEFF

VOLUME 132. Immunochemical Techniques (Part J: Phagocytosis and Cell-Mediated Cytotoxicity)
Edited by GIOVANNI DI SABATO AND JOHANNES EVERSE

VOLUME 133. Bioluminescence and Chemiluminescence (Part B)
Edited by MARLENE DELUCA AND WILLIAM D. MCELROY

VOLUME 134. Structural and Contractile Proteins (Part C: The Contractile Apparatus and the Cytoskeleton)
Edited by RICHARD B. VALLEE

VOLUME 135. Immobilized Enzymes and Cells (Part B)
Edited by KLAUS MOSBACH

VOLUME 136. Immobilized Enzymes and Cells (Part C)
Edited by KLAUS MOSBACH

VOLUME 137. Immobilized Enzymes and Cells (Part D)
Edited by KLAUS MOSBACH

VOLUME 138. Complex Carbohydrates (Part E)
Edited by VICTOR GINSBURG

VOLUME 139. Cellular Regulators (Part A: Calcium- and Calmodulin-Binding Proteins)
Edited by ANTHONY R. MEANS AND P. MICHAEL CONN

VOLUME 140. Cumulative Subject Index Volumes 102–119, 121–134

VOLUME 141. Cellular Regulators (Part B: Calcium and Lipids)
Edited by P. MICHAEL CONN AND ANTHONY R. MEANS

VOLUME 142. Metabolism of Aromatic Amino Acids and Amines
Edited by SEYMOUR KAUFMAN

VOLUME 143. Sulfur and Sulfur Amino Acids
Edited by WILLIAM B. JAKOBY AND OWEN GRIFFITH

VOLUME 144. Structural and Contractile Proteins (Part D: Extracellular Matrix)
Edited by LEON W. CUNNINGHAM

VOLUME 145. Structural and Contractile Proteins (Part E: Extracellular Matrix)
Edited by LEON W. CUNNINGHAM

VOLUME 146. Peptide Growth Factors (Part A)
Edited by DAVID BARNES AND DAVID A. SIRBASKU

VOLUME 147. Peptide Growth Factors (Part B)
Edited by DAVID BARNES AND DAVID A. SIRBASKU

VOLUME 148. Plant Cell Membranes
Edited by LESTER PACKER AND ROLAND DOUCE

VOLUME 149. Drug and Enzyme Targeting (Part B)
Edited by RALPH GREEN AND KENNETH J. WIDDER

VOLUME 150. Immunochemical Techniques (Part K: *In Vitro* Models of B and T Cell Functions and Lymphoid Cell Receptors)
Edited by GIOVANNI DI SABATO

VOLUME 151. Molecular Genetics of Mammalian Cells
Edited by MICHAEL M. GOTTESMAN

VOLUME 152. Guide to Molecular Cloning Techniques
Edited by SHELBY L. BERGER AND ALAN R. KIMMEL

VOLUME 153. Recombinant DNA (Part D)
Edited by RAY WU AND LAWRENCE GROSSMAN

VOLUME 154. Recombinant DNA (Part E)
Edited by RAY WU AND LAWRENCE GROSSMAN

VOLUME 155. Recombinant DNA (Part F)
Edited by RAY WU

VOLUME 156. Biomembranes (Part P: ATP-Driven Pumps and Related Transport: The Na, K-Pump)
Edited by SIDNEY FLEISCHER AND BECCA FLEISCHER

Volume 157. Biomembranes (Part Q: ATP-Driven Pumps and Related Transport: Calcium, Proton, and Potassium Pumps)
Edited by Sidney Fleischer and Becca Fleischer

Volume 158. Metalloproteins (Part A)
Edited by James F. Riordan and Bert L. Vallee

Volume 159. Initiation and Termination of Cyclic Nucleotide Action
Edited by Jackie D. Corbin and Roger A. Johnson

Volume 160. Biomass (Part A: Cellulose and Hemicellulose)
Edited by Willis A. Wood and Scott T. Kellogg

Volume 161. Biomass (Part B: Lignin, Pectin, and Chitin)
Edited by Willis A. Wood and Scott T. Kellogg

Volume 162. Immunochemical Techniques (Part L: Chemotaxis and Inflammation)
Edited by Giovanni Di Sabato

Volume 163. Immunochemical Techniques (Part M: Chemotaxis and Inflammation)
Edited by Giovanni Di Sabato

Volume 164. Ribosomes
Edited by Harry F. Noller, Jr., and Kivie Moldave

Volume 165. Microbial Toxins: Tools for Enzymology
Edited by Sidney Harshman

Volume 166. Branched-Chain Amino Acids
Edited by Robert Harris and John R. Sokatch

Volume 167. Cyanobacteria
Edited by Lester Packer and Alexander N. Glazer

Volume 168. Hormone Action (Part K: Neuroendocrine Peptides)
Edited by P. Michael Conn

Volume 169. Platelets: Receptors, Adhesion, Secretion (Part A)
Edited by Jacek Hawiger

Volume 170. Nucleosomes
Edited by Paul M. Wassarman and Roger D. Kornberg

Volume 171. Biomembranes (Part R: Transport Theory: Cells and Model Membranes)
Edited by Sidney Fleischer and Becca Fleischer

Volume 172. Biomembranes (Part S: Transport: Membrane Isolation and Characterization)
Edited by Sidney Fleischer and Becca Fleischer

Volume 173. Biomembranes [Part T: Cellular and Subcellular Transport: Eukaryotic (Nonepithelial) Cells]
Edited by Sidney Fleischer and Becca Fleischer

Volume 174. Biomembranes [Part U: Cellular and Subcellular Transport: Eukaryotic (Nonepithelial) Cells]
Edited by Sidney Fleischer and Becca Fleischer

Volume 175. Cumulative Subject Index Volumes 135–139, 141–167

Volume 176. Nuclear Magnetic Resonance (Part A: Spectral Techniques and Dynamics)
Edited by Norman J. Oppenheimer and Thomas L. James

Volume 177. Nuclear Magnetic Resonance (Part B: Structure and Mechanism)
Edited by Norman J. Oppenheimer and Thomas L. James

Volume 178. Antibodies, Antigens, and Molecular Mimicry
Edited by John J. Langone

Volume 179. Complex Carbohydrates (Part F)
Edited by Victor Ginsburg

Volume 180. RNA Processing (Part A: General Methods)
Edited by James E. Dahlberg and John N. Abelson

Volume 181. RNA Processing (Part B: Specific Methods)
Edited by James E. Dahlberg and John N. Abelson

Volume 182. Guide to Protein Purification
Edited by Murray P. Deutscher

Volume 183. Molecular Evolution: Computer Analysis of Protein and Nucleic Acid Sequences
Edited by Russell F. Doolittle

Volume 184. Avidin-Biotin Technology
Edited by Meir Wilchek and Edward A. Bayer

Volume 185. Gene Expression Technology
Edited by David V. Goeddel

Volume 186. Oxygen Radicals in Biological Systems (Part B: Oxygen Radicals and Antioxidants)
Edited by Lester Packer and Alexander N. Glazer

Volume 187. Arachidonate Related Lipid Mediators
Edited by Robert C. Murphy and Frank A. Fitzpatrick

Volume 188. Hydrocarbons and Methylotrophy
Edited by Mary E. Lidstrom

Volume 189. Retinoids (Part A: Molecular and Metabolic Aspects)
Edited by Lester Packer

Volume 190. Retinoids (Part B: Cell Differentiation and Clinical Applications)
Edited by Lester Packer

Volume 191. Biomembranes (Part V: Cellular and Subcellular Transport: Epithelial Cells)
Edited by Sidney Fleischer and Becca Fleischer

Volume 192. Biomembranes (Part W: Cellular and Subcellular Transport: Epithelial Cells)
Edited by Sidney Fleischer and Becca Fleischer

Volume 193. Mass Spectrometry
Edited by James A. McCloskey

Volume 194. Guide to Yeast Genetics and Molecular Biology
Edited by Christine Guthrie and Gerald R. Fink

Volume 195. Adenylyl Cyclase, G Proteins, and Guanylyl Cyclase
Edited by Roger A. Johnson and Jackie D. Corbin

Volume 196. Molecular Motors and the Cytoskeleton
Edited by Richard B. Vallee

Volume 197. Phospholipases
Edited by Edward A. Dennis

Volume 198. Peptide Growth Factors (Part C)
Edited by David Barnes, J. P. Mather, and Gordon H. Sato

Volume 199. Cumulative Subject Index Volumes 168–174, 176–194

Volume 200. Protein Phosphorylation (Part A: Protein Kinases: Assays, Purification, Antibodies, Functional Analysis, Cloning, and Expression)
Edited by Tony Hunter and Bartholomew M. Sefton

Volume 201. Protein Phosphorylation (Part B: Analysis of Protein Phosphorylation, Protein Kinase Inhibitors, and Protein Phosphatases)
Edited by Tony Hunter and Bartholomew M. Sefton

Volume 202. Molecular Design and Modeling: Concepts and Applications (Part A: Proteins, Peptides, and Enzymes)
Edited by John J. Langone

Volume 203. Molecular Design and Modeling: Concepts and Applications (Part B: Antibodies and Antigens, Nucleic Acids, Polysaccharides, and Drugs)
Edited by John J. Langone

Volume 204. Bacterial Genetic Systems
Edited by Jeffrey H. Miller

Volume 205. Metallobiochemistry (Part B: Metallothionein and Related Molecules)
Edited by James F. Riordan and Bert L. Vallee

VOLUME 206. Cytochrome P450
Edited by MICHAEL R. WATERMAN AND ERIC F. JOHNSON

VOLUME 207. Ion Channels
Edited by BERNARDO RUDY AND LINDA E. IVERSON

VOLUME 208. Protein–DNA Interactions
Edited by ROBERT T. SAUER

VOLUME 209. Phospholipid Biosynthesis
Edited by EDWARD A. DENNIS AND DENNIS E. VANCE

VOLUME 210. Numerical Computer Methods
Edited by LUDWIG BRAND AND MICHAEL L. JOHNSON

VOLUME 211. DNA Structures (Part A: Synthesis and Physical Analysis of DNA)
Edited by DAVID M. J. LILLEY AND JAMES E. DAHLBERG

VOLUME 212. DNA Structures (Part B: Chemical and Electrophoretic Analysis of DNA)
Edited by DAVID M. J. LILLEY AND JAMES E. DAHLBERG

VOLUME 213. Carotenoids (Part A: Chemistry, Separation, Quantitation, and Antioxidation)
Edited by LESTER PACKER

VOLUME 214. Carotenoids (Part B: Metabolism, Genetics, and Biosynthesis)
Edited by LESTER PACKER

VOLUME 215. Platelets: Receptors, Adhesion, Secretion (Part B)
Edited by JACEK J. HAWIGER

VOLUME 216. Recombinant DNA (Part G)
Edited by RAY WU

VOLUME 217. Recombinant DNA (Part H)
Edited by RAY WU

VOLUME 218. Recombinant DNA (Part I)
Edited by RAY WU

VOLUME 219. Reconstitution of Intracellular Transport
Edited by JAMES E. ROTHMAN

VOLUME 220. Membrane Fusion Techniques (Part A)
Edited by NEJAT DÜZGÜNEŞ

VOLUME 221. Membrane Fusion Techniques (Part B)
Edited by NEJAT DÜZGÜNEŞ

VOLUME 222. Proteolytic Enzymes in Coagulation, Fibrinolysis, and Complement Activation (Part A: Mammalian Blood Coagulation Factors and Inhibitors)
Edited by LASZLO LORAND AND KENNETH G. MANN

Volume 223. Proteolytic Enzymes in Coagulation, Fibrinolysis, and Complement Activation (Part B: Complement Activation, Fibrinolysis, and Nonmammalian Blood Coagulation Factors)
Edited by Laszlo Lorand and Kenneth G. Mann

Volume 224. Molecular Evolution: Producing the Biochemical Data
Edited by Elizabeth Anne Zimmer, Thomas J. White, Rebecca L. Cann, and Allan C. Wilson

Volume 225. Guide to Techniques in Mouse Development
Edited by Paul M. Wassarman and Melvin L. DePamphilis

Volume 226. Metallobiochemistry (Part C: Spectroscopic and Physical Methods for Probing Metal Ion Environments in Metalloenzymes and Metalloproteins)
Edited by James F. Riordan and Bert L. Vallee

Volume 227. Metallobiochemistry (Part D: Physical and Spectroscopic Methods for Probing Metal Ion Environments in Metalloproteins)
Edited by James F. Riordan and Bert L. Vallee

Volume 228. Aqueous Two-Phase Systems
Edited by Harry Walter and Göte Johansson

Volume 229. Cumulative Subject Index Volumes 195–198, 200–227

Volume 230. Guide to Techniques in Glycobiology
Edited by William J. Lennarz and Gerald W. Hart

Volume 231. Hemoglobins (Part B: Biochemical and Analytical Methods)
Edited by Johannes Everse, Kim D. Vandegriff, and Robert M. Winslow

Volume 232. Hemoglobins (Part C: Biophysical Methods)
Edited by Johannes Everse, Kim D. Vandegriff, and Robert M. Winslow

Volume 233. Oxygen Radicals in Biological Systems (Part C)
Edited by Lester Packer

Volume 234. Oxygen Radicals in Biological Systems (Part D)
Edited by Lester Packer

Volume 235. Bacterial Pathogenesis (Part A: Identification and Regulation of Virulence Factors)
Edited by Virginia L. Clark and Patrik M. Bavoil

Volume 236. Bacterial Pathogenesis (Part B: Integration of Pathogenic Bacteria with Host Cells)
Edited by Virginia L. Clark and Patrik M. Bavoil

Volume 237. Heterotrimeric G Proteins
Edited by Ravi Iyengar

Volume 238. Heterotrimeric G-Protein Effectors
Edited by Ravi Iyengar

VOLUME 239. Nuclear Magnetic Resonance (Part C)
Edited by THOMAS L. JAMES AND NORMAN J. OPPENHEIMER

VOLUME 240. Numerical Computer Methods (Part B)
Edited by MICHAEL L. JOHNSON AND LUDWIG BRAND

VOLUME 241. Retroviral Proteases
Edited by LAWRENCE C. KUO AND JULES A. SHAFER

VOLUME 242. Neoglycoconjugates (Part A)
Edited by Y. C. LEE AND REIKO T. LEE

VOLUME 243. Inorganic Microbial Sulfur Metabolism
Edited by HARRY D. PECK, JR., AND JEAN LEGALL

VOLUME 244. Proteolytic Enzymes: Serine and Cysteine Peptidases
Edited by ALAN J. BARRETT

VOLUME 245. Extracellular Matrix Components
Edited by E. RUOSLAHTI AND E. ENGVALL

VOLUME 246. Biochemical Spectroscopy
Edited by KENNETH SAUER

VOLUME 247. Neoglycoconjugates (Part B: Biomedical Applications)
Edited by Y. C. LEE AND REIKO T. LEE

VOLUME 248. Proteolytic Enzymes: Aspartic and Metallo Peptidases
Edited by ALAN J. BARRETT

VOLUME 249. Enzyme Kinetics and Mechanism (Part D: Developments in Enzyme Dynamics)
Edited by DANIEL L. PURICH

VOLUME 250. Lipid Modifications of Proteins
Edited by PATRICK J. CASEY AND JANICE E. BUSS

VOLUME 251. Biothiols (Part A: Monothiols and Dithiols, Protein Thiols, and Thiyl Radicals)
Edited by LESTER PACKER

VOLUME 252. Biothiols (Part B: Glutathione and Thioredoxin; Thiols in Signal Transduction and Gene Regulation)
Edited by LESTER PACKER

VOLUME 253. Adhesion of Microbial Pathogens
Edited by RON J. DOYLE AND ITZHAK OFEK

VOLUME 254. Oncogene Techniques
Edited by PETER K. VOGT AND INDER M. VERMA

VOLUME 255. Small GTPases and Their Regulators (Part A: Ras Family)
Edited by W. E. BALCH, CHANNING J. DER, AND ALAN HALL

Volume 256. Small GTPases and Their Regulators (Part B: Rho Family)
Edited by W. E. Balch, Channing J. Der, and Alan Hall

Volume 257. Small GTPases and Their Regulators (Part C: Proteins Involved in Transport)
Edited by W. E. Balch, Channing J. Der, and Alan Hall

Volume 258. Redox-Active Amino Acids in Biology
Edited by Judith P. Klinman

Volume 259. Energetics of Biological Macromolecules
Edited by Michael L. Johnson and Gary K. Ackers

Volume 260. Mitochondrial Biogenesis and Genetics (Part A)
Edited by Giuseppe M. Attardi and Anne Chomyn

Volume 261. Nuclear Magnetic Resonance and Nucleic Acids
Edited by Thomas L. James

Volume 262. DNA Replication
Edited by Judith L. Campbell

Volume 263. Plasma Lipoproteins (Part C: Quantitation)
Edited by William A. Bradley, Sandra H. Gianturco, and Jere P. Segrest

Volume 264. Mitochondrial Biogenesis and Genetics (Part B)
Edited by Giuseppe M. Attardi and Anne Chomyn

Volume 265. Cumulative Subject Index Volumes 228, 230–262

Volume 266. Computer Methods for Macromolecular Sequence Analysis
Edited by Russell F. Doolittle

Volume 267. Combinatorial Chemistry
Edited by John N. Abelson

Volume 268. Nitric Oxide (Part A: Sources and Detection of NO; NO Synthase)
Edited by Lester Packer

Volume 269. Nitric Oxide (Part B: Physiological and Pathological Processes)
Edited by Lester Packer

Volume 270. High Resolution Separation and Analysis of Biological Macromolecules (Part A: Fundamentals)
Edited by Barry L. Karger and William S. Hancock

Volume 271. High Resolution Separation and Analysis of Biological Macromolecules (Part B: Applications)
Edited by Barry L. Karger and William S. Hancock

Volume 272. Cytochrome P450 (Part B)
Edited by Eric F. Johnson and Michael R. Waterman

Volume 273. RNA Polymerase and Associated Factors (Part A)
Edited by Sankar Adhya

Volume 274. RNA Polymerase and Associated Factors (Part B)
Edited by Sankar Adhya

Volume 275. Viral Polymerases and Related Proteins
Edited by Lawrence C. Kuo, David B. Olsen, and Steven S. Carroll

Volume 276. Macromolecular Crystallography (Part A)
Edited by Charles W. Carter, Jr., and Robert M. Sweet

Volume 277. Macromolecular Crystallography (Part B)
Edited by Charles W. Carter, Jr., and Robert M. Sweet

Volume 278. Fluorescence Spectroscopy
Edited by Ludwig Brand and Michael L. Johnson

Volume 279. Vitamins and Coenzymes (Part I)
Edited by Donald B. McCormick, John W. Suttie, and Conrad Wagner

Volume 280. Vitamins and Coenzymes (Part J)
Edited by Donald B. McCormick, John W. Suttie, and Conrad Wagner

Volume 281. Vitamins and Coenzymes (Part K)
Edited by Donald B. McCormick, John W. Suttie, and Conrad Wagner

Volume 282. Vitamins and Coenzymes (Part L)
Edited by Donald B. McCormick, John W. Suttie, and Conrad Wagner

Volume 283. Cell Cycle Control
Edited by William G. Dunphy

Volume 284. Lipases (Part A: Biotechnology)
Edited by Byron Rubin and Edward A. Dennis

Volume 285. Cumulative Subject Index Volumes 263, 264, 266–284, 286–289

Volume 286. Lipases (Part B: Enzyme Characterization and Utilization)
Edited by Byron Rubin and Edward A. Dennis

Volume 287. Chemokines
Edited by Richard Horuk

Volume 288. Chemokine Receptors
Edited by Richard Horuk

Volume 289. Solid Phase Peptide Synthesis
Edited by Gregg B. Fields

Volume 290. Molecular Chaperones
Edited by George H. Lorimer and Thomas Baldwin

Volume 291. Caged Compounds
Edited by Gerard Marriott

Volume 292. ABC Transporters: Biochemical, Cellular, and Molecular Aspects
Edited by Suresh V. Ambudkar and Michael M. Gottesman

Volume 293. Ion Channels (Part B)
Edited by P. Michael Conn

Volume 294. Ion Channels (Part C)
Edited by P. Michael Conn

Volume 295. Energetics of Biological Macromolecules (Part B)
Edited by Gary K. Ackers and Michael L. Johnson

Volume 296. Neurotransmitter Transporters
Edited by Susan G. Amara

Volume 297. Photosynthesis: Molecular Biology of Energy Capture
Edited by Lee McIntosh

Volume 298. Molecular Motors and the Cytoskeleton (Part B)
Edited by Richard B. Vallee

Volume 299. Oxidants and Antioxidants (Part A)
Edited by Lester Packer

Volume 300. Oxidants and Antioxidants (Part B)
Edited by Lester Packer

Volume 301. Nitric Oxide: Biological and Antioxidant Activities (Part C)
Edited by Lester Packer

Volume 302. Green Fluorescent Protein
Edited by P. Michael Conn

Volume 303. cDNA Preparation and Display
Edited by Sherman M. Weissman

Volume 304. Chromatin
Edited by Paul M. Wassarman and Alan P. Wolffe

Volume 305. Bioluminescence and Chemiluminescence (Part C)
Edited by Thomas O. Baldwin and Miriam M. Ziegler

Volume 306. Expression of Recombinant Genes in Eukaryotic Systems
Edited by Joseph C. Glorioso and Martin C. Schmidt

Volume 307. Confocal Microscopy
Edited by P. Michael Conn

Volume 308. Enzyme Kinetics and Mechanism (Part E: Energetics of Enzyme Catalysis)
Edited by Daniel L. Purich and Vern L. Schramm

Volume 309. Amyloid, Prions, and Other Protein Aggregates
Edited by Ronald Wetzel

Volume 310. Biofilms
Edited by Ron J. Doyle

VOLUME 311. Sphingolipid Metabolism and Cell Signaling (Part A)
Edited by ALFRED H. MERRILL, JR., AND YUSUF A. HANNUN

VOLUME 312. Sphingolipid Metabolism and Cell Signaling (Part B)
Edited by ALFRED H. MERRILL, JR., AND YUSUF A. HANNUN

VOLUME 313. Antisense Technology
(Part A: General Methods, Methods of Delivery, and RNA Studies)
Edited by M. IAN PHILLIPS

VOLUME 314. Antisense Technology (Part B: Applications)
Edited by M. IAN PHILLIPS

VOLUME 315. Vertebrate Phototransduction and the Visual Cycle (Part A)
Edited by KRZYSZTOF PALCZEWSKI

VOLUME 316. Vertebrate Phototransduction and the Visual Cycle (Part B)
Edited by KRZYSZTOF PALCZEWSKI

VOLUME 317. RNA–Ligand Interactions (Part A: Structural Biology Methods)
Edited by DANIEL W. CELANDER AND JOHN N. ABELSON

VOLUME 318. RNA–Ligand Interactions (Part B: Molecular Biology Methods)
Edited by DANIEL W. CELANDER AND JOHN N. ABELSON

VOLUME 319. Singlet Oxygen, UV-A, and Ozone
Edited by LESTER PACKER AND HELMUT SIES

VOLUME 320. Cumulative Subject Index Volumes 290–319

VOLUME 321. Numerical Computer Methods (Part C)
Edited by MICHAEL L. JOHNSON AND LUDWIG BRAND

VOLUME 322. Apoptosis
Edited by JOHN C. REED

VOLUME 323. Energetics of Biological Macromolecules (Part C)
Edited by MICHAEL L. JOHNSON AND GARY K. ACKERS

VOLUME 324. Branched-Chain Amino Acids (Part B)
Edited by ROBERT A. HARRIS AND JOHN R. SOKATCH

VOLUME 325. Regulators and Effectors of Small GTPases
(Part D: Rho Family)
Edited by W. E. BALCH, CHANNING J. DER, AND ALAN HALL

VOLUME 326. Applications of Chimeric Genes and Hybrid Proteins
(Part A: Gene Expression and Protein Purification)
Edited by JEREMY THORNER, SCOTT D. EMR, AND JOHN N. ABELSON

VOLUME 327. Applications of Chimeric Genes and Hybrid Proteins
(Part B: Cell Biology and Physiology)
Edited by JEREMY THORNER, SCOTT D. EMR, AND JOHN N. ABELSON

VOLUME 328. Applications of Chimeric Genes and Hybrid Proteins (Part C: Protein–Protein Interactions and Genomics)
Edited by JEREMY THORNER, SCOTT D. EMR, AND JOHN N. ABELSON

VOLUME 329. Regulators and Effectors of Small GTPases (Part E: GTPases Involved in Vesicular Traffic)
Edited by W. E. BALCH, CHANNING J. DER, AND ALAN HALL

VOLUME 330. Hyperthermophilic Enzymes (Part A)
Edited by MICHAEL W. W. ADAMS AND ROBERT M. KELLY

VOLUME 331. Hyperthermophilic Enzymes (Part B)
Edited by MICHAEL W. W. ADAMS AND ROBERT M. KELLY

VOLUME 332. Regulators and Effectors of Small GTPases (Part F: Ras Family I)
Edited by W. E. BALCH, CHANNING J. DER, AND ALAN HALL

VOLUME 333. Regulators and Effectors of Small GTPases (Part G: Ras Family II)
Edited by W. E. BALCH, CHANNING J. DER, AND ALAN HALL

VOLUME 334. Hyperthermophilic Enzymes (Part C)
Edited by MICHAEL W. W. ADAMS AND ROBERT M. KELLY

VOLUME 335. Flavonoids and Other Polyphenols
Edited by LESTER PACKER

VOLUME 336. Microbial Growth in Biofilms (Part A: Developmental and Molecular Biological Aspects)
Edited by RON J. DOYLE

VOLUME 337. Microbial Growth in Biofilms (Part B: Special Environments and Physicochemical Aspects)
Edited by RON J. DOYLE

VOLUME 338. Nuclear Magnetic Resonance of Biological Macromolecules (Part A)
Edited by THOMAS L. JAMES, VOLKER DÖTSCH, AND ULI SCHMITZ

VOLUME 339. Nuclear Magnetic Resonance of Biological Macromolecules (Part B)
Edited by THOMAS L. JAMES, VOLKER DÖTSCH, AND ULI SCHMITZ

VOLUME 340. Drug–Nucleic Acid Interactions
Edited by JONATHAN B. CHAIRES AND MICHAEL J. WARING

VOLUME 341. Ribonucleases (Part A)
Edited by ALLEN W. NICHOLSON

VOLUME 342. Ribonucleases (Part B)
Edited by ALLEN W. NICHOLSON

VOLUME 343. G Protein Pathways (Part A: Receptors)
Edited by RAVI IYENGAR AND JOHN D. HILDEBRANDT

VOLUME 344. G Protein Pathways (Part B: G Proteins and Their Regulators)
Edited by RAVI IYENGAR AND JOHN D. HILDEBRANDT

Volume 362. Recognition of Carbohydrates in Biological Systems (Part A)
Edited by Yuan C. Lee and Reiko T. Lee

Volume 363. Recognition of Carbohydrates in Biological Systems (Part B)
Edited by Yuan C. Lee and Reiko T. Lee

Volume 364. Nuclear Receptors
Edited by David W. Russell and David J. Mangelsdorf

Volume 365. Differentiation of Embryonic Stem Cells
Edited by Paul M. Wassauman and Gordon M. Keller

Volume 366. Protein Phosphatases
Edited by Susanne Klumpp and Josef Krieglstein

Volume 367. Liposomes (Part A)
Edited by Nejat Düzgüneş

Volume 368. Macromolecular Crystallography (Part C)
Edited by Charles W. Carter, Jr., and Robert M. Sweet

Volume 369. Combinational Chemistry (Part B)
Edited by Guillermo A. Morales and Barry A. Bunin

Volume 370. RNA Polymerases and Associated Factors (Part C)
Edited by Sankar L. Adhya and Susan Garges

Volume 371. RNA Polymerases and Associated Factors (Part D)
Edited by Sankar L. Adhya and Susan Garges

Volume 372. Liposomes (Part B)
Edited by Nejat Düzgüneş

Volume 373. Liposomes (Part C)
Edited by Nejat Düzgüneş

Volume 374. Macromolecular Crystallography (Part D)
Edited by Charles W. Carter, Jr., and Robert W. Sweet

Volume 375. Chromatin and Chromatin Remodeling Enzymes (Part A)
Edited by C. David Allis and Carl Wu

Volume 376. Chromatin and Chromatin Remodeling Enzymes (Part B)
Edited by C. David Allis and Carl Wu

Volume 377. Chromatin and Chromatin Remodeling Enzymes (Part C)
Edited by C. David Allis and Carl Wu

Volume 378. Quinones and Quinone Enzymes (Part A)
Edited by Helmut Sies and Lester Packer

Volume 379. Energetics of Biological Macromolecules (Part D)
Edited by Jo M. Holt, Michael L. Johnson, and Gary K. Ackers

Volume 380. Energetics of Biological Macromolecules (Part E)
Edited by Jo M. Holt, Michael L. Johnson, and Gary K. Ackers

VOLUME 399. Ubiquitin and Protein Degradation (Part B)
Edited by RAYMOND J. DESHAIES

VOLUME 400. Phase II Conjugation Enzymes and Transport Systems
Edited by HELMUT SIES AND LESTER PACKER

VOLUME 401. Glutathione Transferases and Gamma Glutamyl Transpeptidases
Edited by HELMUT SIES AND LESTER PACKER

VOLUME 402. Biological Mass Spectrometry
Edited by A. L. BURLINGAME

VOLUME 403. GTPases Regulating Membrane Targeting and Fusion
Edited by WILLIAM E. BALCH, CHANNING J. DER, AND ALAN HALL

VOLUME 404. GTPases Regulating Membrane Dynamics
Edited by WILLIAM E. BALCH, CHANNING J. DER, AND ALAN HALL

VOLUME 405. Mass Spectrometry: Modified Proteins and Glycoconjugates
Edited by A. L. BURLINGAME

VOLUME 406. Regulators and Effectors of Small GTPases: Rho Family
Edited by WILLIAM E. BALCH, CHANNING J. DER, AND ALAN HALL

VOLUME 407. Regulators and Effectors of Small GTPases: Ras Family
Edited by WILLIAM E. BALCH, CHANNING J. DER, AND ALAN HALL

VOLUME 408. DNA Repair (Part A)
Edited by JUDITH L. CAMPBELL AND PAUL MODRICH

VOLUME 409. DNA Repair (Part B)
Edited by JUDITH L. CAMPBELL AND PAUL MODRICH

VOLUME 410. DNA Microarrays (Part A: Array Platforms and Web-Bench Protocols)
Edited by ALAN KIMMEL AND BRIAN OLIVER

VOLUME 411. DNA Microarrays (Part B: Databases and Statistics)
Edited by ALAN KIMMEL AND BRIAN OLIVER

VOLUME 412. Amyloid, Prions, and Other Protein Aggregates (Part B)
Edited by INDU KHETERPAL AND RONALD WETZEL

VOLUME 413. Amyloid, Prions, and Other Protein Aggregates (Part C)
Edited by INDU KHETERPAL AND RONALD WETZEL

VOLUME 414. Measuring Biological Responses with Automated Microscopy
Edited by JAMES INGLESE

VOLUME 415. Glycobiology
Edited by MINORU FUKUDA

VOLUME 416. Glycomics
Edited by MINORU FUKUDA

Volume 417. Functional Glycomics
Edited by Minoru Fukuda

Volume 418. Embryonic Stem Cells
Edited by Irina Klimanskaya and Robert Lanza

Volume 419. Adult Stem Cells
Edited by Irina Klimanskaya and Robert Lanza

Volume 420. Stem Cell Tools and Other Experimental Protocols
Edited by Irina Klimanskaya and Robert Lanza

Volume 421. Advanced Bacterial Genetics: Use of Transposons and Phage for Genomic Engineering
Edited by Kelly T. Hughes

Volume 422. Two-Component Signaling Systems, Part A
Edited by Melvin I. Simon, Brian R. Crane, and Alexandrine Crane

Volume 423. Two-Component Signaling Systems, Part B
Edited by Melvin I. Simon, Brian R. Crane, and Alexandrine Crane

Volume 424. RNA Editing
Edited by Jonatha M. Gott

Volume 425. RNA Modification
Edited by Jonatha M. Gott

Volume 426. Integrins
Edited by David Cheresh

Volume 427. MicroRNA Methods
Edited by John J. Rossi

Volume 428. Osmosensing and Osmosignaling
Edited by Helmut Sies and Dieter Haussinger

Volume 429. Translation Initiation: Extract Systems and Molecular Genetics
Edited by Jon Lorsch

Volume 430. Translation Initiation: Reconstituted Systems and Biophysical Methods
Edited by Jon Lorsch

Volume 431. Translation Initiation: Cell Biology, High-Throughput and Chemical-Based Approaches
Edited by Jon Lorsch

Volume 432. Lipidomics and Bioactive Lipids: Mass-Spectrometry–Based Lipid Analysis
Edited by H. Alex Brown

Volume 433. Lipidomics and Bioactive Lipids: Specialized Analytical Methods and Lipids in Disease
Edited by H. Alex Brown

Volume 434. Lipidomics and Bioactive Lipids: Lipids and Cell Signaling
Edited by H. Alex Brown

Volume 435. Oxygen Biology and Hypoxia
Edited by Helmut Sies and Bernhard Brüne

Volume 436. Globins and Other Nitric Oxide-Reactive Protiens (Part A)
Edited by Robert K. Poole

Volume 437. Globins and Other Nitric Oxide-Reactive Protiens (Part B)
Edited by Robert K. Poole

Volume 438. Small GTPases in Disease (Part A)
Edited by William E. Balch, Channing J. Der, and Alan Hall

Volume 439. Small GTPases in Disease (Part B)
Edited by William E. Balch, Channing J. Der, and Alan Hall

Volume 440. Nitric Oxide, Part F Oxidative and Nitrosative Stress in Redox Regulation of Cell Signaling
Edited by Enrique Cadenas and Lester Packer

Volume 441. Nitric Oxide, Part G Oxidative and Nitrosative Stress in Redox Regulation of Cell Signaling
Edited by Enrique Cadenas and Lester Packer

Volume 442. Programmed Cell Death, General Principles for Studying Cell Death (Part A)
Edited by Roya Khosravi-Far, Zahra Zakeri, Richard A. Lockshin, and Mauro Piacentini

Volume 443. Angiogenesis: *In Vitro* Systems
Edited by David A. Cheresh

Volume 444. Angiogenesis: *In Vivo* Systems (Part A)
Edited by David A. Cheresh

Volume 445. Angiogenesis: *In Vivo* Systems (Part B)
Edited by David A. Cheresh

Volume 446. Programmed Cell Death, The Biology and Therapeutic Implications of Cell Death (Part B)
Edited by Roya Khosravi-Far, Zahra Zakeri, Richard A. Lockshin, and Mauro Piacentini

Volume 447. RNA Turnover in Bacteria, Archaea and Organelles
Edited by Lynne E. Maquat and Cecilia M. Arraiano

Volume 448. RNA Turnover in Eukaryotes: Nucleases, Pathways and Analysis of mRNA Decay
Edited by Lynne E. Maquat and Megerditch Kiledjian

Volume 449. RNA Turnover in Eukaryotes: Analysis of Specialized and Quality Control RNA Decay Pathways
Edited by Lynne E. Maquat and Megerditch Kiledjian

Volume 450. Fluorescence Spectroscopy
Edited by Ludwig Brand and Michael L. Johnson

Volume 451. Autophagy: Lower Eukaryotes and Non-Mammalian Systems (Part A)
Edited by Daniel J. Klionsky

Volume 452. Autophagy in Mammalian Systems (Part B)
Edited by Daniel J. Klionsky

Volume 453. Autophagy in Disease and Clinical Applications (Part C)
Edited by Daniel J. Klionsky

Volume 454. Computer Methods (Part A)
Edited by Michael L. Johnson and Ludwig Brand

Volume 455. Biothermodynamics (Part A)
Edited by Michael L. Johnson, Jo M. Holt, and Gary K. Ackers (Retired)

Volume 456. Mitochondrial Function, Part A: Mitochondrial Electron Transport Complexes and Reactive Oxygen Species
Edited by William S. Allison and Immo E. Scheffler

Volume 457. Mitochondrial Function, Part B: Mitochondrial Protein Kinases, Protein Phosphatases and Mitochondrial Diseases
Edited by William S. Allison and Anne N. Murphy

Volume 458. Complex Enzymes in Microbial Natural Product Biosynthesis, Part A: Overview Articles and Peptides
Edited by David A. Hopwood

Volume 459. Complex Enzymes in Microbial Natural Product Biosynthesis, Part B: Polyketides, Aminocoumarins and Carbohydrates
Edited by David A. Hopwood

Volume 460. Chemokines, Part A
Edited by Tracy M. Handel and Damon J. Hamel

Volume 461. Chemokines, Part B
Edited by Tracy M. Handel and Damon J. Hamel

Volume 462. Non-Natural Amino Acids
Edited by Tom W. Muir and John N. Abelson

Volume 463. Guide to Protein Purification, 2nd Edition
Edited by Richard R. Burgess and Murray P. Deutscher

Volume 464. Liposomes, Part F
Edited by Nejat Düzgüneş

Volume 465. Liposomes, Part G
Edited by Nejat Düzgüneş

Volume 466. Biothermodynamics, Part B
Edited by Michael L. Johnson, Gary K. Ackers, and Jo M. Holt

Volume 467. Computer Methods Part B
Edited by Michael L. Johnson and Ludwig Brand

Volume 468. Biophysical, Chemical, and Functional Probes of RNA Structure, Interactions and Folding: Part A
Edited by Daniel Herschlag

Volume 469. Biophysical, Chemical, and Functional Probes of RNA Structure, Interactions and Folding: Part B
Edited by Daniel Herschlag

Volume 470. Guide to Yeast Genetics: Functional Genomics, Proteomics, and Other Systems Analysis, 2nd Edition
Edited by Gerald Fink, Jonathan Weissman, and Christine Guthrie

Volume 471. Two-Component Signaling Systems, Part C
Edited by Melvin I. Simon, Brian R. Crane, and Alexandrine Crane

Volume 472. Single Molecule Tools, Part A: Fluorescence Based Approaches
Edited by Nils G. Walter

Volume 473. Thiol Redox Transitions in Cell Signaling, Part A Chemistry and Biochemistry of Low Molecular Weight and Protein Thiols
Edited by Enrique Cadenas and Lester Packer

Volume 474. Thiol Redox Transitions in Cell Signaling, Part B Cellular Localization and Signaling
Edited by Enrique Cadenas and Lester Packer

Volume 475. Single Molecule Tools, Part B: Super-Resolution, Particle Tracking, Multiparameter, and Force Based Methods
Edited by Nils G. Walter

Volume 476. Guide to Techniques in Mouse Development, Part A Mice, Embryos, and Cells, 2nd Edition
Edited by Paul M. Wassarman and Philippe M. Soriano

Volume 477. Guide to Techniques in Mouse Development, Part B Mouse Molecular Genetics, 2nd Edition
Edited by Paul M. Wassarman and Philippe M. Soriano

Volume 478. Glycomics
Edited by Minoru Fukuda

VOLUME 479. Functional Glycomics
Edited by MINORU FUKUDA

VOLUME 480. Glycobiology
Edited by MINORU FUKUDA

VOLUME 481. Cryo-EM, Part A: Sample Preparation and Data Collection
Edited by GRANT J. JENSEN

VOLUME 482. Cryo-EM, Part B: 3-D Reconstruction
Edited by GRANT J. JENSEN

VOLUME 483. Cryo-EM, Part C: Analyses, Interpretation, and Case Studies
Edited by GRANT J. JENSEN

VOLUME 484. Constitutive Activity in Receptors and Other Proteins, Part A
Edited by P. MICHAEL CONN

VOLUME 485. Constitutive Activity in Receptors and Other Proteins, Part B
Edited by P. MICHAEL CONN

VOLUME 486. Research on Nitrification and Related Processes, Part A
Edited by MARTIN G. KLOTZ

VOLUME 487. Computer Methods, Part C
Edited by MICHAEL L. JOHNSON AND LUDWIG BRAND

VOLUME 488. Biothermodynamics, Part C
Edited by MICHAEL L. JOHNSON, JO M. HOLT, AND GARY K. ACKERS

VOLUME 489. The Unfolded Protein Response and Cellular Stress, Part A
Edited by P. MICHAEL CONN

VOLUME 490. The Unfolded Protein Response and Cellular Stress, Part B
Edited by P. MICHAEL CONN

VOLUME 491. The Unfolded Protein Response and Cellular Stress, Part C
Edited by P. MICHAEL CONN

VOLUME 492. Biothermodynamics, Part D
Edited by MICHAEL L. JOHNSON, JO M. HOLT, AND GARY K. ACKERS

VOLUME 493. Fragment-Based Drug Design
Tools, Practical Approaches, and Examples
Edited by LAWRENCE C. KUO

VOLUME 494. Methods in Methane Metabolism, Part A
Methanogenesis
Edited by AMY C. ROSENZWEIG AND STEPHEN W. RAGSDALE

VOLUME 495. Methods in Methane Metabolism, Part B
Methanotrophy
Edited by AMY C. ROSENZWEIG AND STEPHEN W. RAGSDALE

VOLUME 496. Research on Nitrification and Related Processes, Part B
Edited by MARTIN G. KLOTZ AND LISA Y. STEIN

VOLUME 497. Synthetic Biology, Part A
Methods for Part/Device Characterization and Chassis Engineering
Edited by CHRISTOPHER VOIGT

VOLUME 498. Synthetic Biology, Part B
Computer Aided Design and DNA Assembly
Edited by CHRISTOPHER VOIGT

VOLUME 499. Biology of Serpins
Edited by JAMES C. WHISSTOCK AND PHILLIP I. BIRD

VOLUME 500. Methods in Systems Biology
Edited by DANIEL JAMESON, MALKHEY VERMA, AND HANS V. WESTERHOFF

VOLUME 501. Serpin Structure and Evolution
Edited by JAMES C. WHISSTOCK AND PHILLIP I. BIRD

VOLUME 502. Protein Engineering for Therapeutics, Volume A
Edited by K. DANE WITTRUP AND GREGORY L. VERDINE

VOLUME 503. Protein Engineering for Therapeutics, Volume B
Edited by K. DANE WITTRUP AND GREGORY L. VERDINE

VOLUME 504. Imaging and Spectroscopic Analysis of Living Cells
Optical and Spectroscopic Techniques
Edited by P. MICHAEL CONN

VOLUME 505. Imaging and Spectroscopic Analysis of Living Cells
Live Cell Imaging of Cellular Elements and Functions
Edited by P. MICHAEL CONN

VOLUME 506. Imaging and Spectroscopic Analysis of Living Cells
Imaging Live Cells in Health and Disease
Edited by P. MICHAEL CONN

SECTION ONE

VIRUSES, MICROBES, AND PARASITES

CHAPTER ONE

Microbial Cells Analysis by Atomic Force Microscopy

David Alsteens

Contents

1. Introduction 4
2. Immobilization of Microbial Cells 6
   2.1. Isolated cell wall 6
   2.2. Imaging of living cells 7
3. Applications of AFM to Nanomicrobiology 11
   3.1. Imaging dynamic processes 11
   3.2. Elucidating of microbial cell substructures 13
   3.3. Imaging of underlying layers 14
4. Future Challenges 14
Acknowledgments 14
References 15

Abstract

Unraveling the structure of microbial cells is a major challenge in current microbiology and offers exciting prospects in biomedicine. Atomic force microscopy (AFM) appears as a powerful method to image the surface ultrastructure of live cells under physiological conditions and allows real-time imaging to follow dynamic processes such as cell growth, and division and effects of drugs and chemicals.

The following chapter introduces different methods of sample preparation to gain insights into the microbial cell organization. Successful strategies to immobilize microorganisms, including physical entrapment and chemical attachment, are described. This step is a key step and a prerequisite of any analysis and persists as an important limitation to the application of AFM to microbiology due to the wide diversity of microorganisms.

Finally, some applications are depicted which underlie the ability of AFM to explore living microbes with unprecedented resolution.

Institute of Condensed Matter & Nanosciences, Université catholique de Louvain, Louvain-la-Neuve, Belgium

Methods in Enzymology, Volume 506
ISSN 0076-6879, DOI: 10.1016/B978-0-12-391856-7.00025-1

1. Introduction

The nanoscale surface exploration of microbes using atomic force microscopy (AFM) has been increasingly exploited to explore various cells in different areas such as biology, medicine, industry, and ecology. Because they constitute the frontier between the cells and their environment, microbial cell walls play several key functions: supporting the internal turgor pressure of the cell, protecting the cytoplasm from the outer environment, imparting shape to the organism, acting as a molecular sieve, sensing environmental stresses/stimuli, and controlling interfacial interactions and thereby biointerfacial processes (molecular recognition, cell adhesion, and aggregation). These processes have major consequences, which can be either beneficial, for instance in biotechnology (wastewater treatment, bioremediation, immobilized cells in reactors), or detrimental in industrial systems (biofouling, contamination) and in medicine (interactions of pathogens with animal host tissues, accumulation on implants). Hence, there is considerable interest in improving our understanding of the functions of microbial cell surfaces.

The knowledge of these functions requires determination of their structural and biophysical properties. To do so, several techniques have been made available for biologists. For instance, light microscopy has long been a favorite tool of microbiologists, enabling them to count and identify microbial cells and to observe general structural details. Fluorescence techniques provide valuable information on the cell wall organization, assembly, and dynamics, but the resolution remains limited to the wavelength of the light source (Daniel and Errington, 2003; Turner *et al.*, 2010a). Recently, there has been progress in breaking the diffraction limit barrier with the advent of super-resolution microscopy, also named far-field optical nanoscopy, but these methods are not yet well-established in microbiology (Gitai, 2009; Hell, 2007). Our current perception of microbial cell walls owes much to the development of cryo-electron microscopy techniques, which have allowed researchers to obtain high-resolution views of microbial structures in conditions close to their native state (Matias and Beveridge, 2005). Notably, cryo-electron tomography—or three-dimensional electron microscopy—provides images of whole bacterial cells at resolutions that are one to two orders of magnitude higher than those currently attained with light microscopy (Milne and Subramaniam, 2009). However, cryo-electron microscopy methods are demanding and require vacuum conditions during the analysis, implying that live cells cannot be imaged in aqueous solution.

To overcome this limitation, several biophysical methods have been developed to probe microbial surfaces at the nanoscale in liquid condition, including flow chamber experiments, biomembrane force probe, optical and magnetic tweezers, and AFM (Bustamante *et al.*, 2000; Dufrêne, 2008; Evans and Calderwood, 2007; Neuman and Nagy, 2008; Sotomayor and Schulten, 2007). Among these powerful methods, only AFM can both localize and force-probe single molecules on live cells. This technique has recently opened a wide range of novel possibilities for probing microbial surfaces at the nanoscale, going from topographic imaging, mapping of chemical groups, detection of single molecular recognition sites to probing of nanomechanical behavior of individual cell wall proteins.

Briefly, the principle of AFM is based on monitoring the interaction forces between an ultra-sharp tip and a sample surface to generate a topographical surface image. In addition, force measurements make it possible to quantitatively measure forces between interacting moieties. The main parts of the AFM are the cantilever, the tip, the sample stage, and the optical deflection system consisting of a laser diode and a photodetector (Fig. 1.1) (Jena and Hörber, 2002). The sample is mounted on a piezoelectric scanner which ensures three-dimensional positioning with high precision and accuracy. By applying appropriate voltages, piezoelectric materials expand in

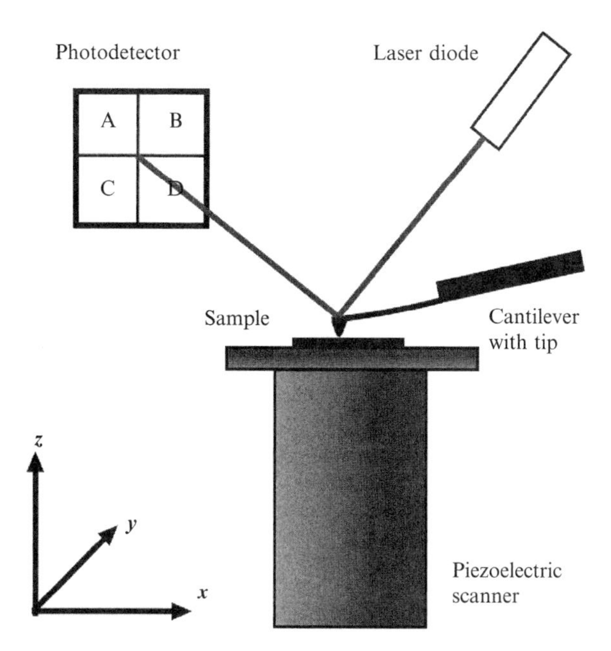

Figure 1.1 AFM setup. The sample is mounted on a three-axis piezoelectric scanner. The topography of the sample is monitored by the deflection of the cantilever measured using a laser beam reflected of its back. The voltage difference between the upper and lower half of the photodetector (A + B − (C + D)) is a measure for the vertical deflection of the cantilever. (For color version of this figure, the reader is referred to the Web version of this chapter.)

some directions in a defined way (Fig. 1.1). Most instruments today use an optical method to measure the cantilever deflection with high resolution by registering the shift of a laser beam on the backside of the cantilever onto a split position-sensitive photodetector (Fig. 1.1). As the cantilever bends, the position of the laser beam on the detector shifts from its initial central point. The resulting voltage difference between the segments of the photodiode gives a direct and linear measure for the deflection of the cantilever. AFMs can be operated in air, vacuum, or liquid (Weisenhorn *et al.*, 1989) and at various temperatures, and therefore microscopists have used it to visualize in real-time how cell surfaces interact with external agents or change during cell growth and division (Dufrêne, 2004).

Typically, the AFM tips are made of silicon (Si) or silicon nitride (Si_3N_4) using microfabrication techniques. Often the backside of the cantilever is covered with a thin gold layer to enhance its reflectivity. The probe shape has a conical or pyramidal apex with a 2–50 nm curvature radius. For biological specimens, best results are generally obtained with cantilevers exhibiting small spring constants, that is, in the range of 0.01–0.10 N/m (Hinterdorfer and Dufrêne, 2006).

Here, different methods useful to gain insights into microbial cell organization will be introduced in the following chapter.

2. Immobilization of Microbial Cells

A unique characteristic of AFM is its ability to reveal structural details cell walls and membranes on single live cells with unprecedented resolution (for a recent review, see Scheuring and Dufrêne, 2010).

2.1. Isolated cell wall

During the past few years, AFM has been extensively used to image isolated membranes. Isolated proteins or fragmented cell wall are generally sufficiently attached by simple adsorption (Müller *et al.*, 1996). The first impressive images of purified membranes of bacteriorhodopsin in buffer solution were obtained in 1990 (Butt *et al.*, 1990) with a resolution down to 1.1 nm. Similar resolutions were obtained on a large variety of other crystalline membrane proteins (Purple membrane of *Halobacterium* (Muller *et al.*, 1995), porins of *Escherichia coli* (Schabert *et al.*, 1995; Scheuring *et al.*, 1999), S-layers from *Corynebacterium glutamicum* (Scheuring *et al.*, 2002), from *Deinococcus radiodurans* (Müller *et al.*, 1996), and from *Bacillus sphaericus* (Gyorvary *et al.*, 2003), etc.). AFM was also able to monitor conformational changes of the individual molecules revealing mechanisms underlying diverse functions (Müller and Engel, 1999; Müller *et al.*, 1996). The development of high-speed AFMs (HS-AFMs) (Ando *et al.*, 2001; Hansma *et al.*, 2006; Humphris *et al.*, 2005) now allows one to acquire images in less than 20 ms. The application of HS-AFM enabled Casuso *et al.* (2009) to record high-resolution movies acquired at a 100 ms frame acquisition time. With the development of this technique, the observations of motions of the trimers in regions between different bacteriorhodopsin have become possible. Dynamic changes in protein conformation in response to illumination were visualized in the bacteriorhodopsin using HS-AFM (Shibata *et al.*, 2010). These findings provide novel perspectives for analyzing diffusion process of membrane proteins.

2.2. Imaging of living cells

An important prerequisite to investigate cell wall architecture is a good sample preparation and immobilization. The growing number of papers published about cell immobilization reveals that this step persists as an important limitation to the application of AFM to microbiology. Because cells dramatically differ in their shape, size, rigidity, and chemical properties, many strategies coexist for their immobilization. Among these, drying microbes onto the substrate before imaging is a popular choice. Nevertheless, this step can induce a cell dehydration leading to the flattening and collapsing of the microorganism. Additionally, imaging in buffer is often crucial to assess the native state of the investigated cell surface. The specimen must also be firmly immobilized to the support to withstand the lateral forces exerted by the scanning tip. While proteins or fragmented cell wall is attached by simple adsorption, stronger attachment is often needed for microbes. Several other difficulties are associated with the imaging of whole microbial cells. First, as opposed to animal cells, microbial cells have a well-defined shape and have no tendency to spread over substrates. As a result, the contact area between cells and substrate is very small, often leading to cell detachment by the scanning probe. A second problem is the vertical motion of the sample (or probe) which is typically limited to a few micrometers (in the range of ~ 10 μm for commercial scanner). This makes it difficult to image the surface of large objects, such as round cells. In this context, some options have been developed to immobilize microorganisms. With these limitations in mind, new successful strategies to immobilize microorganisms have been developed, including physical entrapment and chemical attachment (Fig. 1.2).

2.2.1. Physical entrapment

Rapid advances have been made in imaging live cells. Developed in 1995 (Kasas and Ikai, 1995), a successfully method consists in trapping spherical microbes in polymer membranes with a pore size comparable to the dimension of the cell (Fig. 1.2A).

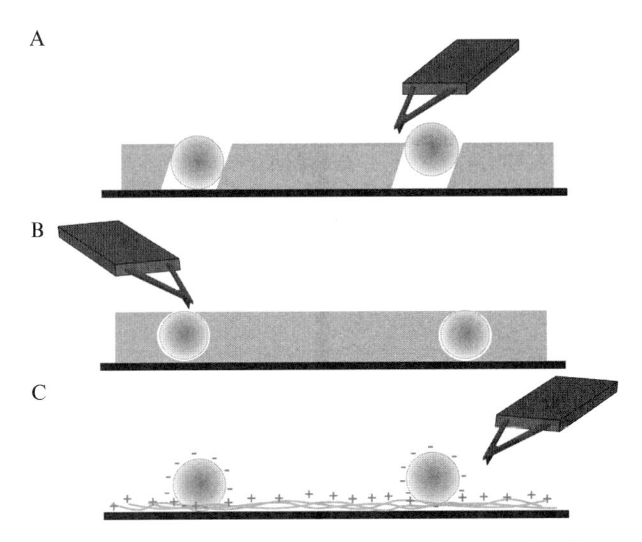

Figure 1.2 Cell immobilization procedures. Live cell imaging implies strong attachment onto a substrate, which can be achieved by (A) physical entrapment into a porous polymer membrane, (B) immobilization in an agar gel matrix, or (C) chemical fixation using glass slides treated with positively charged molecules.

This way offers several advantages:

(i) Mechanical trapping ensures strong attachment to allow live cell imaging at high resolution and the observation of dynamic process.
(ii) The very smooth polycarbonate membrane facilitates the cell localization.
(iii) A large variety of pore size and surface properties of membrane are commercially available.

The main drawbacks of this method are the partial access to the cell surface, a relative selection of the microbe in function of the defined pore size, and the risk of mechanical pressure due to the entrapment which could cause some artifacts in nanomechanical measurements.

Typically, a cell suspension ($\sim 10^6$ cells/ml) is harvested by centrifugation and washed sometimes with buffer. Cells are immobilized by mechanical trapping into porous polycarbonate membrane (Millipore or it4ip) with adequate pore size. After cutting the filter (1 cm $\times$ 1 cm), the lower part is carefully dried on a sheet of tissue and the specimen is then attached to the magnetic sample holder using a double-stick adhesive tape before being mounted in the liquid cell.

The two most crucial characteristics in the choice of the porous membrane are the pore size and the surface properties of the membrane. The pore size has to be slightly smaller than the cell to trap it. Nevertheless, too small pore sizes can deform the cell and induce high pressure which could damage the cell and bias nanomechanical measurements. Therefore, slight enlargement of the pore size, using sodium hydroxide, can be used to circumvent this problem (Turner *et al.*, 2010b). Surface hydrophobicity of polycarbonate membranes also plays a role in cell immobilization. Water contact angle measurements (data not shown) reveal that membranes with a static contact angle above 75° are more effective for trapping yeast cells. A simple approach to increase hydrophobicity is to heat the membranes overnight at 110 °C. Hence, the success of the porous membrane method depends on a combination of mechanical confinement and hydrophobic interactions.

This method was successfully applied to a large series of microbial cells. The technique has allowed observing the supramolecular organization of many cell wall constituents to be directly observed on live cells, including polysaccharides, peptidoglycan, and proteins. As an example, the nanoscale 3D assembly of clustered proteinaceous microfibrils, termed rodlets, was observed on the surface of the human opportunistic pathogen *Aspergillus fumigatus*, which causes several important respiratory diseases. In Fig. 1.3, a

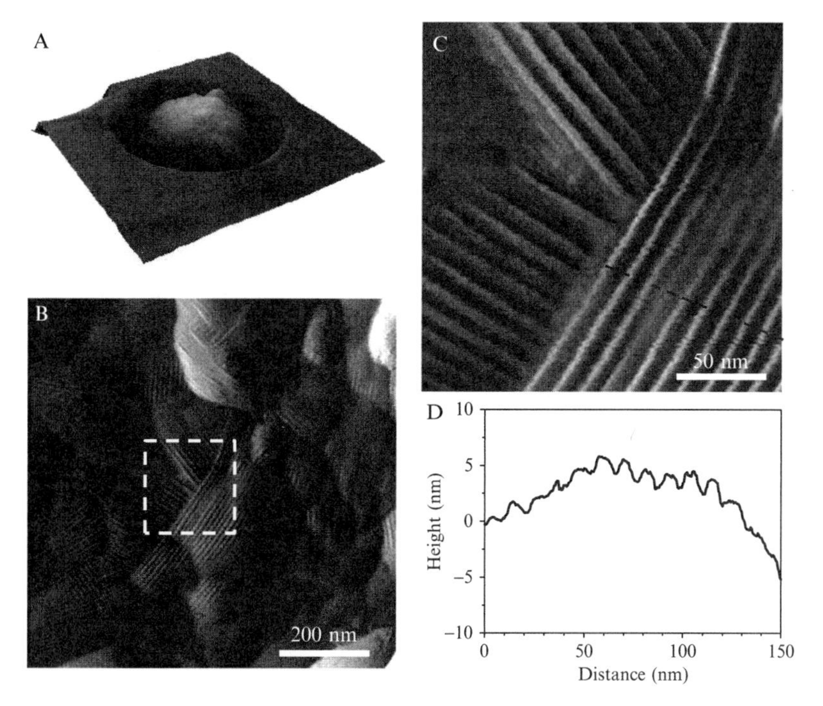

Figure 1.3 Imaging of *Aspergillus fumigatus*. (A) AFM height image of a single conidium trapped into a pore membrane (7 μm × 7 μm). (B) Deflection image showing the crystalline rodlet layer. (C) High-resolution image (see square in B) of the rodlets of hydrophobins. (D) Cross-section (see line in C) showing the hydrophobins height. (For color version of this figure, the reader is referred to the Web version of this chapter.)

single *Aspergillus* fungal spore is trapped in a polycarbonate membrane (Fig. 1.3A) and rodlets can be visualized on higher magnification (Fig. 1.3B and C). Such rodlets are composed of hydrophobic proteins, which are of paramount importance to favor spore dispersion by air currents and mediate adherence to the human tissue causing important respiratory disease such as bronchopulmonary and invasive aspergillosis. Unprecedent good resolution, with a vertical resolution of ~0.5 nm and a lateral resolution of ~5 nm (Fig. 1.3D), was obtained on this rodlets in buffer condition and directly on live cells (Dague *et al.*, 2007). More recently, great efforts have been made to visualize S-layer nanoarrays on living *C. glutamicum* bacterial cells (Dupres *et al.*, 2009). This first *in vivo* visualization of S-layer has allowed a better understanding of the structure of the protein monomolecular array in its native state.

Membrane method is generally not applicable for immobilizing rod-shaped bacteria or cell wall deficient forms of bacteria. An alternative to membrane trapping consists in immobilizing the cells mechanically in an agar gel (Gad and Ikai, 1995) (Fig. 1.2B). In this method, the gel is used as a soft, deformable immobilization matrix, thereby allowing direct visualization of growth process. Nevertheless, this procedure can lead to cell surface contamination by agar and does not ensure sufficient immobilization to allow high-resolution images.

Beside these two strategies, the lithographic patterning of silicon wafers opens new perspectives to better accommodate to the shape of the microorganism (Kailas *et al.*, 2009) without interfering with their cellular physiology. This non-invasive method of trapping has allowed to follow closely a dynamic cellular process, that is, cell division.

2.2.2. Chemical attachment

Other attempts have led to the use of gelatin matrix (Doktycz *et al.*, 2003). For this strategy, microbes need to be suspended in water (the presence of rich media or buffer salts can interfere) and allowed to stand on the gelatin-coated mica surface. Once the cells are immobilized on the surface, they can be imaged in liquid without further alteration. The choice of the gelatin appeared to be crucial. According to Doktycz *et al.* (2003), several bovine gelatins failed while porcine gelatins succeed in immobilizing microorganisms. This stable attachment permits to rinse the cells under a stream of liquid. Specific bonds and electrostatic and hydrophobic interactions take place with gelatin (i.e., denatured collagen), and contribute to retaining microbes. Gelatin has been effective for immobilizing Gram (+) and Gram (−) bacteria, yeasts, and diatoms.

Another strategy consists in using cationic surface coatings (Fig. 1.2C). This approach takes advantage of the negatively charged surface of most cells. Rod-shaped bacteria are often immobilized on a support pretreated with poly-L-lysine or polyethylenimine. For this type of immobilization, the choice of the buffer salts must be considered. In some cases, the addition of divalent cations to the buffer can facilitate binding (Kienberger *et al.*, 2005). Imaging in water can create detrimental osmotic pressure. Therefore, using an isotonic solution like 0.25 *M* sucrose is advisable (Sullivan *et al.*, 2007). Nevertheless, a recent study demonstrated that the use of poly-L-lysine coating for bacterial attachment should probably be avoided due to significant effects on bacterial physiology (Colville *et al.*, 2010). Further, these treatments may induce rearrangement and denaturation of the cell wall, and contamination of the tip. Cationic surface coatings fail to immobilize round-shaped cells due to the too small contact area.

Because of these limitations, covalent bonding strategies appear as a valuable alternative. Substrates are modified with amino groups and subsequently cross-linked to cells using glutaraldehyde. Immobilization by covalent attachment depends on favorable microbe-to-substrate contact, and any repulsion forces that might prevent this contact must be overcome. This technique has been used successfully for AFM imaging with various buffers.

3. Applications of AFM to Nanomicrobiology

Microbial imaging is a crucial challenge for understanding extracellular ultrastructure of single cells, bacterial communities such as biofilms, and microbial behavior to extracellular stresses (for instance growth conditions to antibiotics exposure). The use of AFM in this field will continue to grow particularly due to its ability to study dynamic processes at the nanoscale and on live cells.

3.1. Imaging dynamic processes

AFM appears to be an ideal tool to image dynamic cellular processes. An elegant example is the germination of *A. fumigatus* conidia (Dague *et al.*, 2008), during which progressive disruption of the rodlet layer revealed the underlying inner cell wall structures. Similarly, Kailas *et al.* followed the process of cell division in *Staphylococcus aureus* (Kailas *et al.*, 2009; Touhami *et al.*, 2004). Detailed images of the surface of dividing cells showed ring-like and honeycomb structures at 20 nm resolution (Turner *et al.*, 2010b). Structural analysis of *Bacillus atrophaeus* spores revealed previously unrecognized germination-induced alterations in spore coat architecture (Plomp *et al.*, 2007). The nascent structure of the emerging vegetative cell showed a porous network of peptidoglycan, consistent with a honeycomb model structure.

In pharmacology, real-time AFM imaging is useful to directly visualize the activity of drugs on microbial cell walls (Alsteens *et al.*, 2008; Francius *et al.*, 2008; Verbelen *et al.*, 2006; Yang *et al.*, 2006). There is much interest in studying the molecular mechanism by which the antibiotics affect the cell wall characteristics. Probing the nanoscale surfaces properties of living *Mycobacterium tuberculosis* cells and their modifications upon incubation with four antimycobacterial drugs (i.e., isoniazid (INH), ethionamide (ETH), ethambutol (EMB), and streptomycin (STR)) demonstrated that all drugs induce a substantial increase of surface roughness to an extent that correlates with the localization of the target (Alsteens *et al.*, 2008). As shown in Fig. 1.4, mycobacteria were mechanically

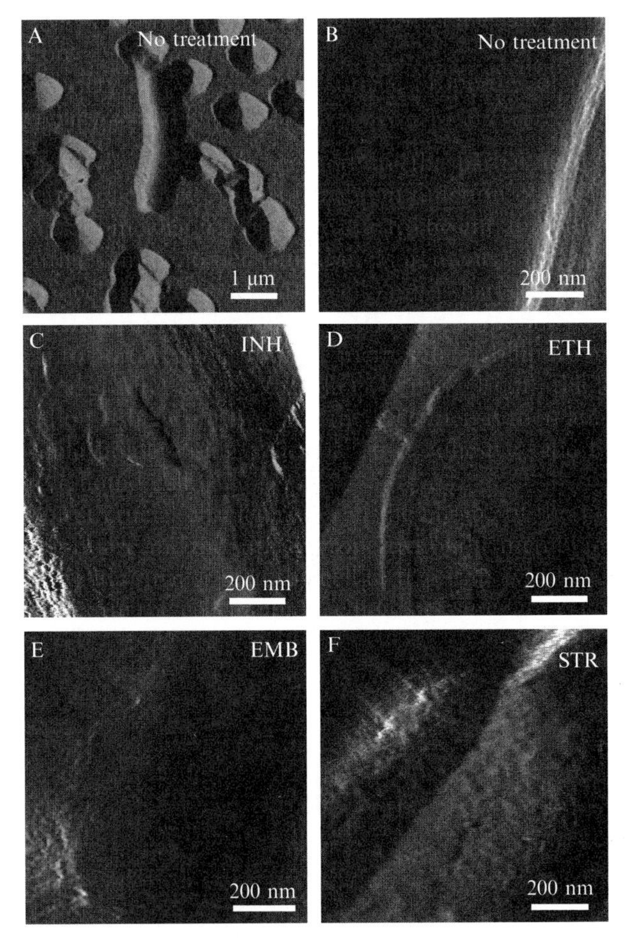

Figure 1.4 Nanoscale imaging of mycobacterial surfaces and their alterations upon treatment with four drugs. (A) AFM deflection image of a single *M. bovis* BCG on a porous membrane. (B) High-resolution deflection images in aqueous solutions of cells prior (B) and after treatment with isoniazid (INH) (C), ethionamide (ETH) (D), ethambutol (EMB) (E), and streptomycin (STR) (F) at the minimum inhibitory concentration. (For color version of this figure, the reader is referred to the Web version of this chapter.)

immobilized onto polycarbonate membrane. For some microbe type cases, the interactions between the cell wall and an untreated substrate are strong enough to ensure immobilization. Here, immobilization results from hydrophobic interaction between polycarbonate and mycolic acid layer. High resolution of *Mycobacterium bovis* BCG cells revealed a very smooth and homogeneous surface, consistent with surface architecture and previous work (Verbelen *et al.*, 2006). The imaging of mycobacteria after incubation for 24 h with INH, ETH, EMB, and STR at concentrations corresponding to the minimal inhibitory concentrations puts in evidence the major ultrastructural alterations: layered structures, striations, and porous morphologies. These modifications result from the inhibition of the synthesis of the three major cell wall constituents, that is, mycolic acids (INH, ETH), arabinans (EMB), and proteins (STR).

Using HS-AFM, Fantner *et al.* (2010) were able to investigate the kinetics of antimicrobial peptide activity. The action of the antimicrobial peptide CM15 on *E. coli* occurred in two steps: a first step or time variable incubation phase (which takes seconds to minutes to complete) was followed by a more rapid execution phase.

3.2. Elucidating of microbial cell substructures

AFM appears also to be an ideal tool to image cell substructures or particular organization at the molecular level with high resolution. Kailas *et al.* have followed the process of cell division in *Staphylococcus aureus* (Kailas *et al.*, 2009; Touhami *et al.*, 2004). Detailed images of the surface of dividing cells showed ring-like and honeycomb structures at 20 nm resolution (Turner *et al.*, 2010b). Structural analysis of *Bacillus atrophaeus* spores revealed previously unrecognized germination-induced alterations in spore coat architecture (Plomp *et al.*, 2007). The nascent structure of the emerging vegetative cell showed a porous network of peptidoglycan, consistent with a honeycomb model structure. Insights into the nanoscale organization of cell wall peptidoglycan were recently revealed by Andre *et al.* (2010). Using mutant strains defective in cell wall polysaccharides, AFM images revealed that peptidoglycan forms periodic bands running parallel to the short axis. Such structures were missing on purified sacculi revealing the importance of imaging directly on live cells and avoiding the use of aggressive treatments.

The study of DNA transfer in *E. coli* during conjugation was also investigated by AFM (Shu *et al.*, 2008). Mating pairs of *E. coli* sharing an extended pilus were imaged, and the AFM tip was used to sever the extended pilus. Further, the area was probed with an antibody to detect ssDNA. This study puts in evidence the ability to image and manipulate a localized region of the cell.

3.3. Imaging of underlying layers

An important drawbacks of AFM as an imaging tool is its ability to image only the superficial layer of the cell wall. To circumvent this, several tricks can be used such as incubation with drug, acid, base, detergent, or enzyme in order to remove the most outer layer.

To gain access to the cytoplasmic membrane, the cell wall components have to be removed. Established protocols, using heat, acid, and enzyme treatments, provide access to the various membrane proteins that reside there (Francius *et al.*, 2008).

The use of genetic manipulation is also a promising tool to investigate the distribution and organization of underlying components by suppressing one particular protein. The recent study of Andre *et al.* (2011) puts in evidence the role played by wall–teichoic acids (WTAs) in *Lactobacillus plantarum*. By comparing wild-type bacteria with mutant defective in the synthesis of WTAs, the nanoscale imaging by AFM showed that WTAs play a central role in controlling cell morphogenesis (surface roughness, cell shape, cell polarization).

4. Future Challenges

A new AFM mode (Scan Asyst by Bruker) enables to control the force applied by the AFM probe on the sample: "vertical and lateral vibrations of the probe during scanning are valuable signals for the characterization of the actual applied force by the tip". In this mode, the set point is continuously and automatically adjusted during the experiment, which enables to obtain high-resolution image of purple membranes at molecular resolution, and cells at high signal-to-noise ratio, for hours without intrinsic force drift (Casuso and Scheuring, 2010). Efforts in applying this technology to live cells are a great challenge for the future.

As previously mentioned, the development of HS-AFM now permits to reduce the image acquisition time to less than 20 ms. These developments provide novel perspectives for analyzing diffusion process of membrane proteins. Nevertheless, transferring this technique to live cell is not an easy task. Although few studies have been performed with HS-AFM on live cell, this technique seems to be very promising to characterize dynamic processes. Such advances will be benefit to better understand biological processes.

ACKNOWLEDGMENTS

This work was supported by the National Foundation for Scientific Research (FNRS). D.A. is a Postdoctoral Researcher of the FRS-FNRS.

REFERENCES

Alsteens, D., Verbelen, C., *et al.* (2008). Organization of the mycobacterial cell wall: A nanoscale view. *Pflugers Arch.* **456,** 117–125.

Ando, T., Kodera, N., *et al.* (2001). A high-speed atomic force microscope for studying biological macromolecules. *Proc. Natl. Acad. Sci. USA* **98,** 12468–12472.

Andre, G., Kulakauskas, S., *et al.* (2010). Imaging the nanoscale organization of peptidoglycan in living *Lactococcus lactis* cells. *Nat. Commun.* **1,** 27.

Andre, G., Deghorain, M., *et al.* (2011). Fluorescence and atomic force microscopy imaging of wall teichoic acids in *Lactobacillus plantarum*. *ACS Chem. Biol.* **6,** 366–376.

Bustamante, C., Macosko, J. C., *et al.* (2000). Grabbing the cat by the tail: Manipulating molecules one by one. *Nat. Rev. Mol. Cell Biol.* **1,** 130–136.

Butt, H. J., Downing, K. H., *et al.* (1990). Imaging the membrane protein bacteriorhodopsin with the atomic force microscope. *Biophys. J.* **58,** 1473–1480.

Casuso, I., and Scheuring, S. (2010). Automated setpoint adjustment for biological contact mode atomic force microscopy imaging. *Nanotechnology* **21**(3), 035104.

Casuso, I., Kodera, N., *et al.* (2009). Contact-mode high-resolution high-speed atomic force microscopy movies of the purple membrane. *Biophys. J.* **97,** 1354–1361.

Colville, K., Tompkins, N., *et al.* (2010). Effects of poly(L-lysine) substrates on attached *Escherichia coli* bacteria. *Langmuir* **26,** 2639–2644.

Dague, E., Alsteens, D., *et al.* (2007). Chemical force microscopy of single live cells. *Nano Lett.* **7,** 3026–3030.

Dague, E., Alsteens, D., *et al.* (2008). High-resolution cell surface dynamics of germinating *Aspergillus fumigatus* conidia. *Biophys. J.* **94,** 656–660.

Daniel, R. A., and Errington, J. (2003). Control of cell morphogenesis in bacteria: Two distinct ways to make a rod-shaped cell. *Cell* **113,** 767–776.

Doktycz, M. J., Sullivan, C. J., *et al.* (2003). AFM imaging of bacteria immobilized on gelatin coated mica surfaces. *Ultramicroscopy* **97,** 209–216.

Dufrêne, Y. F. (2004). Using nanotechniques to explore microbial surfaces. *Nat. Rev. Microbiol.* **2,** 451–460.

Dufrêne, Y. F. (2008). Towards nanomicrobiology using atomic force microscopy. *Nat. Rev. Microbiol.* **6,** 674–680.

Dupres, V., Alsteens, D., *et al.* (2009). *In vivo* imaging of S-layer nanoarrays on *Corynebacterium glutamicum*. *Langmuir* **25,** 9653–9655.

Evans, E. A., and Calderwood, D. A. (2007). Forces and bond dynamics in cell adhesion. *Science* **316,** 1148–1153.

Fantner, G. E., Barbero, R. J., *et al.* (2010). Kinetics of antimicrobial peptide activity measured on individual bacterial cells using high-speed atomic force microscopy. *Nat. Nanotechnol.* **5,** 280–285.

Francius, G., Domenech, O., *et al.* (2008). Direct observation of *Staphylococcus aureus* cell wall digestion by lysostaphin. *J. Bacteriol.* **190,** 7904–7909.

Gad, M., and Ikai, A. (1995). Method for immobilizing microbial cells on gel surface for dynamic AFM studies. *Biophys. J.* **69,** 2226–2233.

Gitai, Z. (2009). New fluorescence microscopy methods for microbiology: Sharper, faster, and quantitative. *Curr. Opin. Microbiol.* **12,** 341–346.

Gyorvary, E. S., Stein, O., *et al.* (2003). Self-assembly and recrystallization of bacterial S-layer proteins at silicon supports imaged in real time by atomic force microscopy. *J. Microsc.* **212,** 300–306.

Hansma, P. K., Schitter, G., *et al.* (2006). High-speed atomic force microscopy. *Science* **314,** 601–602.

Hell, S. W. (2007). Far-field optical nanoscopy. *Science* **316,** 1153–1158.

Hinterdorfer, P., and Dufrêne, Y. F. (2006). Detection and localization of single molecular recognition events using atomic force microscopy. *Nat. Methods* **3,** 347–355.

Humphris, A. D. L., Miles, M. J., *et al.* (2005). A mechanical microscope: High-speed atomic force microscopy. *Appl. Phys. Lett.* **86**(3), 034106–034109.

Jena, B. P., and Hörber, J. K. H. (2002). Atomic Force Microscopy in Cell Biology. Academic Press, San Diego.

Kailas, L., Ratcliffe, E. C., *et al.* (2009). Immobilizing live bacteria for AFM imaging of cellular processes. *Ultramicroscopy* **109,** 775–780.

Kasas, S., and Ikai, A. (1995). A method for anchoring round shaped cells for atomic force microscope imaging. *Biophys. J.* **68,** 1678–1680.

Kienberger, F., Rankl, C., *et al.* (2005). Visualization of single receptor molecules bound to human *Rhinovirus* under physiological conditions. *Structure* **13,** 1247–1253.

Matias, V. R. F., and Beveridge, T. J. (2005). Cryo-electron microscopy reveals native polymeric cell wall structure in *Bacillus subtilis* 168 and the existence of a periplasmic space. *Mol. Microbiol.* **56,** 240–251.

Milne, J. L. S., and Subramaniam, S. (2009). Cryo-electron tomography of bacteria: Progress, challenges and future prospects. *Nat. Rev. Microbiol.* **7,** 666–675.

Müller, D. J., and Engel, A. (1999). Voltage and pH-induced channel closure of porin OmpF visualized by atomic force microscopy. *J. Mol. Biol.* **285,** 1347–1351.

Muller, D. J., Schabert, F. A., *et al.* (1995). Imaging purple membranes in aqueous solutions at subnanometer resolution by atomic force microscopy. *Biophys. J.* **68,** 1681–1686.

Müller, D. J., Baumeister, W., *et al.* (1996). Conformational change of the hexagonally packed intermediate layer of *Deinococcus radiodurans* monitored by atomic force microscopy. *J. Bacteriol.* **178,** 3025–3030.

Neuman, K. C., and Nagy, A. (2008). Single-molecule force spectroscopy: Optical tweezers, magnetic tweezers and atomic force microscopy. *Nat. Methods* **5,** 491–505.

Plomp, M., Leighton, T. J., *et al.* (2007). *In vitro* high-resolution structural dynamics of single germinating bacterial spores. *Proc. Natl. Acad. Sci. USA* **104,** 9644–9649.

Schabert, F. A., Henn, C., *et al.* (1995). Native *Escherichia coli* OMPF porin surfaces probed by atomic force microscopy. *Science* **268,** 92–94.

Scheuring, S., and Dufrêne, Y. F. (2010). Atomic force microscopy: Probing the spatial organization, interactions and elasticity of microbial cell envelopes at molecular resolution. *Mol. Microbiol.* **75,** 1327–1336.

Scheuring, S., Ringler, P., *et al.* (1999). High resolution AFM topographs of the *Escherichia coli* water channel aquaporin Z. *EMBO J.* **18,** 4981–4987.

Scheuring, S., Stahlberg, H., *et al.* (2002). Charting and unzipping the surface layer of Corynebacterium glutamicum with the atomic force microscope. *Mol. Microbiol.* **44,** 675–684.

Shibata, M., Yamashita, H., *et al.* (2010). High-speed atomic force microscopy shows dynamic molecular processes in photoactivated bacteriorhodopsin. *Nat. Nanotechnol.* **5,** 208–212.

Shu, A. C., Wu, C. C., *et al.* (2008). Evidence of DNA transfer through F-pilus channels during *Escherichia coli* conjugation. *Langmuir* **24,** 6796–6802.

Sotomayor, M., and Schulten, K. (2007). Single-molecule experiments *in vitro* and *in silico*. *Science* **316,** 1144–1148.

Sullivan, C. J., Venkataraman, S., *et al.* (2007). Comparison of the indentation and elasticity of *E. coli* and its spheroplasts by AFM. *Ultramicroscopy* **107,** 934–942.

Touhami, A., Jericho, M. H., *et al.* (2004). Atomic force microscopy of cell growth and division in *Staphylococcus aureus*. *J. Bacteriol.* **186,** 3286–3295.

Turner, R. D., Ratcliffe, E. C., *et al.* (2010a). Peptidoglycan architecture can specify division planes in *Staphylococcus aureus*. *Nat. Commun.* **1,** 26.

Turner, R. D., Thomson, N. H., *et al.* (2010b). Improvement of the pore trapping method to immobilize vital coccoid bacteria for high-resolution AFM: A study of *Staphylococcus aureus*. *J. Microsc.* **238,** 102–110.
Verbelen, C., Dupres, V., *et al.* (2006). Ethambutol-induced alterations in *Mycobacterium bovis* BCG imaged by atomic force microscopy. *FEMS Microbiol. Lett.* **264,** 192–197.
Weisenhorn, A. L., Hansma, P. K., *et al.* (1989). Forces in atomic force microscope in air and water. *Appl. Phys. Lett.* **54,** 2651–2653.
Yang, L., Wang, K. M., *et al.* (2006). Atomic force microscopy study of different effects of natural and semisynthetic β-lactam on the cell envelope of *Escherichia coli*. *Anal. Chem.* **78,** 7341–7345.

CHAPTER TWO

Understanding Parasite Transmission Through Imaging Approaches

Mirko Singer *and* Freddy Frischknecht

Contents

1. Why Imaging Parasites? 20
2. *In Vivo* Models: "How Much Wrong Can Be Right?" 20
3. The Choice of the Appropriate Microscope 24
4. New Materials Provide the Basis for New Approaches to Dissect Cell Migration 25
5. Visualizing Cellular Functions in Motile Cells 27
6. Applying Imaging for Drug Discovery 29
Acknowledgments 31
References 31

Abstract

Unicellular parasites are of high medical relevance as they cause such devastating diseases as malaria or sleeping sickness. Besides the search for improved treatments, research on these parasites is valuable as they constitute interesting model cells to study basic processes of life. They can also serve as valuable reality checks for our presumed understanding of biological processes that emerge from the study of human or yeast cells, as our common ancestor with many parasites is much older than the one with yeast. But working with parasites can be tricky and time-consuming, if not outright impossible. Here, we focus on examples from imaging studies investigating the transmission of the malaria parasite. Achieving an understanding of the processes important for malaria transmission necessitates different imaging approaches and new molecular and material technologies. The discussed techniques will include *in vivo* imaging of pathogens in living animals, screening methodologies, and new materials as surrogate 3D environments.

Parasitology, Department of Infectious Diseases, University of Heidelberg Medical School, Heidelberg, Germany

Methods in Enzymology, Volume 506
ISSN 0076-6879, DOI: 10.1016/B978-0-12-391856-7.00026-3

1. Why Imaging Parasites?

Pathogenic bacteria and viruses are brilliant tricksters that can use parts of the cellular machinery of their eukaryotic host cells to their own advantage. This necessitates the study of host–pathogen interactions to uncover potential new interventions for infectious diseases. It also allows the use of these pathogens as living probes to understand the molecular cell biology of their host cells. Parasites are eukaryotic pathogens and thus provide the additional advantage of being model cells for understanding the basic processes of life. This is particularly true for the single-celled protozoa among them. As live cell imaging of nonfluorescent cells within the host is very complicated, the best-studied parasites are those which can be genetically manipulated to express fluorescent proteins. This includes various parasites species of *Plasmodium*, *Toxoplasma*, *Leishmania*, and *Trypanosoma* (Bolhassani *et al.*, 2011; Nishikawa *et al.*, 2008; Pires *et al.*, 2008; Talman *et al.*, 2010). Some of these, like *Plasmodium*, the parasite causing malaria, are intracellular masters of survival and generate whole new organelle systems in their hapless host cells. *Theileria*, a parasite of cattle, enters and transforms macrophages or lymphocytes and divides in perfect synchrony with their host cell. Others, like *Trypanosoma brucei* or *Giardia*, live an extracellular lifestyle and are constantly exposed to the immune system of their host. Microscopy has been used to study these parasites ever since their discovery and has revealed striking insights into the dynamics of parasites in their hosts (Amino *et al.*, 2005, 2006; Coombes and Robey, 2010; Frischknecht, 2010). Imaging studies thus guide biological curiosity as well as medical research.

The aim of this chapter is to highlight a small number of studies that have been performed on parasites focusing on the transmission of parasites from their insect vectors to the mammalian host. We will not do justice to completeness but hope to confer an appreciation of the level of new knowledge that can be gained from studying these remarkable organisms.

2. *In Vivo* Models: "How Much Wrong Can Be Right?"

The gold standard of imaging pathogens at work is to be as close as possible to a natural infection. As our interest is focused mainly on human pathogens, the first corner to be cut is the use of an appropriate surrogate animal model that can be infected, which usually is a mouse. Mice have the advantage of being easy to handle, exist with defined genetic backgrounds and many strains have individual genes deleted or modified. However, mice are not humans and much of what can be learned from infecting them cannot be directly transferred to human biology. Clearly, there is no difference here between studying infectious bacteria, viruses, or parasites. And just like working with viruses or bacteria, the question whether to work with the human parasite in mice or to use a further rodent infecting surrogate arises. This has been answered differently when imaging the first steps of the transmission from the vector to the host of *Plasmodium*, where rodent model parasites were used (Amino *et al.*, 2006; Vanderberg and Frevert, 2004), or *Leishmania*, where a human infecting parasite was imaged (Peters *et al.*, 2008). Both parasites are transmitted into the skin of the host, and as a model site for imaging the ear pinna was used, as it offers easy access for an objective lens on an inverted microscope. This also allows to image cells within the tissue using simple wide-field microscopy (Amino *et al.*, 2006; Hellmann *et al.*, 2011). However, the ear pinna is clearly a special structure with a much thinner dermis as other skin sites, which might in turn lead to problems of generalizing the obtained data.

When imaging parasites transmitted by an arthropod vector, one ideally observes the vector, the parasite and the host during the infectious bite. While this might indeed work for ticks or other slow feeders, to catch a mosquito in the act of biting at a certain skin site calls for much patience and has thus only been rarely observed (Kebaier *et al.*, 2009; Vanderberg and Frevert, 2004). This biological three-body problem can also be dissected to image first the parasite in the vector, for example by fixing the mosquito with superglue onto a glass slide and freeing the stylets (parts of the proboscis that are inserted into the skin) from the surrounding labrum, which should be fixed in a droplet of glue to be kept aside (Frischknecht *et al.*, 2004). Ideally, this operation is carried out at low temperatures so the mosquito does not move and especially does not salivate. Upon transfer to ambient temperatures, the mosquito then awakes rapidly and often immediately starts to salivate, revealing the parasites that are injected (Fig. 2.1A). If a fluorescent parasite is used, it is possible to observe and count the number of parasites that are ejected from the (clearly irritated) mosquito over time (Frischknecht *et al.*, 2004). This has revealed that there is a limited pool of *Plasmodium* sporozoites available for ejection within the salivary glands. Even if there are 20,000 sporozoites resident in the glands, only up to 1000 are ejected with some mosquitoes even ejecting none, despite salivating massively. It has also been described that most mosquitoes eject sporozoites within the first 2 min of salivation (Frischknecht *et al.*, 2004).

To follow the parasites after they are injected into the skin (Sidjanski and Vanderberg, 1997), one can hold the ear of an anesthetized mouse over the edge of a beaker filled with mosquitoes and covered by mosquito gaze (Fig. 2.1B). Ideally, the mosquitoes have been starved over night to increase feeding behavior. With some patience, a mosquito can then be observed to insert its proboscis and take a blood meal. If parasites are imaged later, it is

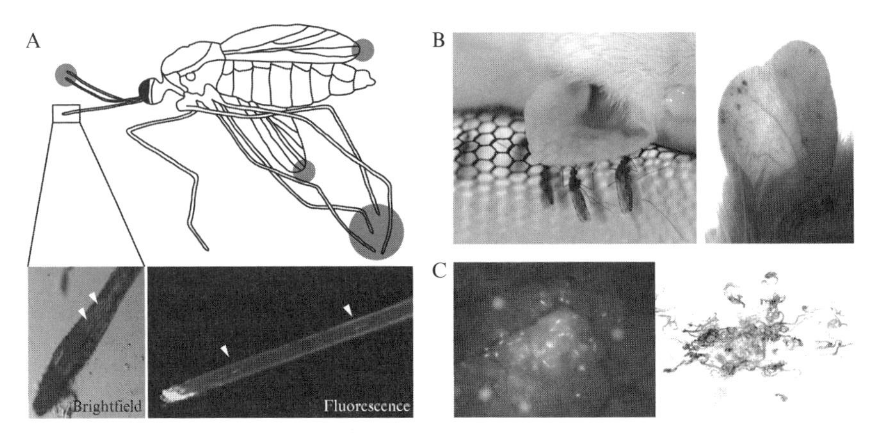

Figure 2.1 Imaging *Plasmodium* sporozoite transmission to the mouse. (A) Schematic drawing of how to immobilize a mosquito on a glass slide using superglue (dark circles) to fix the wings, legs and the labrum. In addition the mosquito is fixed with the thorax (not shown). Enlargement shows at left a brightfield image of the tip of a salivating mosquito proboscis with the salivary canal and at right an image taken with a green filterset on a widefield fluorescence microscope. Note the two sporozoites being ejected by the salivating mosquito. (B) *Anopheles stephensi* mosquitoes feeding on a mouse ear through a mosquito gaze and potentially injecting *Plasmodium* sporozoites into the dermis of the ear pinna (left); right: Haemorrhages after mosquito probing on a mouse ear. (C) Widefield fluorescent image (left) and inverted maximum intensity projection (right) of *Plasmodium berghei* sporozoites expressing GFP in their cytoplasm and deposited in the dermis of the ear pinna. Images modified from Frischknecht *et al.*, 2004; Frischknecht, 2007; Amino *et al.*, 2006.

recommended to remove the hair at the bite site a day or two prior to the bite by hair-removal lotion, duct tape, or shaving to reduce their autofluorescent signal. During the bite, the mosquito moves its stylets rapidly through the skin thus injuring blood capillaries and thereby causing a small hemorrhage. This can be spotted later by holding the mouse against a light bulb such that the light shines through the ear (Fig. 2.1C). One can then place the mouse on the stage of the inverted microscope with the part of the ear showing the hemorrhage positioned over the objective lens. As this takes a few minutes to achieve, one loses information of the very early events after transmission, which can only be obtained by the "painful" three-body approach. The latter indeed revealed that within seconds after being transmitted, the sporozoites, which are largely immotile in the salivary gland (Frischknecht *et al.*, 2004), are rapidly moving within the dermis (Vanderberg and Frevert, 2004) (Fig. 2.1C). The slower approach in turn allowed a quantitative analysis that revealed not just the average speed of the migrating sporozoites but also their slow-down over time and their relative success in entering blood and lymph vessels (Amino *et al.*, 2006). Sporozoites entering blood vessels can eventually reach and infect the liver where they ultimately differentiate into blood-cell invading forms (merozoites), while those sporozoites entering lymph vessels are destroyed in the draining lymph nodes (Amino *et al.*, 2006; Frischknecht, 2007) with their contribution to the outcome of the infection remaining unclear. This appears to be in contrast to *T. brucei*, the causative agents of sleeping sickness, which are entering the blood stream via migration through the lymphatic system (Barry and Emergy, 1984). The careful observations on *Plasmodium* show that quantitative analysis is possible with imaging and can still reveal important new insights into the life cycle of a much-studied parasite. To what degree these observations can be transferred from one species to another, however, remains an open question. At least, the two rodent malaria parasite species *Plasmodium berghei* and *Plasmodium yoelii*, which elicit interesting differences in the immune response of the host and show different molecular requirements for liver cell entry (Silvie *et al.*, 2003, 2007), appear to behave in largely the same way after injection by mosquitoes into the skin (Yamauchi *et al.*, 2007). Importantly, a large proportion of sporozoites of both species remained at the bite site, and a careful study showed that a small proportion of those can invade fibroblasts in the dermis and develop into infectious merozoites (Gueirard *et al.*, 2010). However, no natural infection resulted from these cells, probably as the few infectious parasites failed to enter the blood stream. This shows that one has to be careful when injecting deliberately large numbers of parasites to facilitate observation, as the few parasites injected by a natural bite might behave differently and/or lead to diverse responses by the host.

Interestingly, a sporozoite expressing a mutated form of its major surface antigen, resembling a cleaved version of the circumsporozoite protein (CSP), was subsequently shown to fail to leave the skin, entered skin fibroblasts, and did eventually cause an infection of blood cells, thus completely circumventing the liver (Coppi *et al.*, 2011). This is interesting not only because it highlights an important function of CSP, a major vaccine candidate (Cohen *et al.*, 2010) but also because it gives potential molecular insights into the evolution of the malaria life cycle as *Plasmodium* parasites in birds and reptiles develop within macrophages in the skin, while those infecting mammals somehow have chosen the immune-privileged site of the hepatocyte as their first host cell for replication. Most likely, the difference of invasion preference is not only mediated by recognition of the appropriate host cell that is invaded. *Plasmodium* species infecting mammals display additional behaviors as they need to invade blood capillaries in order to reach the liver, where they are also exposed to antibodies produced in response to previous infections. These imaging studies may also assist in addressing important medical questions for novel whole-parasite vaccine approaches (Hoffman *et al.*, 2002; Matuschewski, 2006), where thousands of sporozoites are administered repeatedly (Roestenberg *et al.*, 2009). For example, could a developing parasite in the skin leak antigen to the draining lymph node and thus prime the immune system in ways that are contributing or counteracting the vaccine-induced immunity of the vaccinated person?

Obviously, the presence of these mosquito-introduced pathogens does not go unnoticed, and immune cells are likely to rush to the site of the mosquito bite (Amino *et al.*, 2008; Frischknecht, 2007). Indeed, using sand flies transmitting *Leishmania*, it was shown that neutrophils migrate to the site of a bite, whether infected or not, and clog up the damaged tissue to allow wound healing (Peters *et al.*, 2008). Interestingly, the parasites are first taken up by neutrophils in the skin but are not yet destroyed. Once neutrophils are taken up by macrophages, the parasite swaps its host cells and starts replicating within the lysosomes of the latter cells (van Zandbergen *et al.*, 2004), a process termed the 'Trojan horse model', of infection (Laskay *et al.*, 2003). Direct observations in the mouse, however, showed that parasites escape from neutrophils (or parasites are released) and are then taken up by macrophages (Peters *et al.*, 2008), a process later termed the 'Trojan rabbit model,' (Ritter *et al.*, 2009). As there are a number of different *Leishmania* species causing diverse clinical symptoms in humans, it remains to be determined which of these models applies to which natural host–parasite combination, or whether one of the two fits all.

3. The Choice of the Appropriate Microscope

Comparing the *Plasmodium* with the *Leishmania* studies, it is notable that the different groups utilized very diverse imaging platforms. All observations of *Plasmodium* sporozoites in mosquitoes were made with wide-field fluorescence microscopes (Akaki and Dvorak, 2005; Frischknecht *et al.*, 2004, 2006). They allow easy and rapid image acquisition and could compromise in terms of the collected out-of-focus light that would be blocked by a confocal system. Clearly, a classical confocal or two-photon system would not be able to visualize the rapidly floating sporozoites that swim around the narrow hemolymph-filled veins in the mosquito wing or are pushed down the proboscis at speeds of up to 300 μm/s with the flow of saliva (Akaki and Dvorak, 2005; Frischknecht *et al.*, 2004).

Curiously, even in the skin of the ear pinna, a wide-field microscope allowed to make the key observations of rapid parasite migration, blood capillary and lymph vessel entry (Amino *et al.*, 2005, 2006; Münter *et al.*, 2009; Vanderberg and Frevert, 2004). Noteworthy, a wide bandwidth emission filter should be used for imaging to increase the fluorescent signal over autofluorescent signal ratio. In this case, the out-of-focus light allows a determination of the tissue depth as autofluorescent patches at the skin surface could be used to judge their distance from the focal plane. Further, as the sporozoites move at 1–2 μm/s, it would be impossible with a classic confocal or two-photon microscope to acquire z-scans at high enough temporal resolution to catch all sporozoites within the necessary volume. Thus, the out-of-focus sporozoites detected by wide-field imaging could still be used for determining the average speed of a population and the total number of transmitted sporozoites. In addition, the weak autofluorescence of the tissue also aided the detection of blood vessels, which appeared darker against the tissue. Nevertheless, subsequent studies used a spinning disk confocal microscope, which allows rapid scanning of a field of view and has nearly the same focal depth as a confocal microscope. This made it possible to sharply visualize the fast moving sporozoites in different z-planes (Amino *et al.*, 2007; Münter *et al.*, 2009) and the lymphocytes clearing up the nonmotile sporozoites in the skin (Amino *et al.*, 2008).

Two-photon microscopy was used to image the practically nonmotile *Leishmania* parasites and to decipher how they interact with slow moving cells of the immune system (Filipe-Santos *et al.*, 2009; Peters *et al.*, 2008). The reason to use two-photon was probably more the availability of and the expertise with the setup as well as the reduced photobleaching and phototoxicity, as the ear pinna is thin enough for conventional one-photon microscopy. Clearly, if other sites like the tail are imaged, where for example *Plasmodium* sporozoites can move deep enough into the dermis to be lost to detection by confocal or wide-field microscopy, a two-photon microscope needs to be considered. As two-photon systems are now available with fast image scanning rates, they might indeed be suitable for revealing such rapid events.

4. New Materials Provide the Basis for New Approaches to Dissect Cell Migration

Motile bacteria use their flagellum or substrate-based gliding motility to search for nutrients (Sun *et al.*, 2011). Similarly, motile parasites, such as *Plasmodium* sporozoites, might sense chemical gradients and move toward their source (Akaki and Dvorak, 2005). After the transmission of sporozoites into the skin, this source is still unclear. It could be constituents of the blood which leak from the wounded capillaries and blood vessels into the tissue. Alternatively, components of the saliva could keep sporozoites at the bite site, as it has been shown for *Borrelia* (Shih *et al.*, 2002). Or sporozoites could "run" away from these components, thus avoiding neutrophils rushing to the bite site. The trajectories of migrating sporozoites in the dermis of the ear appear random. But could this be due to the special architecture of the ear?

A recent study thus investigated sporozoites injected into the tail of a mouse and indeed the trajectories of the parasites appeared in a more linear fashion than in the ear (Hellmann *et al.*, 2011) (Fig. 2.2A). Were these sporozoites attracted to a source deeper within the dermis, where the blood vessels are located? To address this question *in vivo* is difficult, as it would require many parameters to be monitored at the same time, the blood vessels, their rupture, the extracellular matrix, which might be oriented in different ways and thus guide small migrating cells, and the parasites themselves. The question could also be turned upside down and one could ask whether sporozoites need to be chemotactic to cross the short distance from the bite site to the nearest (or one of the nearest) capillaries. Like some lymphocytes, which simply move faster upon stimulation, yielding a disproportional higher number of contacts with their partner cells (Miller *et al.*, 2002), sporozoites might just go fast enough and rely on the surrounding environment to be guided. Like lymphocytes, the malaria parasites might simply rely on deflection by obstacles within the skin tissue to deflect them onto random paths. While this appears on face value possible in the ear, where sporozoites appear to move randomly, it appears less straightforward for the tail, where sporozoites migrate on long trajectories.

To address this, a new approach was taken by placing sporozoites in micropatterned obstacle arrays (Hellmann *et al.*, 2011). These could be manufactured such that the diameter of the individual obstacle could be varied as well as the obstacle-to-obstacle distance (Fig. 2.2B). Previously, such arrays were used to place large cells on top and measure the sites and forces applied by these cells onto the substratum, which was possible as the "obstacles" are flexible and can be bend under force (le Digabel *et al.*, 2010). For the much smaller sporozoites moving at the bottom of the obstacles, no such displacements

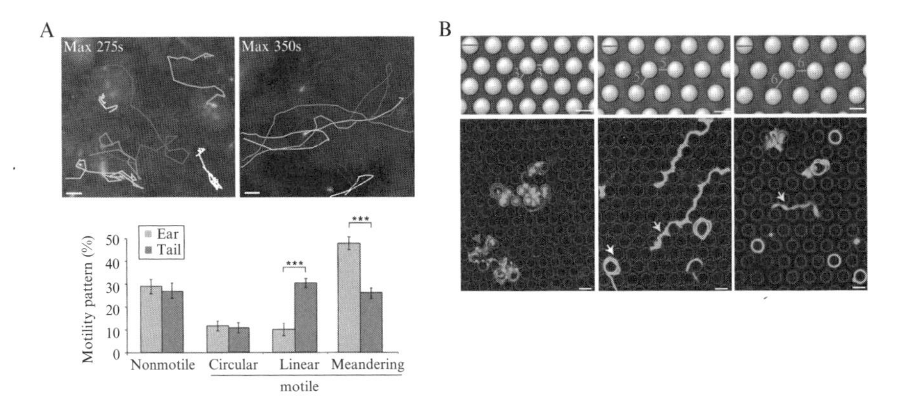

Figure 2.2 Migrating sporozoites appear to be guided by the environment. (A) *Plasmodium* sporozoites migrate on apparent different paths in the dermis of the ear pinna (top left) and the tail of a mouse (top right), which is confirmed by quantitative analysis of the moving patterns (bottom). Scale bars: 25 μm. (B) Micro-patterned obstacles of the same diameter (8 μm, dark grey line) but different obstacle-to-obstacle distance (light grey line) as viewed from the top and projections of motile sporozoites moving in different obstacle arrays. The outline of the obstacles is also depicted. Scale bars: 10 μm. Figure adapted from Hellmann *et al.*, 2011.

were observed. In contrast, sporozoites encountering the obstacles either moved around them in their typical circular fashion or were deflected from their path. Non-circular moving sporozoites have been observed moving in either linear or meandering ways. Quantitative analysis then revealed that with decreasing obstacle-to-obstacle distance, the number of sporozoites moving on either linear or meandering patterns increased (Hellmann *et al.*, 2011). Following the mean square displacements of the sporozoites migrating *in vivo* in the dermis of the ear pinna or the tail and those moving in, the obstacle arrays could be plotted and analyzed. This showed similar parameters for the sporozoites migrating *in vivo* and within some arrays. While this does not disprove that sporozoites might follow a chemical or physical gradient away from the place they were injected into the skin, it shows that the environment can, in principle, guide these rapidly moving pathogens (Hellmann *et al.*, 2011).

Similar approaches could now also be envisaged in 3D environments with collagen or matrigel replacing the fluid between the obstacles or within artificially generated skin cultures. This together demonstrates how novel *in vitro* approaches can provide interesting insights into *in vivo* behavior of parasites. Ultimately, a combination of 2D, 3D, and *in vivo* approaches is needed for dissecting many of the currently open questions about host–parasite interactions.

5. Visualizing Cellular Functions in Motile Cells

In classical cell biological research, fluorescent approaches to understand cellular functions *in vivo* are used in a broad range of research topics like cytoskeleton dynamics, protein interaction, or membrane dynamics (see Chapters 18 & 19 of MIE Volume 504 and Chapters 1 & 15 of MIE Volume 505). Most of these studies have in common that the reference point, the cell, is relatively stationary in respect to the more rapidly changing fluorescent signal. If dynamics of fluorescent proteins are followed in fast moving cells, the two trajectories of the protein under investigation and the cell have to be separated. If the cell does not change its shape while it is motile, a kymograph can be used as has been done before for movement of protein in motile bacteria or adhesion site dynamics in motile sporozoites (Münter *et al.*, 2009; Sun *et al.*, 2011) by aligning the cell at each time point in respect to each other (Fig. 2.3). If the cell also changes its shape during motility, as has been most intensely studied in *Dictyostelium* (Heinrich *et al.*, 2008), analysis is limited to analogous readout and quantification (except fluorescence intensity) become very laborious.

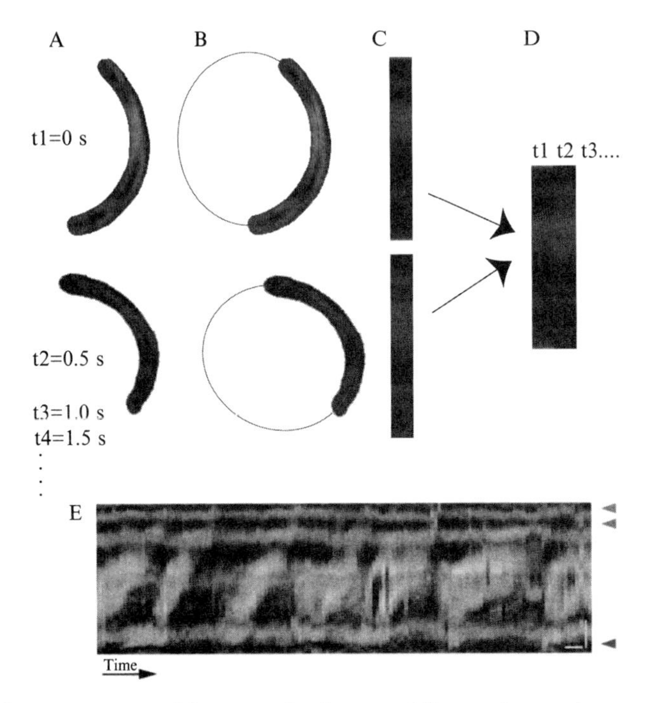

Figure 2.3 Generation of kymographs from rapidly moving and crescent-shaped *Plasmodium berghei* sporozoites. (A) Sporozoites are imaged at 2 Hz with reflection interference contrast optics to reveal their substrate adhesion sites (black). (B) Using an ellipse, the central axis of a sporozoite is defined for each frame. (C) The intensities of the central axis along the ellipse are visualized by expanding the value of the individual profile-pixel to a rectangle. (D) A kymograph is generated by stringing together a series of rectangles. (E) A kymograph of a rapidly motile sporozoite shows the dynamic turnover of adhesion sites at front (bottom), center and rear (top). Scale bars: 2 s (horizontal), 2.5 μm (vertical). Arrowheads indicate distinct adhesion sites at the parasite front and rear. Figure adapted from Münter *et al.*, 2009.

Imaging of protein dynamics during transmission of sporozoites has, so far, only been investigated *in vitro*. *In vivo* imaging of subcellular processes is further hampered due to the challenges of imaging relatively weak signals within a motile cell, which is surrounded by thousands of host cells and extracellular matrix. In order to reach sufficient t- and z-resolution, the field of view would have to be very small, which in turn limits the cells observed, in case of a natural transmission, only a few in total. However, it would be very interesting to visualize, for example, the dynamics and fate of parasite proteins that are important for motility or are currently in clinical trials as vaccine candidates like the CSP. In this respect, recent advances in the field of adaptive optics (Planchon *et al.*, 2011) might prove essential for doing such studies.

6. Applying Imaging for Drug Discovery

The above section illustrated some of the current findings that imaging studies have revealed about our understanding of the transmission process of malaria parasites. Imaging also has a long tradition in drug screening approaches and should thus, in principle, also be applicable for finding novel inhibitors against this early transmission stage. The question arises whether a small molecule or an antibody could be used to block *Plasmodium* sporozoite motility in the skin or in the liver prior to infection of a hepatocyte? Such a molecule could be a valuable addition to the limited set of malaria prophylactic drugs. Different approaches are possible to screen for inhibitors of parasite motility.

First, as sporozoite motility appears to be necessary for invasion of hepatocytes, one could infect cultured or primary hepatocytes with sporozoites in the presence of potential inhibitors and score for the success of invasion by investigating the numbers of intracellular parasites (i.e., liver stages) after certain time points during liver stage development. For ease of handling 24 h could be used, where *P. berghei* or *P. yoelii* parasites have rounded up and can be clearly distinguished from autofluorescent particles that sometimes cause havoc with quantitative imaging approaches. As both parasites are available in fluorescent forms, no fixation and no staining are necessary and the culture could be directly imaged on a 384-(or more) well plate and the results analyzed by signal detection software (Mahmoudi *et al.*, 2003; Mota and Rodriguez, 2008; Prudencio *et al.*, 2008). The drawback of this approach is that only about 2% of hepatocytes are ever invaded by sporozoites, the reasons for this low infectivity not being clear. The great advantage, however, is that inhibitors against motility, invasion, and intrahepatic growth, which constitute another important new drug target (Friesen *et al.*, 2010; Mazier *et al.*, 2009), could be isolated.

A second approach would be to stain the molecules left behind the moving parasite on the substrate, which for sporozoites yield characteristic circular patterns. These are classically stained after fixation of a population of sporozoites that were placed on a substrate (e.g., the glass bottom of a 386-well plate) by an antibody against the major surface protein of the sporozoites, CSP (Stewart and Vanderberg, 1988). Thus, parasites incubated with a molecule that inhibits motility can be detected by the absence of circular trails, which again could be analyzed by automated software. An important drawback of this methodology is the fixation and antibody staining. Also, this makes it hard to discriminate whether a parasite has moved in just one or more circles, that is, whether a potential inhibition was partially successful.

A third option is to film the sporozoites in the presence of inhibitors in a multi-well plate for a short period of time, determining the number of motile sporozoites and their average speed. Such an approach would necessitate the least preparatory steps (no host cells, no fixation or staining) and has the additional advantage that each parasite contributes to the read out. On the downside, the imaging takes longer as not just one but at least 3–10 images would have to be taken per sample and the analysis is likely more cumbersome as it necessitates to distinguish from few images whether sporozoites are motile or not. Nevertheless, this was accomplished by generating an analysis tool based on the open source software ImageJ (Hegge *et al.*, 2009). This now allows the acquisition of short movies from 100 to 250 sporozoites with a 10× objective lens on a 384-well format and to process them rapidly (Fig. 2.4). It is, however, still not a high-throughput format, as sporozoites only stay comparatively motile for about 40 min at room temperature (Hegge *et al.*, 2009). Afterward, the number of motile sporozoites drops and their speed decreases. Thus, only about 40 conditions can currently be assessed in parallel. Nevertheless, it has allowed the screening of a small library of potential inhibitors and resulted in the identification of efficient new molecules that arrest sporozoite gliding (Frischknecht lab, unpublished data).

Importantly, whenever a potential inhibitor of sporozoite motility or parasite intracellular development has been identified by *in vitro* screenings, it is possible to perform rapid *in vivo* testing using bioluminescent parasites

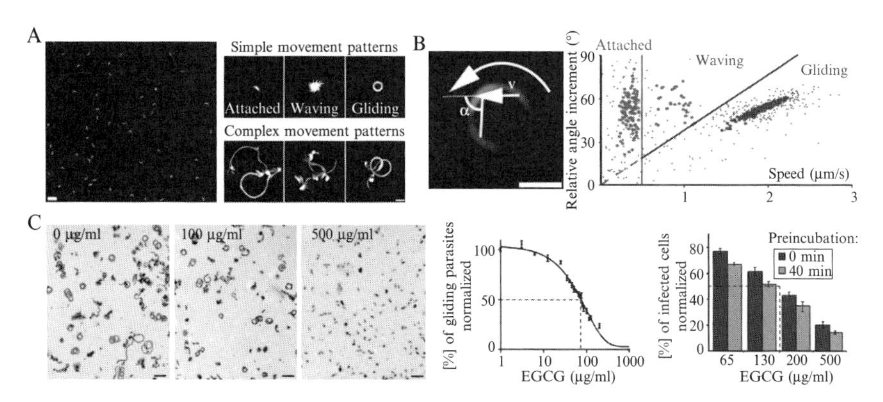

Figure 2.4 Quantitative screening of *Plasmodium* sporozoite motility. (A) *Plasmodium berghei* sporozoites expressing GFP in the cytoplasm as viewed through a 10x objective lens (left) and the patterns of movement they can perform while gliding on a flat support (right). (B) Rationale for automatically separating sporozoites into attached, waving and gliding parasites by plotting their angular change over their speed. (C) Arrest of sporozoite motility by the active component of green tea, EGCG used as a proof of concept test. Inverted fluorescent maximum projections show that with increasing EGCG concentrations fewer sporozoites move. Quantification over a large number of concentrations allows the determination of the IC50. Follow-up manual quantification shows that EGCG also inhibits infection of liver cells but with a lower IC50 (right panel). Images modified from Hegge *et al.*, 2009 and Hellmann *et al.*, 2011.

and directly score their success of replicating in the host (Mwakingwe *et al.*, 2009; Ploemen *et al.*, 2009). Clearly, the bioluminescent signal emitted by individual pathogens is not nearly strong enough to be detected in an intact mouse. However, at the end of the intrahepatic development, which in rodent malaria parasites can be determined with the necessary precision, several thousand parasites have been formed and thus produce a strong focal signal that can readily be detected with the use of a 3D imaging setup (Ploemen *et al.*, 2009). This should thus allow to rapidly move from determining inhibitor efficiency *in vitro* to doing so *in vivo*.

ACKNOWLEDGMENTS

We thank Ann-Kristin Mueller, Leandro Lemgruber, and Simone Lepper for comments; Stephan Hegge, Janina Hellmann, Christine Selhuber-Unkel and Sylvia Münter for figures; the Federal German Ministry for Education and Research (BMBF), the German Research Foundation (DFG, SFB 544 and SPP1464), the European Union FP7 (EVIMalaR) and the Chica and Heinz Schaller Foundation for funding.

REFERENCES

Akaki, M., and Dvorak, J. A. (2005). A chemotactic response facilitates mosquito salivary gland infection by malaria sporozoites. *J. Exp. Biol.* **208**(Pt 16), 3211–3218.

Amino, R., Menard, R., *et al.* (2005). In vivo imaging of malaria parasites—Recent advances and future directions. *Curr. Opin. Microbiol.* **8**(4), 407–414.

Amino, R., Thiberge, S., *et al.* (2006). Quantitative imaging of Plasmodium transmission from mosquito to mammal. *Nat. Med.* **12**(2), 220–224.

Amino, R., Thiberge, S., *et al.* (2007). Imaging malaria sporozoites in the dermis of the mammalian host. *Nat. Protoc.* **2**(7), 1705.

Amino, R., Giovannini, D., *et al.* (2008). Host cell traversal is important for progression of the malaria parasite through the dermis to the liver. *Cell Host Microbe* **3**(2), 88.

Barry, J. D., and Emergy, D. L. (1984). Parasite development and host responses during the establishment of Trypanosoma brucei infection transmitted by tsetse fly. *Parasitology* **88**(Pt 1), 67–84.

Bolhassani, A., Taheri, T., *et al.* (2011). Fluorescent Leishmania species: Development of stable GFP expression and its application for in vitro and in vivo studies. *Exp. Parasitol.* **127**(3), 637–645.

Cohen, J., Benns, S., *et al.* (2010). The malaria vaccine candidate RTS, S/AS is in phase III clinical trials. *Ann. Pharm. Fr.* **68**(6), 370–379.

Coombes, J. L., and Robey, E. A. (2010). Dynamic imaging of host-pathogen interactions in vivo. *Nat. Rev. Immunol.* **10**(5), 353–364.

Coppi, A., Natarajan, R., *et al.* (2011). The malaria circumsporozoite protein has two functional domains, each with distinct roles as sporozoites journey from mosquito to mammalian host. *J. Exp. Med.* **208**(2), 341.

Filipe-Santos, O., Pescher, P., *et al.* (2009). A dynamic map of antigen recognition by CD4 T cells at the site of Leishmania major infection. *Cell Host Microbe* **6**(1), 23–33.

Friesen, J., Silvie, O., *et al.* (2010). Natural immunization against malaria: Causal prophylaxis with antibiotics. *Sci. Transl. Med.* **2**(40), 40ra49.

Frischknecht, F. (2007). The skin as interface in the transmission of arthropod-borne pathogens. *Cell. Microbiol.* **9**(7), 1630–1640.

Frischknecht, F. (2010). Imaging parasites at different scales. *Cell Host Microbe* **8**(1), 16–19.

Frischknecht, F., Baldacci, P., *et al.* (2004). Imaging movement of malaria parasites during transmission by Anopheles mosquitoes. *Cell. Microbiol.* **6**(7), 687.

Frischknecht, F., Martin, B., *et al.* (2006). Using green fluorescent malaria parasites to screen for permissive vector mosquitoes. *Malar. J.* **5**, 23.

Gueirard, P., Tavares, J., *et al.* (2010). Development of the malaria parasite in the skin of the mammalian host. *Proc. Natl. Acad. Sci. USA* **107**(43), 18640.

Hegge, S., Kudryashev, M., *et al.* (2009). Automated classification of Plasmodium sporozoite movement patterns reveals a shift towards productive motility during salivary gland infection. *Biotechnol. J.* **4**(6), 903–913.

Heinrich, D., Youssef, S., *et al.* (2008). Actin-cytoskeleton dynamics in non-monotonic cell spreading. *Cell Adh. Migr.* **2**(2), 58–68.

Hellmann, J. K., Münter, S., *et al.* (2011). Environmental constraints guide migration of malaria parasites during transmission. *PLoS Pathog.* **7**(6), e1002080.

Hoffman, S. L., Goh, L. M., *et al.* (2002). Protection of humans against malaria by immunization with radiation-attenuated Plasmodium falciparum sporozoites. *J. Infect. Dis.* **185**(8), 1155–1164.

Kebaier, C., Voza, T., *et al.* (2009). Kinetics of mosquito-injected Plasmodium sporozoites in mice: Fewer sporozoites are injected into sporozoite-immunized mice. *PLoS Pathog.* **5**(4), e1000399.

Laskay, T., van Zandbergen, G., *et al.* (2003). Neutrophil granulocytes—Trojan horses for Leishmania major and other intracellular microbes? *Trends Microbiol.* **11**(5), 210–214.

le Digabel, J., Ghibaudo, M., *et al.* (2010). Microfabricated substrates as a tool to study cell mechanotransduction. *Med. Biol. Eng. Comput.* **48**(10), 965–976.

Mahmoudi, N., Ciceron, L., *et al.* (2003). In vitro activities of 25 quinolones and fluoroquinolones against liver and blood stage Plasmodium spp. *Antimicrob. Agents Chemother.* **47**(8), 2636–2639.

Matuschewski, K. (2006). Getting infectious: Formation and maturation of Plasmodium sporozoites in the Anopheles vector. *Cell. Microbiol.* **8**(10), 1547–1556.

Mazier, D., Renia, L., *et al.* (2009). A pre-emptive strike against malaria's stealthy hepatic forms. *Nat. Rev. Drug Discov.* **8**(11), 854–864.

Miller, M. J., Wei, S. H., *et al.* (2002). Two-photon imaging of lymphocyte motility and antigen response in intact lymph node. *Science* **296**(5574), 1869–1873.

Mota, M. M., and Rodriguez, A. (2008). New pieces for the malaria liver stage puzzle: where will they fit? *Cell Host Microbe* **3**(2), 63–65.

Münter, S., Sabass, B., *et al.* (2009). Plasmodium sporozoite motility is modulated by the turnover of discrete adhesion sites. *Cell Host Microbe* **6**(6), 551.

Mwakingwe, A., Ting, L. M., *et al.* (2009). Noninvasive real-time monitoring of liver-stage development of bioluminescent Plasmodium parasites. *J. Infect. Dis.* **200**(9), 1470–1478.

Nishikawa, Y., Zhang, H., *et al.* (2008). Construction of Toxoplasma gondii bradyzoite expressing the green fluorescent protein. *Parasitol. Int.* **57**(2), 219–222.

Peters, N. C., Egen, J. G., *et al.* (2008). In vivo imaging reveals an essential role for neutrophils in leishmaniasis transmitted by sand flies. *Science* **321**(5891), 970–974.

Pires, S. F., DaRocha, W. D., *et al.* (2008). Cell culture and animal infection with distinct Trypanosoma cruzi strains expressing red and green fluorescent proteins. *Int. J. Parasitol.* **38**(3–4), 289–297.

Planchon, T. A., Gao, L., *et al.* (2011). Rapid three-dimensional isotropic imaging of living cells using Bessel beam plane illumination. *Nat. Methods* **8**(5), 417–423.
Ploemen, I. H., Prudencio, M., *et al.* (2009). Visualisation and quantitative analysis of the rodent malaria liver stage by real time imaging. *PLoS One* **4**(11), e7881.
Prudencio, M., Rodrigues, C. D., *et al.* (2008). Kinome-wide RNAi screen implicates at least 5 host hepatocyte kinases in Plasmodium sporozoite infection. *PLoS Pathog.* **4**(11), e1000201.
Ritter, U., Frischknecht, F., *et al.* (2009). Are neutrophils important host cells for Leishmania parasites? *Trends Parasitol.* **25**(11), 505–510.
Roestenberg, M., McCall, M., *et al.* (2009). Protection against a malaria challenge by sporozoite inoculation. *N. Engl. J. Med.* **361**(5), 468–477.
Shih, C. M., Chao, L. L., *et al.* (2002). Chemotactic migration of the Lyme disease spirochete (Borrelia burgdorferi) to salivary gland extracts of vector ticks. *Am. J. Trop. Med. Hyg.* **66**(5), 616–621.
Sidjanski, S., and Vanderberg, J. P. (1997). Delayed migration of Plasmodium sporozoites from the mosquito bite site to the blood. *Am. J. Trop. Med. Hyg.* **57**(4), 426–429.
Silvie, O., Rubinstein, E., *et al.* (2003). Hepatocyte CD81 is required for Plasmodium falciparum and Plasmodium yoelii sporozoite infectivity. *Nat. Med.* **9**(1), 93–96.
Silvie, O., Franetich, J. F., *et al.* (2007). Alternative invasion pathways for Plasmodium berghei sporozoites. *Int. J. Parasitol.* **37**(2), 173–182.
Stewart, M. J., and Vanderberg, J. P. (1988). Malaria sporozoites leave behind trails of circumsporozoite protein during gliding motility. *J. Protozool.* **35**(3), 389–393.
Sun, M., Wartel, M., *et al.* (2011). From the cover: Motor-driven intracellular transport powers bacterial gliding motility. *Proc. Natl. Acad. Sci. USA* **108**(18), 7559–7564.
Talman, A. M., Blagborough, A. M., *et al.* (2010). A Plasmodium falciparum strain expressing GFP throughout the parasite's life-cycle. *PLoS One* **5**(2), e9156.
van Zandbergen, G., Klinger, M., *et al.* (2004). Cutting edge: Neutrophil granulocyte serves as a vector for Leishmania entry into macrophages. *J. Immunol.* **173**(11), 6521–6525.
Vanderberg, J. P., and Frevert, U. (2004). Intravital microscopy demonstrating antibody-mediated immobilisation of Plasmodium berghei sporozoites injected into skin by mosquitoes. *Int. J. Parasitol.* **34**(9), 991–996.
Yamauchi, L. M., Coppi, A., *et al.* (2007). Plasmodium sporozoites trickle out of the injection site. *Cell. Microbiol.* **9**(5), 1215–1222.

CHAPTER THREE

Multiphoton Microscopy Applied for Real-Time Intravital Imaging of Bacterial Infections *In Vivo*

Ferdinand X. Choong,* Ruben M. Sandoval,† Bruce A. Molitoris,† *and* Agneta Richter-Dahlfors*

Contents

1. Introduction 36
2. Selection of Reporters for a Complex Site of Infection 37
    2.1. Visualizing bacteria 38
    2.2. Visualizing tissues 39
3. Setup of a Live Imaging Platform of Infection 41
    3.1. Selecting and customizing a multiphoton imaging platform and analysis 41
    3.2. Pathogen and host 43
    3.3. Anesthesia and surgical creation of retroperitoneal window 44
    3.4. Introducing uropathogenic *E. coli* to initiate infection 46
4. Studying Tissue Responses in Real-Time During Infection 50
    4.1. Probing histological changes of the mucosal lining 50
    4.2. Imaging vascular features such as rate, clotting, and integrity 51
    4.3. Local tissue oxygen tension 54
    4.4. Systemic dissemination 56
    4.5. Imaging obstruction of a perfused environment 56
5. Postimaging Analysis 57
    5.1. Immunohistochemistry 57
6. Extending Real-Time Imaging into Molecular Analysis 58
    6.1. Molecular view of host responses 58
    6.2. Molecular details based on bacterial genetics 59

Acknowledgments 60

References 60

* Department of Neuroscience, Swedish Medical Nanoscience Center, Karolinska Institutet, Stockholm, Sweden

† Indiana University School of Medicine, Indianapolis, Indiana, USA

Methods in Enzymology, Volume 506

ISSN 0076-6879, DOI: 10.1016/B978-0-12-391856-7.00027-5

Abstract

To understand the underlying mechanisms of bacterial infections, researchers have for long addressed the molecular interactions occurring when the bacterium interacts with host target cells. In these studies, primarily based on *in vitro* systems, molecular details have been revealed along with increased knowledge regarding the general infection process. With the recent advancements in *in vivo* imaging techniques, we are now in a position to bridge a transition from classical minimalistic *in vitro* approaches to allow infections to be studied in its native complexity—the live organ. Techniques such as multiphoton microscopy (MPM) allow cellular-level visualization of the dynamic infection process in real time within the living host. Studies in which all interplaying factors, such as the influences of the immune, lymphatic, and vascular systems can be accounted for, are likely to provide new insights to our current understanding of the infection process. MPM imaging becomes extra powerful when combined with advanced surgical procedure, allowing studies of the illusive early hours of infection. In this chapter, our intention is to provide a general view on how to design and carry out intravital imaging of a bacterial infection. While exemplifying this using a spatiotemporally well-controlled uropathogenic *Escherichia coli* (UPEC) infection in rat kidneys, we hope to provide the reader with general considerations that can be adapted to other bacterial infections in organs other than the kidney.

1. Introduction

Interactions occurring when pathogens encounter their host are characterized by a dynamic and complex interplay. Classical approaches to probing the process of infection has relied heavily on *in vitro* cell culture models. While this method has yielded substantial valuable findings, and has essentially molded our current understanding of host–pathogen interactions, the native state of infection *in vivo* is inherently much more complex than what any *in vitro* model can simulate. Host tissues are heterogeneous organizations of a diverse population of cell types whose composition changes during the time-course of infection. As tissue homeostasis is dynamically altered during infection, bacteria must adapt simultaneously to cope with the changing microenvironment. The physiology of the bacterial population is thus likely to be altered during the infection process. With these dynamic, multifactorial processes in mind, the infection process is preferentially studied in real-time within the live animal. Recent advances in imaging platforms and data analysis tools have been beneficial in greatly improving the data quality of live visualization of an infection as it progresses within the organ of an animal (Dunn *et al.*, 2002, 2007; Månsson *et al.*, 2007a,b). Noninvasive multiphoton imaging is one such tool.

MPM is a noninvasive technique developed by biophysicists (Goeppert-Mayer, 1931; Weissman *et al.*, 2007) and adapted by biologists to study dynamic processes within living organs. In essence, MPM enables both high resolution and high-sensitivity fluorescence microscopy of intact tissues over time. Since its initial application by Denk *et al.* (1990), it has been adapted to explore a wide variety of research fields, including studies of Ca^{2+} fluctuations in individual synapses (Svoboda and Yasuda, 2006; Tian *et al.*, 2006), the role of astrocytes in the brain (Tian *et al.*, 2006), tumor vascularization (Brown *et al.*, 2001), embryonic development (Squirrell *et al.*, 1999), kidney physiology (Molitoris and Sandoval, 2005), and bacterial infections (Månsson *et al.*, 2007a,b; Melican and Richter-Dahlfors, 2009a,b).

The following sections will cover both concepts and considerations behind the design of a model for real-time imaging of bacterial infections *in vivo*. Host–pathogen interactions when *in vivo* can be highly unpredictable. Depending on the anatomy and physiology of the organ, along with its relation to the vasculature and other organs, there will be a wide variation in the dynamics of pathogen clearance, persistence, or dissemination. That being so, all possible tissue compartments that may be involved in the host response or that the pathogen may access to have to be sampled. In our *in vivo* model, a wealth of information at a range of complexity is generated. This implies that one might not fully comprehend the significance of the data when it is encountered. Therefore, clear definition of the pathology of disease and physiology of the organ of interest must be fed into the design of the selected animal model.

Our group has primarily employed intravital imaging on UPEC infection of the upper urinary tract in live rats. The kidney is an organ containing a good variety of microenvironments for the study of host–pathogen interactions and is, in addition, well suited for surgical procedures and MPM studies using an inverted microscope setup (Dunn *et al.*, 2003, 2007). The structure of the nephron, such as the close proximity between the tubular epithelium and peritubular capillaries, allows the bacterial infection to be initiated in the mucosal lining, while monitoring the onset of the inflammatory response. The dynamic nature of the model enables us to perform kinetic studies of bacterial colonization and host responses, as well as monitor any accompanying effects infection may have on vascular and renal flow. Thus, this model can provide new data not possible to achieve in static end-point analysis of infected tissues (Melican and Richter-Dahlfors, 2009a,b).

2. Selection of Reporters for a Complex Site of Infection

To follow the progression of infection *in vivo*, bacterial and host components need to be visible under the microscope. Under fluorescence-based imaging, the ironclad rule is that only fluorophore-associated molecules are visible. Fluorophores with appropriate spectral features must thus be selected and applied to the experiment using the appropriate vectors.

2.1. Visualizing bacteria

Bacteria are commonly visualized by methods such as specific dyes, antibody-based systems, and genetic approaches. While all these work well for *in vitro* studies, the live setting provides further challenges. There is a shortage of dyes for prokaryotes to be used in intravital studies. Use of antibodies is also limited due to the risk of them being immunogenic and the difficulty to predict their biodistribution after systemic delivery. As an alternative, bacteria can be genetically engineered to express a fluorescent protein, from either plasmid-bornc or chromosomally inserted genes. Plasmid-borne reporter systems while feasible *in vitro* are not fully compatible with *in vivo* models, as the necessary absence of antibiotics in the animal may lead to a variation in plasmid copy number. By inserting a single-copy gene onto the bacterial chromosome, drawbacks of plasmid-based systems can be circumvented, including the risk of expressing unnecessarily high levels of GFP that may exert cytotoxic effects. The chromosomal site for gene insertion must be selected with care to avoid any polar effects. Similarly, the gene must be cloned under a promoter whose activity is not affected by the changing tissue microenvironment. Thus, promoters regulating ribosomal protein gene expression should be avoided as bacteria rapidly adjust ribosome numbers according to nutrient availability.

2.1.1. Required materials

- Bacterial strain
- UPEC strain CFT073, serotype O6:K2:H1
- *Oligonucleotides constructing UPEC strains with chromosomal encoding gfp*
- $P_{LtetO\text{-}1}$: TTCGTCTTC ACCTCGAGTC CCTATCAGTG ATAGAGATTG ACATCCCTAT CAGTGATAGA GATACTGAGC ACATCAGCAG GACGCACTGA CGAATTCATT AAAGAGGAG AAAGGTACCC ATGGG
- Ptet_F1: CATGCGACCC GGGTTCGTCT TCACCTCGAG TCC
- Ptet_R1: GTCGCCATT CTAGACCCAT GGGTACCTTT CTCCT
- P1cob: gtgacgggag gcgcacggag cgggaagagt cgccacgcaG CCTGGGGTAA TGACTCTCTA GC
- P2cob: cagcgtatc gcccgtttgc ccgcccagcg tacgtttgag aCGTCATTTC TGCCATTCAT CC
- *Kits for genetic manipulation*
- Techniques and kits for genetic manipulation of bacteria vary significantly between research groups and are thus not listed here. This section will only give a brief overview of the major steps to generate a bacterium with chromosomally encoded GFP^{+}.

2.1.2. Construction of UPEC strains with chromosomal encoding of GFP$^+$

GFP$^+$-producing bacteria are constructed by site-specific integration of the gene-encoding enhanced green fluorescent protein (*gfp*$^+$) into the *cobS* site on the chromosome of uropathogenic *Escherichia coli* (UPEC) strain CFT073 according to procedures described by Hautefort *et al.* (2003). The tetracycline promoter $P_{LtetO\text{-}1}$ was used in the absence of the Tet repressor, resulting in constitutive activation of the promoter.

To begin cloning, amplify the tetracycline promoter $P_{LtetO\text{-}1}$ using primers Ptet_R1 and Ptet_F1 by PCR. Digest both amplified $P_{LtetO\text{-}1}$ and the *gfp*$^+$-*cm*^{r}-containing plasmid pZEP08 (Hautefort *et al.*, 2003) with *Xba*I and *Sma*I (Sigma, Germany). Purify and ligate the digestion products to give the plasmid pKM001 ($P_{LtetO\text{-}1}$-*gfp*$^+$-*cm*^{r}). Amplify pKM001 ($P_{LtetO\text{-}1}$-*gfp*$^+$-*cm*^{r}) by PCR amplification with promoters P1cob and P2cob. These promoters contain a 40 nt 5′ region homologous to the *cobS* locus at 44 min on the CFT073 chromosome, and a 21–23 nt sequence-binding site to the *gfp*$^+$-*cm*^{r}-containing plasmid. Perform chromosomal recombination of PCR products by λ Red recombination (Datsenko and Wanner, 2000), deleting 1143 bp (positions 2319458–2320600). Verify accurate clones by sequencing. Perform a bacterial growth assay to ensure that *gfp*$^+$ is successfully expressed and does not alter bacterial physiology.

Before applying a GFP$^+$-producing bacterial strain to the intravital model, it would be prudent to first conduct a preliminary test of *gfp*$^+$ expression *in vivo*. This is because bacteria will experience different microenvironments during the infection, and it is important to ensure that the GFP$^+$ promoter activity is independent of bacterial location in the tissue. For UPEC infections, perform a traditional, ascending UPEC infection (see Section 3.4.3), sacrifice the rat 4 days postinfection, and isolate the kidneys. After fixation, use cryosections to compare the proportion of GFP$^+$-expressing UPEC in the tissue against the total bacteria population as detected by fluorophore-conjugated (Rhodamin or Texas red) antibody staining for lipopolysaccharide (LPS). Optimally, there should be a 100% overlap.

2.2. Visualizing tissues

Tissue autofluorescence is often regarded a problem during microscopy as it lowers the signal-to-noise ratio. During live imaging, however, it is often considered advantageous as it facilitates orientation in the tissue (Dunn *et al.*, 2003, 2007). In addition, fluorescent probes are used to increase information obtained from the tissue. When delivered, probes are, however, constantly cycled between different bodily compartments and may not necessarily appear in the tissue of interest. Biodistribution is hence an important factor when administering probes during live imaging. Dextran is a versatile glucan polysaccharide with uses ranging from biomaterials to drugs and probes. It can be synthesized and purified to produce samples of a specific molecular mass, and it rarely exhibits toxicity at functional doses. Creative use of fluorophore-conjugated dextran allows the visualization of different tissue compartments. Store all fluorophore-conjugated dextrans wrapped in foil ≤1 month at 4 °C.

2.2.1. Required materials

Fluorophore-conjugated dextrans

- 10 kDa fluorophore-conjugated dextran in 0.9% sterile saline (20 mg/ml): A bulk probe that, when injected intravenously, is freely filtered by the glomerulus and is somewhat permeant in the vasculature. Used for assays of glomerular permeability, proximal tubule endocytosis, and vascular permeability.
- 500 kDa rhodamine-conjugated dextran in 0.9% sterile saline (8 mg/ml): A bulk probe that, when injected intravenously, is not filtered by the kidney but is retained in the vasculature. Used for assays of glomerular permeability, vascular flow, and vascular permeability.

Reagents and disposables

- Sterile 0.9% saline solution
- 10,000 MWCO membrane
- Surgical syringe

2.2.2. Visualizing filtrate flow in the nephron

Contents of blood are actively filtered by the glomerulus into the luminal compartment of the nephron. To visualize this, slowly inject 10 kDa fluorophore-conjugated dextran (20 mg/ml) from a surgical syringe via an access line in the femoral or jugular vein. When deciding which particular fluorophore to use, one must consider whether other fluorophores will be used in combination. With the rat positioned on the MPM stage for imaging (see Section 3.3), systemically introduced dextran first appears as a flash within the vasculature before it is rapidly filtered by the glomerulus. For detailed protocols on how to image renal filtration, readers are referred to Wang *et al.* (2010).

2.2.3. Visualizing the renal vasculature

Fluorophore-conjugated molecules of molecular weight (Mw) larger than the upper glomerular filtration limit remains in the vasculature and are therefore useful to visualize the blood plasma. Prepare an 8 mg/ml sample of the rhodamine-conjugated 500 kDa dextran in 0.9% sterile saline. Dialyze 5–10 ml of this solution using a 10,000 MWCO membrane against 0.9% (w/v) sterile saline (5 l) overnight at room temperature. After systemic

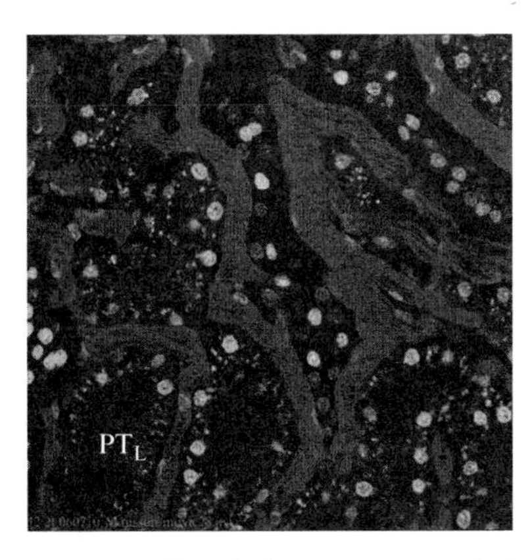

Figure 3.1 Fluorescence image of live kidney tissue morphology under multiphoton. Blood plasma (red) is labeled by fluorophore-conjugated 500 kDa dextran, black streaks represent erythrocytes. Epithelial lining of the proximal tubules exhibits autofluorescence and appears dull green, PT$_L$ indicates the lumen of the proximal tubule, and cell nuclei are stained by Hoechst 33342 (blue). Image adapted from Movie S1 in Månsson *et al.* (2007a,b). (See Color Insert.)

delivery, this fluid-phase marker remains in the vasculature for several hours (Fig. 3.1). As erythrocytes exclude the dye, these cells can be observed as black streaks within the red plasma.

3. Setup of a Live Imaging Platform of Infection

Real-time live imaging of bacterial infections requires a variety of competences ranging across microbiology, renal physiology, and biophysics. A typical study would involve several phases covering bacterial cloning and cultivation; animal handling, surgery and maintenance; microscopy and data analysis. This section will provide optimum conditions and procedures covering preimaging preparations, initiation of infection, and the basics of imaging of the infected tissue.

3.1. Selecting and customizing a multiphoton imaging platform and analysis

Multiphoton imaging systems have become increasingly easy to setup, and ready-made all-in-one systems are available from several commercial suppliers, for example, Leica, Thorlabs, and Zeiss. Alternatively, "homemade" systems, which offer more freedom of expansion, can be built from scratch or by modifying confocal microscope systems (Majewska *et al.*, 2000;

Mller *et al.*, 2003; Nguyen *et al.*, 2001). A major decision for choosing a system is to decide between an upright and an inverted system, being that the animal preparation and use defers between the two systems.

In an inverted system, the exposed tissue, in our case the kidney, is wedged between the animal and the bottom of the dish facing the objective. This helps to reduce tissue movement as the position of the kidney is stabilized by the weight of the animal. Alternatively, for an upright system, the exposed kidney is presented and immobilized using a custom designed kidney cup.

Image capture: The quality of captured images decreases dramatically with movement of the sample. While proper surgical preparation and mounting of the animal can minimize sample movement, there is a limit to how much can be done. The chance of mortality of the animal increases with the degree of invasiveness of surgical procedures, and intrinsic pulsations of the live sample from contractions of the heart and respiration cannot be avoided. High image capture rates can accommodate sample movement by acquiring a series of clear "still" shots or a fluid video of the sample as it moves. However, such systems tend to be costly and therefore not easily available.

The system we primarily have used is limited to one frame per second for a 512-by-512 frame and one frame per 2 ms for a line scan. This has been sufficient for tracking dynamic processes of host–pathogen interactions in the kidney, as well as more rapid physiological processes such as blood flow and glomerular filtration.

Digital image analysis: Live imaging of any process in real-time generates enormous datasets/image volumes of multiples planes. Often datasets involving time series are large files stored in multiple copies and variations, which then requires massive amounts of digital real estate. Aside from the resolution capability of the imaging platform and the acquisition speed of the system, processing power and memory capacity of the data analysis platform are of equal importance. With continuous turnover of powerful personal computer systems, CPUs capable of handling most forms of imaging analysis are being increasingly easy to obtain, as are image analysis software capable of handling live imaging data. Image analysis software can be purchased commercially or obtained for free via shareware. Though sharewares may have limited capabilities, they tend to be sufficient for basic image analysis and enhancement. For image analysis, we have primarily used ImageJ (a free image-processing tool from the NIH with an extensive library of plug-ins; http://rsb.info.nih.gov/ij/) and Voxx (a voxel-based 3D imaging program; http://www.nephrology.iupui.edu/imaging/voxx/index.html), followed by Adobe Photoshop (Adobe, CA, USA) for final adjustments for presentation. In case quantitative analysis shall be performed, commercial software such as Metamorph (Molecular Devices, Sunnyvale, CA) and Amira (Visage Imaging, San Diego, CA) are better calibrated. The authors suggest beginners to first experiment and familiarize with shareware before exploring other options.

3.2. Pathogen and host

3.2.1. Required materials

Rats and bacterial strains for infection model

- UPEC strain LT004 (CFT073 *cobS::*$\Phi(P_{LtetO\text{-}1}\text{-}gfp^{+})$, *cm*^{r})
- Standard Sprague-Dawley (264 ± 16 g) and Munich-Wistar (255 ± 22 g) rats

Culture medium and additional equipment

- Luria–Bertani (LB) agar and LB broth
- 37 °C shaking incubator for liquid culture growth
- 37 °C incubator for agar plates
- 100-ml conical flask or 15-ml Falcon tube
- Centrifuge
- 1.5-ml eppendorf tubes
- Glass spreader

3.2.2. Pathogen

Our studies are primarily based on the clinical, uropathogenic *E. coli* isolate CFT073 (Mobley *et al.*, 1990; Welch *et al.*, 2002) engineered to produce GFP^{+}. Isogenic mutants have been constructed and applied when testing the role of bacterial virulence factors *in vivo*. Preparation of bacterial culture for subsequent use in animal studies is relatively straightforward. Prepare a shaking overnight culture of bacteria in a 100-ml conical flask or a 15-ml Falcon tube containing LB broth at 37 °C. To ensure sufficient aeration, the volume of the culture should not exceed 20% of the total volume of the receptacle. The following morning, perform a 100-fold dilution into fresh LB broth and dispense an aliquot of suitable volume into a fresh sterile receptacle. Cultivate the culture according to above until it reaches a density of $OD_{600} = 0.6$. To minimize the volume that will be infused into the animal, concentrate the culture to an approximate density of 10^{9} CFU/ml. In general, UPEC cultures with $OD_{600} = 0.6$ have a cell density of 10^{8} CFU/ml. To concentrate the culture, harvest the bacteria by centrifugation at $5000 \times g$ for 5 min. Discard the supernatant and resuspend the pellet in 0.9% sterile normal saline of 10% the original volume. Add Fast Green FCF (1 mg/ml) (in order to visualize the direction and speed of fluid flow during the infusion to the nephron) and Cascade blue-conjugated 10 kDa dextran (0.2 mg/ml). The latter accumulates rapidly in endosomes/lysosomes of proximal tubule cells, thereby selectively outlining the injected tubule. This is important in perfused environments, where the bulk of bacteria are expected to be flushed out by the filtrate flow. Maintain the bacterial suspension on ice until used. After infusion, determine the actual live cell density of the bacterial culture by performing dilution plating. This involves performing serial 10-fold dilutions in 1.5-ml eppendorf tubes, and dispensing 100 μl aliquots of each dilution onto separate LB agar plates. Prepare triplicates for each dilution made. Evenly spread the 100 μl aliquot over the agar using a glass spreader and incubate plates in a 37 °C incubator overnight. The following morning, perform a visual count of all colonies on each plate. Select the dilution that gives a count of between 50 and 300 colonies per plate and determine the average colony count from all plates in the triplicate. The cell density of the original culture is estimated by multiplying this average number by the dilution factor.

3.2.3. Host

We applied Sprague–Dawley (264 ± 16 g) and Munich–Wistar (255 ± 22 g) rats in our studies. The latter has surface-localized glomeruli, which enable easy visualization of the proximal convoluted tubules. Before start of the experiment, rats were given free access to chow and water.

3.3. Anesthesia and surgical creation of retroperitoneal window

In order to perform live imaging, the animal must be anesthetized and surgically prepared for infection and orientation on an imaging platform. With the organ surgically exposed, the animal is more fragile, and vital parameters such as blood pressure, pulse rate, and breathing become prone to fluctuations. Such fluctuations not only increase the risk of mortality but also impact host responses to the induced infection. As such, much attention needs to be placed on how to optimally prepare and maintain the animal. The following section describes an optimized set of such surgical procedures.

3.3.1. Required materials

Mice strains for surgical preparation

- Standard Sprague-Dawley (264 ± 16 g) and Munich-Wistar (255 ± 22 g) rats

Surgical consumables

- Isoflurane/oxygen mixtures (5% (v/v) and 2% (v/v))
- Pentabarbital (optional)
- Buprinorphine
- Germicidal soap
- 0.9% sterile saline, prewarmed to 37 °C
- Vascular catheters (PE-60 tubing for rats and PE-50 tubing for mice; Becton Dickinson)

Surgical equipment

- Anesthesia induction chamber
- Homeothermic table
- Rectal probe for temperature recording
- Electric clippers
- Kidney cup
- Surgical scissors (Braintree Scientific)
- Appropriate animal temperature control devices (e.g., circulating water blanket attached to a temperature-controlled circulating water bath)

Fluorescent labels and probes

- See Section 2

3.3.2. Presurgery preparation

Place the animal in an anesthesia induction chamber containing a 5% isoflurane/oxygen mixture. After initial anesthesia is obtained, transfer the animal to a clean surgical area on a homeothermic table. Maintain anesthesia with a 2% isoflurane/oxygen mixture, titrate to effect. Inject 0.05 g/kg buprinorphine subcutaneously in preparation for the subsequent surgery. Shave areas to be incised. This is typically performed at the left flank area (access the kidney), neck (jugular vein and artery access line), and inner thigh (femoral vein access line) using electric clippers. Cleanse the respective areas with germicidal soap and water and then towel dry. Thoroughly remove any cut hairs. In our experience, minute presence of stray hairs can greatly interfere with imaging. Prior to surgery, insert a rectal probe for temperature monitoring.

3.3.3. Surgery

Small incisions are preferred when performing live imaging of the kidney, and it is easier to progressively increase the incision rather than stitch a large one. Large incisions also allow more movement of the exteriorized kidney, which is counterproductive for imaging. In our studies, a small incision (using surgical scissors) is typically made. First, make a small incision over the femoral vein and insert a venous access line (PE-60 tubing) for subsequent dye infusion. Next, expose the kidney by making a 0.5–1 cm incision (using surgical scissors) in the left flank through the retroperitoneum. We have found it beneficial to palpate the left flank for the kidney first to get an idea of its size and location. Using a pair of forceps, draw back the adipose tissues which might be covering the kidney. Lift the kidney out of the incision by holding on to the hilar fat pad. Place a Petri dish on the microscope stage and fill it with prewarmed normal saline. Move the animal to the microscope stage and position the animal such that the left kidney, immersed in saline, is facing down toward the objective. Maintain the temperature of the animal with a heating blanket and supply appropriate anesthesia.

3.4. Introducing uropathogenic *E. coli* to initiate infection

In the classical retrograde model of upper urinary tract infection, bacteria are delivered via a catheter into the bladder, from where they ascend to the kidney (Fig. 3.2). While this model resembles the natural route of infection, it suffers from drawbacks such as poor spatial and temporal control. This is because of the difficulty in predicting *where* and *when* bacteria enter into the kidney. Accordingly, this model is best suited for imaging end-stage infections when macroscopic identification of cortical abscesses directs the researcher to the site-of-interest to be imaged. To overcome these complications and generate an infection model with high spatial and temporal precision, we applied intratubular micropuncture (Fig. 3.2). Here, bacteria are slowly infused directly into the lumen of the proximal tubule in a single nephron (spatial control) at a defined time point (temporal control). For details, see below.

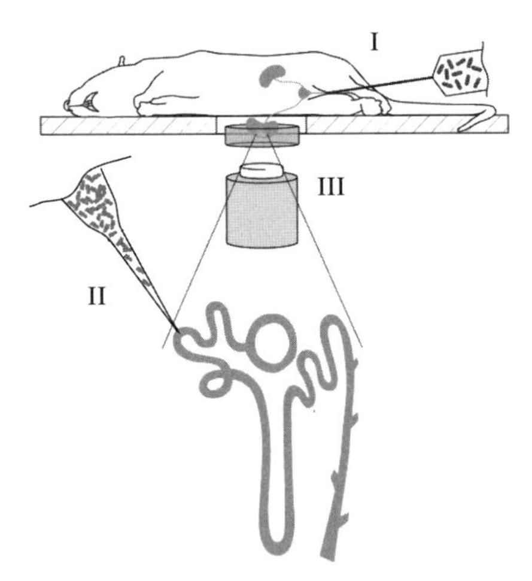

Figure 3.2 Graphical depiction showing the positioning of a live rodent on an inverted multiphoton microscope. The left kidney is stabilized in a cell culture dish and immersed in isotonic saline. The microscope stage is heated, and the rat is wrapped in a heating pad (not shown). Green fluorescent bacteria can be introduced either via (I) retrograde infection where bacteria are infused into the bladder or (II) injected by micropuncture directly into the proximal tubule of one nephron (yellow) in an exposed kidney. Infection is imaged by multiphoton microscopy (III). Image adapted from Månsson *et al.* (2007a,b). (See Color Insert.)

3.4.1. Required materials

Rats and bacterial strains for infection model and respective preparations

- See Section 3.2

Consumables

- Sterile PBS
- Heavy mineral oil
- 2% isofluran/oxygen anesthesia
- Halothane/oxygen anesthesia
- Sudan black-stained castor oil

Appropriate fluorescent probes

- 1 mg/ml Fast Green Fisher
- See Section 2

Infection and Imaging equipment

- Sharpened micropipettes
- Stereoscopic microscope
- Leitz micromanipulator
- Mercury leveling bulb
- Imaging Platform (MPM)

3.4.2. Intratubular micropuncture

Aspirate the bacterial suspension (prepared as outlined in Section 3.2) into sharpened micropipettes (tip diameter 7–8 μm) filled with heavy mineral oil. Prepare the animal as described in Section 3.3 and position it on the stereoscopic microscope stage. Wean the animal off the isoflurane/oxygen mixture used during preparation, to halothane/oxygen anesthesia, allowing fine-tuning and recovery. Focus on the projected site of infection under observation at 96-fold magnification. Using a Leitz micromanipulator and mercury leveling bulb, inject the bacterial suspension as prepared in Section 3.2.2 into the proximal tubule. To avoid increased shear stress in the tubule during delivery, an infusion rate of approximately 50 nl/min is used for 10 min. This results in total delivery of ca. 5×10^5 CFU of bacteria. Take notice of the time point for infusion, as this marks the initiation of the infection process. We find it useful to inject one to three proximal tubules for each rat, either with bacteria or with sterile PBS. The latter serves as a control for any effects of the surgical procedure. Also inject Sudan black-stained castor oil in a nephron near each of the infection sites. This will aid in the localization of injection sites once the animal is positioned on the MPM stage.

Following infusion, the animal is moved to the multiphoton microscope stage and positioned for imaging as described in Section 3.3.3. For gross identification of the infection site, use nearby tubules marked with Sudan black-stained castor oil (Fig. 3.3). To identify the bacterially injected nephron, look for the characteristic blue fluorescent outline of the tubuloepithelium, which originates from endocytosed cascade-blue-conjugated dextran coinfused with bacteria. Though the original bacterial load was ca. 5×10^5 CFU, a majority of bacteria has been flushed out by the filtrate flow, only few remain bound to the epithelium. To find these bacteria, carefully focus on the injected tubule, expect to see only few green fluorescent bacteria attached to the blue lining. In contrast, if the site is imaged at later time points, or if infection sites are used that lack perfusion, one might expect to see large masses of green fluorescent bacteria.

Next, introduce appropriate probes via intravenous injection, making sure to first flush dead space of catheter. In the kidney, we usually image the infection sites for 1–8 h after bacterial delivery (Fig. 3.4). While the animal is maintained on the microscope stage for image acquisition, the depth of anesthesia, core body temperature, and blood pressure (if desired) is monitored. Repeat experiments on different days, each time using freshly prepared bacterial inoculums. If desired, the rat may be taken off the stage, housed overnight, and imaged on subsequent days. Again, nephrons injected with Sudan black-stained castor oil will help in guiding the microscopist to find the infection sites.

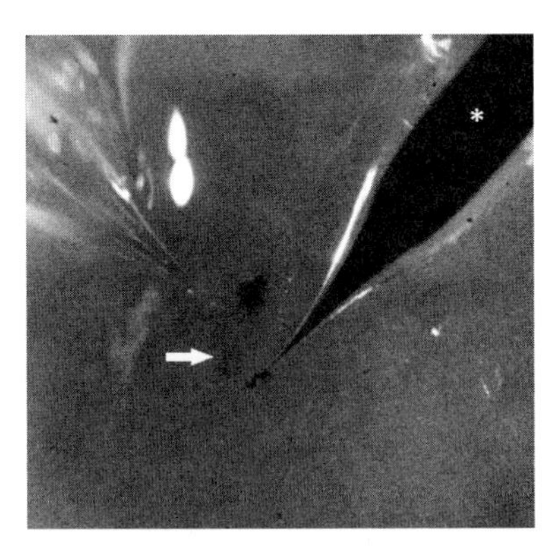

Figure 3.3 Photograph showing the microperfusion of bacteria into a tubule with a micropipette of tip diameter 8 μm (left). (Arrow) indicates the infused tubule. (*) Identical micropipette filled with Sudan black-stained castor oil is used for injection of nearby nephron to aid in orientation of the infection site during imaging. (For color version of this figure, the reader is referred to the Web version of this chapter.)

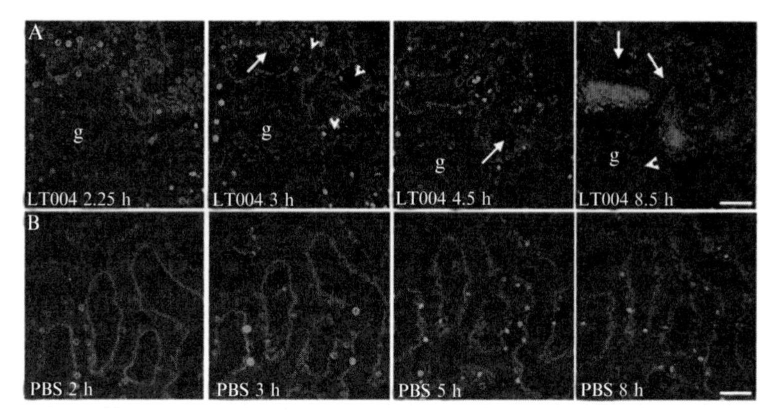

Figure 3.4 Selected time points obtained from real-time multiphoton imaging of UPEC infected (A, top panel) and PBS control-injected (B, lower panel) proximal tubules. The injected tubules are outlined by endocytosed blue dextran while tubules in uninfected nephrons exhibit green autofluorescence. In (A), normal blood flow, visualized by infusion of large Mw fluorescently labeled dextran (red), is observed in peritubular and glomerular (g) capillaries 2.25 h after onset of infection. Erythrocytes are seen as black streaks within vessels. At 3 h, initiation of capillary collapse and altered blood flow (arrowheads) occur as a consequence of infection. At 4.5 h, bacteria fill the tubule lumen (arrow). At 8.5 h, with persistent multiplication of the pathogen, fluorescence signal of proximal tubule-specific labeling disappears indicative of epithelial linings disintegration. A single glomerular loop with slowed flow of erythrocytes is shown (arrowhead). In (B), a PBS sham-injected nephron (blue outline) shows no indication of abnormal function. Image adapted from Melican *et al.* (2008). (See Color Insert.)

3.4.3. Imaging using the retrograde infection model of acute pyelonephritis

Prepare an overnight culture of bacteria in LB broth. The following morning, prepare a fresh 100-fold diluted culture in LB broth and cultivate it to $OD_{600} = 0.6$ as described in Section 3.2.2. Harvest the cells by centrifugation of the culture at $5000 \times g$. Discard the supernatant and resuspend the pellet in an equal volume of PBS to achieve a culture density of approximately 10^8 CFU/ml. We commonly use four to six rats in each series of experiments due to the inherited randomness of onset of pyelonephritis in the ascending model. Female Sprague-Dawley rats are anesthetized using isofluran. Via a catheter, 1 ml of the bacterial suspension is slowly infused into the bladder of each rat. Infusion of PBS is used in control rats. Allow the infection to progress for 4 days, then prepare rats for live multiphoton imaging as described in Section 3.3. Use ocular inspection to identify sites of cortical abscesses, which then are imaged on the MPM. Due to the inherited poor spatiotemporal control of this model, it lends itself best to the study of late-stage infection.

4. Studying Tissue Responses in Real-Time During Infection

In the presence of infectious agents, the host defends itself by onset of inflammatory responses, meaning that tissue homeostasis is inevitably altered during the time-course of infection. When studying these processes in real-time, a general knowledge of the organ architecture and function is required to ensure that both expected and unexpected deviations from native anatomy and physiology, occurring as result of the infection process, will be noticed. To verify whether observed effects relate to the infection process, comparative studies, controlling for the surgical procedure (PBS sham injections), should be performed. Time-course imaging is also useful to identify infection-associated effects, as such effects may become more pronounced as the infection progresses. Alternatively, infection of the animal using isogenic bacteria harboring mutations in specific virulence genes can be used to verify their role in an infection, and if the lack of such genes abrogates or alters respective contributions to pathogenesis and host responses.

4.1. Probing histological changes of the mucosal lining

With careful selection and introduction, fluorescent probes can be used to cleanly define respective tissue compartments. We have demonstrated that bacteria colonizing the tubule lumen can be visualized by bacterially expressed GFP^{+}, at the same time as differently sized dextrans are used to mark the epithelial lining and the vasculature (Fig. 3.1), with no visible leakage of probes when initially introduced (Månsson *et al.*, 2007a,b; Melican *et al.*, 2008). As such, boundaries between mutually exclusive fluorescent signals correspond to the boundary between respective tissue compartments. By then tracking changes of the position of these target-specific fluorescent probes, changes in histology of tissues arising from the progression of UPEC infection can be detected. Several examples are described in the following section.

4.1.1. Required materials

Rats and bacterial strains for infection model and their respective preparations

- See Sections 3.2 and 3.4

Fluorescent labels and probes

- See Section 2

4.1.2. Integrity of the mucosal epithelium

Drawing from Månsson *et al.* (2007a,b) as an example, at early time points postinfection, the luminal face of the proximal tubular epithelium is seen as a continuous blue lining, arising from the internalization of coinfused small Mw cascade blue-conjugated dextran (Månsson *et al.*, 2007a,b). Bacterium which attaches to the epithelial surface and colonize then presents as green fluorescence located directly against the blue epithelial layer (Månsson *et al.*, 2007a,b). Within the next few hours, as bacteria multiply and their numbers escalate, green-fluorescing bacteria begin to breach the continuous blue fluorescent luminal lining of the epithelial layer, moving between individual blue units to the basal lamina (Månsson *et al.*, 2007a,b). This illustrates how, during UPEC infection, bacteria gradually disrupt the integrity of the renal epithelium (Månsson *et al.*, 2007a,b).

4.1.3. Exfoliation of epithelial cells

As a compound effect of infection-induced local ischemia and contact with bacterial virulence factors, exfoliation of proximal tubule epithelial lining could be observed approximately 6 h postinfusion (Melican *et al.*, 2008). Due to the original cascade-blue staining of the epithelial lining, exfoliated cells can be identified as isolated blue fluorescent units within the lumen of the tubule with no apparent contact with the basal lamina.

4.1.4. Neutrophil Infiltration

When long-term survival of the rat is not a study objective, cytotoxic and mutagenic dyes can be introduced into the animal for short-term imaging of the tissue. In our studies, we have applied the DNA-binding fluorescent stain Hoechst 33342 to probe for the presence and location of neutrophils during UPEC infection. Neutrophils are identified based on the characteristic polymorphonuclear shape outlined in blue. As Hoechst disrupts normal cell functions by interfering with DNA replication, rat infused with Hoechst is not maintained after imaging.

4.2. Imaging vascular features such as rate, clotting, and integrity

Large Mw fluorophore-conjugated dextran injected into the vasculature remains within the systemic circulation, since glomeruli cannot filter it. This red dye, contained in the lumen of the vessels, can be used to probe several vascular features. This includes if the area under investigation has an appropriate blood supply can be visualized, the rate of the blood flow can be quantified, and the integrity of the vessel walls can be studied, as well as the formation of blood clots.

4.2.1. Imaging and estimation of vascular flow

Prepare the animal as described in Section 3. Inject rhodamine-labeled 500 kDa dextran intravenously via a jugular vein access line. Erythrocytes, which do not internalize the red dextran, remain unlabeled and appear as dark streaks rapidly moving within the vessels (Dunn *et al.*, 2003, 2007). Using a line-scan method, the axial view of a capillary is recorded every 1–2 ms using the multiphoton microscope, for 2 s. Erythrocytes moving within the vessel are identified as diagonal black streaks in the line-scan images (Fig. 3.5) from which the velocity (μm s^{-1}) of each erythrocyte can be determined by calculating $\Delta d/\Delta t$ where d represents the distance (μm) and t represents time (s) (Brown *et al.*, 2001; Kang *et al.*, 2006; Kleinfeld *et al.*, 1998; Ogasawara *et al.*, 2000; Yamamoto *et al.*, 2002). Several vessels can be analyzed to obtain an averaged estimation of the blood flow; however, it is essential to choose vessels of identical dimensions.

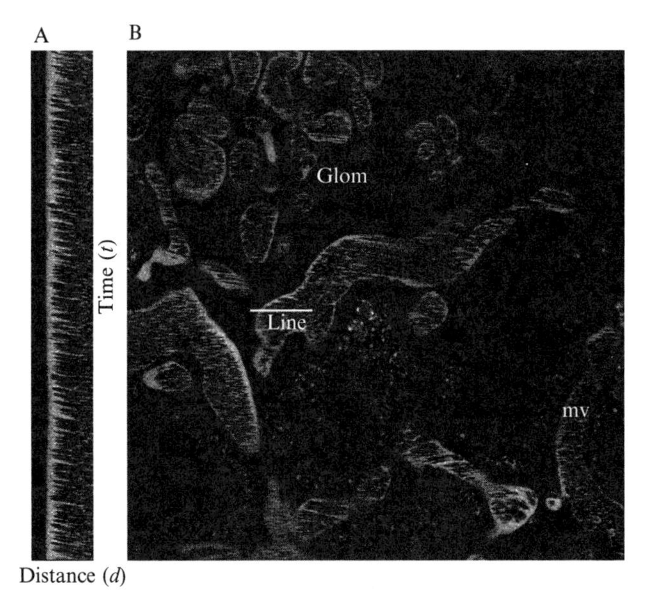

Figure 3.5 Vascular blood flow as measured by MPM. Texas Red-labeled Rat Serum Albumin (TR-RSA) injected i.v. fills the plasma volume within the vasculature denoting the surrounding microvasculature (mv) and a glomerulus (Glom). A line scan shown in panel (A) was performed along the center of the vessel shown in panel (B, line). Because erythrocytes flowing throughout the systemic circulation are impermeable to TR-RSA, they appear as black streaks within the (mv). The single line show in panel B was repeatedly imaged to form the tall montage shown in (A) where the vertical axis represents time (in this case, 2 ms per line), while the horizontal axis represents distance (in this case, 330 nm per pixel). The passage of RBCs along the scan form a speed-dependent motion artifact from which velocity can be determined by using $\Delta d/\Delta t$; here average RBC velocity was measured to be 890 μm/s ± 134.

When analyzing the effect, an infection may have on the blood flow in nearby capillaries, line-scan data should be compared to those obtained when analyzing flow rate in capillaries located next to a PBS sham-injected tubule.

4.2.2. Imaging platelets and clots with and without anticoagulant therapy

The size and shape of the black streaks seen in the red fluid-phase dextran marker varies according to cell type. Small black silhouettes, whose size indicates they are too small to be red blood cells or infiltrating immune cells, typically represent platelets. These silhouettes can be seen aggregating in capillaries, thereby forming large black nonfluorescent aggregates adhering to the vessel wall. Next to these obstructing masses, plasma, essentially devoid of red blood cells, can be identified by the intense solid red color of the fluid-phase marker. Collectively, these signs are typical for local clot formation (Fig. 3.6; Melican *et al.*, 2008). The role of clotting, whether in blood-borne infection, or as in our model when bacteria are localized at the mucosal lining, can be analyzed by applying anticoagulant therapy.

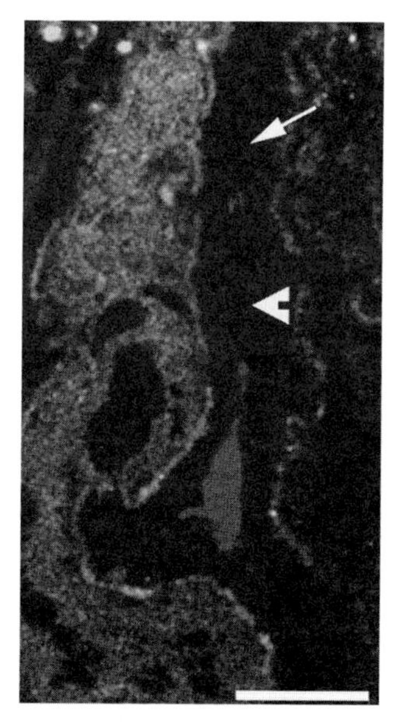

Figure 3.6 Images obtained during live multiphoton analysis of an UPEC (strain LT004, green) infected proximal tubule (blue) and adjacent blood vessel (red). A lack of erythrocyte movement is seen in the area. Black silhouettes within vessels are indicative of platelets (arrow). Aggregation of black masses adhering to the vessel wall (arrowhead) suggests platelet aggregates. Image adapted from Melican *et al.* (2008). (See Color Insert.)

Immediately following bacterial infusion, rats are treated with 400 U/kg of heparin sodium via the jugular venous access line. After 4 h, a second dose (200 U/kg) is administered. The progression of infection is imaged on the multiphoton microscope as usual with the status of the animal being monitored by body temperature and blood pressure. The latter recording is important as a sudden, sharp drop in blood pressure is indicative of the onset of sepsis. If this occurs, animal should be sacrificed, and internal organs such as the liver, spleen, and heart should be aseptically removed, homogenized in PBS, and analyzed for the presence of bacteria. This is performed by serial dilution and plating, see description in Section 3.2. Blood samples shall also be taken throughout the experiment to confirm lack of clotting in animals on anticoagulant therapy as well as to analyze for the presence of bacteria (see Section 3.2).

4.2.3. Imaging the integrity of the endothelial lining

Increased endothelial permeability causing perivascular leakage of the large Mw red dextran can be observed as a local effect of infection. In case immune cells have extravasated, they can be observed as black silhouettes within the area of leakage (Melican *et al.*, 2008).

4.3. Local tissue oxygen tension

In our UPEC infection studies, we noted an apparent cessation of blood flow in peritubular capillaries within 3–4 h of infection. As this was accompanied by typical signs of ischemia, it suggested that the tissue was suffering from a lack of vital oxygen (Melican *et al.*, 2008). By measuring the local oxygen tension, the highly targeted and localized ischemia was demonstrated as a host defense mechanism preventing systemic dissemination of the pathogen. Measuring local-tissue oxygen tension may also provide interesting insights into the role of ROS, which is an important host innate immune response against infections. Production of reactive oxygen species is regulated by oxygen tension (Wu *et al.*, 2007). Specifically, the rate of superoxide production by phagocytic oxygenase in leukocytes from glucose and oxygen is directly proportional to the local oxygen concentration (Hunt and Aslam, 2004). It would thus be highly appreciated if the imaging data on blood flow cessation could be complemented with actual recordings of the tissue oxygen.

4.3.1. Required materials

Equipment

- Modified Clark-type microelectrode system
- Stereomicroscope

Reagents

- 120 mg/kg Thiobutabarbital
- Water saturated with $Na_2S_2O_5$

Additional materials for animal preparation and infection

- See Sections 3.2 and 3.4 for animal preparation and infection

Bacterial growth and enumeration

- LB agar and LB broth
- 37 °C shaking incubator for liquid culture growth
- 37 °C incubator for agar plates

4.3.2. Surgical preparation of the animal

Anesthetize animals with an intraperitoneal injection of thiobutabarbital at 120 mg/kg. Transfer the animal to a temperature-controlled operating table at 37 °C. Perform a tracheotomy and insert polyethylene catheters into both femoral arteries and the right femoral vein. Use one arterial catheter to monitor blood pressure and the other for blood sampling. Maintain the fluid levels of the animal by saline infusion into the vein. Catheterize the bladder to allow urinary drainage. Expose the left kidney by a left subcostal flank incision and immobilize the kidney in a plastic cup surrounded by saline-soaked cotton wool. Cover the surface of the kidney with paraffin oil and then catheterize the left ureter for urine collection. Retain an ample amount for bacterial enumeration when required. Allow a 60-min window for the animal to stabilize. Infection can then be performed by microperfusion as described in Section 3.4.2.

4.3.3. Measurement of renal PO_2

Tissue oxygen tension (PO_2) is commonly measured using modified Clark-type microelectrodes (Severinghaus, 2002). Before applying the electrodes to the tissue, perform a two-point calibration in water saturated with either $Na_2S_2O_5$ or air at 37 °C. Next, initiate the infection by infusing bacteria according to previous description. In parallel, PBS should be infused into a neighboring site in the same kidney to serve as internal control. While the animal still is under stereomicroscopic observation, use the micromanipulator to insert one microelectrode into each of the tubule lumens immediately after infusion of bacteria and PBS. In our setup, simultaneous recording of PO_2 in both sites was conducted for a total of 240 min. At this stage, PO_2 in the infected nephron had plummeted to 0 mmHg. During the whole procedure, blood pressure should be monitored either via tail cuff or via a direct arterial line with the use of a pressure transducer. By presenting recorded PO_2 over time for both sites in the same graph, together with the blood pressure, a clear picture emerges whether the infection causes any local or global effects in the animal.

4.4. Systemic dissemination

In patients, bacteria sometimes find their ways from the urinary tract into the systemic circulation, thereby giving rise to urosepsis. When simulating the natural course of infection in an animal model, it is important to control different tissue compartments for the presence of the pathogen. In each experiment, blood and urine samples shall be collected from the animal at appropriate time points and analyzed for the presence of bacteria by direct enumeration of cell density by dilution plating, see Section 3.2.2. A typical sign that urosepsis has occurred is when a sharp drop in blood pressure is observed.

4.5. Imaging obstruction of a perfused environment

By infusing the vasculature with low Mw fluorophore-conjugated dextrans, the functionality of an infected nephron can be assessed. After being filtered, the fluorophore signal within the nephron is short lived as it is immediately channeled into distal reaches of the nephron. Månsson *et al.* observed the accumulation of a low Mw rhodamine-conjugated dextran in an UPEC infected nephron after the bacterium had established itself in the proximal tubule (Månsson *et al.*, 2007a,b). The persistence of the otherwise transient signal here suggested that UPEC presence in the proximal tubule ultimately leads to the stoppage of renal flow.

The high flexibility inherent to our intravital imaging model then allowed the introduction of various UPEC isogenic mutants to define the root cause of the above observed obstruction. Specifically, the synergistic role of major adhesion factors Type 1 and P fimbriae in PCT obstruction was shown by applying respective knockout strains (Fig. 3.7). By imaging the site of infection and analyzing the functionality of the nephron in relation to the clearance of the intravenously infused fluorophore-conjugated low Mw dextran, the chain of events leading to the obstruction could be identified. Applying this method, Melican *et al.* clearly showed the synergistic role of P and Type 1 fimbriae in colonization of the luminal surface and the center of the tubule, respectively, and that the cessation of filtrate flow was the result of obstruction by large bacterial aggregate bacteria residing in the center of the proximal tubule lumen, rather than a host response to epithelium-bound colonies.

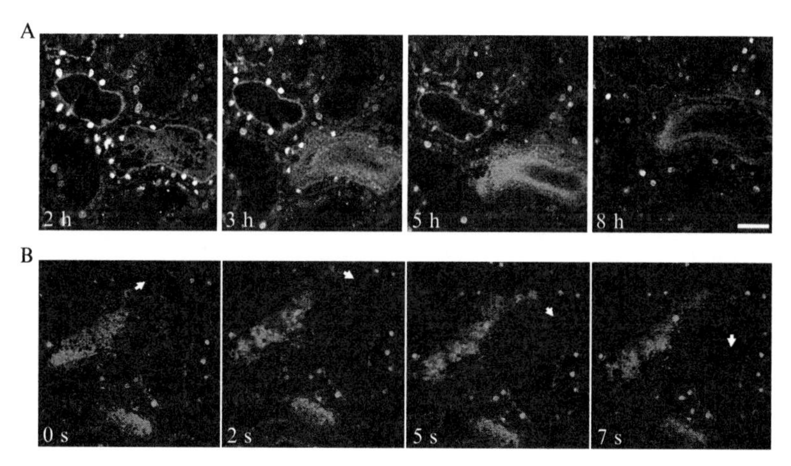

Figure 3.7 Synergistic action of attachment organelles during colonization in presence of renal flow. (A) Mutation of the UPEC Type 1 fimbriae tip adhesion negatively impacts the ability of this strain to colonize the center of the tubule lumen. This is shown from selected time points during 8 h. The same bacterium expresses P fimbriae, which promotes normal binding to the epithelium. Scale bar = 30 μm. (B) UPEC lacking P fimbriae-mediated attachment shows an inability for epithelial attachment. Increased susceptibility to the shear stress is shown by the large bacteria aggregates that are seen "flushing" in the direction of filtrate flow (arrow) with a rate of 1 $\mu m\ s^{-1}$. Image adapted from Melican *et al.* (2011). (For color version of this figure, the reader is referred to the Web version of this chapter.)

5. Postimaging Analysis

While real-time study of the infection process *in vivo* is an attractive concept, many important techniques in the studying of host–pathogen interactions remain "endpoint" measures. By applying our spatiotemporally controllable infection model, endpoint studies can be made accurately and in parallel to the real-time setting, thereby increasing the relevance of non-real-time data in describing the natural progression of host–pathogen interactions.

5.1. Immunohistochemistry

5.1.1. Required materials

Reagents and equipment for sample preparation

- Paraformaldehyde in PBS (4%)
- Sucrose in sterile PBS (20%)
- OCT Tissue-tek
- 3-methyl-butane
- Microtome

Antibodies and fluorescent dyes

- Phalloidin conjugated with TRITC or Texas Red
- Hoechst 33342 dye
- RbaCollagen IV
- GtaRb-Cy3

When ending the multiphoton imaging, aseptically remove the infected kidney and fix with freshly prepared 4% paraformaldehyde in PBS pH 7.4 (Merck, Germany) for 1 h. Wash the kidney three times in PBS and immerse it overnight in 20% sucrose in PBS at 4 °C for cryoprotection of the organ. Mount pieces of the organ in OCT Tissue-tek and snap freeze the samples using 3-methyl-butane at −50 to −60 °C. In a cryostat, make 10 μm thick renal cortical slices for staining with dyes and/or antibodies and perform confocal laser scanning microscopy. In our setting, phalloidin-conjugated TRITC or Texas Red was used to visualize actin, whereas Hoechst 33342 dye stained for cell nuclei. Antibody Rb anti-Collagen IV was used to stain the basal lamina for the major extracellular matrix protein collagen IV, which was detected using the secondary antibody Gt-anti-Rb-Cy3. Bacteria were visualized using their endogenous GFP^+.

6. Extending Real-Time Imaging into Molecular Analysis

Several features of the model for real-time imaging of infection have been found to be very useful in generating a molecular understanding of the infection process. Here, we summarize such features from the host as well as microbe perspectives.

6.1. Molecular view of host responses

Initiating the infection by microinfusing, one defined nephron gives this model an outstanding spatial and temporal precision. We find this very useful when supplementing the imaging data with transcriptional profiling to indicate by which molecular details the host orchestrates its tissue response. As the infection site is known, it can be precisely dissected from the kidney and used for RNA preparation. In the absence of the bulk of noninfected tissue from the entire organ, the ratio between infected and uninfected tissue enables RNA to be prepared at a quality suitable for microarray analysis. By obtaining tissue at time point of interests defined from parallel imaging experiments, and by comparing to the transcriptome obtained from noninfected tissue, a dynamic description of the host's response to infection emerges (Boekel *et al.*, 2011). The study can be extended by comparative tissue transcriptomics, allowing conserved trends in host responses to be identified. By doing so, we identified the importance of IFN-γ in interorgan communication already within the first hours of infection. These data illustrate that host responses at the local site of infection may result from systemic influences.

6.2. Molecular details based on bacterial genetics

Similar to the above reasoning, the spatiotemporal precision of infection provides a nice opportunity to identify bacterial gene expression *in vivo*. Preparing bacterial RNA from tissues dissected at selected time points, use quantitative reverse-phase RT-PCR to analyze whether bacteria express a gene-of-interest, that is, a virulence factor, during organ colonization (Månsson *et al.*, 2007a,b). Traditionally, virulence factors have been associated to infections in specific organs. For bacterial attachment organelles, the dogma associates P fimbriae to renal infections and Type 1 fimbriae to cystitis. Using the model described herein, we tested this hypothesis and found that both fimbriae are being expressed in the kidney during colonization (Melican *et al.*, 2011).

Using an isogenic mutant, which is unable to produce a specific virulence factor, the role of this factor for disease progression *in vivo* can be tested. When knocking out expression of the UPEC-associated toxin *a*-hemolysin (HlyA), the kinetics of the host response, rather than the actual outcome of infection at 24 h, was altered, suggesting a role for this factor in modulating the inflammatory response (Månsson *et al.*, 2007a,b).

Intravital imaging allows the function of bacterial proteins to be tested in contexts not previously attainable. Many tissue compartments expose bacteria to mechanical challenges, such as shear stress from the filtrate flow. By comparing the infection process using strains expressing or not expressing the above-mentioned attachment organelles, we studied the mechanisms that help bacteria maintain themselves in perfused environments. A synergistic function of the P and Type 1 fimbriae was identified, with P being important for attachment to the tubule epithelium, while Type 1 is engaged in interbacterial binding and biofilm formation (Melican *et al.*, 2011). The latter enables bacterial colonization across the central part of the tubule lumen where there is no epithelium to hold on to (Fig. 3.7). In dynamic environments, the infectious niche can thus be defined to be as small as center or periphery of the tubule lumen, rather than bladder and kidney as previously suggested.

In conclusion, we hope this chapter illustrates that real-time intravital imaging holds great potential in studying the dynamics of host–pathogen interactions. The concerns we have raised are in principle valid no matter which organ is studied. Hopefully, this chapter can assist others in defining systems for future analysis.

ACKNOWLEDGMENTS

The research relevant to this chapter was supported by the Swedish Research Council, the Swedish Foundation for Strategic Research, and the Swedish Royal Academy of Sciences (to A. Richter-Dahlfors), and NIH Grant DK061594 awarded to B. A. Molitoris to establish a "George M. O'Brien Center for Advanced Renal Microscopic Analysis" at the Indiana Center for Biological Microscopy, Indianapolis, Indiana.

REFERENCES

Boekel, J., Källskog, O., Rydén-Aulin, M., Rhen, M., and Richter-Dahlfors, A. (2011). Comparative tissue transcriptomics reveal prompt inter-organ communication in response to local bacterial kidney infection. *BMC Genomics* **12,** 123.

Brown, E. B., Campbell, R. B., Tsuzuki, Y., Xu, L., Carmeliet, P., Fukumura, D., and Jain, R. K. (2001). In vivo measurement of gene expression, angiogenesis and physiological function in tumors using multiphoton laser scanning microscopy. *Nat. Med.* **7**(7), 864–868.

Datsenko, K. A., and Wanner, B. L. (2000). One-step inactivation of chromosomal genes in Escherichia coli K-12 using PCR products. *Proc. Natl. Acad. Sci. USA* **97**(12), 6640–6645.

Denk, W., Strickler, J. H., and Webb, W. W. (1990). Two-photon laser scanning fluorescence microscopy. *Science* **248**(4951), 73–76.

Dunn, K. W., Sandoval, R. M., Kelly, K. J., Dagher, P. C., Tanner, G. A., Atkinson, S. J., Bacallao, R. L., and Molitoris, B. A. (2002). Functional studies of the kidney of living animals using multicolor two-photon microscopy. *Am. J. Physiol. Cell Physiol.* **283**(3), C905–C916.

Dunn, K. W., Sandoval, R. M., and Molitoris, B. A. (2003). Intravital imaging of the kidney using multiparameter multiphoton microscopy. *Nephron Exp. Nephrol.* **94**(1), e7–e11.

Dunn, K., Sutton, T., and Sandoval, R. (2007). Live-animal imaging of renal function by multiphoton microscopy. *Curr. Protoc. Cytom.* 12.9.1–12.9.18.

Goeppert-Mayer, M. (1931). Über Elementarakte mit zwei Quantensprüngen. *Ann. Phys.* **9**(3), 273–295.

Hautefort, I., Proença, M. J., and Hinton, J. C. D. (2003). Single-copy green fluorescent protein gene fusions allow accurate measurement of Salmonella gene expression in vitro and during infection of mammalian cells. *Appl. Environ. Microbiol.* **69**(12), 7480–7491.

Hunt, T., and Aslam, R. (2004). Oxygen 2002: Wounds. *Undersea Hyperb. Med.* **31,** 147–153.

Kang, J. J., Toma, I., Sipos, A., McCulloch, F., and Peti-Peterdi, J. (2006). Quantitative imaging of basic functions in renal (patho)physiology. *Am. J. Physiol. Renal Physiol.* **291**(2), F495–F502.

Kleinfeld, D., Mitra, P. P., Helmchen, F., and Denk, W. (1998). Fluctuations and stimulus-induced changes in blood flow observed in individual capillaries in layers 2 through 4 of rat neocortex. *Proc. Natl. Acad. Sci. USA* **95**(26), 15741–15746.

Majewska, A., Yiu, G., and Yuste, R. (2000). A custom-made two-photon microscope and deconvolution system. *Pflugers Arch.* **441**(2–3), 398–408.

Månsson, L. E., Melican, K., Boekel, J., Sandoval, R. M., Hautefort, I., Tanner, G. A., Molitoris, B. A., and Richter-Dahlfors, A. (2007a). Real-time studies of the progression of bacterial infections and immediate tissue responses in live animals. *Cell. Microbiol.* **9**(2), 413–424.

Månsson, L. E., Melican, K., Molitoris, B. A., and Richter-Dahlfors, A. (2007b). Progression of bacterial infections studied in real time—Novel perspectives provided by multiphoton microscopy. *Cell. Microbiol.* **9**(10), 2334–2343.

Melican, K., and Richter-Dahlfors, A. (2009a). Multiphoton imaging of host-pathogen interactions. *Biotechnol. J.* **4**(6), 804–811.
Melican, K., and Richter-Dahlfors, A. (2009b). Real-time live imaging to study bacterial infections in vivo. *Curr. Opin. Microbiol.* **12**(1), 31–36.
Melican, K., Boekel, J., Månsson, L. E., Sandoval, R. M., Tanner, G. A., Källskog, O., Palm, F., Molitoris, B. A., and Richter-Dahlfors, A. (2008). Bacterial infection-mediated mucosal signalling induces local renal ischaemia as a defence against sepsis. *Cell. Microbiol.* **10**(10), 1987–1998.
Melican, K., Sandoval, R. M., Kader, A., Josefsson, L., Tanner, G. A., Molitoris, B. A., and Richter-Dahlfors, A. (2011). Uropathogenic Escherichia coli P and Type 1 fimbriae act in synergy in a living host to facilitate renal colonization leading to nephron obstruction. *PLoS Pathog.* **7**(2), e1001298.
Mller, M., Schmidt, J., Mironov, S., and Richter, D. (2003). Construction and performance of a custom-built two-photon laser scanning system. *J. Phys. D Appl. Phys.* **36,** 1747.
Mobley, H. L., Green, D. M., Trifillis, A. L., Johnson, D. E., Chippendale, G. R., Lockatell, C. V., Jones, B. D., and Warren, J. W. (1990). Pyelonephritogenic Escherichia coli and killing of cultured human renal proximal tubular epithelial cells: Role of hemolysin in some strains. *Infect. Immun.* **58**(5), 1281–1289.
Molitoris, B. A., and Sandoval, R. M. (2005). Intravital multiphoton microscopy of dynamic renal processes. *Am. J. Physiol. Renal Physiol.* **288**(6), F1084–F1089.
Nguyen, Q., Callamaras, N., Hsieh, C., and Parker, I. (2001). Construction of a two-photon microscope for video-rate Ca2+ imaging. *Cell Calcium* **30**(6), 383–393.
Ogasawara, Y., Takehara, K., Yamamoto, T., Hashimoto, R., Nakamoto, H., and Kajiya, F. (2000). Quantitative blood velocity mapping in glomerular capillaries by in vivo observation with an intravital videomicroscope. *Methods Inf. Med.* **39**(2), 175–178.
Severinghaus, J. W. (2002). The invention and development of blood gas analysis apparatus. *Anesthesiology* **97**(1), 253–256.
Squirrell, J. M., Wokosin, D. L., White, J. G., and Bavister, B. D. (1999). Long-term two-photon fluorescence imaging of mammalian embryos without compromising viability. *Nat. Biotechnol.* **17**(8), 763–767.
Svoboda, K., and Yasuda, R. (2006). Principles of two-photon excitation microscopy and its applications to neuroscience. *Neuron* **50**(6), 823–839.
Tian, G. F., Takano, T., Lin, J. H. C., Wang, X., Bekar, L., and Nedergaard, M. (2006). Imaging of cortical astrocytes using 2-photon laser scanning microscopy in the intact mouse brain. *Adv. Drug Deliv. Rev.* **58**(7), 773–787.
Wang, E., Sandoval, R., Campos, S., and Molitoris, B. (2010). Rapid diagnosis and quantification of acute kidney injury using fluorescent ratio-metric determination of glomerular filtration rate in the rat. *Am. J. Physiol. Renal Physiol.* **299**(5), F1048.
Weissman, S. J., Warren, J. W., Mobley, H. L., and Donnenberg, M. S. (2007). Host-pathogen interactions and host defense mechanisms. *Dis. Kidney Urin. Tract* **1,** 3776 8th edition.
Welch, R. A., Burland, V., Plunkett, G., Redford, P., Roesch, P., Rasko, D., Buckles, E. L., Liou, S. R., Boutin, A., Hackett, J., Stroud, D., Mayhew, G. F., *et al.* (2002). Extensive mosaic structure revealed by the complete genome sequence of uropathogenic Escherichia coli. *Proc. Natl. Acad. Sci. USA* **99**(26), 17020–17024.
Wu, W., Platoshyn, O., Firth, A. L., and Yuan, J. X. J. (2007). Hypoxia divergently regulates production of reactive oxygen species in human pulmonary and coronary artery smooth muscle cells. *Am. J. Physiol. Lung Cell. Mol. Physiol.* **293**(4), L952–L959.
Yamamoto, H., Lee, C. E., Marcus, J. N., Williams, T. D., Overton, J. M., Lopez, M. E., Hollenberg, A. N., Baggio, L., Saper, C. B., Drucker, D. J., and Elmquist, J. K. (2002). Glucagon-like peptide-1 receptor stimulation increases blood pressure and heart rate and activates autonomic regulatory neurons. *J. Clin. Invest.* **110**(1), 43–52.

CHAPTER FOUR

Analysis of Virus Entry and Cellular Membrane Dynamics by Single Particle Tracking

Helge Ewers* *and* Mario Schelhaas†

Contents

1. Introduction 64
2. Fluorescent Labeling of Viruses 66
   2.1. Chemical labeling of viral surface proteins 67
   2.2. Labeling the viral envelope with lipid tracers 71
3. Data Acquisition and Analysis 73
   3.1. Microscopy 73
   3.2. Tracking virus particles 74
   3.3. Cotracking with cellular markers 76
   3.4. Extracting information from tracks 77
4. Conclusions and Outlook 78
Acknowledgments 79
References 79

Abstract

Viruses have evolved to mimic cellular ligands in order to gain access to their host cells for replication. Since viruses are simple in structure, they rely on host cells for all their transportation needs. Following single virus particles during the initial phase of infection, that is, virus entry into target cells, can reveal crucial information on the mechanism of pathogen infections and likewise cellular transport and membrane dynamics.

Here, we give an overview on how to fluorescently label virus particles for live cell microscopy, and on how virus entry can be analyzed by single particle tracking experiments. Highlighted are strategies, on how to *chemically* introduce fluorophores into virions, and on how to extract *quantitative* information from live cell data.

* Institute of Biochemistry, ETH Zurich, Schafmattstrasse, Zurich, Switzerland
† Emmy-Noether Group, Virus Endocytosis', Institutes of Molecular Virology and Medical Biochemistry, ZMBE, University of Münster, Münster, Germany

Methods in Enzymology, Volume 506
ISSN 0076-6879, DOI: 10.1016/B978-0-12-391856-7.00028-7

1. Introduction

Virus entry is the process during which viruses gain access to viral replication sites within uninfected cells, a multistep course of events that starts with binding to target cells. Since viruses are simple in structure/composition and lack any locomotive capacity, viruses depend on cellular trafficking mechanisms during entry, which most often involves endocytosis (Schelhaas, 2010). This makes virus entry a feasible tool to explore membrane dynamics by following virus particle dynamics. Virus entry is highly dynamic: For one, the subsequent interactions of the virus with cellular structures during the various entry steps (binding, plasma membrane dynamics, internalization/endocytosis, intracellular trafficking, penetration/membrane fusion) are transient, and they require motion of the virion. For two, the virus interacts with cellular structures that are dynamic themselves. Probing the dynamic interactions and movements of viral particles as they enter cells by live cell imaging is not only satisfying in a "seeing-is-believing" way but also yields novel and crucial information on the mechanism of this process, which is important for the development of antiviral strategies and for our understanding of cellular membrane dynamics (Burckhardt and Greber, 2009).

Historically, the visualization of virus particles by direct fluorophore labeling and microscopy during entry started 1980, when Helenius and colleagues described the binding of rhodamine-labeled Semliki Forrest Virus to cells (Helenius *et al.*, 1980). From the early 1990s, the detection of viruses in live cells by many groups followed.

After pioneering work on the attachment and mobility of incoming adeno-associated virus (Seisenberger *et al.*, 2001), the analysis of viral entry by live cell microscopy has increasingly focused on quantitative aspects of viral motion on the cell surface and within cells by single particle tracking (Table 4.1; Brandenburg and Zhuang, 2007; Burckhardt and Greber, 2009).

Single particle tracking of viruses can provide data that are inaccessible in ensemble measurements, and only by single particle tracking we are able to dissect two or more processes that occur in parallel or in a timely overlay. To understand the cell biology of viral infection, it is desirable to identify and characterize the different patterns of movement and interactions that occur during infectious entry. For this, quantification and classification of a large number of trajectories is required. This has been exemplified in work by the Zhuang lab (Lakadamyali *et al.*, 2003) that followed mobility and fluorescence intensity of influenza virions during cell entry. When influenza virions bound to cells, they underwent a period of random motion presumably on the cell surface. Subsequently, virions exhibited a burst of directed motion at about 2 μm/s toward the perinuclear area indicating transport toward the minus end of microtubules by dynein motors. Afterward, the fluorescence of the lipid envelope of the

Table 4.1 Single particle tracking studies of various virus families

Virus	Reference
Adeno-associated virus	Seisenberger *et al.* (2001)
Adenovirus	Gazzola *et al.* (2009), Helmuth *et al.* (2007)
Dengue virus	van der Schaar *et al.* (2008)
Hepatitis C virus	Coller *et al.* (2009)
HIV	Baumgartel *et al.* (2011), Endress *et al.* (2008)
Human papillomavirus	Schelhaas *et al.* (2008)
Influenza virus	Lakadamyali *et al.* (2003)
Murine leukemia virus	Lehmann *et al.* (2005)
Murine polyoma virus	Ewers *et al.* (2005)
Poliovirus	Brandenburg and Zhuang (2007)
Simian Virus 40	Ewers *et al.* (2007), Kukura *et al.* (2009)

virus rose quickly before vanishing, suggesting fusion of the viral envelope with the membrane of an endosomal vacuole.

In this simple but telling experiment, the combination of the temporal organization of the events with the quantitative information on viral movement already provides a testable model on the infectious entry pathway of influenza virus.

Thus, quantifying the motion of viral particles from time-lapse fluorescence microscopy can provide a wealth of information. The speed of directed motion may provide information on the kind of motors that may transport viruses along distinct cytoskeletal tracks. The track length, whether stops occur, or whether directionality changes can help to identify the mechanism of motion along cytoskeletal tracks. If the viral receptor is linked to the cytoskeleton, this will lead to deviation of viral movement from random Brownian motion in the plasma membrane, for example, to directed motion by actin retrograde flow (Lehmann *et al.*, 2005; Schelhaas *et al.*, 2008) or to confinement to small areas in the plane of the plasma membrane (Ewers *et al.*, 2005). In this way, quantitative information on viral mobility can tell us about the cellular mechanisms underlying membrane receptor motion within the plasma membrane and the events leading to viral internalization.

If single virus tracking is then combined with cotracking of cellular components, further information on the cell biology of virus entry is obtained. This approach revealed, for example, that influenza viruses may use two different endocytic internalization mechanisms in parallel (Rust *et al.*, 2004), and that clathrin-mediated endocytosis requires actin polymerization for the internalization of infectious but not noninfectious Vesicular Stomatitis Virus particles simply due to their different size (Cureton *et al.*, 2010).

Several challenges are associated with the analysis of viral entry in live cells on both, the experimental level and the level of data analysis and interpretation: Practical challenges lie within the detection noise of the sample, that is, how reliably viral particles or cellular structures can be discerned from background noise, within the spatial and temporal resolution of live cell movies, and within the accurate and quantitative analysis of the large amounts of data obtained during acquisition. Data interpretation is challenging due to "biological noise," which is based upon random fluctuations in or superimposition of biological processes. These processes, particularly quick, transient events, occur never synchronously. Different virions within the population may follow different itineraries at the time of acquisition. Or, a single virus may exhibit several different types of motion during the time of data acquisition such as diffusion, directed motion, confinement (on the plasma membrane) followed by fast long range and possibly saltatory intracellular movement. In addition, the ratios of overall virus particles to infectious particles can be high, and this may make it difficult to define the relevant motion patterns of infectious particles as opposed to the "biological noise" of noninfectious particles.

To interpret data correctly, it is of utmost importance, to *quantitatively* analyze the motion of viruses by multiple single particle tracks and to dissect and to reconstruct the subsequent motion types in virus ensembles. Thus, the analysis needs to include a routine to define and extract the different patterns computationally from a huge amount of potentially heterogeneous data. Cell biological and virological controls are required to differentiate between data and noise.

The practical challenges faced in the tracking of single particles during virus entry will be discussed in the beginning of this chapter, whereas the techniques for data extraction and analysis that facilitate the separation of relevant information from noise will be discussed in the second part.

2. Fluorescent Labeling of Viruses

To follow viruses by live cell microscopy, virions need to be fluorescently labeled. Two main approaches have been used to follow viruses by fluorescent microscopy in live cells (Brandenburg and Zhuang, 2007): (i) tagging of viral structural proteins with fluorescent proteins or with small peptides that have a high affinity for exogenously supplied small fluorescent compounds and (ii) chemical conjugation or introduction of fluorophores to or into viral structures.

One experimental challenge is to introduce a minimal number of fluorophores necessary to detect virus particles against the fluorescent background of cells. This is particularly important when viral assembly/egress is studied due to the strong out-of-focus fluorescence of highly overexpressed, tagged

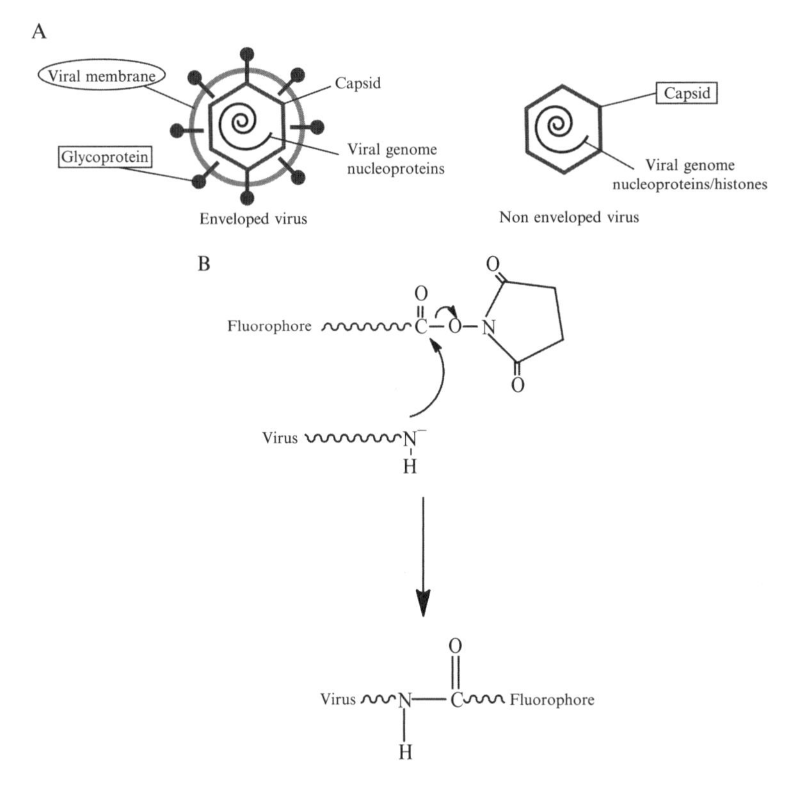

Figure 4.1 Labeling virus particles. (A) Schematic depiction of a typical enveloped versus nonenveloped virus. Boxed are components accessible for chemical conjugation of fluorophores, whereas encircled is the viral membrane accessible for lipophilic dye insertion. (B) Schematic depiction of the reaction of a succinimidyl ester-coupled fluorophore with amine residues within a virus. (For color version of this figure, the reader is referred to the Web version of this chapter.)

viral proteins. As with all proteins, attaching dyes or tags to viral structures can alter their biological function, may render the particles noninfectious, and needs to be carefully controlled.

Here, we focus on how to chemically label viral surface proteins or the viral envelope to study virus entry and endocytosis (see Fig. 4.1A), since the strategies for the introduction of tags into viral structural proteins are highly diverse.

2.1. Chemical labeling of viral surface proteins

There are many different chemical reactions that can be used to covalently attach fluorophores to amino acids/proteins. Mostly, amine- and thiol-reactive reagents are used with which virtually any protein on the virus surface can be labeled, that is, viral glycoproteins (envelope proteins) or

Table 4.2 Chemical labeling reagents and some suppliers

Target	Reactive chemical compound	Supplier
Surface proteins Amine reactive	Alexa fluor succinimidyl esters	Invitrogen (www.invitrogen.com)
Surface proteins Amine reactive	Atto succinimidyl esters	Atto Tec (www.atto-tec.com)
Surface proteins Amine-reactive	Oyster succinimidyl esters	Denovo Biolabels (www.biolabels.com)
Surface proteins Thiol-reactive	Alexa fluor maleimide	Invitrogen (www.invitrogen.com)
Surface proteins Thiol-reactive	Atto maleimide	Atto Tec (www.atto-tec.com)
Viral membrane Dialkylcarbocyanines	DiI, DiO, DiD, DiR	Invitrogen (www.invitrogen.com)
Viral membrane Phospholipids	Atto DPPE, DOPE, PPE, DMPE	Atto Tec (www.atto-tec.com)

capsid proteins for nonenveloped viruses (e.g., Schelhaas *et al.*, 2008; Vonderheit and Helenius, 2005). A summary of labeling reagents from various sources is given in Table 4.2. Based on our experience, we recommend to initially use succinimidyl ester-conjugated fluorophores.

The following protocol is optimized for nonenveloped viruses but has been successfully used for enveloped viruses such as Semliki Forrest Virus.

2.1.1. General considerations

2.1.1.1. Virus Importantly, the virus particles need to be purified prior to labeling and should be free of proteins or cellular debris at least on a coomassie gel. Any addition of BSA or stabilizing proteins should be avoided during the labeling procedure.

The concentration of the viral protein(s) to be labeled should be 1–2 mg/ml for optimal results. However, lower concentrations of as little as 0.2 mg/ml typically give satisfactory results, too. Higher concentrations increase the tendency of viruses to aggregate during labeling. The optimal viral protein concentration for the labeling procedure will have to be determined once.

2.1.1.2. Reactive dye Succinimidyl esters form an amide bond between the dye and the viral protein. The reactive dyes should be dissolved in high quality, anhydrous dimethyl sulfoxide (DMSO, 10 mg/ml), ideally directly prior to use. However, dye solutions can be stored at −80 °C for a number of months to obtain satisfactory results. Keep away from light to avoid photobleaching of the fluorophore.

It is important to consider that the number and surface position of the amines (i.e., lysine residues that are exposed to the surface of both the protein itself and the capsid) will vary greatly among the surface proteins of different viruses, as will the reactivity of the dyes. We recommend to try three different degrees of labeling whenever possible, using different molar ratios of the reactive reagent to viral protein, and base future protocols on the amount of reagent that gives the most satisfactory results for a particular virus.

2.1.1.3. Reaction buffer Succinimidyl esters react with nonprotonated aliphatic amine groups, including the amine terminus of proteins and the ϵ-amino group of lysines (see Fig. 4.1B). The ϵ-amino group has a pK_a of around 10.5. To maintain it in the nonprotonated form, the conjugation must take place in a buffer with slightly basic pH. It is important to avoid buffers that contain primary amines (e.g., Tris), since these will compete for conjugation. In general, a 0.1 to 0.2 *M* sodium bicarbonate buffer, pH 8.3 is favorable for conjugation of succinimidyl esters with proteins. However, the reaction often also works satisfactory for PBS–based buffers, pH 7.4 (see example below). In this buffer primarily the amine terminus of proteins will be labeled.

2.1.1.4. Purification The labeled virus will have to be separated from free dye with a purification method of your choice. Use the optimized protocol for purification of your viral particles such as gradient ultracentrifugation, size exclusion columns or extensive dialysis. Viruses are usually large compared to free dyes, so buffer exchange columns with size exclusion of >100 kDa often work fine, as long as your virus does not bind to the resin.

2.1.1.5. Analysis of the labeling procedure The labeling efficiency may require optimization in order to visualize your particles. Determine the relative efficiency of a labeling reaction by measuring the absorbance of the protein at 280 nm and the absorbance of the dye at its excitation maximum (λ_{max}). Using the Beer–Lambert law ($A = \epsilon \times$ path length $\times$ concentration), the approximate number of dye molecules per protein molecule can be calculated.

2.1.1.6. Aggregate formation It is important to avoid aggregation of virus particles during the labeling procedure, and the handling will have to be optimized for this purpose. To test the dispersity of the fluorescent virus solution, one may use light scattering techniques or electron microscopy. A fast and satisfactory way to analyze the sample is to bind the labeled virus particles to coated or uncoated glass coverslips and to visualize them by fluorescence microscopy. The particles should exhibit a single Gaussian intensity distribution of your fluorescent spots with similar signal intensity and dimension. Clusters typically exhibit high intensity and spatial variations and do not exhibit single Gaussian profiles.

2.1.2. Example protocol: Labeling of HPV-16 virions

This protocol allows the fluorescent labeling of human papillomavirus type 16 (HPV-16) pseudovirions. HPV-16 is a nonenveloped virus that consists of two structural proteins, the major capsid protein, L1 and the mostly internally located minor capsid protein, L2, which form an icosahedral ($T = 7$) capsid that is 55 nm in diameter.

1. (*Optional*: Dialyse a suitable amount of purified virus (1–2 mg) into 0.1 *M* $NaHCO_3$, pH 8.3, or in another suitable buffer).

   For labeling, dilute HPV-16 virions purified over a linear 5–20% Optiprep gradient in PBS, 0.8 *M* NaCl, pH 7.4 (we avoid the carbonate buffer, since the elevated pH affects the viral structure).
2. For HPV-16, the optimal viral protein concentration for labeling is 0.5 mg/ml. Determine the *accessible* protein concentration, that is, the viral surface proteins. A protein assay with intact virions will give a good estimation of the general protein amount, then estimate the molar concentration of accessible viral proteins, here L1, the major capsid protein, as based on a coomassie gel.
3. Take 0.5 ml of your virus suspension and pipet the appropriate amount of dye (stock = 10 mg/ml in DMSO, 10× molar ratio of dye/L1) *dropwise* into the virus suspension *while vortexing the tube*!!!

   Note: If the complete amount of dye is pipetted at once into the tube without mixing, virus particles will aggregate considerably. When aggregates are being formed, too much dye was used, or your virus concentration was too high. Scaling down the dye may result in inefficient labeling, so concentration of virus is the one to adjust at this moment.
4. Close the tube, wrap with aluminum foil and rotate side-over-side on a rotator for 1 h at RT.

   (*Optional*: Stop reaction by adding 40 μl of freshly made 1.5 *M* hydroxylamine, pH 8.5, and incubate for another hour. Or proceed immediately with next steps. For size exclusion, we actually recommend not to block the labeling reaction. The unbound, still-reactive dye will then bind to the column and reduces the chance that some unbound dye remains in the final virus suspension.)
5. *Note*: The fastest method to remove unbound dye is to carry out a size exclusion chromatography step. Be sure, however, that your virus does not tend to bind unspecifically to Sephadex G25 columns. A safer method (actually we use this method first until we have found the optimal conditions for a particular virus) is to reband the virus by gradient centrifugation using CsCl or Sucrose. We then isolate the band with a needle and dialyze in the particular virion buffer.

We use a NAP5 column (Amersham Pharmacia) to separate unbound dye, and equilibrate the column with 10 ml virion buffer. Apply the labeling mixture, discard the flow through. Elute with 0.5–1 ml virion buffer. The virus will be diluted up to two times.

Note: Use any buffer you like, but preferably one that does not contain any protein or free amine groups. However, if the column needs to be saturated with BSA to eliminate any unspecific binding of the virus to the column, the reaction has to be stopped with hydroxylamine as described above.

6. Determine the virus protein concentration again (if there is BSA in the preparation, because it was used during size exclusion chromatography, assume that the virus protein concentration to be diluted twice, which is true for HPV-16). Measure the amount of absorption of the specific dye of the labeled virus solution with a spectrophotometer and observe the amount of absorption of the specific dye. Calculate the amount of dye molecules per virus particle. Check, which proteins are typically labeled by the fluorophore after SDS-PAGE gel using a transilluminator for this purpose.
7. Test the infectivity of your labeled virus preparation in comparison to the unlabeled input. We simply use reporter gene expression from a construct delivered by HPV-16, but any other infectivity assay such as plaque assays or viral gene expression can be used, too.

2.2. Labeling the viral envelope with lipid tracers

A number of lipophilic tracers could be applied to label enveloped viral particles. Of those, the long chain lipophilic dialkylcarbocyanines (DiI, DiO, DiD, DiR, Table 4.3) and fluorophore-coupled phospholipids have most often been used (for manufacturers, see Table 4.2). Dialkylcarbocyanines are weakly fluorescent in water but highly fluorescent and quite photostable when incorporated into membranes. The spectral properties of the dialkylcarbocyanines are largely independent of the lengths of the alkyl chains. They have extremely high extinction coefficients, moderate fluorescence quantum yields, and short excited state lifetimes in lipid environments.

Table 4.3 Excitation and emission maxima of dialkylcarbocyanines

Dye	Excitation max.	Emission max.
DiI	484	501
DiO	549	565
DiD	644	663
DiR	748	780

2.2.1. General considerations

2.2.1.1. Virus Again, the virus particles need to be purified prior to labeling and should be free of proteins or cellular debris at least on a coomassie gel. The following protocol works only for enveloped viruses such as Vesicular Stomatitis Virus, Semliki Forrest Virus, Influenza viruses, since the tracer molecules require a viral membrane to insert into. The ratio of virus particles to amount of dye needs to be established once.

2.2.1.2. Reactive dye The low solubility of the dye in aqueous solution leads to high variance in free concentration and this may lead to inconsistent dye incorporation during labeling. We recommend to use commercially available solutions of the dye for staining such as the Vybrant^{TM} solutions (Molecular Probes). Dissolving the dye crystal works reasonably well in anhydrous DMSO (preferred 1 mg/ml), but then the labeling protocol may require optimization for any new batch of dissolved dye. For initial tries to stain enveloped viruses, DiI is to be preferred over DiO, DiD, as staining of the latter tend to be rather weak. However, DiD may be advantageous for imaging, as cells typically exhibit less background fluorescence toward longer wavelengths.

2.2.2. Example protocol: Influenza A virus (Strain X31) labeling by DiD

1. Dialyse purified influenza A viruses from a sucrose gradient overnight against PBS +/+. After dialysis, adjust the protein concentration to 0.5 mg/ml.
2. Take 0.5 ml of the virus suspension and pipet the appropriate amount of dye, here 100 μl (stock = 1 mg/ml in DMSO), *dropwise* into the virus suspension *while vortexing the tube*!!! To test the right concentration of dye, initially try 1, 10, 100, 200 μl of dye. Add PBS +/+ to a final volume of 1.5 ml.
3. Close the tube, wrap with aluminum foil and rotate side-over-side on a rotator for 2 h at RT.
4. To remove dye that has not been incorporated into virus, we recommend a step or linear gradient *ultracentrifugation*. Purification over a 30%/60% sucrose step gradient yields a distinct blue band on the cushion. Extract by syringe (total volume about 1 ml).
5. Dialyse over night against PBS +/+, 0.2% BSA at 4 °C.
6. As a result, up to 5 ml dialyzed virus solution are obtained. To remove aggregates, we pass the influenza virus solution through a 0.45-μm sterile filter, that we wetted with virion buffer in advance. After passing the virus through the filter, wash it with an additional 5 ml PBS +/+, 0.2% BSA. Plenty of particles are lost this way, but in exchange a monodisperse particle solution is obtained.

7. Determine again your virus protein concentration (if you have BSA in your preparation, you may simply assume the dilution factor). Measure the amount of absorption of the specific dye of your labeled virus solution with a spectrophotometer and observe the absorption of the specific dye. Calculate the ratio of dye molecules per virus particle.
8. Test the infectivity of the labeled virus preparation in comparison to the unlabeled input.

3. Data Acquisition and Analysis

3.1. Microscopy

To visualize and detect viruses, a number of different microscopy setups are used. However, epifluorescence microscopy, confocal techniques, and total internal reflection microscopy (TIRFM) are being employed in most cases.

Epifluorescence microscopy has the advantage of large imaging depth and relatively low cost. Disadvantageous may be the high background due to out-of-focus fluorescence and autofluorescence of the cell, so that virions with low fluorophore labeling may not be reliably distinguished from noise during image analysis. The large imaging depth may also increase the risk of particles crossing paths during tracking which results in dismissing the respective information (see below).

Confocal techniques have the advantage that the out-of-focus background is considerably reduced, and that three-dimensional data can be acquired. This advantage, however, comes at the cost of lower signal intensity. A multiplexed detection by a spinning disk setup reduces the time of illumination necessary for detection as compared to a laser scanning setup. Here, the price is the loss of adjustable confocal depth present in the laser scanning method.

TIRFM has minimal signal loss, since it is a wide-field technique. However, the fluorescence excitation by the evanescent field close to the coverslip (about 200 nm deep) limits the technique to events that occur close to the bottom surface of a cell such as virus motion on the plasma membrane or internalization.

Advantageous for live cell tracking are high quantum yield charge-coupled device cameras that allow rapid acquisition of data (in the ms range) to follow the fast events during virus membrane trafficking. For cotracking, illumination wavelengths have to be switched very fast, to acquire the signals of virus versus cellular marker proteins within a reasonable time frame in particular, if colocalization of potentially fast moving structures needs to be captured. This can be achieved by employing fast excitation filter changers, monochromators, or better laser-based illumination sources coupled with an acousto-optical tunable filter. Split view devices can be employed for simultaneous detection of two fluorophores at the cost of a smaller field of view. There, the reflected light is split into two light paths that are filtered before they are each imaged simultaneously on separate halves of the camera chip each.

3.2. Tracking virus particles

The investigation of lipids and membrane proteins by single particle tracking has lead to important discoveries about the structure and organization of the plasma membrane. With technological development, more delicate properties of membrane molecules such as receptor dimerization and the interplay between signaling and receptor mobility could be addressed. However, these techniques usually address processes at the plasma membrane only.

For viruses, the entire entry process can be followed by single particle tracking from attachment and plasma membrane mobility to endocytosis and intracellular trafficking. The third dimension, although important, is typically neglected due to the limit in z-resolution of current microscopy techniques. The quantitative representation of the viral movement contains information about the infectious and noninfectious pathways of viral movement as well as information on the cell biology of the organelles the virus interacts with. Since viruses are generally thought to bind to more than one receptor molecule, single particle tracking can additionally be used to investigate the consequences of receptor cross-linking.

3.2.1. General considerations for virus tracking

To track single viruses, monodisperse solutions of fluorescence-labeled virions are required that are still infectious or at least mimic the behavior of infectious virus. If TIRF illumination is used to reduce cellular background, the viruses need to be sufficiently small to be able to move into the narrow space between the coverglass and the bottom membrane of the cell. Alternatively, cells can be seeded on micropillars.

3.2.1.1. Identification Most viral particles will be smaller than the resolution limit of fluorescence microscopy and hence, will exhibit the pattern of a point source on the camera chip. This point spread function can be fit by a Gaussian function to determine the subresolution position of the particle. The peak of the fit Gaussian function is considered to be the place the particle is at and the quality of the fit gives the localization error. The localization error depends on the number of photons emitted by the labeled virus and the resulting signal to noise ratio. Viruses are generally labeled with many fluorophores and hence relatively bright. This is why the localization precision is typically better than 50 nm. A number of fast algorithms are available for particle detection and localization (Cheezum *et al.*, 2001; Sbalzarini and Koumoutsakos, 2005). Typically, particles are detected by a watershed mark as brightest spots within a given frame and are then fitted as mentioned above. Here, the parameters must be carefully adjusted to distinguish clearly between viral particles and spots resulting from background or noise.

3.2.1.2. Tracking To link particles in subsequent frames, a circular extension is computationally created around the localized particle. In the subsequent frame, particles within this circle are linked. The size of this circle thus determines how close particles can be in respect to one another and how far they can move between frames without being lost. This parameter needs to be carefully controlled such that it does not lead to a bias in the sampling of particle mobility. Particles that cross path cannot be distinguished intervention-free and must be dismissed. Hence, the density of the particles on the cell is important (Fig. 4.2).

Sometimes, particles are manually linked or approximations are being made based on the directionality of movement in prior frames. Bear in mind, that this constitutes a direct interference with data analysis and a potential introduction of bias. This may lead to misinterpretation. Also, it may be considered falsifying results, if not carefully documented and discussed.

Tracking algorithms are freely available in platforms such as ImageJ (rsbweb.nih.gov/ij) or Fiji (http://fiji.sc/Fiji) or on the websites of developers

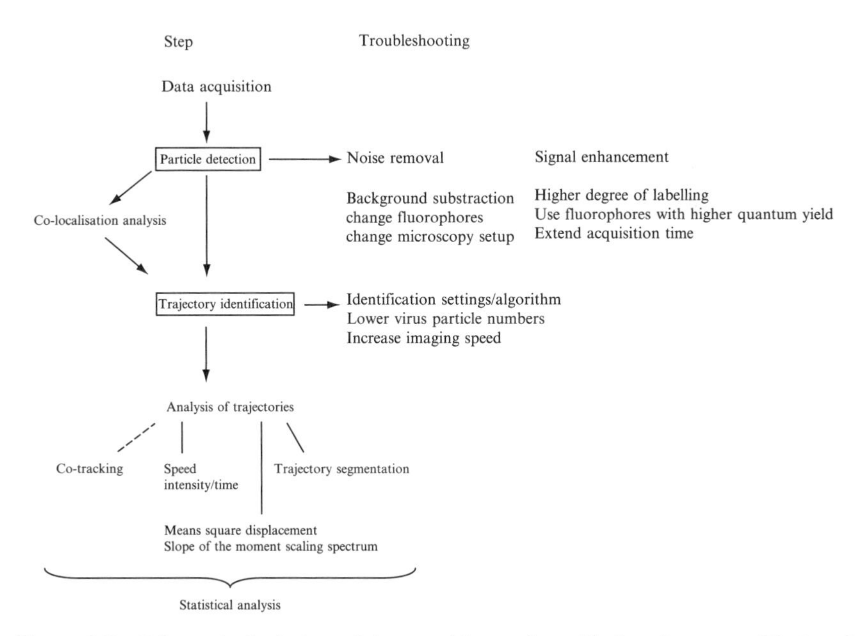

Figure 4.2 Schematic depiction of virus tracking and troubleshooting steps. Displayed on the left is the typical workflow of a single virus tracking experiment that starts with the acquisition of data. Typical troubleshooting steps in this workflow are boxed and suggestions for troubleshooting are depicted on the right.

(e.g., www.mosaic.ethz.ch/Downloads/matlabtracker). Several commercially available image analysis packages also contain tracking applications.

3.2.2. Experimental example: Single particle tracking of Simian Virus 40 at the plasma membrane

1. Plate CV-1 cells grown in indicator-free Dulbeccos's Modification of Eagles Medium (DMEM) supplemented with 10% FCS and 4 m*M* Glutamax on sterile coverglass 1 day prior to the experiment.
2. On the day of the experiment prewarm indicator-free medium supplemented with 20 m*M* HEPES, pH 7.4 and install the sample on the TIRF microscope.
3. Add an appropriate dilution of fluorescently labeled virions. The density of viral particles in the field of view should allow unambiguous tracking of individual virions. Note the time. Take reference images, a fluorescent image of the bottom surface of the cell, and one of the cell in phase contrast to document cell morphology.
4. When the first virions appear under the cell in the TIRF field, select a region of interest and acquire images at an appropriate frame rate, depending on the maximal mobility of the particles, here 1000 frames at 20 Hz.
5. Continue data acquisition for 45 min at most.
6. Take an image of the cell in phase contrast at the end of the experiment to ascertain by cell morphology that the cell has not sustained photo damage.
7. Repeat experiment, if desired with different condition.
8. Take a control movie of viruses stuck to the cover glass under experimental conditions.
9. Use particle localization and tracking algorithms to extract trajectories of particles. Important: optimize localization and tracking parameters on a number of time series and use the same parameters for the analysis of all data eventually. Analyze and calculate the diffusion coefficient and other parameters for the particles from the control movie (see step 8.). All particles imaged on cells that are below this value are rejected from analysis.
10. Visually inspect all acquired trajectories for obviously false linkages.

3.3. Cotracking with cellular markers

For the functional analysis of viral particle motion in cells, the correlation of incoming particles with known organelles of endocytic mechanisms is very important, since it contains information on the infectious entry pathway of the virus. An overview of typical cellular markers can be found in Brandenburg and Zhuang (2007) and Mercer *et al.* (2010).

For analysis, typically the amount of overlap of fluorescent virions with fluorescent markers of organelles is determined. Since incoming virions may show overlap with different organelles of the endocytic pathway over time during the infectious process, such analysis must be done carefully. One common mistake is the quantification of viral localization to large intracellular organelles by counting individual particles in fluorescence images that show overlap with fluorescent staining of intracellular organelles. This is very error prone, due to the high image depth (0.5–several μm) and small object size (30–200 nm in diameter) that leaves hundreds of nanometer space in which a virion can be located without being functionally associated with the organelle itself. Also, a reliable correlation with extensive organelles such as the endoplasmatic reticulum is very difficult to ascertain in fluorescence microscopy.

At best, functional association of viruses with organelles is tested by time-lapse acquisition, since only the subsequent overlap in several timeframes ideally accompanied by concomitant movement of fluorescent signals assures true colocalization of point-like objects. There are two ways of analyzing a potential concomitant movement of viruses and organelles. For one, a colocalizing object can be tracked. For this, colocalization pixels based on thresholding are defined. From these thresholded pixels, colocalizing objects are defined that match size and intensity criteria (e.g., one virus signal). Then, the colocalizing objects are tracked. For two, detection of tracking of the objects (virus, organelles) is carried out separately. Cotracks are then defined, if the two tracks match in directionality and speeds of motion within a stringent spatial limit.

3.4. Extracting information from tracks

From viral trajectories, several parameters can be quantified as exemplified in Fig. 4.3. First, the diffusion coefficient (D in $\mu m^2/s$) gives the speed of motion of a particle in the plane of the membrane. This will differ between the motility of a lipid in the plasma membrane ($< 5\ \mu m^2/s$) and an immobile particle (detection limit usually around $0.001\ \mu m^2/s$). Second, the mean square displacement plot, a statistical analysis of all individual steps of a trajectory can tell, whether a particle is confined, undergoing purely Brownian or directed motion. The shape of the mean square displacement plot (see Fig. 4.3) and even more reliably the slope of the moment-scaling spectrum (S_{MSS}; Ewers *et al.*, 2005) define these three prototypic behaviors.

Typically, viruses exhibit not only one type of behavior over time, but switch between different modes of motion. Such diverse trajectories are of great interest, because these switches in behavior often represent events in virus–host interaction, they need to be analyzed carefully. One possible way is the so-called moving window analysis of the trajectory. Here, for example, a portion of 100 time frames is analyzed according to parameters such as

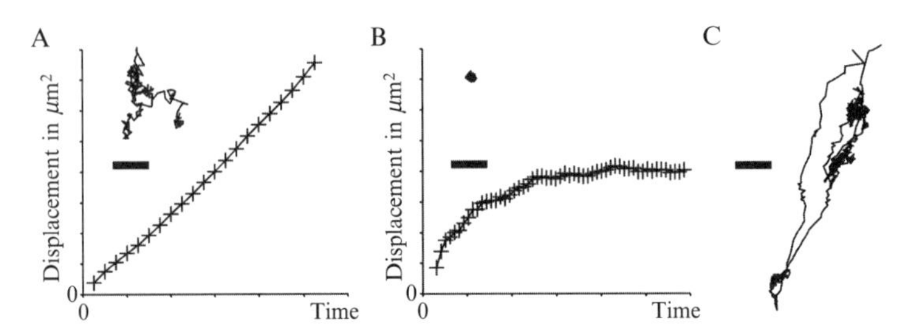

Figure 4.3 Examples of viral trajectories on the plasma membrane. (A) A single SV40 virion undergoing random motion at the cell surface. The trajectory shows a random variety of step orientation and a statistical distribution of step length. The plot of the mean square displacement versus time shows linear growth indicative of a random walk. The virus and its receptors undergo Brownian motion in the plane of the membrane. (B) Another trajectory of an SV40 virion. The particle is confined to a small area in the cell membrane. The mean square displacement plot is asymptotic indicating that the virion cannot escape the point of confinement. (C) A trajectory of an HPV-16 virion on the cell surface. The virion shows a complex trajectory alternating between points of confinement and directed motion. The speed of the directed movement contains information about what cellular mechanism may control the movement (i.e., motor-mediated transport on microtubule or actin tracks, attachment to actin fibers which undergo treadmilling). Scale bars are 2 μm.

the diffusion coefficient (D) and the slope of the moment-scaling spectrum (S_{MSS}). This analysis window is then moved frame by frame along the trajectory. The values of D and S_{MSS} will vary over time and define the transition points (Ewers *et al.*, 2005).

In addition, the transition point between motion types (e.g., random motion and directed motion) can be computationally extracted to unambiguously identify characteristic motion patterns in virus–host interaction by segmenting whole trajectories using trajectory segmentation algorithms based on supervised vector classification (Helmuth *et al.*, 2007).

4. Conclusions and Outlook

Here, we have introduced and summarized several means of how to employ single particle tracking of viruses for the analysis of viral entry and vesicular trafficking. These techniques allow to access information unavailable to ensemble measurements. Hence, they provide new insights into virus entry mechanisms, which, in turn would be fruitful for the analysis of membrane trafficking and the development of antiviral strategies.

Since the interactions of viruses with cellular proteins occurs typically on a nanometer scale, it may be necessary to develop new techniques to go beyond current resolution and detection limits. It will be interesting to see how newly developed superresolution techniques can be employed for this purpose. Among those are photoactivated localization microscopy (PALM), structured illumination microscopy (SIM), stimulated emission depletion microscopy (STEDM), and stochastic optical reconstruction microscopy (STORM; Betzig *et al.*, 2006; Gustafsson, 2005; Klar *et al.*, 2000; Rust *et al.*, 2006). In addition, new mathematical algorithms are being developed to extract subresolution information from images of non-single point light sources (e.g., Helmuth *et al.*, 2009). The development of new fluorescent probes that are brighter and smaller may allow to visualize the interactions of viruses and cells when only few molecules interact. An especially fascinating perspective is the investigation of single virus–host interactions in tissue models or animals with novel techniques potentially involving endoscopy or molecular imaging.

ACKNOWLEDGMENTS

We would like to thank all members of the Schelhaas lab for critical comments on the manuscript and proof reading. We apologize to all those great individuals who have advanced the field of virus imaging and who were not mentioned in the manuscript due to space limitations. M.S. is supported by the German science foundation (DFG, Emmy-Noether grant SCHE 1552/2-1, collaborative research center grant SFB629). H.E. is supported by the NCCR Neural Plasticity and Repair.

REFERENCES

Baumgartel, V., Ivanchenko, S., Dupont, A., Sergeev, M., Wiseman, P. W., Krausslich, H. G., Brauchle, C., Muller, B., and Lamb, D. C. (2011). Live-cell visualization of dynamics of HIV budding site interactions with an ESCRT component. *Nat. Cell Biol.* **13,** 469–474.

Betzig, E., Patterson, G. H., Sougrat, R., Lindwasser, O. W., Olenych, S., Bonifacino, J. S., Davidson, M. W., Lippincott-Schwartz, J., and Hess, H. F. (2006). Imaging intracellular fluorescent proteins at nanometer resolution. *Science* **313,** 1642–1645.

Brandenburg, B., and Zhuang, X. (2007). Virus trafficking—Learning from single-virus tracking. *Nat. Rev. Microbiol.* **5,** 197–208.

Burckhardt, C. J., and Greber, U. F. (2009). Virus movements on the plasma membrane support infection and transmission between cells. *PLoS Pathog.* **5,** e1000621.

Cheezum, M. K., Walker, W. F., and Guilford, W. H. (2001). Quantitative comparison of algorithms for tracking single fluorescent particles. *Biophys. J.* **81,** 2378–2388.

Coller, K. E., Berger, K. L., Heaton, N. S., Cooper, J. D., Yoon, R., and Randall, G. (2009). RNA interference and single particle tracking analysis of hepatitis C virus endocytosis. *PLoS Pathog.* **5,** e1000702.

Cureton, D. K., Massol, R. H., Whelan, S. P., and Kirchhausen, T. (2010). The length of vesicular stomatitis virus particles dictates a need for actin assembly during clathrindependent endocytosis. *PLoS Pathog.* **6,** e1001127.

Endress, T., Lampe, M., Briggs, J. A., Krausslich, H. G., Brauchle, C., Muller, B., and Lamb, D. C. (2008). HIV-1-cellular interactions analyzed by single virus tracing. *Eur. Biophys. J.* **37,** 1291–1301.

Ewers, H., Smith, A. E., Sbalzarini, I. F., Lilie, H., Koumoutsakos, P., and Helenius, A. (2005). Single-particle tracking of murine polyoma virus-like particles on live cells and artificial membranes. *Proc. Natl. Acad. Sci. USA* **102,** 15110–15115.

Ewers, H., Jacobsen, V., Klotzsch, E., Smith, A. E., Helenius, A., and Sandoghdar, V. (2007). Label-free optical detection and tracking of single virions bound to their receptors in supported membrane bilayers. *Nano Lett.* **7,** 2263–2266.

Gazzola, M., Burckhardt, C. J., Bayati, B., Engelke, M., Greber, U. F., and Koumoutsakos, P. (2009). A stochastic model for microtubule motors describes the in vivo cytoplasmic transport of human adenovirus. *PLoS Comput. Biol.* **5,** e1000623.

Gustafsson, M. G. (2005). Nonlinear structured-illumination microscopy: Wide-field fluorescence imaging with theoretically unlimited resolution. *Proc. Natl. Acad. Sci. USA* **102,** 13081–13086.

Helenius, A., Kartenbeck, J., Simons, K., and Fries, E. (1980). On the entry of Semliki forest virus into BHK-21 cells. *J. Cell Biol.* **84,** 404–420.

Helmuth, J. A., Burckhardt, C. J., Koumoutsakos, P., Greber, U. F., and Sbalzarini, I. F. (2007). A novel supervised trajectory segmentation algorithm identifies distinct types of human adenovirus motion in host cells. *J. Struct. Biol.* **159,** 347–358.

Helmuth, J. A., Burckhardt, C. J., Greber, U. F., and Sbalzarini, I. F. (2009). Shape reconstruction of subcellular structures from live cell fluorescence microscopy images. *J. Struct. Biol.* **167,** 1–10.

Klar, T. A., Jakobs, S., Dyba, M., Egner, A., and Hell, S. W. (2000). Fluorescence microscopy with diffraction resolution barrier broken by stimulated emission. *Proc. Natl. Acad. Sci. USA* **97,** 8206–8210.

Kukura, P., Ewers, H., Muller, C., Renn, A., Helenius, A., and Sandoghdar, V. (2009). High-speed nanoscopic tracking of the position and orientation of a single virus. *Nat. Methods* **6,** 923–927.

Lakadamyali, M., Rust, M. J., Babcock, H. P., and Zhuang, X. (2003). Visualizing infection of individual influenza viruses. *Proc. Natl. Acad. Sci. USA* **100,** 9280–9285.

Lehmann, M. J., Sherer, N. M., Marks, C. B., Pypaert, M., and Mothes, W. (2005). Actin- and myosin-driven movement of viruses along filopodia precedes their entry into cells. *J. Cell Biol.* **170,** 317–325.

Mercer, J., Schelhaas, M., and Helenius, A. (2010). Virus entry by endocytosis. *Annu. Rev. Biochem.* **79,** 803–833.

Rust, M. J., Lakadamyali, M., Zhang, F., and Zhuang, X. (2004). Assembly of endocytic machinery around individual influenza viruses during viral entry. *Nat. Struct. Mol. Biol.* **11,** 567–573.

Rust, M. J., Bates, M., and Zhuang, X. (2006). Sub-diffraction-limit imaging by stochastic optical reconstruction microscopy (STORM). *Nat. Methods* **3,** 793–795.

Sbalzarini, I. F., and Koumoutsakos, P. (2005). Feature point tracking and trajectory analysis for video imaging in cell biology. *J. Struct. Biol.* **151,** 182–195.

Schelhaas, M. (2010). Come in and take your coat off—How host cells provide endocytosis for virus entry. *Cell. Microbiol.* **12,** 1378–1388.

Schelhaas, M., Ewers, H., Rajamaki, M. L., Day, P. M., Schiller, J. T., and Helenius, A. (2008). Human papillomavirus type 16 entry: Retrograde cell surface transport along actin-rich protrusions. *PLoS Pathog.* **4,** e1000148.

Seisenberger, G., Ried, M. U., Endress, T., Buning, H., Hallek, M., and Brauchle, C. (2001). Real-time single-molecule imaging of the infection pathway of an adeno-associated virus. *Science* **294,** 1929–1932.

van der Schaar, H. M., Rust, M. J., Chen, C., van der Ende-Metselaar, H., Wilschut, J., Zhuang, X., and Smit, J. M. (2008). Dissecting the cell entry pathway of dengue virus by single-particle tracking in living cells. *PLoS Pathog.* **4,** e1000244.

Vonderheit, A., and Helenius, A. (2005). Rab7 associates with early endosomes to mediate sorting and transport of Semliki forest virus to late endosomes. *PLoS Biol.* **3,** e233.

CHAPTER FIVE

Imaging of Live Malaria Blood Stage Parasites

Christof Grüring *and* Tobias Spielmann

Contents

1. Introduction 82
2. Imaging Malaria Blood Stages at Ambient Temperature 83
   2.1. Required materials 85
   2.2. Preparation and viewing of parasites at ambient temperature 85
   2.3. Staining of parasites with Bodipy-TR-C_5-ceramide 86
3. Long-Term 4D Imaging of *P. falciparum* Blood Stages 86
   3.1. Required materials 87
   3.2. Preparation of parasites for long-term imaging 88
   3.3. Long-term time-lapse imaging 89
   3.4. Evaluation of the experiment and troubleshooting 90
   3.5. Use of Dendra2 conversion in time-lapse imaging to track specific protein pools 90
Acknowledgments 91
References 91

Abstract

Life cell imaging is a tool for cell biology that has provided invaluable insights into many dynamic processes such as cell division, morphogenesis, or endo- and exocytosis. While observing cells by time-lapse imaging is a standard procedure in many systems, this technique was until recently not available for blood stages of *Plasmodium falciparum*, the causative agent of the most severe form of human malaria. Here, we provide a detailed description of the procedure for time-lapse-based four-dimensional microscopy in blood stages of this important pathogen. With the widespread use of *P. falciparum* transfection to fluorescently tag proteins of interest, this technique provides a new tool to study the biology of malaria blood stages that is hoped to lead to a better appreciation of the dynamic processes in this life cycle phase.

Bernhard Nocht Institute for Tropical Medicine, Parasitology Section, Hamburg, Germany

Methods in Enzymology, Volume 506
ISSN 0076-6879, DOI: 10.1016/B978-0-12-391856-7.00029-9

1. Introduction

The phylum Apicomplexa contains several important pathogens such as *Plasmodium falciparum*, the causative agent of the most dangerous form of human malaria. With approximately 225 million cases and 781,000 deaths per year, malaria remains an enormous health problem in developing countries (WHO, 2010). Using a light microscope, the French military surgeon Alfonse Laveran discovered the malaria parasite in 1881 during the examination of fresh blood samples of sick soldiers. In these samples, he noted the presence of pigmented spherical bodies of variable size displaying amoeboid movement (Laveran, 1881). With the advent of staining procedures such as the still ubiquitously used Giemsa stain and the use of electron microscopy, the different stages of the parasite were described in much greater detail. However, because these procedures require fixed samples, this led to a largely static perception of the parasite's life cycle. Nevertheless, seminal live cell studies were already conducted in the mid-1970s when Dvorak and colleagues firstly visualized the infection of a red blood cell by a *Plasmodium* parasite (Dvorak *et al.*, 1975). With the emergence of new microscopy technologies, important aspects of the parasite's life cycle in the human host were uncovered (Amino *et al.*, 2006; Mota *et al.*, 2001; Sturm *et al.*, 2006). However, no live cell imaging studies covered the important disease causing blood stage.

In blood stages, the recent appliance of cryo-electron tomography allowed for the detailed investigation of the ultrastructure of the parasite and the modifications of the host cell (Hanssen *et al.*, 2008), extending on a large body of previous electron microscopy studies (Bannister *et al.*, 2000). Although very informative regarding morphological details, it is impossible to infer dynamic processes from these static snapshots in the life cycle. The absence of live cell techniques did not only hamper the study of processes such as protein transport but so far also prevented the visualization of parasite growth during its progression through the asexual development in erythrocytes.

We recently established long-term time-lapse imaging in *P. falciparum* blood stages, creating three-dimensional (3D) representations of growing parasites in regular time intervals (four-dimensional (4D) imaging) to show the morphological changes associated with parasite growth and to provide new insights into protein export and host cell modifications (Grüring *et al.*, 2011). The capacity for time-lapse imaging is an important asset to study the biology of this parasite, and we give here a detailed description of this method. We also provide some general recommendations for viewing live parasites at ambient temperature, a technique that is widely used.

2. Imaging Malaria Blood Stages at Ambient Temperature

A convenient method to visualize *P. falciparum* blood stages is to image them at ambient temperature directly after removing them from *in vitro* culture. This has the advantage of simplicity and is widely used, especially with parasites expressing fluorescently tagged proteins (Tilley *et al.*, 2007). Although these parasites show some differences in morphology to actively growing parasites (Grüring *et al.*, 2011), this method nevertheless provides a good starting point for a first analysis of protein localization and to characterize new transgenic cell lines.

For many experimental questions, it is important to be able to visualize as many of the structures in infected red blood cells as possible. Ideally, this should be possible with a single dye of spectral properties compatible with the simultaneous use of parasites expressing fluorescent proteins. Further, the dye should not affect the morphology and viability of the cells. Among the large number of commercially available options, Bodipy-C_5-ceramide found frequent use in malaria research. For instance, it was used to visualize the tubovesicular network extending from the parasitophorous vacuole surrounding the parasite in the host cell (Elmendorf and Haldar, 1994), parasite-induced host cell modifications termed Maurer's clefts (Hanssen *et al.*, 2008) or the cavity in the parasite body (Grüring *et al.*, 2011). It was also used for short-term time-lapse imaging to follow parasite morphology (Grüring *et al.*, 2011). Originally, this membrane label was used to stain the Golgi in mammalian cells (Pagano *et al.*, 1991) but was found to also visualize intracellular pathogens (see, e.g., Boleti *et al.*, 2000). The Texas Red-conjugated version (Bodypi-TR-C_5-ceramide) is especially useful because it can be used alongside GFP which is most commonly used to fluorescently tag proteins of interest in transgenic parasites. A variant possessing spectral properties with an emission spectrum in the green range is also available, but the concentration-dependent shift of the spectral properties of this label leading to emission in the red spectrum (Pagano *et al.*, 1991) limits the range of this dye for colocalization experiments (Fig. 5.1A). Examples of Bodipy-TR-C_5-ceramide stained parasites are shown in Fig. 5.1B.

A further useful and ubiquitously used stain in parasites viewed at ambient temperature is DAPI (or Hoechst) which can be added to visualize the nuclei of the parasite. Its emission in the blue range is well suited for colocalizations. However, the nuclei of young stage parasites are not or only poorly stained with these reagents (see Fig. 5.1A, top panel) and excess or prolonged staining should be avoided as this can lead to artifactual emission in the green spectral range (Fig. 5.1C).

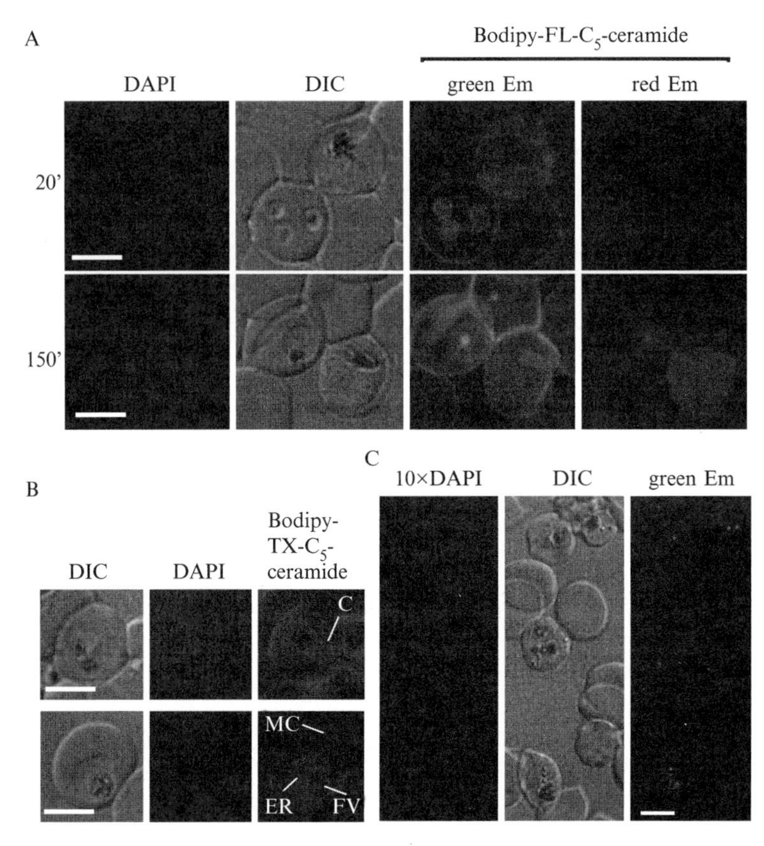

Figure 5.1 Imaging of stained *P. falciparum* parasites at ambient temperature. (A) Top row shows a schizont and a red blood cell infected with three ring stages after incubation with Bodipy-FL-C_5-ceramide for 20 min (indicated on the left). Note the lack of DAPI staining in the three ring-stage parasites. The bottom row shows a trophozoite and a schizont parasite after incubation with Bodipy-FL-C_5-ceramide for 150 min. Increased incubation time leads to increased emission of the dye in the red spectrum. This is more pronounced in later stage parasites (compare the schizont stage parasites in the two samples). (B) Two examples of Bodipy-TX-C_5-ceramide stained infected red blood cells. C, cavity; ER, endoplasmatic reticulum; FV, food vacuole; MC, Maurer's clefts. (C) Overexposure of untransfected parasites with DAPI leads to focal emission in the green spectrum. Em, emission. Size bars: 5 μm. (For interpretation of the references to color in this figure legend, the reader is referred to the Web version of this chapter.)

Due to the limited optical contrast, except for the hemozoin in the food vacuole, phase contrast or preferentially differential interference contrast (DIC) is of advantage to visualize the parasite in the transmission light channel.

2.1. Required materials

2.1.1. Parasites

- *P. falciparum* blood stage parasites grown *in vitro* according to standard conditions (Trager and Jensen, 1976) or tail blood of rodents infected with appropriate malaria species

2.1.2. Devices

- Fluorescence microscope equipped with 100× or 63× magnification oil immersion lenses and if possible the capacity for DIC

2.1.3. Reagents

- Bodipy-TR-C_5-ceramide 5 μ*M* stock solution (Invitrogen), either in 1× PBS complexed to BSA according to the manufacturer's specifications or purchased in BSA complexed form
- 1 mg/ml stock DAPI (4′,6-diamidino-2-phenylindole) in dH_2O (Roche)
- RPMI medium for parasite culture: 10.43% (w/v) RPMI 1640 (or as specified for the lot used), 25 m*M* HEPES, 12 m*M* $NaHCO_3$, 6 m*M* D-glucose, 0.5% (w/v) albumax, 0.2 m*M* hypoxanthine

2.1.4. Disposables

- Coverslips (for instance, 24 × 65 mm, 0.13–0.16 mm thickness, R. Langenbrinck, Emmendingen, Germany)
- Microscopy slides (for instance 76 × 26 mm, Engelbrecht, Edermünde, Germany)
- 1.5 ml tubes (Eppendorf)

2.2. Preparation and viewing of parasites at ambient temperature

To image *P. falciparum* blood stages, remove a small aliquot (50–500 μl) from a *P. falciparum in vitro* culture and transfer it into a 1.5-ml tube. If required, add DAPI to 1 μg/ml final and incubate at 37 °C for 10 min. To avoid overstaining with DAPI, centrifuge the cells at 1000×*g* for 30 s and resuspend the pellet in 500 μl of RPMI. Spin once more and remove the supernatant to leave approximately 1–2 pellet volumes of medium. If no DAPI was added, this washing step can be omitted. Resuspend the pellet in the remaining RPMI culture medium and transfer 5–15 μl of the suspension as a single drop to the center of a microscopy glass slide. Immediately add a coverslip and image the cells. The hematocrit and total volume of the suspension added to the slide will determine how suitable the cells will be for imaging. It is advisable to scan the slide to find appropriate areas, avoiding the borders of the slide. The cells should be nearly confluent without extensive touching to avoid distorting their shape. Further they should not be lying on top of each other or appear squashed by the coverslip. If cells move, this usually will stop after a short while. Watch for signs of drying out or osmotic stress which is apparent from lysed cells nearby or erythrocytes showing echinocytosis, respectively. Generally, care should be taken to complete imaging within ~30 min or less if the cells start to deteriorate.

2.3. Staining of parasites with Bodipy-TR-C_5-ceramide

Remove 500 μl from a *P. falciparum in vitro* culture and transfer it to a 1.5-ml tube. Centrifuge the cells at 1000×*g* for 30 s and resuspend the pellet in 25 μl of RPMI culture medium. Add 50 μl of a 5 μ*M* Bodipy-TR-C_5-ceramide stock solution (2.5 μ*M* final), mix gently, and incubate for 15–30 min at 37 °C. If desired, include DAPI at 1 μg/ml final concentration. The duration of staining with Bodipy-TR-C_5-ceramide will determine the extent of labeling and which structures are stained. Generally, longer incubation periods are required to achieve labeling of the Maurer's clefts, whereas the parasite periphery is rapidly labeled. Extended labeling periods with high concentrations of Bodipy-TR-C_5-ceramide may be detrimental to the morphology of the cells. After staining, wash the cells once in 500 μl RPMI culture medium and image as described in Section 2.2. Infected red blood cells labeled with Bodipy-TR-C_5-ceramide can also be imaged over time under culture conditions (Section 3) as the label in the parasites (as opposed to the host cell membrane) only very slowly (over hours) extracts back into the medium. However, in this case, it should be considered that depending on the extent of labeling, parasite viability may be affected.

Bodipy-TR-C_5-ceramide can also be applied to *Plasmodium berghei* blood stages (Grüring *et al.*, 2011). In this case, care should be taken to add RPMI culture medium slowly to the mouse blood and to wash the cells once before the addition of Bodipy-TR-C_5-ceramide as described above. If PBS is to be used, this should be added slowly after the first resuspension of the cells in RPMI medium, aiming for a gradual increase of the proportion of PBS. Direct addition of PBS to mouse blood should be avoided because it results in grave echinocytosis of erythrocytes.

3. Long-Term 4D Imaging of *P. falciparum* Blood Stages

In this section, we provide the basic requirements for 4D imaging of *P. falciparum* parasites. First, parasites are coated on a dish with a translucent bottom suitable for microscopy (Section 3.2). These cells are then imaged over time under culture conditions (Section 3.3). In the setup described a programmable motor stage is used. This makes it possible to do multi-area time-lapse imaging, that is, instead of gathering data from a single area per time point only, the stage moves to several preprogrammed positions for each time point to acquire data from many individual cells in a single experiment. A *z*-drift correction unit is used to maintain focus over extended periods of time. In each position, a stack is collected that can be reconstructed into a 3D representation. Assembly of the 3D reconstruction of all time points of a region of interest then leads to a 4D dataset. During the establishment phase of the method, the state of the culture and the imaged areas should be assessed after completion of the experiment (Section 3.4). As a method to follow specific pools of protein over time, the use of the photoconvertible protein Dendra2 is described in Section 3.5.

The acquired data can be viewed using the microscope software or other suitable programs. We routinely use the commercial Imaris package (Bitplane) as well as the freely available ImageJ (http://rsb.info.nih.gov/ij/) to display and analyze the 4D data. A description of data analysis is however beyond the scope of this chapter.

3.1. Required materials

3.1.1. Parasites

P. falciparum parasites grown according to standard procedures (Trager and Jensen, 1976) are used for live cell imaging. We use RPMI medium supplemented with 0.5% albumax. Although not tested, it is likely that parasites grown in medium supplemented with serum work equally well. If you intend to image parasites expressing a protein of interest fused to a fluorescent protein, it is advisable to first verify the fluorescence levels in a sample of these parasites using a standard fluorescence microscope. The level of fluorescence will determine the laser intensities required for the time-lapse imaging with the confocal microscope. Poorly fluorescent cell lines are not suitable for 4D imaging.

3.1.2. Devices and materials

We use a commercially available FV1000 confocal microscope (Olympus) equipped with a 100× lens with an aperture of 1.4, a motor stage, an Olympus Cellcubator with climate control unit and an Olympus *z*-drift correction autofocus system. It is likely that other systems of inverted microscopes can be used to achieve long-term 4D imaging of *P. falciparum* parasites, but we recommend that they are carefully tested to make sure images of acceptable quality can be obtained while illumination intensities are kept sufficiently low to not affect parasite viability. A microscope equipped with a 63× lens is similarly suitable, as in principle equal or even better resolution can be achieved with these objectives compared to lenses of a higher magnification.

3.1.3. Disposables

Dishes with a translucent bottom suitable for imaging are needed. Evaporation represents a major harm for the parasites. We therefore recommend to use dishes with firm sitting lids and to further seal the dish using parafilm. One example of appropriate dishes is the sterile, glass bottom, uncoated, hydrophobic, high, 35 mm μ-Dishes distributed by ibidi (Martinsried, Germany). Attention should also be paid that the microscopy dishes and the immersion oil are compatible.

3.1.4. Media and solutions

- RPMI culture medium (see Section 2)
- Imaging medium: to reduce phototoxicity we strongly recommend the use of phenol red-free RPMI medium for long-term life cell microscopy (recipe as for the standard RPMI culture medium outlined in Section 2). As growth of *P. falciparum* parasites in this medium is comparable to the standard medium, there is no need to adapt the parasites to this medium before the experiment
- 0.5 mg/ml cell culture grade concanavalin A (Sigma) in dH_2O, filter sterilized
- Sterile 1× PBS

3.2. Preparation of parasites for long-term imaging

Four to six hours before the planned experiment, the incubator of the confocal microscope should be started to equilibrate the system to 37 °C. Failure to do this in time may result in *z*-shifts due to temperature changes during programming and subsequent imaging. Next prepare the glass bottom microscopy dish by adding enough of a 5 mg/ml concanavalin A solution to completely cover the bottom (at least 200 μl for ibidi 35 mm μ-Dishes). This will allow you to immobilize the erythrocytes from the parasite culture on the bottom of the dish. For glass surfaces, it is however essential that the concanavalin A is dissolved in dH_2O rather than PBS (Spielmann *et al.*, 2003). Seal the microscopy dish and carry out the incubation with concanavalin A at 37 °C for at least 20 min or until the parasites are ready to be added (extending this step to hours does not appear to be detrimental).

To prepare the parasites, remove 500 μl from the culture of interest that has been fed the day before using normal RPMI. Ideally, the parasitemia should be between 5% and 10% to avoid long search times in order to minimize exposure of the culture to illumination during programming of the cells (Section 3.3). The following steps should be carried out without unnecessary delay to reduce the time parasites are exposed to none culture conditions. Transfer the cells to a 1.5-ml tube and centrifuge for 30 s at 1000×*g*. Remove the supernatant and wash the cells twice in 1 ml of PBS using 30 s centrifugations at 1000×*g*. Thereafter, gently resuspend the cells in 500 μl of PBS. During the last centrifugation step, remove the concanavalin A from the microscopy dish and gently rinse the bottom of the dish with PBS twice. Remove excess PBS and immediately apply ~300 μl of the washed cell suspension to the bottom of the dish and allow the cells to settle for 10 min. Note that while in principle excess cells are not problematic and are washed off in the next step, addition of a suspension with a very high hematocrit will cause the erythrocytes to settle upright which is not advisable for most imaging questions. Remove unbound cells by gently rinsing the bottom of the dish with PBS until a faint red-golden layer is left. Immediately fill up (8 ml for ibidi 35 mm μ-Dishes) the microscopy dish containing the coated cells with pre-warmed (37 °C) phenol red-free RPMI medium. Close the lid and seal the dish with parafilm. Immediately place the dish into the incubator of the microscope.

3.3. Long-term time-lapse imaging

The actual programming of the cells for multi-area time-lapse imaging will vary with the microscopy system used. As a general guideline, use the brightfield/DIC to find the focal plane containing the cells. After this, scan sections using the laser to excite your fluorophore of interest using no or low zoom levels and the 100× lens. The zoom levels required to find cells of interest will depend on how easily the fluorescence pattern is detectable. If possible avoid the use of the epifluorescence lamp to search for cells to limit exposure of the culture to excessive illumination. Also try to avoid too many passes of the laser for a given area. Once a cell is detected, zoom in on it and optimize the zoom, laser, and detector settings. Then set the autofocus, the borders of the stack and the stack step size and feed these data into the multi-area time-lapse program. We routinely use a zoom level of 3–8, laser intensities of 0.5–5%, dwell time of 4 μs and stack step size of 0.33–0.38 μm. Generally, the laser intensity should be kept as low as possible. In principle, it would also be possible to acquire only a single optical section per time point; however, minimal shifts in *z* may lead to imaging of different sections of the cell of interest in different time points, limiting the usefulness of this approach.

After a region of interest was programmed, continue to search for a new area to add more cells to your experiment. After programming a suitable number of cells, set the interval time (we use 5 min to 1 h for long-term time lapse) and the total duration of the experiment (we use 12–80 h), then start the program.

3.4. Evaluation of the experiment and troubleshooting

Especially during the initial phase of establishing the method, it is crucial to carefully assess the imaged culture before removing it from the microscope. Firstly, the imaged areas should be analyzed to see signs of photodamage. If dead cells are apparent in this area, it should be checked in other areas of the dish whether this was a general problem in the imaged culture or if this was indeed laser induced. If laser damage occurred, try to use lower laser levels by increasing detectors, lower zoom levels, increase the time interval between time points, and attempt to reduce the total number of laser passes to program the settings of an area. If there was a general problem with the imaged culture, try placing dishes containing the phenol red-free medium and normal culture medium in the microscope incubator and compare their growth with normally grown parasites. We found that on occasions the much lower number of cells in the microscope dish caused the culture to be very sensitive to otherwise unproblematic situations (e.g., addition of 1% DMSO). This can often be amended by preincubating the medium to be used for imaging with a final concentration of 5% uninfected red blood cells that are removed before adding the medium to the dish containing the cells for imaging. A further problem affecting all cells in the imaged culture is evaporation that will lead to an increased concentration of salts and osmotic stress. Make sure the microscopy dish is filled up and properly sealed with parafilm to avoid this problem.

If the culture in the microscopy dish is fine and the imaged parasites show growth, it is crucial to look for reinvasion after schizont rupture of imaged parasites. Reinvasion is a good indicator for appropriate parasite development and should be used as a gold standard.

3.5. Use of Dendra2 conversion in time-lapse imaging to track specific protein pools

The development of photoconvertible proteins enabled researches to specifically follow a protein population. Recently, we applied this technique together with long-term 4D imaging in *P. falciparum* blood stages (Grüring *et al.*, 2011) using the photoconvertible protein Dendra2 (Chudakov *et al.*, 2007). Dendra2 can be excited using the 488 nm laser line and emits in the green spectrum (maximum at 507 nm). The protein is converted using a 405 nm laser, leading to a change in its excitation maximum to 553 nm and an emission maximum of 573 nm (Chudakov *et al.*, 2007). The very low intensities required for conversion alleviate the drawback of using a laser in the UV range that normally is highly detrimental to living cells.

3.5.1. Additional materials

- Parasites need to be generated that express your protein of interest fused to Dendra2. Expression of the transgene should be assessed under a normal fluorescence microscope using filter sets suitable for GFP. Only if the fluorescence is well detectable, it is worth to attempt conversion and time-lapse imaging.
- The confocal microscope for time-lapse imaging should be equipped with a laser in the UV range (e.g., 405 nm) for the conversion, and laser lines near 490 and 553 nm for excitation of the unconverted and the converted form of Dendra2, respectively.

3.5.2. Achieving conversion of Dendra2 in *P. falciparum* parasites

Prepare the parasites expressing the protein of interest fused to Dendra2 for long-term imaging as described in Section 3.2. Search for cells using the settings normally used for GFP. Once a suitable cell is found, select a region of interest and convert it by scanning it with the UV-laser (405 nm) with 0.1–0.3% intensity and a dwell time of 2000 μs. Assess the level of conversion by comparing the proportion of green signal (excited with 488 nm; detection around 507 nm) that is left compared to the newly apparent red signal (we use a 559 nm laser for excitation; emission around 573 nm). The conversion can be repeated to completely convert a pool of protein or to increase the proportion of converted versus unconverted protein. Thereafter follow the converted cell (or multiple cells if multi-area time lapse is possible with your setup) as described in Section 3.3. It is advisable to also have cells without converted regions in the imaged area, as inefficient conversion can occur with the 488 nm laser line used to visualize the green pool of Dendra2 (Chudakov *et al.*, 2007). These cells will serve as control to exclude this possibility, although so far this did not appear to be a problem in our hands. It is notable that several regions of interest can be converted within a single cell. Despite prolonged exposure to the UV-laser during conversion, cells proved to be viable and suitable for 4D imaging (Grüring *et al.*, 2011).

ACKNOWLEDGMENTS

The research to establish 4D imaging was funded by DFG grants SP1209/1 and SP1209/1–2. We thank Dr. Monica Hagedorn and the members of our lab for critically reading the chapter.

REFERENCES

Amino, R., Thiberge, S., Martin, B., Celli, S., Shorte, S., Frischknecht, F., and Menard, R. (2006). Quantitative imaging of *Plasmodium* transmission from mosquito to mammal. *Nat. Med.* **12,** 220–224.

Bannister, L. H., Hopkins, J. M., Fowler, R. E., Krishna, S., and Mitchell, G. H. (2000). A brief illustrated guide to the ultrastructure of *Plasmodium falciparum* asexual blood stages. *Parasitol. Today* **16,** 427–433.

Boleti, H., Ojcius, D. M., and Dautry-Varsat, A. (2000). Fluorescent labeling of intracellular bacteria in living host cells. *J. Microbiol. Methods* **40,** 265–274.

Chudakov, D. M., Lukyanov, S., and Lukyanov, K. A. (2007). Using photoactivatable fluorescent protein Dendra2 to track protein movement. *Biotechniques* **41,** 553, 555, 557 passim.

Dvorak, J. A., Miller, L. H., Whitehouse, W. C., and Shiroishi, T. (1975). Invasion of erythrocytes by malaria merozoites. *Science* **187,** 748–750.

Elmendorf, H. G., and Haldar, K. (1994). *Plasmodium falciparum* exports the Golgi marker sphingomyelin synthase into a tubovesicular network in the cytoplasm of mature erythrocytes. *J. Cell Biol.* **124,** 449–462.

Grüring, C., Heiber, A., Kruse, F., Ungefehr, J., Gilberger, T. W., and Spielmann, T. (2011). Development and host cell modifications of *Plasmodium falciparum* blood stages in four dimensions. *Nat. Commun.* **2,** 165.

Hanssen, E., Sougrat, R., Frankland, S., Deed, S., Klonis, N., Lippincott-Schwartz, J., and Tilley, L. (2008). Electron tomography of the Maurer's cleft organelles of *Plasmodium falciparum*-infected erythrocytes reveals novel structural features. *Mol. Microbiol.* **67,** 703–718.

Laveran, A. (1881). Nature parasitaire des accidents de l'impaludism: Description d'un nouveau parasite trouvé dans le sang des maladesatteints de fieèvre palustre. J.-B. Baillière, Paris.

Mota, M. M., Pradel, G., Vanderberg, J. P., Hafalla, J. C., Frevert, U., Nussenzweig, R. S., Nussenzweig, V., and Rodriguez, A. (2001). Migration of *Plasmodium* sporozoites through cells before infection. *Science* **291,** 141–144.

Pagano, R. E., Martin, O. C., Kang, H. C., and Haugland, R. P. (1991). A novel fluorescent ceramide analogue for studying membrane traffic in animal cells: Accumulation at the Golgi apparatus results in altered spectral properties of the sphingolipid precursor. *J. Cell Biol.* **113,** 1267–1279.

Spielmann, T., Fergusen, D. J., and Beck, H. P. (2003). *etramps*, a new *Plasmodium falciparum* gene family coding for developmentally regulated and highly charged membrane proteins located at the parasite-host cell interface. *Mol. Biol. Cell* **14,** 1529–1544.

Sturm, A., Amino, R., van de Sand, C., Regen, T., Retzlaff, S., Rennenberg, A., Krueger, A., Pollok, J. M., Menard, R., and Heussler, V. T. (2006). Manipulation of host hepatocytes by the malaria parasite for delivery into liver sinusoids. *Science* **313,** 1287–1290.

Tilley, L., McFadden, G., Cowman, A., and Klonis, N. (2007). Illuminating *Plasmodium falciparum*-infected red blood cells. *Trends Parasitol.* **23,** 268–277.

Trager, W., and Jensen, J. B. (1976). Human malaria parasites in continuous culture. *Science* **193,** 673–675.

WHO (2010). Malaria. World Health Organization, Geneva Fact Sheet NO. 94.

CHAPTER SIX

Escherichia coli K1 Invasion of Human Brain Microvascular Endothelial Cells

Lip Nam Loh *and* Theresa H. Ward

Contents

1. Introduction 94
2. HBMEC Cell Culture and Transfection 95
   2.1. Cell culture and transfection 95
   2.2. Evaluation of HBMEC transfection efficiency 97
3. Making Fluorescent *E. coli* K1 that Retain Virulence 101
   3.1. Surface labeling of *E. coli* K1 with FITC 101
   3.2. Construction of fluorescent protein-expressing *E. coli* K1 103
   3.3. Screening for virulence of fluorescent protein-expressing *E. coli* K1 103
4. Live Imaging of *E. coli* K1-Infected HBMEC 107
Acknowledgments 110
References 111

Abstract

The pathogenic *Escherichia coli* strain *E. coli* K1 is a primary causative agent of neonatal meningitis. Understanding how these bacteria cross the blood–brain barrier is vital to develop therapeutics. Here, we describe the use of live-cell imaging techniques to study *E. coli* K1 interactions with cellular markers following infection of human brain microvascular endothelial cells, a model system of the blood–brain barrier. We also discuss optimization of endothelial cell transfection conditions using nonviral transfection technique, bacterial labeling techniques, and *in vitro* assays to screen for fluorescent bacteria that retain their ability to invade host cells.

Department of Immunology and Infection, Faculty of Infectious and Tropical Diseases, London School of Hygiene and Tropical Medicine, London, United Kingdom

Methods in Enzymology, Volume 506
ISSN 0076-6879, DOI: 10.1016/B978-0-12-391856-7.00030-5

1. Introduction

Escherichia coli strains possessing the K1 capsular polysaccharide (*E. coli* K1) are one of the most common causative agents for neonatal bacterial meningitis (NBM), with a mortality rate of 40% in developing countries (Bonacorsi and Bingen, 2005; Holt *et al.*, 2001; Mulholland, 1998; Polin and Harris, 2001; Robbins *et al.*, 1974). Infants are believed to acquire the pathogen from the maternal commensal flora during delivery or from the environment, leading to colonization of the infant's intestinal tract (Bonacorsi and Bingen, 2005). Following intestinal colonization, the bacteria translocate into the bloodstream, where they survive intravascularly and multiply to high levels of bacteremia to initiate an infection (Glode *et al.*, 1977; Kim *et al.*, 1992; Pluschke *et al.*, 1983). For meningitis to develop, the bacteria must passage through the blood–brain barrier (BBB) into the central nervous system (CNS) and stimulate the infiltration of immune cells (Kim, 2003).

Over the past two decades, the availability of *in vitro*-cultured human brain microvascular endothelial cells (HBMEC) has allowed identification of bacterial and host factors that are required for bacterial binding, invasion, and intracellular survival (Huang *et al.*, 1995; Khan *et al.*, 2002, 2007; Kim *et al.*, 2003). The signaling events that lead to actin polymerization to enable *E. coli* K1 uptake into the nonphagocytic HBMEC have been well investigated (Maruvada *et al.*, 2008; Prasadarao, 2002; Prasadarao *et al.*, 1999; Reddy *et al.*, 2000a,b), but, beyond bacterial entry, the bacterial intracellular survival and trafficking remain poorly understood.

The discovery of fluorescent proteins and the technological advancement of imaging hardware have revolutionized every aspect of cellular and molecular biology research in recent years (Remington, 2006; Ward and Lippincott-Schwartz, 2006). Due to the inert chemical properties of these proteins, cytoplasmic expression of fluorescent proteins has been routinely applied to label bacteria for microscopy studies (Drecktrah *et al.*, 2007; Knodler *et al.*, 2005; Lamberti *et al.*, 2010; Qazi *et al.*, 2001; Ruthel *et al.*, 2004; Weingart *et al.*, 1999). The availability of fluorescent protein-tagged cellular markers also allows examination of the host factors involved in a pathogen's intracellular life cycle (Berón *et al.*, 2002; Guignot *et al.*, 2004; Rzomp *et al.*, 2003). In this chapter, we discuss strategies for labeling bacteria and the potential pitfalls of cytoplasmic fluorescent protein expression, and the application of live-cell imaging to study the intracellular fate of *E. coli* K1 post-invasion of HBMEC.

2. HBMEC Cell Culture and Transfection

2.1. Cell culture and transfection

As with other endothelial cells, HBMEC are difficult to transfect with nonviral transfection techniques (Kang *et al.*, 2009; Kovala *et al.*, 2000; Segura *et al.*, 2001). We have tested various proprietary cationic polymer- and lipid-based transfection reagents and electroporation techniques, and the resulting transfection efficiency in HBMEC was almost always below 30%. In this section, we provide the optimal transfection conditions of HBMEC with jetPRIME™ transfection reagent, which has enabled significant improvement in HBMEC transfection efficiency to approximately 50%.

Materials

- HBMEC complete growth medium: RPMI-1640 (Sigma-Aldrich, Poole, Dorset, UK) supplemented with 20% (v/v) fetal bovine serum (FBS) (Biosera, Ringmer, UK), 2 m*M* L-glutamine (Sigma-Aldrich), 1 m*M* sodium pyruvate (Sigma-Aldrich), 100 μg/ml streptomycin (Sigma-Aldrich), 100 units/ml penicillin (Sigma-Aldrich), nonessential amino acids (Gibco, Invitrogen, Paisley, UK), and vitamins (Gibco). Prewarmed to 37 °C.
- Unsupplemented RPMI-1640 medium (Sigma-Aldrich). Prewarmed to 37 °C.
- Trypsin–EDTA solution: trypsin (0.5 g/l) and ethylenediamine tetraacetic acid (EDTA) (0.5 m*M*) (Sigma-Aldrich).
- jetPRIME™ transfection reagent (Polyplus-transfection SA, France).
- HBMEC.
- pN1-EGFP plasmid DNA (Clontech, BD Biosciences, Oxford, UK) (Table 6.1): endotoxin-free plasmid DNA purified with Pureyield™ Midiprep kit (Promega, Southampton, UK) and eluted in 10 m*M* Tris–HCl, pH 8.5, buffer.

  Note: To ensure good transfection efficiency, the ratio of spectrophotometric absorbance of the purified DNA at 260 nm to that of 280 nm (A_{260}/A_{280}) must be at least 1.8.
- Tissue culture CO_2 incubator.
- Tissue culture disposables:
  - Microscope glass cover slips (No. 1 13 mm diameter) (VWR, Lutterworth, UK).

    Note: Glass cover slips are cleaned with 70% ethanol and then air-dried before use.
  - Choice of cell culture flask according to requirements: 75 cm^2 tissue culture flasks (Sarstedt, Leicester, UK) for routine maintenance; 6-well tissue culture dishes (Corning, VWR) for Western blotting experiments;

Table 6.1 Plasmids used in this study

Plasmid	Relevant characteristics	Source or reference
pN1-EGFP	Mammalian expression vector encoding a red-shifted variant of wild-type GFP, kanamycin resistant.	Clontech, BD Biosciences.
pFPV25.1	Bacterial expression vector carrying *gfpmut3a* gene downstream of a *Salmonella typhimurium* promoter, *rpsM*, ampicillin resistant. K1-GFP is the transformed *E. coli* K1 strain.	Valdivia and Falkow (1996); a kind gift from Dr. Olivier Marchés.
pFPV-mCherry	Bacterial expression vector carrying mCherry cDNA, cloned by PCR from pRSET-B mCherry, downstream of a *S. typhimurium* promoter, *rpsM*, ampicillin resistant. K1-Cherry is the transformed *E. coli* K1 strain.	Shaner *et al.* (2004); pRSET-B mCherry a kind gift from Professor Roger Tsien. Constructed in this study.
rpsM$^-$	pFPV25.1 with *rpsM* deleted, ampicillin resistant. K1rpsM$^-$ is the transformed *E. coli* K1 strain.	Constructed in this study.
GFP-Rab7	pC1-EGFP vector containing wild-type canine Rab7 cDNA, kanamycin resistant.	Bucci *et al.* (2000); a kind gift from Professor Albert Haas and Dr. Bianca Schneider.
GFP-Rab5wt	pC1-EGFP vector containing wild-type canine Rab5 cDNA, kanamycin resistant.	Nichols *et al.* (2001); a kind gift from Dr. Ben Nichols.
GFP-Rab5S34N	pC1-EGFP vector containing Rab5S34N mutant cDNA, kanamycin resistant.	Nichols *et al.* (2001); a kind gift from Dr. Ben Nichols.
GFP-Rab5Q79L	pC1-EGFP vector containing Rab5Q79L mutant cDNA, kanamycin resistant.	Nichols *et al.* (2001); a kind gift from Dr. Ben Nichols.

24-well tissue culture dishes (Corning) for *in vitro* infection experiments; LabTek 8-well glass cover slip chambers (Nalge Nunc, VWR) for time-lapse imaging experiments.

Methods

1. HBMEC are grown at 37°C with 5% CO_2 in HBMEC complete growth medium. At near confluence, they are passaged with trypsin–EDTA to maintain in 75 cm^2 flasks. A 1:20 split provides sufficient cells to approach confluency in 4–5 days. For transient transfections, seed 1×10^5 cells/ml of HBMEC on cleaned glass cover slips in 24-well tissue culture dishes 2 days before transfections, in order to achieve approximately 80% confluency at the time of transfection.
2. jetPRIME™ transfection reagent is used according to the manufacturer's protocol. In brief, prior to transfection, add 0.5–1.0 μg pN1-EGFP to 50 μl jetPRIME™ buffer. Mix the tube gently by tapping. Depending on the amount of plasmid DNA used for transfection, add 1 or 2 μl of jetPRIME™ transfection reagent to the diluted DNA. Mix the tube by tapping immediately and incubate at room temperature for 10 min. Pipette the DNA–lipid complex mixture dropwise onto the cells in a 24-well tissue plate.

   Note: To ensure good distribution of the DNA–lipid complex on cells throughout the well, the medium is repeatedly pipetted up and down with a P200 micropipette.
3. After 5–6 h, remove medium containing DNA–lipid complex and wash cells once with prewarmed unsupplemented RPMI-1640. Replenish the cells with fresh HBMEC complete growth medium and transfer to the tissue culture incubator. Cells can be screened for expression 20–24 h following transfection.

   Note: HBMEC can also be grown into a polarized monolayer on a collagen-coated Transwell-Clear™ filter insert (Corning). Upon complete confluency, the culture medium is replaced with HBMEC complete growth medium containing 500 n*M* hydrocortisone (Sigma-Aldrich) to induce tight junction formation (Khan and Siddiqui, 2009).

2.2. Evaluation of HBMEC transfection efficiency

Cell transfection efficiency can be evaluated in several ways, such as by Western blot, FACS analysis, or fluorescence microscopy imaging. Western blots allow identification and quantification of the total protein expression in a population of transfected cells and can identify whether a chimeric protein is being expressed at the expected size, or if free GFP is expressed by the construct. In order to perform Western blots, it is necessary to have an antibody that is able to detect the protein of interest, for example, anti-GFP if checking chimeric protein expression. However, it can be time-consuming and no information can be gathered about individual cell expression. For FACS analysis, the percentages of positive cells can be determined in a population of transfected cells, and can also enable cell sorting if needed. Levels of expression in individual cells can be analyzed but this approach does not give information on protein localization. Fluorescence microscopy imaging is another routine approach applied to evaluate cell transfection efficiency. This approach allows direct visualization of the individual transfected cells and the chimeric protein localization can be assessed. If the transfected construct does not express a fluorescent protein, then alternative approaches must be used to check for its expression, for example, epitope-tagging, or morphological changes to the cell caused by exogenous protein expression that can be scored. A fluorescence microscope and image analysis software are required for this approach.

In this section, we describe the evaluation of HBMEC transfection efficiency by live-cell imaging and Western blotting. The live-cell imaging approach is preferred to fixed cells as autofluorescent signal from fixatives (such as formaldehyde) is avoided.

2.2.1. Microscopy scoring

Materials

- Unsupplemented RPMI-1640 medium (Sigma-Aldrich). Prewarmed to 37 °C.
- HBMEC complete growth medium. Prewarmed to 37 °C.
- Imaging medium:

  RPMI-1640 without phenol red (Sigma-Aldrich) supplemented with 5% (v/v) FBS, 2 m*M* L-glutamine, 25 m*M* HEPES (pH 7.4). Pre-equilibrated in tissue culture incubator.
- Live-cell nuclear stain, such as DRAQ5™ (Biostatus, Leicestershire, UK) or Hoechst 33342 (Invitrogen).
- Axioplan 2 upright fluorescent microscope (Zeiss, Welwyn Garden City, UK) with CCD camera, operated with Volocity imaging software (Improvision, Perkin-Elmer, Coventry, UK), or LSM510 confocal microscope (Zeiss).
- Low-tech rubber gasket imaging chamber as described in Ward (2007).

  Note: For short-term imaging, this simple imaging chamber is sufficient for imaging purposes as it affords the cells some media to keep them healthy for short amounts of time (generally up to 2 h). It can be washed in appropriate detergent and/or 70% ethanol and reused.
- Sharp point tweezers.

Methods

1. Twenty-four hours after transfection on cover slips in a 24-well tissue culture plate as described in Section 2.1, wash cells with unsupplemented RPMI-1640 once.
2. Add 250 μl HBMEC complete growth medium containing 5 μ*M* DRAQ5™ to the cells, and incubate in CO_2 incubator for 5 min.
3. Remove the DRAQ5™-containing medium from cells, wash cells with unsupplemented RPMI-1640 three times.
4. Place a drop (~35 μl) of imaging medium in the hole of the imaging chamber.
5. Lift cover slip with previously transfected cells from the well, and invert onto the liquid (cells facing media). Dry excess liquid on the cover slip with tissue.
6. Image the cells with a confocal microscope or a fluorescence microscope. Imaging parameters, such as gain and offset levels, and line averaging, are optimized to avoid oversaturation of pixels and to improve signal:noise ratio. To collect data from a large area of the cover slip automatically, acquire tile images by setting the tile dimensions [$n(x)$, $n(y)$], and single plane images are collected.

   Note: Tiling requires a motorized stage under software control, so that images are automatically acquired from neighboring fields over large areas of the cover slip. This avoids potential bias where the experimenter may inadvertently select areas positive for GFP-transfected cells.
7. Count the total number of nuclei and GFP-expressing cells in a field manually, and the percentage of GFP expression is calculated (Fig. 6.1A).

   Note: Total fluorescence intensity can be quantified using imaging software. However, to compare the total fluorescence intensity between different areas or samples, all images must be acquired with similar imaging parameters, and pixel oversaturation must be avoided.

2.2.2. Western blot analysis

Manual microscopy scoring is the cheapest and fastest approach to assess cell transfection efficiency. However, this approach can be prone to an experimenter's bias during image acquisition. Therefore, in a parallel approach to confirm the results from microscopy scoring, the total GFP protein expressed in cells can be assessed by Western blotting. Western blotting protocols are beyond the scope of this chapter, and will be described briefly in this section. For detailed protocols, see Gallagher (2006), Gallagher *et al.* (2008), and Ward (2007).

Materials

- Sample buffer: 7.0 ml 0.5 *M* Tris–Cl, pH 6.8, 3.0 ml glycerol, 1.0 g SDS, 1.2 mg bromophenol blue, 0.6 ml β-mercaptoethanol.

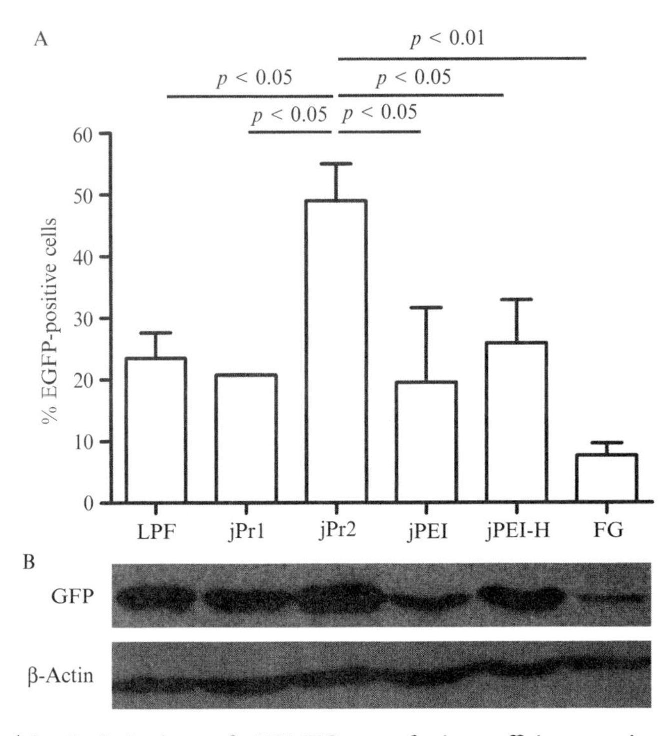

Figure 6.1 Optimization of HBMEC transfection efficiency using different transfection reagents. (A) HBMEC were transfected with pN1-EGFP plasmid using different transfection reagents according to the manufacturers' instructions, and according to the procedure detailed above for jetPRIME™. LPF, Lipofectamine™ 2000 and PLUS™ Reagent (Invitrogen) + 0.5 μg plasmid DNA; jPr1, 1 μl jetPRIME™ + 0.5 μg plasmid DNA; jPr2, 2 μl jetPRIME™ + 1 μg plasmid DNA; jPEI, jetPEI™ (Polyplus-transfection SA) + 1 μg plasmid DNA; jPEI-H, jetPEI™-HUVEC (Polyplus-transfection SA) + 2 μg plasmid DNA; FG, FuGENE® HD (Roche Diagnostics, Burgess Hill, UK) + 0.5 μg plasmid DNA. After 24 h, transfected cells were stained with DRAQ5™, a live-cell nuclear stain, and cells were imaged using the tiling technique. Transfection efficiency is expressed as the percentage of GFP-expressing cells over total DRAQ5™-positive nuclei. Results are representative of two independent experiments performed in duplicate and presented as mean ± standard deviation. Statistical significance was evaluated with analysis of variance and Newman–Keuls *post hoc* analysis. (B) Total GFP protein in the transfected cells was determined by Western blotting analysis, lanes equivalent to chart above.

- PBS, pH 7.2.
- Cell scrapers.
- 10% SDS-PAGE gel.
- PVDF membrane (Millipore, Sigma-Aldrich).
- PBS-T: 1× PBS (pH 7.2) and 0.1% (v/v) Tween 20.
- Blocking buffer: 5% (w/v) nonfat dry milk powder in PBS-T.

- Antibodies: rabbit anti-GFP (Molecular Probes, Invitrogen), mouse anti-β-actin clone AC-15 (Sigma–Aldrich), horseradish peroxidase (HRP)-conjugated anti-rabbit, and anti-mouse antibodies (GE Healthcare, Buckinghamshire, UK).

Methods

1. Twenty-four hours posttransfection, wash cells in a 6-well tissue culture plate once with 1× PBS (pH 7.2). Add 50 μl sample buffer to cells, scrape the cells off the dish, and transfer to a microcentrifuge tube.
2. Incubate the lysates at 98 °C, 10 min, and then on ice for 1 min. Centrifuge the tubes at 13,000 rpm with a microcentrifuge at room temperature for 5 min and remove the supernatant to avoid the contaminating pellet although this is not always present.
3. Resolve the lysates on a 10% SDS-PAGE gel (Gallagher, 2006) and transfer onto a PVDF membrane (Gallagher *et al.*, 2008).
4. Block the membrane using standard protocol and incubate sequentially with primary antibodies (rabbit anti-GFP diluted 1:1000 and mouse anti-β-actin diluted 1:20,000) and secondary antibodies (HRP anti-rabbit antibody at 1:3000 dilution and HRP anti-mouse antibody at 1:3000). Bands were detected with enhanced chemiluminescent (ECL) Western blotting substrate (GE Healthcare) (Fig. 6.1B).

Our results show that the transfection efficiency of HBMEC was significantly improved by increasing the concentration of the plasmid DNA to 1 μg and the volume of jetPRIME™ to 2 μl per transfection reaction (Fig. 6.1A). Western blot analysis confirmed the scoring (Fig. 6.1B).

3. Making Fluorescent *E. coli* K1 that Retain Virulence

To visualize live bacteria microscopically, bacteria can be either surface labeled with fluorescent dye or by expressing cytoplasmic fluorescent proteins. Bacterial labeling with either approach requires careful assessment, to ensure no detrimental biological effect on the bacteria during infection of eukaryotic cell lines. In this section, we present the methods for labeling bacteria and *in vitro* screening of fluorescently labeled bacteria.

3.1. Surface labeling of *E. coli* K1 with FITC

Surface labeling of bacteria with fluorescent dyes, such as fluorescein isothiocyanate (FITC) and NHS Rhodamine, has been widely used to study bacteria–host cell interaction (Hazenbos *et al.*, 1994; Kim *et al.*, 2003;

Schneider *et al.*, 2000; Steele-Mortimer *et al.*, 1999; Weingart *et al.*, 1999). This labeling approach is rapid and cheap. One downside is that the dye will be diluted over each generation during bacterial replication. Further, during surface labeling with FITC, the dye binds to the amines of amino acid side chains, and this has been found to interfere with *Bordetella pertussis* extracellular virulence factors (Weingart *et al.*, 1999); therefore, the usefulness of this approach must be assessed for any particular pathogen.

Materials

- Luria–Bertani (LB) broth: 1% (w/v) tryptone, 0.5% (w/v) yeast extract, 0.5% (w/v) NaCl.
- *E. coli* K1 colonies on LB agar plate.
- 1× PBS (pH 7.2).
- Labeling buffer: 50 m*M* sodium carbonate, 100 m*M* sodium chloride, pH 8.0. The solution is sterilized by passing through 0.22 μm syringe filter.
- FITC (Sigma-Aldrich): Dissolve FITC in DMSO to make 10 mg/ml stock solution. Aliquot in single use aliquots and store at −20 °C.
- Spectrophotometer.
- Shaking incubator.
- Fluorescence microscope.
- Disposables:
  - 1.5 ml disposable polystyrene cuvette, 20 ml sterile universal, microscope glass slide, glass cover slip (No. 1 18 mm × 18 mm) (VWR).

Methods

1. Inoculate a single colony of *E. coli* K1 from LB agar plate into 5 ml LB broth in a 20 ml universal. Incubate the bacterial culture overnight in a shaking incubator at 37 °C, 200 rpm. In order to obtain bacteria at active log phase, subculture the overnight bacterial culture at 50 times dilution in fresh LB on the day of experiment and grow for approximately 2 h.
2. Bacterial culture concentration is measured by absorbance at 595 nm in a spectrophotometer. The bacterial culture is adjusted to A_{595} of 0.22, which is equivalent to approximately 1×10^9 cfu/ml of *E. coli* K1 bacteria.
3. Centrifuge 1 ml of the log phase bacterial culture at 13,000 rpm for 5 min.
4. Remove the supernatant carefully without disturbing the bacterial pellet. Then, wash the bacterial pellet once in 1 ml PBS, and centrifuge at 13,000 rpm for 5 min.
5. Remove the supernatant carefully and resuspend the pellet in 1 ml labeling buffer containing 0.5 mg/ml FITC. The FITC-containing labeling buffer must be freshly prepared on the day of experiment. Incubate the bacterial suspension for 30 min at room temperature in the dark.

Note: The labeling buffer itself may have adverse effects on bacterial virulence.

6. Wash the FITC-labeled bacteria three times with 1× PBS and each time centrifuge at 7000 rpm for 5 min.
7. Resuspend the bacterial pellet in 1 ml PBS.
8. Check for bacterial fluorescence under a fluorescence microscope. Briefly, drop 5 μl of the labeled bacterial culture on a microscope glass slide, lay a cover slip onto the culture, and examine the culture under a fluorescence microscope.
9. Screen for bacterial virulence as described in Section 3.3, Method 2.

3.2. Construction of fluorescent protein-expressing *E. coli* K1

In choosing the fluorescent protein, several criteria must be taken into consideration, such as photostability, brightness, protein oligomerization, and cross talk in excitation and emission channels (Shaner *et al.*, 2005). It is also important to note that not all of the fluorescent proteins are properly expressed in all strains of bacteria. Variations in the bacterial codon usage may also have significant effects on fluorescent protein expression in the bacteria and codon optimization may be necessary (Qazi *et al.*, 2001; Sastalla *et al.*, 2009). Further, fluorescent protein expression from a plasmid may be unstable and integration into the genome may be a preferred option (Andreu *et al.*, 2010; Karlyshev and Wren, 2005).

3.2.1. Transformation of *E. coli* K1 with bacterial expression plasmid DNA

Generally, plasmid DNA can be introduced into bacterial cells either by heat shock or by electroporation. We transformed *E. coli* K1 with bacterial fluorescent protein-expressing constructs (Table 6.1) by electroporation, since plasmid DNA transformation using heat shock of chemically competent cells was very inefficient for this strain. Bacterial transformation protocols are beyond the scope of this chapter. Detailed discussions and protocols are described in Seidman *et al.* (2001). After transformation, some bacterial colonies have visible coloration. For confirmation, the transformed bacteria are screened with a fluorescence microscope as described in Section 3.1, step 8.

3.3. Screening for virulence of fluorescent protein-expressing *E. coli* K1

Cytoplasmic fluorescent protein expression in bacteria provides continuously fluorescent live bacteria, but some adverse outcomes from the burden on protein synthesis to bacterial pathogenicity have been observed (Knodler *et al.*, 2005). In addition to the production of fluorescent proteins, the presence of plasmid, and the antibiotic resistance marker in the plasmid can affect the bacterial physiology during infection (Abromaitis *et al.*, 2005; Knodler *et al.*, 2005). We describe below *in vitro* assays used to screen for virulence in fluorescent protein-expressing *E. coli* K1. First, the transformed bacterial growth kinetics are compared to the native strain, and then the gentamicin protection assay enables quantification of viable intracellular bacteria protected from external antibiotic.

Materials

- HBMEC complete growth medium. Prewarmed to 37 °C.
- Experimental medium: RPMI-1640 (Sigma-Aldrich) supplemented with 5% (v/v) FBS (Biosera), and 2 m*M* L-glutamine (Sigma-Aldrich). Prewarmed to 37 °C.
- Unsupplemented RPMI-1640 (Sigma-Aldrich). Prewarmed to 37 °C.
- LB broth.
- LB agar plates.
- 0.3% (w/v) SDS solution in PBS (pH 7.2).
- Gentamicin solution, 50 mg/ml (Sigma–Aldrich).
- HBMEC.
- Bacterial cultures:
  - *E. coli* K1 and transformed strains (from Section 3.2.1).
  - *E. coli* K-12, HB101 strain (nonpathogenic control).
- Spectrophotometer.
- Refrigerated centrifuge.
- Tissue culture CO_2 incubator.
- Incubator at 37 °C.
- Disposables:
  - 1.5 ml disposable polystyrene cuvette, 24-well tissue culture dish, 20 ml sterile universals.

Method 1: Bacterial growth kinetics

1. Inoculate a single colony of *E. coli* K1 or transformed bacteria from a LB agar plate into 5 ml LB broth (with antibiotic for selection of transformed strains as applicable) in a 20 ml universal. Incubate the bacterial culture overnight in a shaking incubator at 37 °C, 200 rpm.
2. Subculture the overnight bacterial culture at 50 times dilution in fresh LB. Sample 1 ml of the diluted bacterial culture, load the culture into a disposable polystyrene cuvette and read A_{595}.
3. Incubate the diluted bacterial culture in a shaking incubator at 37 °C, 200 rpm. At 0.5, 1, and 2 h after incubation, sample 1 ml of the culture and read A_{595}.

Our results show that the growth kinetics of transformed *E. coli* K1 expressing GFP (K1-GFP) and *E. coli* K1 expressing mCherry (K1-Cherry) are slightly lower than native *E. coli* K1 (K1) (Fig. 6.2A and B).

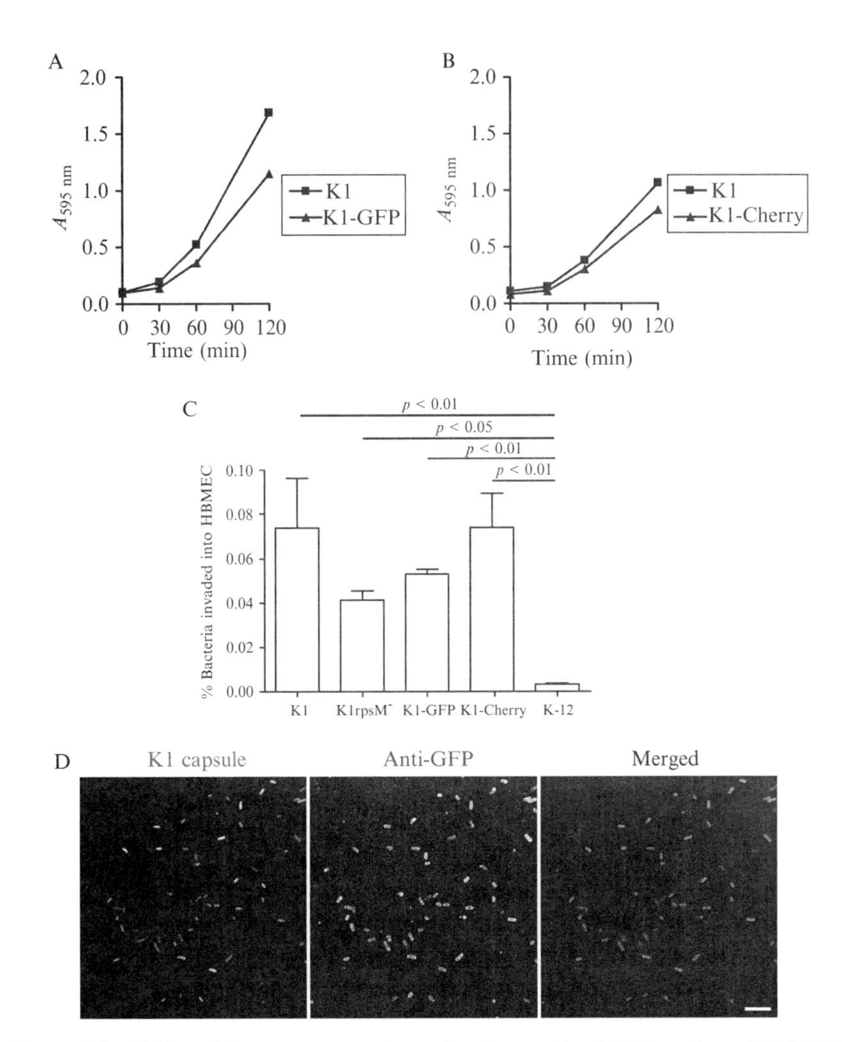

Figure 6.2 Effect of fluorescent protein production on *E. coli* K1 invasion of HBMEC. (A, B) The growth kinetics of *E. coli* K1 expressing GFP (K1-GFP) and *E. coli* K1 expressing mCherry (K1-Cherry), respectively, in LB broth were monitored at the indicated time points with a spectrophotometer at A_{595}. The growth kinetics of transformed *E. coli* K1 were marginally lower than the native *E. coli* K1 (K1). (C) HBMEC were infected with untransformed *E. coli* K1 or plasmid-containing strains at m.o.i. 100 for 2 h, and then infected cells were incubated in medium containing 100 μg/ml gentamicin for 1 h. Cells were lysed and plated on LB agar for CFU enumeration (intracellular bacteria). The result represents percentage of bacteria that invaded into HBMEC. All transformed strains were invasive compared to the laboratory nonpathogenic K-12 strain, HB101. Results are representative of two independent experiments performed in triplicate and presented as mean ± standard deviation. Statistical significance was evaluated with analysis of variance and Newman–Keuls *post hoc* analysis. (D) K1-GFP culture was smeared on a glass cover slip, air-dried, and fixed with 2% (v/v) formaldehyde. The bacterial smear was stained with 10 μg/ml PK1A-GFP probe in wash buffer (PBS with 5%, v/v, FBS). To differentiate from cytoplasmic GFP in the bacterial strain, the probe was counterstained

Method 2: Gentamicin protection assay

1. Seed HBMEC at 1 × 10^5 cells/ml into 24-well tissue culture dish 3 days before experiment as described in Section 2.1. For infection, the cells are cultured until confluent.
2. At least 30 min before infection, wash cells three times with unsupplemented RPMI-1640 media to remove antibiotics. Replenish with 0.5 ml experimental medium.
3. Add log phase bacteria at a m.o.i. of 100 to HBMEC. This is calculated by trypsinizing an equivalent chamber of HBMEC and counting on a hemocytomer. Bacterial numbers are counted by adjusting culture concentration to 1 × 10^9 bacteria/ml (Section 3.1). After the addition of bacteria, centrifuge the plate at 500×*g* at 15 °C for 5 min to bring the bacteria into contact with the cells and incubate together for 2 h at 37 °C with 5% CO_2.
4. Reserve bacteria inoculum and plate out in serial dilutions to confirm the CFU of inoculum introduced to the cells, as the absorption is only an estimate.
5. At the end of the infection period, wash cells three times with unsupplemented RPMI-1640 media, and incubate infected cells with experimental medium containing 100 μg/ml gentamicin to kill extracellular bacteria for a further 1 h at 37 °C with 5% CO_2.
6. Wash cells three times with unsupplemented RPMI-1640 media, and lyse cells with 0.5 ml 0.3% SDS.
7. Plate serial dilutions of lysates directly on to LB agar, and incubate plates at 37 °C for 16–18 h, and score colonies.

We show that bacterial strains carrying either the promoterless plasmid DNA (K1rpsM$^-$) or GFP-expressing plasmid DNA (K1-GFP) exhibited decreased HBMEC invasion compared to the K1 wild type, although they remain significantly more invasive than the nonpathogenic *E. coli* K-12 laboratory strain (Fig. 6.2C). This demonstrates that even the plasmid alone can create a burden that affects bacterial pathogenicity.

As Kim *et al.* (2003) showed that the bacterial K1 polysialic capsule is the crucial virulence factor for *E. coli* K1 intracellular survival by evading lysosomal degradation, we speculated that expression of GFP in the bacteria might affect the bacterial K1 capsule expression. The K1 capsule can be stained with a GFP-tagged catalytically inactive endosialidase probe (PK1A-GFP) (Jokilammi *et al.*, 2004) and this revealed the presence of the K1 capsule on K1-GFP (Fig. 6.2D). Therefore, fluorescent protein production did not affect K1 capsule production in the transformed bacteria, and the

---

with mouse anti-GFP clone 3E6 (1:500 dilution; Invitrogen), then FluoProbes® 642-conjugated donkey anti-mouse antibody (1:200 dilution; Interchim, Cheshire Science Ltd., Chester, UK). The images show the presence of the bacterial K1 capsule on K1-GFP. Scale bar = 10 μm. (For interpretation of the references of color in this figure legend, the reader is referred to the Web version of this chapter.)

reduction in K1-GFP invasion efficiency is unlikely to be caused by intracellular lysis but may reflect a problem in the invasion process. In contrast, the invasion efficiency of K1-Cherry was very similar to the native bacteria (Fig. 6.2C), providing a useful tool with which to investigate invasion pathways of *E. coli* K1 into HBMEC by live-cell imaging.

In conclusion, the level of the bacterial virulence can be affected by the presence of plasmids and cytoplasmic production of fluorescent proteins. Our data demonstrate that although the growth kinetics of the fluorescent protein-expressing bacteria are slightly lower than the growth kinetics of the native *E. coli* K1, bacterial virulence is retained.

4. Live Imaging of *E. coli* K1-Infected HBMEC

Having optimized the HBMEC transfection efficiency by using jetPRIME™ transfection reagent, and successfully obtained an *E. coli* K1 strain expressing fluorescent protein that retains virulence, we are able to look at direct interaction between the bacteria and different cellular compartments by confocal live-cell imaging techniques. We describe below two live imaging protocols of *E. coli* K1-infected HBMEC. Method 1 is only suitable for short-duration imaging experiments, whereas Method 2 is for time-lapse imaging experiments. First, the interactions of *E. coli* K1 with various GFP-tagged cellular markers are screened, and time-lapse imaging then allows tracking of the bacterial postinvasion events over time.

Materials

- Unsupplemented RPMI-1640 (Sigma-Aldrich). Prewarmed to 37 °C.
- Experimental medium. Prewarmed to 37 °C.
- Imaging medium. Pre-equilibrated in a tissue culture incubator.
- Plasmid DNA (Table 6.1) purified with Midiprep kit (Promega) and eluted in 10 m*M* Tris–HCl, pH 8.5, buffer.
- Low-tech rubber gasket imaging chamber (Ward, 2007).
- Sharp point tweezers.
- Disposables:
  - Cleaned microscope glass cover slips.
  - LabTek 8-well glass cover slip chambers (Nalge Nunc).

Method 1: Live imaging of E. coli *K1 interaction with early endosomes*

1. Seed HBMEC on glass cover slips and transfect with GFP-Rab5 chimera expression vectors (Table 6.1) as described in Section 2.1.
2. Infect cells with log phase *E. coli* K1 expressing mCherry at a m.o.i. of 100. After the addition of bacteria, centrifuge the plate at 500×*g* at 15 °C for 5 min.

3. After 2 h incubation at 37 °C with 5% CO_2, wash the cells three times with unsupplemented RPMI-1640 and incubate the infected cells in experimental medium containing 100 μg/ml gentamicin for 1 h at 37 °C with 5% CO_2.
4. At the end of gentamicin treatment, wash cells three times with unsupplemented RPMI-1640, lift the cover slip from the well, and mount the cover slip onto a low-tech rubber gasket imaging chamber with approximately 30 μl imaging medium (Ward, 2007).
5. Prior to imaging, preheat the microscope imaging chamber to 37 °C to avoid focus problems resulting from expanding metal components of the stage.
6. Optimize imaging parameters, such as gain and offset levels, and line averaging. To acquire 3D images, set the z-stack parameters. For acquisition of tile-z stacks images, set tile dimensions [$n(x)$, $n(y)$] in addition to the z-stack parameters. A sample result is shown in Fig. 6.3A.

   Note: For imaging performed with any short wavelength laser, such as 405 nm often used for CFP illumination, the laser power must be kept as low as possible to avoid phototoxicity.

Method 2: Time-lapse imaging of E. coli *K1 interaction with late endosomes*

1. Seed 5 × 10^4 cells/ml of HBMEC in a LabTek 8-well glass cover slip chamber 3 days prior to experiment and transfect with GFP-Rab7 (Table 6.1) as described in Section 2.1.
2. Infect cells with log phase *E. coli* K1 expressing mCherry at a m.o.i. of 100, and incubate the cells for 2 h at 37 °C with 5% CO_2.

   Note: To image invasion, infect the cells for 30 min, and then process the cells as described in *step 3* but without adding gentamicin into the imaging medium.
3. Following infection, wash the cells three times with unsupplemented RPMI-1640. Replenish the infected cells with imaging medium containing 100 μg/ml gentamicin.
4. Prior to imaging, in addition to preheating the microscope imaging chamber to 37 °C as described in *Method 1*, switch on CO_2 control to zero the CO_2 in the system from previous usage. Approximately 15 min before imaging, supply CO_2 to the system.
5. Place LabTek chamber on the stage holder, and perform condenser alignment to achieve Köhler illumination.

   Note: The presence of the LabTek chamber lid changes the microscope light path. Therefore, the microscope condenser must be aligned to ensure even illumination of cells in the chamber when tracking the cells in phase contrast.
6. To acquire a time series at multiple locations, mark all the locations of interest in the software.

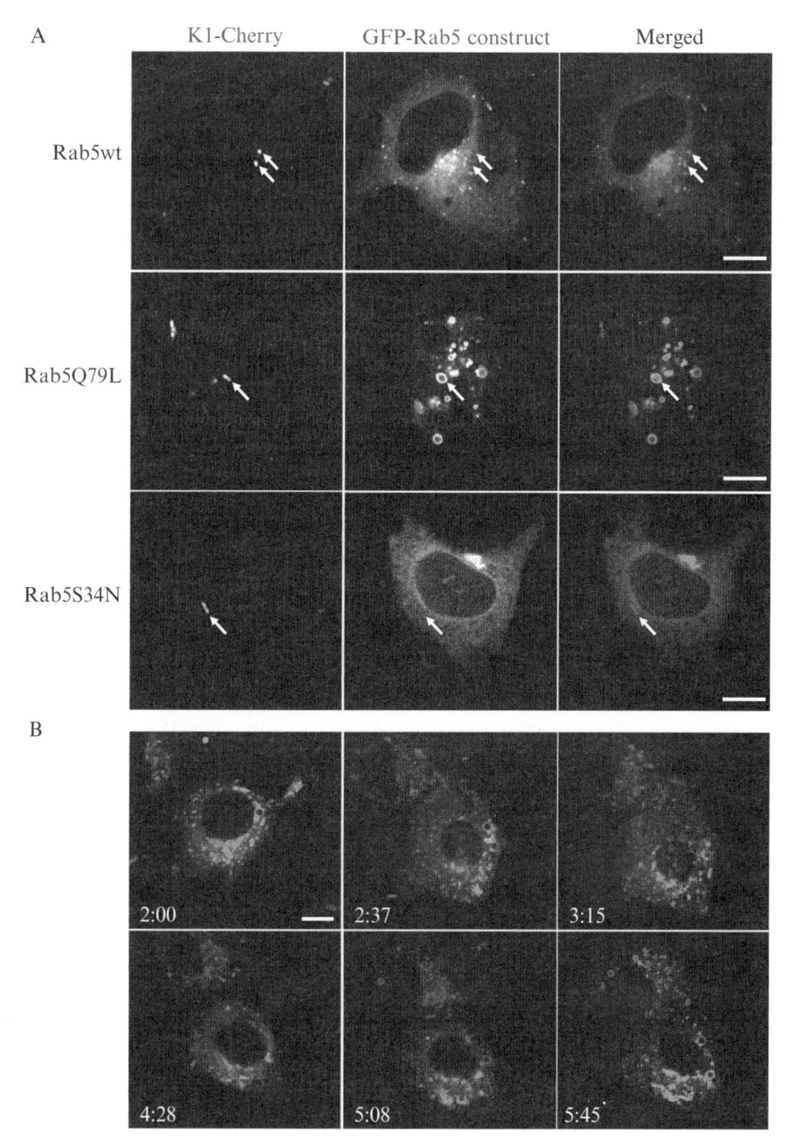

Figure 6.3 Interaction of *E. coli* K1 with endosomal markers. (A) HBMEC were transfected with GFP-Rab5 chimera expression plasmids: either wild-type (Rab5wt), the dominant positive Rab5Q79L mutant, or the dominant negative Rab5S34N mutant, as indicated. Transfected cells were infected with K1-Cherry at m.o.i. 100 for 2 h, and then infected cells were incubated in medium containing 100 μg/ml gentamicin for 1 h. Cells were imaged with a point-scanning confocal microscope, LSM510 (Zeiss). A single confocal slice is shown. Chimeric Rab5 (green) and *E. coli* K1 (red). Arrows indicate the location of bacteria. (B) HBMEC were transiently transfected with wild-type GFP-Rab7 and were infected with K1-Cherry at m.o.i. 100 for 2 h. At the end of the 2 h incubation, cells were washed with unsupplemented

7. Set up imaging parameters. For multicolor imaging, select line scanning. It is also advised to use faster scan speeds ($\geq$8 μs/pixel) to reduce the time spent by the laser at each pixel (Frigault *et al.*, 2009). However, it should be noted that the image resolution may be reduced by increasing the scanning speed. To acquire high-quality images, line averaging is set to 4.
8. Set the number of imaging cycles and the time interval between imaging cycle. Select autofocus function for long time courses to avoid focus drift. A sample result of time-lapse live-cell imaging is shown in Fig. 6.3B.

Rab GTPases are members of the Ras superfamily of small GTPases, which regulate each stage of membrane trafficking in the secretory and endocytic pathways (Schwartz *et al.*, 2007). Pathogens have been found to acquire these proteins on pathogen-containing vacuoles during infection (Guignot *et al.*, 2004; Rzomp *et al.*, 2003; Steele-Mortimer *et al.*, 1999). Our results (Fig. 6.3A) show that *E. coli* K1 did not colocalize with wild-type Rab5 (Rab5wt) or dominant negative Rab5 (Rab5S34N), but localized in an enlarged dominant positive Rab5 compartment (Rab5Q79L). These results indicate that *E. coli* K1 did not arrest the endosomal maturation at the early endosomal stage. In contrast, long-term association of a late endosomal marker, Rab7, with *E. coli* K1 containing vacuoles was observed (Fig. 6.3B).

ACKNOWLEDGMENTS

We thank Dr. Olivier Marchés (Queen Mary, University of London) for pFPV25.1 plasmid DNA; Dr. Naveed Khan (University of Nottingham) for HBMEC and *E. coli* K1 and K-12 strains; Professor Jukka Finne and Dr. Anne Jokilammi (University of Turku, Finland) for the PK1A-GFP probe; Professor Roger Tsien (University of California, San Diego) for mCherry cDNA; Professor Albert Haas (University of Bonn, Germany) and Dr. Bianca Schneider (Research Center Borstel, Germany) for Rab7 plasmid DNA; Dr. Ben Nichols (MRC LMB, Cambridge, UK) for Rab5 plasmid DNA. Lip Nam Loh was supported by Bloomsbury Colleges PhD Studentship. This work was supported by Central Research Fund, University of London.

RPMI-1640 to remove extracellular bacteria and replenished with imaging medium containing 100 μg/ml gentamicin. Cells were imaged over time after the initial 2 h infection with a point-scanning confocal microscope, LSM510 (Zeiss). A single confocal slice is shown. Representative time points (hour:minutes) are presented. Chimeric Rab7 (green) and *E. coli* K1 (red). Scale bar = 10 μm. (For interpretation of the references to color in this figure legend, the reader is referred to the Web version of this chapter.)

REFERENCES

Abromaitis, S., Faucher, S., Béland, M., Curtiss, R., III, and Daigle, F. (2005). The presence of the *tet* gene from cloning vectors impairs *Salmonella* survival in macrophages. *FEMS Microbiol. Lett.* **242,** 305–312.

Andreu, N., Zelmer, A., Fletcher, T., Elkington, P. T., Ward, T. H., Ripoll, J., Parish, T., Bancroft, G. J., Schaible, U., Robertson, B. D., and Wiles, S. (2010). Optimisation of bioluminescent reporters for use with mycobacteria. *PLoS One* **5,** e10777.

Berón, W., Gutierrez, M. G., Rabinovitch, M., and Colombo, M. I. (2002). *Coxiella burnetii* localizes in a Rab7-labeled compartment with autophagic characteristics. *Infect. Immun.* **70,** 5816–5821.

Bonacorsi, S., and Bingen, E. (2005). Molecular epidemiology of *Escherichia coli* causing neonatal meningitis. *Int. J. Med. Microbiol.* **295,** 373–381.

Bucci, C., Thomsen, P., Nicoziani, P., McCarthy, J., and van Deurs, B. (2000). Rab7: A key to lysosome biogenesis. *Mol. Biol. Cell* **11,** 467–480.

Drecktrah, D., Knodler, L. A., Howe, D., and Steele-Mortimer, O. (2007). *Salmonella* trafficking is defined by continuous dynamic interactions with the endolysosomal system. *Traffic* **8,** 212–225.

Frigault, M. M., Lacoste, J., Swift, J. L., and Brown, C. M. (2009). Live-cell microscopy—Tips and tools. *J. Cell Sci.* **122,** 753–767.

Gallagher, S. R. (2006). One-dimensional SDS gel electrophoresis of proteins. *Curr. Protoc. Mol. Biol.* **75,** 10.2.A.1. Chapter 10, Unit 10.2A.

Gallagher, S., Winston, S. E., Fuller, S. A., and Hurrell, J. G. (2008). Immunoblotting and immunodetection. *Curr. Protoc. Mol. Biol.* Chapter 10, Unit 10.8.

Glode, M. P., Sutton, A., Moxon, E. R., and Robbins, J. B. (1977). Pathogenesis of neonatal Escherichia coli meningitis: Induction of bacteremia and meningitis in infant rats fed *E. coli* K1. *Infect. Immun.* **16,** 75–80.

Guignot, J., Caron, E., Beuzón, C., Bucci, C., Kagan, J., Roy, C., and Holden, D. W. (2004). Microtubule motors control membrane dynamics of *Salmonella*-containing vacuoles. *J. Cell Sci.* **117,** 1033–1045.

Hazenbos, W. L. W., van den Berg, B. M., van't Wout, J. W., Mooi, F. R., and van Furth, R. (1994). Virulence factors determine attachment and ingestion of nonopsonized and opsonized *Bordetella pertussis* by human monocytes. *Infect. Immun.* **62,** 4818–4824.

Holt, D. E., Halket, S., de Louvois, J., and Harvey, D. (2001). Neonatal meningitis in England and Wales: 10 years on. *Arch. Dis. Child. Fetal Neonatal Ed.* **84,** F85–9.

Huang, S.-H., Wass, C., Fu, Q., Prasadarao, N. V., Stins, M., and Kim, K. S. (1995). *Escherichia coli* invasion of brain microvascular endothelial cells in vitro and in vivo: Molecular cloning and characterization of invasion gene *ibe10*. *Infect. Immun.* **63,** 4470–4475.

Jokilammi, A., Ollikka, P., Korja, M., Jakobsson, E., Loimaranta, V., Haataja, S., Hirvonen, H., and Finne, J. (2004). Construction of antibody mimics from a noncatalytic enzyme—Detection of polysialic acid. *J. Immunol. Methods* **295,** 149–160.

Kang, J., Ramu, S., Lee, S., Aguilar, B., Ganesan, S. K., Yoo, J., Kalra, V. K., Koh, C. J., and Hong, Y.-K. (2009). Phosphate-buffered saline-based nucleofection of primary endothelial cells. *Anal. Biochem.* **386,** 251–255.

Karlyshev, A. V., and Wren, B. W. (2005). Development and application of an insertional system for gene delivery and expression in *Campylobacter jejuni*. *Appl. Environ. Microbiol.* **71,** 4004–4013.

Khan, N. A., and Siddiqui, R. (2009). *Acanthamoeba* affects the integrity of human brain microvascular endothelial cells and degrades the tight junction proteins. *Int. J. Parasitol.* **39,** 1611–1616.

Khan, N. A., Wang, Y., Kim, K. J., Chung, J. W., Wass, C. A., and Kim, K. S. (2002). Cytotoxic necrotizing factor-1 contributes to *Escherichia coli K1* invasion of the central nervous system. *J. Biol. Chem.* **277,** 15607–15612.

Khan, N. A., Kim, Y., Shin, S., and Kim, K. S. (2007). FimH-mediated *Escherichia coli* K1 invasion of human brain microvascular endothelial cells. *Cell. Microbiol.* **9,** 169–178.

Kim, K. S. (2003). Pathogenesis of bacterial meningitis: From bacteraemia to neuronal injury. *Nat. Rev. Neurosci.* **4,** 376–385.

Kim, K. S., Itabashi, H., Gemski, P., Sadoff, J., Warren, R. L., and Cross, A. S. (1992). The K1 capsule is the critical determinant in the development of *Escherichia coli* meningitis in the rat. *J. Clin. Invest.* **90,** 897–905.

Kim, K. J., Elliott, S. J., Di Cello, F., Stins, M. F., and Kim, K. S. (2003). The K1 capsule modulates trafficking of *E. coli*-containing vacuoles and enhances intracellular bacterial survival in human brain microvascular endothelial cells. *Cell. Microbiol.* **5,** 245–252.

Knodler, L. A., Bestor, A., Ma, C., Hansen-Wester, I., Hensel, M., Vallance, B. A., and Steele-Mortimer, O. (2005). Cloning vectors and fluorescent proteins can significantly inhibit *Salmonella enterica* virulence in both epithelial cells and macrophages: Implications for bacterial pathogenesis studies. *Infect. Immun.* **73,** 7027–7031.

Kovala, A. T., Harvey, K. A., McGlynn, P., Boguslawski, G., Garcia, J. G. N., and English, D. (2000). High-efficiency transient transfection of endothelial cells for functional analysis. *FASEB J.* **14,** 2486–2494.

Lamberti, Y. A., Hayes, J. A., Perez Vidakovics, M. L., Harvill, E. T., and Rodriguez, M. E. (2010). Intracellular trafficking of *Bordetella pertussis* in human macrophages. *Infect. Immun.* **78,** 907–913.

Maruvada, R., Argon, Y., and Prasadarao, N. V. (2008). *Escherichia coli* interaction with human brain microvascular endothelial cells induces signal transducer and activator of transcription 3 association with the C-terminal domain of Ec-gp96, the outer membrane protein A receptor for invasion. *Cell. Microbiol.* **10,** 2326–2338.

Mulholland, K. (1998). Serious infections in young infants in developing countries. *Vaccine* **16,** 1360–1362.

Nichols, B. J., Kenworthy, A. K., Polishchuk, R. S., Lodge, R., Roberts, T. H., Hirschberg, K., Phair, R. D., and Lippincott-Schwartz, J. (2001). Rapid cycling of lipid raft markers between the cell surface and Golgi complex. *J. Cell Biol.* **153,** 529–541.

Pluschke, G., Mercer, A., Kusécek, B., Pohl, A., and Achtman, M. (1983). Induction of bacteremia in newborn rats by *Escherichia coli* K1 is correlated with only certain O (lipopolysaccharide) antigen types. *Infect. Immun.* **39,** 599–608.

Polin, R. A., and Harris, M. C. (2001). Neonatal bacterial meningitis. *Semin. Neonatol.* **6,** 157–172.

Prasadarao, N. V. (2002). Identification of *Escherichia coli* outer membrane protein A receptor on human brain microvascular endothelial cells. *Infect. Immun.* **70,** 4556–4563.

Prasadarao, N. V., Wass, C. A., Stins, M. F., Shimada, H., and Kim, K. S. (1999). Outer membrane protein A-promoted actin condensation of brain microvascular endothelial cells is required for *Escherichia coli* invasion. *Infect. Immun.* **67,** 5775–5783.

Qazi, S. N. A., Rees, C. E. D., Mellits, K. H., and Hill, P. J. (2001). Development of *gfp* vectors for expression in *Listeria monocytogenes* and other low G+C gram positive bacteria. *Microb. Ecol.* **41,** 301–309.

Reddy, M. A., Prasadarao, N. V., Wass, C. A., and Kim, K. S. (2000a). Phosphatidylinositol 3-kinase activation and interaction with focal adhesion kinase in *Escherichia coli* K1 invasion of human brain microvascular endothelial cells. *J. Biol. Chem.* **275,** 36769–36774.

Reddy, M. A., Wass, C. A., Kim, K. S., Schlaepfer, D. D., and Prasadarao, N. V. (2000b). Involvement of focal adhesion kinase in *Escherichia coli* invasion of human brain microvascular endothelial cells. *Infect. Immun.* **68,** 6423–6430.

Remington, S. J. (2006). Fluorescent proteins: Maturation, photochemistry and photophysics. *Curr. Opin. Struct. Biol.* **16,** 714–721.

Robbins, J. B., McCracken, G. H., Jr., Gotschlich, E. C., Ørskov, F., Ørskov, I., and Hanson, L. A. (1974). *Escherichia coli* K1 capsular polysaccharide associated with neonatal meningitis. *N. Engl. J. Med.* **290,** 1216–1220.

Ruthel, G., Ribot, W. J., Bavari, S., and Hoover, T. A. (2004). Time-lapse confocal imaging of development of *Bacillus anthracis* in macrophages. *J. Infect. Dis.* **189,** 1313–1316.

Rzomp, K. A., Scholtes, L. D., Briggs, B. J., Whittaker, G. R., and Scidmore, M. A. (2003). Rab GTPases are recruited to chlamydial inclusions in both a species-dependent and species-independent manner. *Infect. Immun.* **71,** 5855–5870.

Sastalla, I., Chim, K., Cheung, G. Y. C., Pomerantsev, A. P., and Leppla, S. H. (2009). Codon-optimized fluorescent proteins designed for expression in low-GC gram-positive bacteria. *Appl. Environ. Microbiol.* **75,** 2099–2110.

Schneider, B., Gross, R., and Haas, A. (2000). Phagosome acidification has opposite effects on intracellular survival of *Bordetella pertussis* and *B. bronchiseptica*. *Infect. Immun.* **68,** 7039–7048.

Schwartz, S. L., Cao, C., Pylypenko, O., Rak, A., and Wandinger-Ness, A. (2007). Rab GTPases at a glance. *J. Cell Sci.* **120,** 3905–3910.

Segura, I., González, M. A., Serrano, A., Abad, J. L., Bernad, A., and Riese, H. H. (2001). High transfection efficiency of human umbilical vein endothelial cells using an optimized calcium phosphate method. *Anal. Biochem.* **296,** 143–147.

Seidman, C. E., Struhl, K., Sheen, J., and Jessen, T. (2001). Introduction of plasmid DNA into cells. *Curr. Protoc. Mol. Biol.* Chapter 1, Unit 1.8.

Shaner, N. C., Campbell, R. E., Steinbach, P. A., Giepmans, B. N. G., Palmer, A. E., and Tsien, R. Y. (2004). Improved monomeric red, orange and yellow fluorescent proteins derived from *Discosoma* sp. red fluorescent protein. *Nat. Biotechnol.* **22,** 1567–1572.

Shaner, N. C., Steinbach, P. A., and Tsien, R. Y. (2005). A guide to choosing fluorescent proteins. *Nat. Methods* **2,** 905–909.

Steele-Mortimer, O., Méresse, S., Gorvel, J.-P., Toh, B.-H., and Finlay, B. B. (1999). Biogenesis of *Salmonella typhimurium*-containing vacuoles in epithelial cells involves interactions with the early endocytic pathway. *Cell. Microbiol.* **1,** 33–49.

Valdivia, R. H., and Falkow, S. (1996). Bacterial genetics by flow cytometry: Rapid isolation of *Salmonella typhimurium* acid-inducible promoters by differential fluorescence induction. *Mol. Microbiol.* **22,** 367–378.

Ward, T. H. (2007). Trafficking through the early secretory pathway of mammalian cells. *In* "Protein Targeting Protocols," (M. van der Giezen, ed.), pp. 281–296. Humana Press, New Jersey.

Ward, T. H., and Lippincott-Schwartz, J. (2006). The uses of green fluorescent protein in mammalian cells. *In* "Green Fluorescent Protein: Properties, Applications, and Protocols," (M. Chalfie and S. R. Kain, eds.), pp. 305–337. John Wiley & Sons, Inc., New Jersey.

Weingart, C. L., Broitman-Maduro, G., Dean, G., Newman, S., Peppler, M., and Weiss, A. A. (1999). Fluorescent labels influence phagocytosis of *Bordetella pertussis* by human neutrophils. *Infect. Immun.* **67,** 4264–4267.

SECTION TWO

DISEASE STATES

CHAPTER SEVEN

Real-Time Imaging After Cerebral Ischemia: Model Systems for Visualization of Inflammation and Neuronal Repair

Pierre Cordeau *and* Jasna Kriz

Contents

1. Introduction 118
   1.1. Brain response to ischemic injury: A short overview 118
   1.2. The value of bioluminescence/biophotonic imaging in stroke research 119
2. Design and Generation of Dual Reporter Mouse Models for *In Vivo* Imaging of Brain Response to Ischemic Injury 119
   2.1. Mouse model for visualization of microglial activation and innate immune response: TLR2/luc/gfp mouse model 119
   2.2. Mouse model for real-time imaging of neuronal responses to brain injuries: GAP-43/luc/gfp mouse model 120
   2.3. Genotyping 122
   2.4. Advantages of multireporter approach in optical imaging: Why bicistronic constructs? 122
3. Induction of Cerebral Ischemia 124
   3.1. Material/equipment 124
   3.2. Surgical procedures: Experimental ischemia 124
4. *In Vivo* Bioluminescence Imaging 125
   4.1. Preparation of the material: Luciferin solution for *in vivo* imaging 126
   4.2. Preparation of the imaging session 127
   4.3. Relevant control groups 127
   4.4. Protocol for 2D *in vivo* bioluminescence imaging 128
   4.5. Protocol for 3D reconstruction of bioluminescent sources 129
5. Additional Tips for Successful *In Vivo* Bioluminescence Imaging 130
   5.1. Impact of mouse fur color 130
   5.2. Covering ectopic and/or saturated signals 131

Department of Psychiatry and Neuroscience, Laval University, Centre de Recherche du Centre Hospitalier de l'Université Laval, 2705 boul. Laurier, Québec, Québec City, Canada

Methods in Enzymology, Volume 506
ISSN 0076-6879, DOI: 10.1016/B978-0-12-391856-7.00031-7

Abstract

Brain response to ischemic injury is characterized by initiation of a complex pathophysiological cascade comprising the events that may evolve over hours or several days and weeks after initial attack. At present, spatial and temporal dynamics of these events is not completely understood. To enable better understanding of the brain response to ischemic injury we developed and validated several novel transgenic mouse models of bioluminescence and fluorescence, allowing the noninvasive and time-lapse imaging of neuroinflammation, neuronal damage/stress and repair. These mice represent a powerful analytical tool for understanding *in vivo* pathology as well as the evaluating pharmacokinetics and longitudinal responses to drug therapies. Here, we describe the basic procedures of generating biophotonic mouse models for live imaging of microglial activation and neuronal stress and recovery, followed by a detailed description of *in vivo* bioluminescence imaging protocols used after experimental stroke.

1. Introduction

1.1. Brain response to ischemic injury: A short overview

Stroke is the leading cause of death and disabilities in industrialized countries (Dirnagl *et al.*, 1999; Lo *et al.*, 2003). In practice, the term "stroke" refers to an umbrella of conditions caused by the occlusion or hemorrhage of blood vessel supplying the brain. In all instances, stroke ultimately involves dysfunction and death of brain cells. The neurological deficits will reflect the location and size of the compromised brain area. At present, although progress has been made in understanding the molecular pathway that lead to ischemic cell death, the current clinical treatments remain poorly effective (Dirnagl *et al.*, 1999; Lo *et al.*, 2003). Here, it is important to emphasize that brain response to ischemic injury is characterized by initiation of a complex pathophysiological cascade. Once activated the pathological cascade comprises events that may evolve over hours or several days and weeks. This will be followed by the induction of many genes that will initiate postischemic inflammation as well as brain recovery. However, at present, a spatial and temporal dynamics of these events is not completely understood. To enable better understanding of the brain response to ischemic injury, we developed and validated several novel transgenic mouse models of bioluminescence and fluorescence, allowing the noninvasive and time-lapse imaging of neuroinflammation, neuronal damage/stress and repair. The series of our recent studies suggest that biophotonic/bioluminescence signals imaged from the live animals can be used as valid biomarkers to visualize distinct pathological events after ischemic injury (Cordeau *et al.*, 2008; Gravel *et al.*, 2011; Lalancette-Hebert *et al.*, 2009), screen for novel biocompatible molecules (Lalancette-Hebert *et al.*, 2010; Maysinger *et al.*, 2007), and/or analyze early pathogenesis and therapeutic approaches in models of chronic neurological disorders (Keller *et al.*, 2009, 2011; Swarup *et al.*, 2011).

1.2. The value of bioluminescence/biophotonic imaging in stroke research

Bioluminescence is a naturally occurring form of chemiluminescence where energy is released by a chemical reaction in the form of light emission. To date, Firefly luciferase systems have been and are widely used in the field of genetic engineering as reporter genes (for review, see Close *et al.*, 2011; Contag and Bachmann, 2002; Luker and Luker, 2010). In our experiment, we generated and used transgenic mice expressing dual reporters, the firefly luciferase (Fluc) and the green fluorescence reporter GFP, whose transcription is dependent upon the selected gene promoter (Gravel *et al.*, 2011; Lalancette-Hebert *et al.*, 2009). The Fluc bioluminescence is measured and quantified in living intact animals by using high sensitivity/high-resolution cooled, charged-coupled device (CCD) camera, whereas high-resolution fluorescence imaging can be done with microscopes equipped with two-photon laser scanning capabilities. These mice represent a powerful analytical tool for understanding *in vivo* pathology of acute and chronic neurological disorders and may be used for evaluation of pharmacokinetics and longitudinal responses to drug therapies. Here, we describe the procedures of generating biophotonic mouse models for live imaging of microglial activation/innate immune response and neuronal stress and recovery, followed by a detailed description of bioluminescence imaging protocols after the experimental stroke.

2. Design and Generation of Dual Reporter Mouse Models for *In Vivo* Imaging of Brain Response to Ischemic Injury

2.1. Mouse model for visualization of microglial activation and innate immune response: TLR2/luc/gfp mouse model

Activated microglial cells are the main effectors of the innate immune response in the brain following injuries including ischemia. Although the role of activated microglial cells in ischemia is not yet clear, the activation of microglial cells is characterized by a robust induction of the transmembrane receptor family of Toll-like receptor 2 (TLR2). To selectively image activated microglial cells from the brains of living animals, we generated a transgenic mouse model bearing the dual reporter system luciferase and green fluorescent protein under the transcriptional control of a murine TLR2 promoter (Lalancette-Hebert *et al.*, 2009). In this mouse model, transcriptional activation of TLR2 can be visualized in the brains of live animals using biophotonic/bioluminescence imaging and high-resolution CCD camera.

Plasmids and vectors

- Internal Ribosome Entry Site (IRES) vector (BD Biosciences, Mississauga, ON, Canada)
- Luciferase reporter gene (*luc2*) from pGL4 (Promega, Madison, WI, USA)
- *Aequorea coerulescens* Green Fluorescent Protein (AcGFP) (or any other fluorescent reporter protein of choice) gene from pAcGFP1 (BD Biosciences)

A fragment of 1548 base pair (bp) from the murine TLR2 promoter was amplified by polymerase chain reaction (PCR) from the 158H14 clone isolated from the RPCI-24 mouse (C57Bl/6J male) bacterial artificial chromosome (BAC) library under standard conditions by using the Expand HiFi PCR System (Roche, Mississauga, ON, Canada) with oligonucleotides corresponding to pGL3-1486 and pGL3 antisense (Musikacharoen, *et al.*, 2001). The PCR-amplified fragments were inserted into the pCR2.1 vector (TA cloning kit; Invitrogen, Burlington, ON, Canada) and completely sequenced to verify their integrity. The *Spe1/Pst1* promoter fragments of 1548 bp (TLR2/S) were inserted into the IRES vector (BD Biosciences), previously double digested with Spe1 and Pst1 to remove the immediate early promoter of cytomegalovirus ($PCMV_{IE}$). Both the Nhe1/Sal1 1.7 kb fragment corresponding to the luciferase reporter gene (*luc2*) from pGL4 (Promega) and the Sal1/Not1 0.7 kb fragment corresponding to the AcGFP reporter gene from pAcGFP1 (BD Biosciences) were inserted into the IRES recombinant vector. The integrity of the final construct designated IRES-TLR2-LUC2-AcGFP-p(A) was verified by sequencing. The TLR2 promoter-luciferase-AcGFP transgenes were isolated as an ~5.2 kb Spe1/Cla1 fragment of IRES-TLR2-LUC2-AcGFP-p(A) and microinjected into the male pronucleus of fertilized C57BL/6 oocytes.

2.2. Mouse model for real-time imaging of neuronal responses to brain injuries: GAP-43/luc/gfp mouse model

Plasmids and vectors

- IRES vector (BD Biosciences)
- Luciferase reporter gene (*luc2*) from pGL4 (Promega)

- AcGFP (or any other fluorescent reporter protein of choice) gene from pAcGFP1 (BD Biosciences)

Taking a similar approach as in generation of the TLR2 reporter mouse model (see Fig. 7.1), to study neuronal responses to injury and repair, longitudinally and in real time, we generated a transgenic mouse carrying reporter genes *luc* and *gfp* under the transcriptional control of the murine growth associated protein-43 (GAP-43), a neuron specific promoter. We selected the GAP-43 gene promoter as a good candidate because of the following characteristics: (i) In physiological conditions and in the adult brain, it is expressed at very low levels. (ii) It is highly upregulated following brain ischemia and in other types of neuronal injuries. (iii) After brain injury, its induction have been mostly restricted to injured neurons, an injury induced induction. Altogether, this mouse model represents a novel and valuable transgenic tool for *in vivo* bioluminescence and fluorescence imaging of neurite outgrowth in development and in the neuronal responses to injury and repair in adult nervous system (Gravel *et al.,* 2011).

The Nhe1/Sal1 1.7-kb fragment corresponding to the luciferase reporter gene (*luc2*) from pGL4 (Promega) and the Sal1/Not1 0.7-kb fragment corresponding to the AcGFP reporter from pAcGFP1 (BD Biosciences) were both inserted into the IRES vector (BD Biosciences). A 10 kb fragment containing the murine GAP-43 promoter (Zhu and Julien, 1999) was cloned into the IRES-LUC2-AcGFP recombinant vector. The integrity of the final construct was verified by sequencing. The GAP43-luciferase-GFP transgene was isolated as a Spe1/Cla1 fragment and microinjected into the male pronucleus of fertilized C57BL/6 oocytes. Transgenic mice were generated

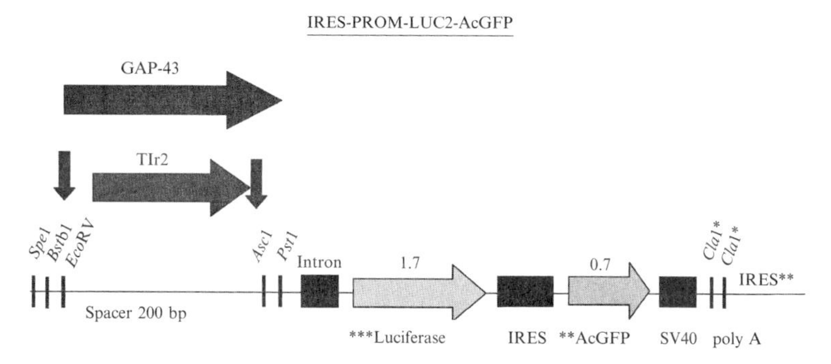

Figure 7.1 Schematic representation of the DNA constructs used to generate TLR2-Fluc-GFP and the GAP-43-luc/gfp transgenic mice. (For color version of this figure, the reader is referred to the Web version of this chapter.)

in the Transgenic and Knockout Facility of the Research Center of the Centre Hospitalier de l'Université Laval (CHUL).

2.3. Genotyping

Transgenic animals were genotyped by PCR detection of luciferase reporter gene with HotStar Taq Master mix Kit (Quiagen, Mississauga, ON, Canada) in 15 m*M* $MgCl_2$ PCR buffer with the following primers: (primer JK3: 5′-GGCGCAGTAGGCAAGGTGGT-3′ and JK4: 5′CAGCAGGATGCTCTCCAGTTC-3′). The PCR conditions were as follows: 95 °C–15 min, 30 cycles (94 °C–30 s, 65 °C–30 s, 72 °C–1 min, 72 °C–7 min.) (Cordeau *et al.*, 2008; Lalancette-Hebert *et al.*, 2009).

2.4. Advantages of multireporter approach in optical imaging: Why bicistronic constructs?

Majority of mouse models generated and used for bioluminescence imaging are designed by driving expression of reporter gene luciferase under the transcriptional control of the gene promoter of interest. Therefore, in these mouse models there is a delay between initial activation of the promoter and the expression of a sufficient amount of enzyme necessary for detection by *in vivo* visualization technique. Although *in vivo* bioluminescence imaging and the use of luciferases as reporter molecules in whole animal imaging has significant advantages including inherently low background, the light from luciferase enzymes cannot be detected by microscopy or flow cytometry. In addition, luciferase is not readily detectable by available antibodies using standard immunohistochemistry approach. In the majority of available transgenic mouse models, the luciferase reporter gene is induced and expressed in multiple tissues; however, without means for cell-type specific detection, it would be difficult to determine potential functional implications of the bioluminescent signal. For example, the TLR2 upregulation may not have the same functional significance in microglial cells as compared to neurons. It has been well established that the induction of the TLR2 in microglial cells is a marker of cellular activation, while in neurons the TLR2 induction has been limited to the subset of caspase-3 positive apoptotic cells. To overcome this problem in our reporter mouse models, we used bicistronic plasmids coexpressing luciferase together with GFP. As shown in the Fig. 7.1, the two reporter genes were connected via an IRES. Because the previous evidence (Martinez-Salas, 1999) suggests that the use of IRES may lead to lower expression of the coding sequence placed downstream of the IRES therefore in our bicistronic constructs the GFP has been used as a second reporter system. The great advantage of the dual reporter transgenic approach is that the luciferase light emission can be used for *in vivo* whole animal imaging while the parallel induction of the GFP reporter, in the same subset of cells, will greatly facilitate the cell-specific detection and will enable more detailed and meaningful signal analysis (see Fig. 7.2). Contrary to luciferase, the GFP is readily detectable by different types of microscopy and the signal can be further increased by the use of commercial anti-GFP monoclonal or polyclonal antibodies. Because in our transgenic models the GFP is induced in parallel with luciferase, these models can also be used for intravital multiphoton imaging. Once transgenic lines are generated, we strongly recommend validating transgenic lines in controlled conditions. Superficial analysis of the induced signal can lead to serious mistakes in the interpretation of the obtained results and/or creation of the artifacts. Before engaging in experimental imaging protocols, we strongly recommend:

- To investigate whether transgene expression follows the expression pattern of the endogenous protein in at least two known experimental models.
- To verify whether spatial and temporal dynamics of reporter gene (luciferase) induction follows the predicted pattern of gene promoter activation to a known and controlled stimuli.

Once transgenic mouse models have been properly validated they can be used in more complex *in vivo* imaging experiments. Here, we will describe in detail the protocols and procedures for induction of cerebral ischemia followed by *in vivo* bioluminescence imaging of neuroinflammation and neuronal stress/recovery after experimental stroke.

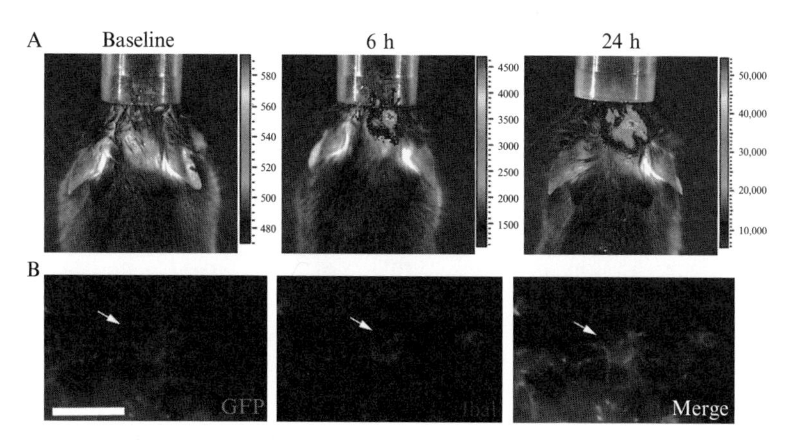

Figure 7.2 The TLR2-luc/gfp mouse is a bicistronic reporter system in (A) the TLR2 induction after LPS injection in the brain of a male mouse at baseline, 6 and 24 h is measured using the luciferase reporter. While in (B) to achieve microscope resolution and to identify the induction of the transgene in microglial cells using the microglia specific Iba1 marker the GFP reporter was used. Note colocalization of the Iba1 with the TLR2 driven transgene GFP. Scale bar = 25 μm. (See Color Insert.)

3. Induction of Cerebral Ischemia

3.1. Material/equipment

Precoated filaments for artery occlusion

- Silicon rubber-coated monofilament for MCAO model. Filament size 6–0, diameter 0.09–0.11 mm, length 20 mm; diameter with coating 0.23 ± 0.02 mm; coating length ≥ 5 mm, Doccol Corporation, www.doccol.com.

Surgical threat

- 4–0, 1.5 Metric SOFSILK (waxcoated, braided silk nonabsorbable sutures), Tyco Healthcare Group, LP Norwalk, CT, USA

- Sterilized microsurgery set
- Operating microscope

3.2. Surgical procedures: Experimental ischemia

Unilateral transient focal cerebral ischemia was induced by intraluminal filament occlusion of the left middle cerebral artery (MCA) during 1 h followed by a reperfusion period (Beaulieu *et al.*, 2002; Belayev *et al.*, 1999; Lalancette-Hebert *et al.*, 2007). The middle cerebral artery occlusion (MCAO) is carried out on 2–3-month-old transgenic or wild-type (WT) male (20–25 g). The animals are normally anesthetized with 2% isoflurane in 100% oxygen at a flow rate of 2 l/min. To avoid cooling, the body temperature has to be regularly checked and maintained at 37 °C with a heating pad. Under an operating microscope, the left common carotid artery and ipsilateral external carotid artery (ECA) are exposed through a midline neck incision and are carefully isolated from surrounding tissue. The internal carotid artery (ICA) is than isolated and carefully separated from the adjacent tissue so monofilament suture can inserted via the proximal ECA into the ICA and then into the circle of Willis, thus occluding the MCA. To induce moderate ischemic brain lesion, the MCA was occluded during 1 h. Depending on the type of imaging protocol, the occlusion period is then followed by various reperfusion times, starting from 24 to 72 h to several months after initial ischemic injury. After surgery, to avoid cooling for the first 72 h the body temperature has to be regularly checked and maintained at 37 °C with a heating pad. As previously described (Weng and Kriz, 2007), to confirm successful MCAO at 6 and 24h following surgery, the animals were examined for early neurological deficits. Only the mice expressing "positive phenotypes" including circling behavior, slight motor deficits of contralateral front paw and reduced spontaneous activity were selected for the study. The induction of early TLR2 biophotonic signals can be used as an additional *in vivo* imaging control.

Comment: In our laboratory, we use the MCA occlusion/reperfusion model. However, described transgenic mice and *in vivo* imaging protocols can be applied to several other experimental models of ischemia and several other types of acute and more chronic brain injuries.

4. *In Vivo* Bioluminescence Imaging

Bioluminescence is a naturally occurring form of chemiluminescence where energy is released by a chemical reaction in the form of light emission (Fig. 7.3). The luciferase used in our studies belongs to a family of light producing enzymes called Firefly luciferase. In our experiments, we used a novel generation of reporters, a pGL4 luciferase reporter vector. These new generation synthetic enzymes are optimized for better expression in mammalian cells, reduced background and due to a removal of cryptic DNA regulatory elements a reduced risk of anomalous transcription. As shown in Fig. 7.3, these enzymes can generate visible light in the presence of enzyme-specific substrate (in case of Firefly luciferase it is D-luciferin), oxygen, and ATP as a source of energy. In this chemical reaction, D-luciferin is converted into oxyluciferin and part of the chemical energy is converted into visible light. Unlike insects, certain bacteria and some marine organisms, genetically modified mammals do not produce luciferin. Therefore, before every imaging session, the substrate has to be injected into mice. D-luciferin is

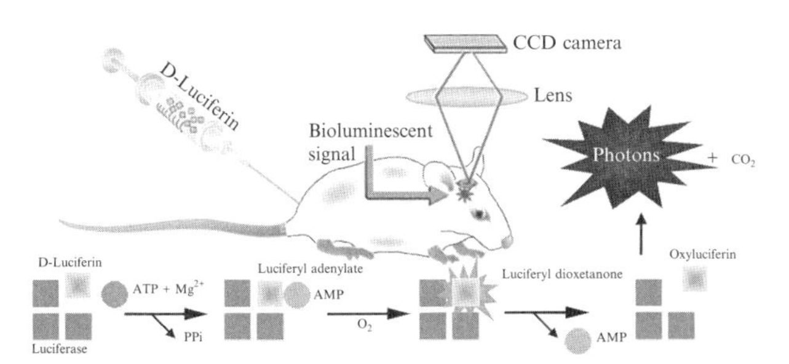

Figure 7.3 Schematic presentation of the bioluminescent source detection from the brain of living mice. In the presence of ATP and magnesium, the luciferin will bind to the luciferase. Afterward, the release of pyrophosphate (PPi) after the creation of AMP will generate luciferyl adenylate. This adenylate will be oxidized to an unstable molecular form after several intermediate stages. To reach molecular stability, a photon of light will be release and captured by the CCD camera. (For color version of this figure, the reader is referred to the Web version of this chapter.)

nontoxic, and after intraperitoneal (i.p.) injection, it is distributed in all tissues including brain. However, when considering steady-state bioluminescent imaging from the brain structures, to avoid artifacts, it is very important to obtain stable concentration of the substrate D-luciferin in the brain. From our experience and in our experimental model, it takes between 18 and 20 min after i.p. injection to obtain steady-state bioluminescent signal from the brains of stroked mice. Because these values can be system and model dependent, we provide here the detailed description of experimental procedures and some tips for successful and valid bioluminescence imaging.

4.1. Preparation of the material: Luciferin solution for *in vivo* imaging

D-luciferin is a small molecule generated in bioluminescent organisms. It diffuses easily across the membranes including the blood brain barrier, the blood placenta barrier, and the blood testis barrier. Further, because of its size, the activation of the immune system seems unlikely. Here, it is important to mention that D-luciferin is a molecule sensitive to light, to oxygen, and in its powder form to moisture, thus minimizing the exposure of the powder and solution to direct light is important. For these reasons, we recommend to work with amber eppendorfs and/or before experiments to wrap a regular transparent eppendorf with aluminum foil to keep the luciferin stable. Further, to keep the quality and intensities of the bioluminescent signals stable it is recommended to prepare a fresh solution of luciferin before every experiment. However, in our hands aliquots stored at −80 °C have been proven to work really well. They can be kept for at least 6 months without changing the effectiveness. To avoid repeated cycles of freezing and thawing, small aliquots of 1 ml are kept at −80 °C. After thawing, the luciferin solution is kept at 4 °C and used within following day or two.

Material needed to prepare the luciferin solution:

- Vial of D-luciferin firefly potassium salt.
- D-phosphate buffered saline (PBS) (with no Mg^{2+} and Ca^{+}).
- Syringe filter, 0.2 μm.
- Syringe
- Dark eppendorf

Procedure: Dissolve the D-luciferin in the D-PBS making a stock solution at a final concentration of 15 mg/ml. For example, dissolve 1 g of D-luciferin in 66.6 ml of D-PBS. Filter the solution with the syringe filter and divide the stock solution in small aliquots (recommended volume is 1 ml) and store at −80 °C, in dark eppendorf until use. Before imaging, thaw the aliquots, on ice or on the bench (both seem to work equally well), and inject 10 μl/g of body weight i.p. A 20 g mouse will receive 200 μl of solution. In general, an adult mouse can receive i.p. up to 1 ml of solution without any problem. A recommended dose is 150 mg/kg. To avoid injuries, the luciferin should be injected (i.p.) holding the animal abdomen side up, head pointing downward. This will force the moveable organs up the animal cavity, toward the diaphragm, minimizing the risk of injecting or puncturing an organ. In our experiments, the animals are injected in the lower left abdominal quadrant using a 1 cc syringe. Avoiding injury and minimizing stress is very important element, especially in long-term imaging experiment where same animals are imaged (and injected) in repetitive cycles. In our long-term experiments (3–4 months), we normally image the same animal once per week. To date, although we performed several long-term imaging studies (Gravel *et al.*, 2011; Keller *et al.*, 2009, 2011; Lalancette-Hebert *et al.*, 2009), we did not observed any luciferin induced toxicity.

4.2. Preparation of the imaging session

In bioluminescence imaging, planning is a key to success. Before every imaging session, especially when dealing with more than one group of experimental animals, we suggest to prepare a log book containing the following information:

- mice identification number and gender
- time of the substrate D-luciferin injection
- time of induction of anesthesia
- the acquisition time

Why all these elements are important? First, some of the tissue responses such as neuroinflammation after stroke are gender and estrogen dependent (Cordeau *et al.*, 2008). Next, when imaging responses from the brain, it takes between 15 and 20 min after i.p. injection for the luciferin to cross the blood brain barrier and it usually takes approximately 5 min to anesthetize one mouse. It is also important to note that programming an acquisition time of 2 min will take a little bit more than 2 min. For example, with the IVIS 200, if you take a 2 min bioluminescence picture at field of view A (4 × 4 cm, a high-resolution field), it will take approximately 150 s before you can open the door of the imaging system and place another experimental animal. Because in certain experiments the peak luciferase expression time used in steady-state bioluminescence imaging may be relatively short (5–10 min), all of these parameters are extremely important for timely, valid, and efficient *in vivo* bioluminescence imaging protocols.

4.3. Relevant control groups

- Identifying luciferin kinetic curve and peak luciferase expression time
- Establishing a baseline curve

To perform quality steady-state *in vivo* bioluminescence imaging for each new transgenic reporter mice and novel experimental paradigm, it is necessary to establish the luciferin kinetic curve and the peak luciferase expression time. For example, when imaging microglial activation/innate immune response after stroke in TLR2 reporter mice (Lalancette-Hebert *et al.*, 2009), the luciferin kinetic curve and luciferase peak expression time will not be the same if we use different stroke models, that is, permanent MCA occlusion versus transient MCA occlusion. Further, these important parameters would change if we image animals after stroke using different reporter mouse lines such as GFAP-luc and/or GAP-43-luc/gfp (Cordeau *et al.*, 2008; Gravel *et al.*, 2011). Therefore, we recommend establishing the luciferin kinetic curve and peak luciferase expression time before every experiment. Namely, before introduction to any new experimental paradigm at least 3–5 transgenic reporter mice should be injected with luciferin and imaged every 5 min for 40–45 min. Once the imaging data is collected and properly analyzed it will be easy to calculate the kinetics and the peak luciferase expression and estimate the duration of the steady-state that will be used in imaging experiments.

In addition to luciferin kinetic controls, for each transgenic mouse model, it is important to analyze the level of baseline reporter gene expression. Depending on the selected gene promoter, there is a possibility that even in control conditions (without stroke) we can observe a baseline bioluminescent signal. For that reason in every experiment employing *in vivo* bioluminescence imaging, it is important to establish a baseline curve and compare all the obtained imaging results to appropriate controls.

4.4. Protocol for 2D *in vivo* bioluminescence imaging

In our imaging protocols, we use IVIS® 200 Imaging System (CaliperLS, Alameda, CA, USA); however, the sequence of events described here can be applied to any of the imaging systems adapted for whole small animals *in vivo* bioluminescence imaging. As previously described (Cordeau, *et al.*, 2008), 25 min prior to imaging session, the mice will receive i.p. injection of the luciferase substrate D-luciferin (150 mg/kg—for mice between 20 and 25 g, 150–187.5 μl of a solution of 20 mg/ml of D-luciferin dissolved in 0.9% saline was injected) (CaliperLS). In our experiment, we always use D-luciferin from the same provider (*note*: although the source of luciferin is decided by the end user, once experiments are initiated, we would not recommend changing the source). 10–15 min after luciferin injection the mice were anesthetized with 2% isoflurane in 100% oxygen at a flow rate of 2 l/min and placed in the heated, light-tight imaging chamber and maintained anesthetized by constant delivery of 2% isoflurane/oxygen mixture at 1 l/min through an IVIS® anesthesia manifold (the anesthesia manifold and adaptors are usually system dependent but equally efficient). Images are collected using high sensitivity (CCD) camera with wavelengths ranging from 300 to 660 nm. Exposition time for imaging is usually 1–2 min using different field of views and f/1 lens aperture. We recommend initiating the protocol with default settings for *in vivo* bioluminescence, which are usually 1 min/medium binning. The first bioluminescent images and measurements will normally provide us with initial information about the signal intensities and these parameters will be then adjusted accordingly. For imaging of bioluminescence signals from brain, we normally use higher resolution/ sensitivity, the 4 × 4 cm field of view. Focusing on the head provides more detailed images of the signal arising from the brain than imaging a whole mouse (Cordeau *et al.*, 2008; Gravel *et al.*, 2011; Lalancette-Hebert *et al.*, 2009). As previously described, bioluminescence emission is normalized and displayed in physical units of surface radiance, photons per second per centimeter squared per steradian (photons/s/cm^2/sr), as is common in bioluminescence imaging (Kadurugamuwa *et al.*, 2005; Zhu *et al.*, 2004). The light output is quantified by determining the total number of photons emitted per second using the Living Image 2.5 acquisition and imaging software (CaliperLS). Region of interest (ROI) measurements on the images are used to convert surface radiance (photons/s/cm^2/sr) to source flux or total flux of photons expressed in photons/seconds (p/s). *Note*: To standardize your data acquisition, it is very important to use the same ROI measurements over time. In biolumienscence imaging, the data are usually represented as pseudo-color images indicating light intensity (red and yellow, most intense, or any other color-code), and overlaid over grayscale reference photographs. After stroke, we usually recommend to image/analyze acute phase (up to 72 h after initial stroke), followed by the later stages of tissue response to ischemic injury. The great advantage of *in vivo* bioluminescence and in general optical imaging is that the same animal could be imaged over time. Moreover, using this approach we can visualize the spatial and temporal dynamics of the different elements of the brain response to ischemic injury.

4.5. Protocol for 3D reconstruction of bioluminescent sources

While bioluminescence imaging is a powerful tool for small animal imaging studies, one should be aware of some limitation using this approach. We already discussed here the difficulties of luciferase detection at cellular levels. Another potential problem is accurate detection of the signal from deeper brain structures. Namely, the bioluminescence images are typically two-dimensional (2D) and because the light is attenuated approximately 10-fold/cm of tissue, the light from superficial sources would be detected in greater extent than the light from deeper tissue structures (Contag and Bachmann, 2002; Luker and Luker, 2010). In stroke imaging this may affect the signals arising from cortical infarction versus striatum. To overcome this problem in our imaging experiments, we use a protocol for 3D representation of our signals. The technique is called the diffuse luminescent imaging tomography (DLIT). As previously described (Cordeau *et al.*, 2008), for the acquisition of 3D images, we acquired grayscale photographs and structured light images followed by series of bioluminescent images using different wavelengths (560–660 nm). The 3D reconstruction of bioluminescent sources in the brain is normally accomplished by using DLIT algorithms (Living Image 3D Analysis Software, CaliperLS).

Diffuse tomography imaging is a technique that analyzes the light generated and projected on the surface of the mice to recreate a 3D reconstruction of the zone generating that same light. To generate the 3D image, a structured light image of the mouse has to be taken. The structured light image consists of a series of parallel lasers lines projected on the mouse to determine the surface of the mouse. Afterward the software will analyze the displacement of the laser light to recreate the 3D mouse. This displacement is used to determine the surface topography of the mouse using polygons. Therefore, to remove any artifacts that could be generated by the software, we recommend to comb or to shave the mice before structure light images. After the 3D surface of the mouse is created, the software can now localize and quantify the light source in the mouse. To achieve this, minimum of two emissions filter (560–660 nm) is needed to localize the light source. Using the different wavelengths, the enzyme with the defined emission spectra (in our case genetically modified Firefly luciferase) and the known tissue properties (bones, muscles, brain, etc.) the software can calculate and map the surface radiance to the photons density for every wavelength. The next step is to divide the interior of the mouse in voxels and to attribute to each voxel the right source strength to generate the photon density measured at each surface element of the mouse (Contag and Bachmann, 2002). In our previous work and as shown here in Fig. 7.4, we tested several protocols for 3D bioluminescence imaging. Our analysis was further confirmed by immunohistochemistry measurements and revealed that the 3D reconstruction using DLIT algorithms is rather accurate methodology (Cordeau *et al.*, 2008).

5. Additional Tips for Successful *In Vivo* Bioluminescence Imaging

5.1. Impact of mouse fur color

Here, it is important to mention that mouse fur color will affect the capacity of the signal generated by the luciferase chemical reaction with the luciferin to reach the camera. While hemoglobin absorbs the wavelengths below 600 nm, the fraction of light above 600 nm can be blocked by black fur. The pigment that gives the fur its color, the melanin, will absorb light and scatter the signal. Because the majority of the transgenic and reporter animals used

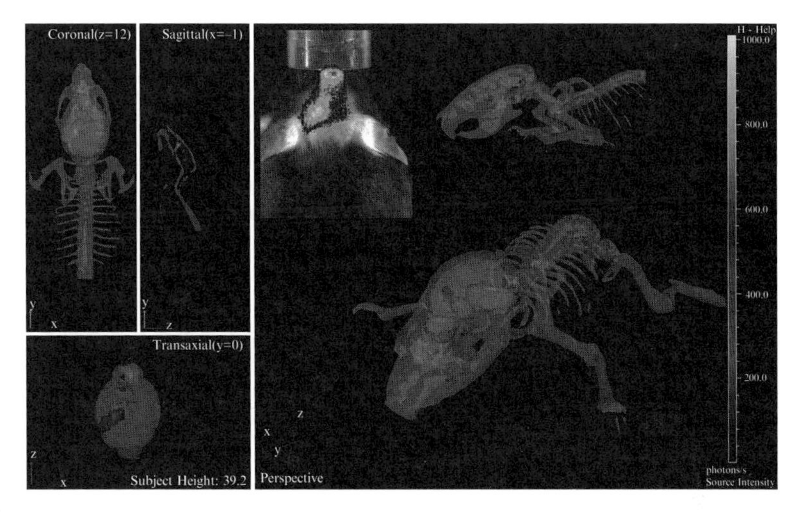

Figure 7.4 Three-dimensional reconstruction of bioluminescent signals emitted from the brain of a 3-month-old TLR2-luc/gfp mouse 24 h after MCAO. Reconstruction is build using three different wavelengths (600, 620, and 640 nm) across the emission spectrum of the bioluminescent source (Living Image software, Xenogen). Green area, concentrated in the ischemic lesion, represents areas of the brain with the highest photons emission thus the biggest TLR2 activation. Scales bar on the right are the color maps for source intensity in photons per second. Insert image was taken with the A field of view with a 2 min exposition prior to the 3D sequence image. (See Color Insert.)

in neuroscience/brain imaging are generated in BL/6 background this may represent a potential problem. To alleviate this problem before imaging session, we normally shave the ROI. When removing the fur, precautions must be taken to not harm the skin of the mouse, since cuts or local inflammation will affect the bioluminescence acquisition or emission when working with a mouse model that respond to inflammation. When shaved, over time, some C57BL/6 mice will develop dark skin pigmentation that if covering the ROI can create problems. For that reason, it may be advantageous to derive your transgenic lines in C57BL/6 albino background or if possible use any other strain with light fur color.

5.2. Covering ectopic and/or saturated signals

When imaging, it is important to keep in mind that some transgenic mice may have ectopic and/or constitutively high transgene (luciferase) expression at the areas potentially interfering with your ROI. For example, the GFAP-luc mouse expresses a high level of luciferase in ears. Because during generation of bioluminescent images, the majority of the imaging software will generate a default scales based on the highest recorded signal if your expected signals are lower than the baseline transgene signal close to your ROI's you may not observe it. Therefore, it is useful to prepare a piece of black paper to cover the area on the mouse that expresses really high baseline level of the signal. In case of GFAP-luc mouse, if you cover the ears with the black paper you will be able to image a more subtle induction of the signals from the brain area (Keller *et al.,* 2009). Importantly, this procedure has no impact on the amount of light emitted in the non-covered area, it's mainly cosmetic.

5.3. Imaging necrotic and hypoxic tissue

Luciferase needs oxygen to complete its reaction, that's why the use of luciferases technology is limited to aerobic environments and living organisms. In an environment where the oxygen levels are really low, the light produced may not correlate with the luciferase expression. It is a thing to keep in mind if you are working with a model that implicates large necrosis and/or hypoxia.

6. Summary

Unlike many current animal studies that are based on single end point data, the presented reporter mice together with the use of *in vivo* biophotnic/bioluminescence imaging represents novel, powerful, analytical tools for understanding *in vivo* pathology of acute and chronic neurological injuries. In addition, use of *in vivo* imaging in stroke research may significantly improve the quality of preclinical drug testing and may lead to more efficient translation of experimental therapies to clinic.

ACKNOWLEDGMENTS

This work was supported by the Canadian Institutes of Health Research (CIHR) and the Heart and Stroke Foundation–Canada. J. K. is a recipient of the Career Award from the R&D and CIHR. P. C. received CIHR Canada Doctoral Scholarships.

REFERENCES

Beaulieu, J. M., Kriz, J., and Julien, J. P. (2002). Induction of peripherin expression in subsets of brain neurons after lesion injury or cerebral ischemia. *Brain Res.* **946,** 153–161.

Belayev, L., Busto, R., Zhao, W., Fernandez, G., and Ginsberg, M. D. (1999). Middle cerebral artery occlusion in the mouse by intraluminal suture coated with poly-L-lysine: Neurological and histological validation. *Brain Res.* **833,** 181–190.

Close, D. M., Hahn, R. E., Patterson, S. S., Baek, S. J., Ripp, S. A., and Sayler, G. S. (2011). Comparison of human optimized bacterial luciferase, firefly luciferase, and green fluorescent protein for continuous imaging of cell culture and animal models. *J. Biomed. Opt.* **16,** 047003.

Contag, C. H., and Bachmann, M. H. (2002). Advances in *in vivo* bioluminescence imaging of gene expression. *Annu. Rev. Biomed. Eng.* **4,** 235–260.

Cordeau, P., Jr., Lalancette-Hebert, M., Weng, Y. C., and Kriz, J. (2008). Live imaging of neuroinflammation reveals sex and estrogen effects on astrocyte response to ischemic injury. *Stroke* **39,** 935–942.

Dirnagl, U., Iadecola, C., and Moskowitz, M. A. (1999). Pathobiology of ischaemic stroke: An integrated view. *Trends Neurosci.* **22,** 391–397.

Gravel, M., Weng, Y. C., and Kriz, J. (2011). Model system for live imaging of neuronalresponses to injury and repair. *Mol. Imaging.* April 27, (Epub ahead of print).

Kadurugamuwa, J. L., Modi, K., Coquoz, O., Rice, B., Smith, S., Contag, P. R., and Purchio, T. (2005). Reduction of astrogliosis by early treatment of pneumococcal meningitis measured by simultaneous imaging, *in vivo*, of the pathogen and host response. *Infect. Immun.* **73,** 7836–7843.

Keller, A. F., Gravel, M., and Kriz, J. (2009). Live imaging of amyotrophic lateral sclerosis pathogenesis: Disease onset is characterized by marked induction of GFAP in Schwann cells. *Glia* **57,** 1130–1142.

Keller, A. F., Gravel, M., and Kriz, J. (2011). Treatment with minocycline after disease onset alters astrocyte reactivity and increases microgliosis in SOD1 mutant mice. *Exp. Neurol.* **228,** 69–79.

Lalancette-Hebert, M., Gowing, G., Simard, A., Weng, Y. C., and Kriz, J. (2007). Selective ablation of proliferating microglial cells exacerbates ischemic injury in the brain. *J. Neurosci.* **27,** 2596–2605.

Lalancette-Hebert, M., Phaneuf, D., Soucy, G., Weng, Y. C., and Kriz, J. (2009). Live imaging of Toll-like receptor 2 response in cerebral ischaemia reveals a role of olfactory bulb microglia as modulators of inflammation. *Brain* **132,** 940–954.

Lalancette-Hebert, M., Moquin, A., Choi, A. O., Kriz, J., and Maysinger, D. (2010). Lipopolysaccharide-QD micelles induce marked induction of TLR2 and lipid droplet accumulation in olfactory bulb microglia. *Mol. Pharm.* **7,** 1183–1194.

Lo, E. H., Dalkara, T., and Moskowitz, M. A. (2003). Mechanisms, challenges and opportunities in stroke. *Nat. Rev. Neurosci.* **4,** 399–415.

Luker, K. E., and Luker, G. D. (2010). Bioluminescence imaging of reporter mice for studies of infection and inflammation. *Antiviral Res.* **86,** 93–100.

Martinez-Salas, E. (1999). Internal ribosome entry site biology and its use in expression vectors. *Curr. Opin. Biotechnol.* **10,** 458–464.

Maysinger, D., Behrendt, M., Lalancette-Hebert, M., and Kriz, J. (2007). Real-time imaging of astrocyte response to quantum dots: *In vivo* screening model system for biocompatibility of nanoparticles. *Nano Lett.* **7,** 2513–2520.

Musikacharoen, T., Matsuguchi, T., Kikuchi, T., and Yoshikai, Y. (2001). NF-kappa B and STAT5 play important roles in the regulation of mouse toll-like receptor 2 gene expression. *J. Immunol.* **166,** 4516–4524.

Swarup, V., Phaneuf, D., Bareil, C., Robertson, J., Kriz, J., and Julien, J. P. (2011). Pathological hallmarks of amyotrophic lateral sclerosis/frontotemporal lobar degeneration in transgenic mice produced with genomic fragments encoding wild-type or mutant forms of human transactive response DNA-binding protein 43. *Brain* **134,** 2610–2626.

Weng, Y. C., and Kriz, J. (2007). Differential neuroprotective effects of a minocycline-based drug cocktail in transient and permanent focal cerebral ischemia. *Exp. Neurol.* **204,** 433–442.

Zhu, Q., and Julien, J. P. (1999). A key role for GAP-43 in the retinotectal topographic organization. *Exp. Neurol.* **155,** 228–242.

Zhu, L., Ramboz, S., Hewitt, D., Boring, L., Grass, D. S., and Purchio, A. F. (2004). Non-invasive imaging of GFAP expression after neuronal damage in mice. *Neurosci. Lett.* **367,** 210–212.

CHAPTER EIGHT

In Vivo Real-Time Visualization of Leukocytes and Intracellular Hydrogen Peroxide Levels During a Zebrafish Acute Inflammation Assay

Luke Pase,[*,†,‡] Cameron J. Nowell,[§] *and* Graham J. Lieschke[*,†,‡]

Contents

1. Introduction 136
2. Visualizing Leukocyte Behavior During Acute Inflammation in Zebrafish 137
   2.1. Inducing an acute inflammatory response in zebrafish 139
3. Things to Consider When Planning *In Vivo* Visualization of the Inflammatory Response 143
   3.1. General considerations for multichannel fluorescence imaging 143
   3.2. Specific considerations for determining changes in H_2O_2 levels using HyPer 148
Acknowledgments 154
References 154

Abstract

Following injury, the inflammatory response directs the host immune cells to the wound to maintain tissue integrity and combat pathogens. The recruitment of immune cells to inflammatory sites is achieved through the establishment of a variety of signal gradients. Using a zebrafish embryo injury model, it was recently demonstrated that, upon injury, cells at the wound margin rapidly produce hydrogen peroxide (H_2O_2) which serves as an early paracrine signal to leukocytes. This chapter provides a method for performing *in vivo* time-lapse fluorescence microscopy to visualize leukocyte behaviors and wound-produced H_2O_2 simultaneously in single zebrafish embryos during an acute inflammatory

[*] Australian Regenerative Medicine Institute, Monash University, Clayton, Victoria, Australia
[†] Cancer and Haematology Division, Walter and Eliza Hall Institute of Medical Research, Parkville, Victoria, Australia
[‡] Department of Medical Biology, University of Melbourne, Parkville, Victoria, Australia
[§] Centre for Advanced Microscopy, Ludwig Institute for Cancer Research Melbourne-Parkville Branch, Parkville, Victoria, Australia

Methods in Enzymology, Volume 506
ISSN 0076-6879, DOI: 10.1016/B978-0-12-391856-7.00032-9

response. Protocols are included for inducing a robust, reproducible acute inflammatory response, for rapidly mounting immobilized embryos for time-lapse imaging, and for computing ratiometric data from the images of embryos expressing the genetically encoded H_2O_2 sensor fluorophore HyPer. General issues to consider when designing multichannel fluorescent imaging are discussed, including particular considerations to note when monitoring intracellular H_2O_2 concentration dynamics using HyPer.

1. Introduction

The inflammatory response is a multicellular process coordinated by many factors in order to protect a host from infectious agents, maintain tissue integrity, and promote healing of damaged tissue. The many triggers of the inflammatory response include exogenous triggers such as microorganisms and endogenous triggers such as cell death or damage. These triggers release various chemical inducers of inflammation such as microbial products (e.g., lipopolysaccharide (LPS)) or intracellular compounds (e.g., adenosine triphosphate (ATP)) that may be detected by cells of the surrounding tissues through various sensing mechanisms including specific receptors. Once the initial signal is sensed, this information is translated into inflammatory mediators (e.g., chemokines, cytokines, vasoactive amines, and hydrogen peroxide (H_2O_2)) which play a role in sensitizing pain receptors, dilating local vessels, and importantly promote the recruitment of innate immune cells (Medzhitov, 2008; Thelen and Stein, 2008).

The first innate immune cells to migrate to a site of inflammation are neutrophils and macrophages. Both of these leukocytes are phagocytes, which play a critical role in containing, killing, and removing microbial threats. Macrophages are also particularly important for effective wound healing and clearance of cell corpses (Martin and Leibovich, 2005). Failure to initiate an inflammatory response and recruit leukocytes risks uncontrolled proliferation of invading microorganisms and severe tissue damage that may be fatal to the host. Conversely, an over-exuberant inflammatory response or a failure to resolve an immune response can also cause severe tissue damage and may lead to chronic inflammation, which is also detrimental to the host. Therefore, the inflammatory response is only beneficial to the host when its intensity is balanced in proportion to the inflammatory trigger and is adequately resolved.

To understand the inflammatory response in its entirety, it is necessary to study it *in vivo*. The zebrafish is an excellent model for studying the inflammatory response *in vivo* and in real time (Ellett and Lieschke, 2010). The optical transparency and small size of zebrafish embryos and larvae and their genetic tractability combine to permit fluorescence-based methods for simultaneous visualization of leukocyte behaviors and inflammatory mediators. Several transgenic lines with fluorescently marked leukocytes are available (Table 8.1) and the genetically encoded H_2O_2 sensor fluorophore HyPer has recently been employed to follow tissue-scale H_2O_2 concentration dynamics in living zebrafish embryos (Belousov *et al.*, 2006; Niethammer *et al.*, 2009). It was demonstrated that immediately following wounding, epithelial cells rapidly produce H_2O_2 catalyzed by dual oxidase (Duox) with the resulting tissue-scale gradient of H_2O_2 serving as an early paracrine signal to leukocytes (Niethammer *et al.*, 2009).

In this chapter, we discuss methods and highlight considerations for *in vivo* real-time imaging in zebrafish embryos of two dynamic components of the inflammatory response: the generation of the leukocyte chemoattractant H_2O_2 and leukocyte behaviors. Firstly, a method is detailed for generating a reliable and reproducible trigger of an acute inflammatory response and for immobilizing embryos for time-lapse imaging. Secondly, several important considerations for optimal *in vivo* real-time imaging and detecting changes in intracellular H_2O_2 concentration using HyPer fluorescence are outlined. A detailed method and MetaMorph journal for generating HyPer ratiometric images specific to HyPer-expressing tissue is provided.

2. Visualizing Leukocyte Behavior During Acute Inflammation in Zebrafish

Zebrafish embryos have a number of anatomical features that make them an excellent model for visualizing cellular behaviors in real time and *in vivo*. First and foremost, zebrafish embryos are optically transparent, making them highly amenable to visual light microscopy. Secondly, their small size allows for them to be easily mounted and anesthetized for extended periods of time as they maintain sufficient oxygen supply by diffusion and the nutrients are provided from the yolk sac. Lastly, their fin is an excellent anatomical location for visualizing migrating leukocytes during acute inflammation as it is only a few cells thick, hence the majority of the tissue can be captured in focus in single or just several Z sections. With these inherent features, migratory cells can be easily distinguished during the inflammatory response using just transmitted light (Herbomel *et al.*, 1999; Redd *et al.*, 2006).

Although time-lapse imaging of leukocytes under brightfield illumination has some advantages (e.g., minimal photodamage to the specimen), using brightfield conditions alone constrains the information that can be obtained: it can be difficult to identify leukocyte types unequivocally; the cellular definition of a leukocyte can be lost within the irregular cellular morphology of the wound edge; cell tracking can only be accurately performed manually. To overcome these limitations, fluorescence microscopy using transgenic zebrafish expressing fluorophores that distinguish

Table 8.1 Selected transgenic zebrafish lines that mark hematopoietic lineages

Hematopoietic cell marked by transgene	Gene promoter	Transgenes	References
Neutrophils	*mpx*	Tg(*mpx*:EGFP)^{i114} Tg(*mpx*:EGFP)^{uwm1} Tg(*mpx*:mCherry)^{uwm7} Tg(*mpx*:Dendra2)^{uwm4}	Mathias *et al.* (2006), Renshaw *et al.* (2006), Walters *et al.* (2010), and Yoo and Huttenlocher (2011)
	lyz	Tg(*lyz*:EGFP)^{nz50} Tg(*lyz*:DsRED)^{nz117} Tg(*lyz*:EGFP)^{rj1} Tg(*lyz*:EGFP)^{ko2} Tg(*lyz*:EGFP)^{ko3} Tg(*lyz*:EGFP)^{ko4}	Hall *et al.* (2007), Kitaguchi *et al.* (2009), and Zhang *et al.* (2008)
	myc enhancer	Et(CLG:YFP)^{smb383} Et(CLG:YFP)^{smb463}	Meijer *et al.* (2008)
Macrophages	*mpeg1*	Tg(*mpeg1*:GAL4) Tg(*mpeg1*:EGFP) Tg(*mpeg1*:mCherry)	Ellett *et al.* (2011)
	fms	Tg(*fms*:GAL4VP16)	Gray *et al.* (2011)
Leukocytes	*myd88*	Tg(*myd88*:EGFP)^{zf163} Tg(*myd88*:DsRED)^{zf164}	Hall *et al.* (2009)
Myeloid progenitors Primitive macrophages	*spi1*	Tg(*spi1*:EGFP)^{gl21} Tg(*spi1*:EGFP)^{zdf11} Tg(*spi1*:GAL4)^{zf149}	Hsu *et al.* (2004), Peri and Nusslein-Volhard (2008), and Ward *et al.* (2003)
	fli1a	Tg(*fli1a*:EGFP)^{y1}	Lawson and Weinstein (2002)
Eosinophils	*gata2*	Tg(*gata2*:EGFP)^{zf35} Tg(*gata2*:EGFP)^{la3}	Balla *et al.* (2010) and Traver *et al.* (2003)
Antigen-presenting mononuclear phagocytes	*mhc2*	Tg(*mhcdab*:GFP)^{sd6} Tg(*mhcdab*:mCherry)^{sd7}	Wittamer *et al.* (2011)
Thrombocytes	*itga2b*	Tg(*itga2b*:EGFP)^{la2}	Lin *et al.* (2005)
Erythrocytes	*gata1*	Tg(*gata1*:EGFP)^{sd2}	Long *et al.* (1997)

different leukocyte types can be used. Some currently available transgenic lines that mark different hematopoietic lineages are listed in Table 8.1.

Following is a method for visualizing leukocyte behavior during an acute inflammation model in a 3 days postfertilization (dpf) zebrafish embryos. 3 dpf embryos combine several useful features: they have an ample number of leukocytes; they are still amenable to manipulation by transient reverse genetic techniques (including protein overexpression from mRNA injection and gene knockdown from antisense morpholino oligonucleotide injection); their pigmentation from melanization can be suppressed by 1-phenyl-2-thiourea (PTU) treatment.

2.1. Inducing an acute inflammatory response in zebrafish

There are multiple environmental and cellular triggers that can initiate the inflammatory response in zebrafish including mechanical injury (Hall *et al.*, 2007; Lieschke *et al.*, 2001; Renshaw *et al.*, 2006), chemical insult, cell death (d'Alencon *et al.*, 2010; Olivari *et al.*, 2008), and the presence of transformed cancer cells (Feng *et al.*, 2010). Here, the induction of an acute inflammatory response by transection of the tip of the caudal fin is described in detail (Fig. 8.1). Caudal fin injury is preferred for the following reasons: (1) it

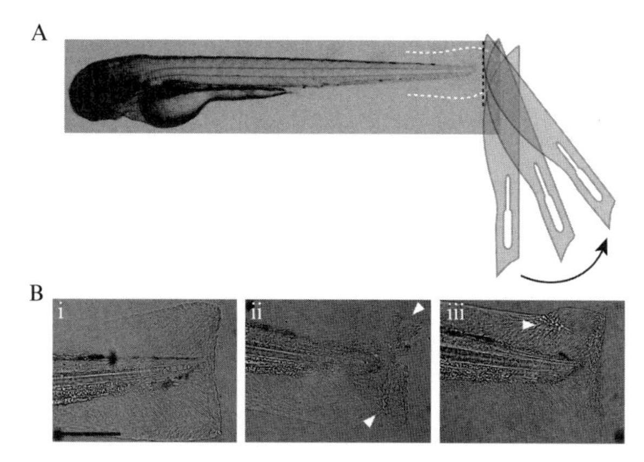

Figure 8.1 Tail fin transection—a reproducible mechanical injury for inducing an acute inflammatory response. (A) Dissecting microscope brightfield image of a 3 days postfertilization embryo and a schematic of a curved scalpel blade demonstrating the anatomical location of amputation (black dashed line). Arrow indicates the motion of the scalpel blade while cutting the caudal fin. Outline of the caudal fin (white dashed line). (B) Transmitted light images of transected caudal fins. (i) A representative embryo with an ideal injury that can be easily reproduced across multiple experiments. (ii and iii) Two injuries that are difficult to replicate as they have multiple insults inflicted during wounding and mounting (arrowheads). Although these injuries will induce an inflammatory response, the magnitude and zone of injury is variable. Scale bar: 200 μm, also applies to panels (B)ii and iii.

provides a single linear wound edge producing a unidirectional chemoattractant gradient for migrating leukocytes; (2) the injury is easy to generate; (3) the magnitude of injury is reproducible between embryos as the wound edge is constrained by the diameter of the caudal fin; (4) as mentioned above, the thinness of the caudal tail and fin provides the anatomic advantage that the majority of tissue can be captured in a single or just a few Z sections, depending on the microscope used; (5) the caudal fin injury does not rupture any vessels, providing for consistency of leukocyte behaviors such as interstitial migratory speed and directionality, endothelial adhesion, and diapedesis.

2.1.1. Protocol for caudal fin amputation

Required materials

- E3 embryo medium (without methylene blue): 5 m*M* NaCl; 0.17 m*M* KCl; 0.33 m*M* $MgSO_4$; 0.33 m*M* $CaCl_2$; pH 7.5.
- Ethyl-*m*-aminobenzoate methanesulfonate (tricane; Sigma-Aldrich). Use at 0.168 mg/mL in E3. Prepare a stock solution of 0.4 g/100 mL, using 1 *M* Tris–HCl (pH 9) to adjust pH to 7.0; use 4.2 mL of stock per 100 mL E3.
- PTU (Sigma-Aldrich). Prepare 100× stock of 0.3% PTU (w/v) in distilled water; use at 0.003% PTU in E3.
- Tissue culture Petri dish (Corning Inc.). We find the use of bacterial Petri dishes often results minor tissue tears along the caudal fin.
- Sterile curved surgical blade (Swann Morton).
- Dissecting microscope.
- Dechorionated 3 dpf transgenic embryos, raised in 0.003% PTU in E3 from 12 hpf.

Method

1. Place transgenic embryos into a clean Petri dish containing 28 °C E3 medium with tricane.
2. Wait several minutes until the embryos are anesthetized.
3. Place Petri dish under the objective lens of a dissecting microscope and adjust the contrast of the microscope to best define the caudal fin.
4. Place the middle of a clean curved scalpel blade onto the Petri dish perpendicular to (but not touching) the center of the caudal fin (Fig. 8.1A).
5. With a smooth continuous motion, roll the blade over the fin.

2.1.2. Agarose mounting of wounded embryos

In order to perform time-lapse imaging, live zebrafish embryos must be immobilized within their liquid medium. A common and effective method is to anesthetize embryos in tricane and mount in 1.5% low-melting point (LMP) agarose. The mounting method described below can immobilize embryos for 10 h while imaging the inflammatory response without adverse

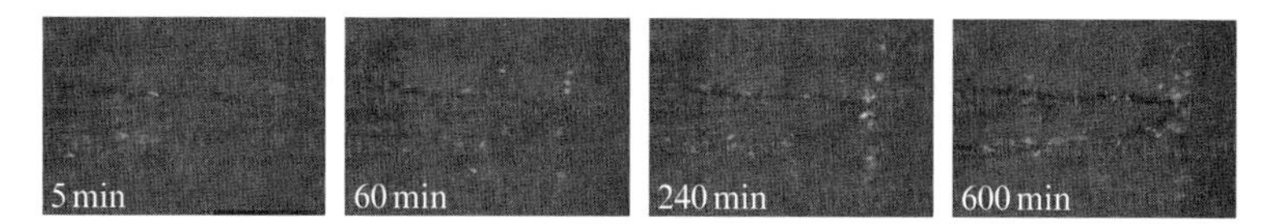

Figure 8.2 *In vivo* visualization of fluorescent macrophages and neutrophils during acute inflammation in zebrafish. Serial still images showing photoconverted Kaede-marked macrophages (red cells) and EGFP-marked neutrophils (green cells) in an injured 3 days postfertilization compound transgenic zebrafish embryo (*mpeg1*:Gal4/UAS:Kaede/*mpx*:EGFP). Times indicate minutes after injury. Note that in this compound transgenic zebrafish, there are no double fluorescent leukocytes, permitting accurate distinction between neutrophil and macrophage behaviors during the inflammatory response. Scale bar (applies to all panels): 200 μm. Frames extracted from a movie originally published in *Blood*. Ellett *et al.* (2011), © The American Society of Hematology. (For interpretation of the references to color in this figure legend, the reader is referred to the Web version of this chapter.)

effects (Fig. 8.2) (Ellett *et al.*, 2011). However, if longer time lapses are required, several precautions should be taken. The agarose mold needs to be adapted to accommodate embryo growth and the prolonged use of tricane eventually slows the embryo's heartbeat leading to death. As an alternative to tricane, paralyzed strains such as the nicotinic receptor mutant, *chrna1*, can be used (Westerfield *et al.*, 1990). Homozygous *chrna1* mutants show no skeletal muscle contraction yet importantly have a normal circulation and heart development.

To capture the initial phase of the inflammatory response, it is essential to minimize the time between wounding and the start of image acquisition. This section provides an efficient method for preparing, mounting, and imaging multiple embryos in a single experiment.

Required materials

- All materials required for caudal fin amputation.
- 1.5% LMP agarose (Scientifix) in E3 medium.
- Sterile microcentrifuge 1.5 mL tubes.
- Heat block set at 34 °C.
- Inverted fluorescence microscope equipped with an incubator.
- Glass-bottom dish (World Precision Instruments, Inc.).
- Eyelash manipulator.
- Timer.
- Sterile plastic transfer pipettes.

Preparation

1. Prewarm E3 medium containing tricane in a Petri dish.
2. Melt 1.5% LMP agarose in E3 medium and aliquot 1 mL into 1.5 mL microcentrifuge tubes. Place the tubes into a heat block set at 34 °C to cool, but still preventing the LMP agarose from setting.

3. Turn on inverted microscope and heat chamber to 28 °C.
4. Set filter configuration to that optimally suited for the fluorophore to be imaged.
5. Select embryos that express the transgene of interest and place into the prewarmed E3 medium with tricane. The number of embryos able to be imaged depends on the time-lapse interval and speed of acquisition per embryo. We generally succeed in commencing imaging 4–8 min after injury and image five embryos, each every 30 s.
6. Use a test embryo to adjust microscope focus and optimize both transmitted light and fluorescent exposure times.

Agarose mounting

1. For maximum efficiency, place 34 °C melted agarose, scalpel blade, glass-bottom Petri dish, eyelash manipulator, stop watch, plastic transfer pipette within easy reach of the dissecting microscope.
2. Place anesthetized embryos on the dissecting microscope. To quickly cut multiple embryos, we align them side-by-side head to tail perpendicular to the user using an eyelash manipulator. This ordering minimizes the need to rotate the embryos for wounding and minimizes the need to search for the next embryo while under the dissecting microscope.
3. Cut the caudal fin for each embryo as described in Section 2.1.1.
4. Start timer.
5. Take two to three embryos with the plastic transfer pipette. Embryos are divided into smaller groups to permit sufficient time to manipulate embryos flat before the agarose hardens.
6. Raise the transfer pipette upright and allow embryos to sink to the tip of the pipette.
7. Transfer embryos into LMP agarose, minimizing the transfer of E3 medium.
8. Expel remaining liquid from the pipette.
9. Use pipette to mix embryos within the agarose.
10. Aspirate the embryos together with the agarose and keep the embryos at the tip of the pipette.
11. Transfer embryos with a small amount of agarose onto a glass-bottom Petri dish.
12. Under the dissecting microscope, use the eyelash manipulator to quickly and gently orientate the embryos flat on the base of the glass-based dish taking particular attention to positioning the caudal fin evenly flat.
13. Repeat steps 5–12 with the remaining embryos. Importantly, do not add the second agarose drop to the first as this may rotate the prepared samples.

14. While the agarose sets, place a glass-based Petri dish on the inverted microscope and set multipoint coordinates of each embryo using transmitted light.
15. Gently submerge embryos and agarose in 28 °C E3 medium with tricane. For time-lapse imaging series longer than 5 h, place lid on top of the glass-based Petri dish to reduce evaporation.
16. Adjust focus for each embryo and reset multipoint coordinates.
17. Acquire time lapse.
18. Stop timer and record the time after injury that the imaging acquisition series commenced.

3. Things to Consider When Planning *In Vivo* Visualization of the Inflammatory Response

3.1. General considerations for multichannel fluorescence imaging

3.1.1. Microscopes and camera/detector types

For fluorescence microscopy, there are several types of microscope systems available and imaging detector choices are generally limited to either cameras (widefield imaging) or photon multiplier tubes (confocal imaging). Tables 8.2 and 8.3 list various options with their advantages and disadvantages.

3.1.2. Acquisition speed

Ideally, acquisition speed should always be as fast as possible to minimize exposure to potentially photodamage inducing light and to maximize temporal resolution. The attainable speed of image acquisition will be governed by several factors:

- Sensitivity of the detector system (camera/PMT)
  - High sensitivity detection allows for shorter exposures, which on a system with fast shutters will equate to dramatically reduced exposure of the sample to potentially damaging light.
  - Additionally, a high sensitivity detector may also allow longer exposures of very low level excitation light which can sometimes be less damaging than a short intense burst.
- Turning on and off the light sources
  - In general, the default inbuilt microscope shutters are not very fast, taking 500 ms or more to respond. This can result in unnecessarily long exposure of the sample when a shorter exposure would suffice.

Table 8.2 Advantages and disadvantages of several different microscope systems

System	Advantages	Disadvantages	Best suited for
Standard widefield	• Fast • Sensitive (with appropriate camera) • Cheap	• Limited to thin samples (<100 μm)	• Whole cell imaging • Macroanimal imaging
Deconvolution	• Fast—if processing is done offline • Sensitive • Improved resolution over standard widefield	• Expensive (for full commercial system) • Limited to thin samples (<100 μm) • Can introduce deconvolution artifacts	• Single cell • Subcellular localization
Point scan confocal	• Works on thick specimens (<400 μm) • Moderate speed at low scan resolutions	• Expensive • Slow at high resolutions	• Subcellular localization • Thick samples
Line scan confocal	• Fast • Works on thick specimens	• Expensive • Can require high levels of excitation light	• Rapidly moving objects in thick samples
Spinning disk confocal	• Relatively inexpensive • Sensitive • Fast • Very low phototoxicity	• Limited to thin specimens (<100 μm)	• Light sensitive samples • Subcellular localization • Whole cells
Theta confocal	• Fast—in certain configurations • Sensitive • Works on very thick specimens (>1 mm)	• Not commercially available • Expensive to maintain	• Macroimaging (e.g., whole embryos)

- Movement of stage for multiple samples
  - Stage movement speed can limit how many samples can be imaged in a given time period. To maximize temporal resolution, samples should be placed as close together as possible to minimize the distance the stage has to move between them.

Table 8.3 Advantages and disadvantages of several different camera/detector types

Camera/ detector type	Advantages	Disadvantages	Best suited for
Color camera (CMOS)	• Inexpensive • Fast color imaging	• Low sensitivity	• Brightfield
Monochrome camera (CCD)	• Affordable • Sensitive (~65% QE) • Fast • High resolution • Some models can do three shot color	• Monochrome • Slow in three shot color mode	• Fluorescence • Slow brightfield
Monochrome camera (EM-CCD)	• Fast • Very sensitive (~95% QE)	• Expensive • Low resolution (512 × 512 or 1024 × 1024)	• Low light fluorescence
PMT (confocal)	• Confocal imaging • Can be fast depending on settings	• Potential for high noise • Low sensitivity (~20% QE)	• Thick fluorescence samples
GaAsP detector (confocal)	• Confocal imaging • Increased sensitivity of standard PMT (~40% QE)	• Expensive • Limited availability	• Thick fluorescence samples

Abbreviations: CCD, charge-coupled device; CMOS, complementary metal oxide semiconductor; EM-CCD, electron multiplying charge-coupled device; GaAsP, gallium arsenide phosphide; PMT, photon multiplier tube; QE, quantum efficiency.

- Changing of fluorescent filters
  - During multichannel fluorescence imaging, temporal resolution can be greatly compromised by slow moving filter turrets. This can be alleviated by using a fast turret or and excitation/emission filter wheel combination that allows rapid changing of filters.

3.1.3. Environmental control

For any live imaging, the environmental conditions need to be tightly controlled for the results to have physiological relevance. Temperature needs to be maintained at the correct physiological level; additionally, other factors such as gas composition and humidity also may need to be kept consistent.

Control of temperature is also important for the stability of the imaging system hardware itself. Fluctuations in the temperature can result in samples drifting out of focus or the system in general performing below expectations. Several options exist for controlling the temperature of the imaging environment. Heated stages (or objective heaters) work to some extent, but are notorious for creating gradients and hot/cold spots. Ideally, the whole microscope platform (objectives, stage, and part of the body) should be contained within a temperature-controlled incubator that is left running even when not imaging. These whole microscope incubators provide the best thermal stability of the control systems available.

Control of gas and humidity can be achieved in two ways. Gas can be supplied to the sample directly, or to a small chamber surrounding the sample, from a premixed source (e.g., 5% CO_2 in compressed air). This has limitations if different conditions are required for different experiments and tends to be expensive as custom gas mixes need to be made and bottled. Alternatively, gas can be supplied through a gas mixer. Gas mixers vary in complexity from ones that draw in room air and mix it with a bottle gas (e.g., CO_2) to create the final mix that is required up to multichannel mixers than can control multiple gas and humidity levels. Mixers allow for increased flexibility and in the long run reduce costs as the individual pure gas bottles are cheaper than premixed ones.

3.1.4. Objective choice

The objective used is key to not just image size/magnification but also governs how much light can be collected (affecting exposure times) and the resolution of the image captured. In general, it is best to use the objective with the highest numerical aperture (NA) and lowest magnification that will give the required field of view. High NA objectives have higher resolution and collect much more light from the sample. High magnification objectives give a larger image but without a high NA will not provide any more resolution. Increases in magnification also lead to a decrease in the light gathering capability of the objective; for example, a 60× 1.4 NA objective will collect around four times as much light as a 100× 1.4 NA objective.

High powered objectives are generally not used for long-term live imaging, due either to their low light gathering ability (e.g., 60× dry objective), or to their requirement for an immersion medium such as oil or water. Keeping the immersion media on the objective for the length of the imaging run can be very difficult. Solutions exist for this if required, but can offer variable reliability. Using an immersion objective may be possible for single position imaging if oil or 'sticky water' is used, but for multi-position time-lapse imaging, this is not easily achieved.

In general, long working distance objectives with correction collars (allowing for correction of the thickness of the dish/slide the sample is mounted on) come with a trade off in NA but are easily adapted to different experimental conditions.

Correction of aberrations (planar and chromatic) is also very important to consider. Most modern research grade objectives are corrected from chromatic aberration at green and red wavelengths and are planar corrected to some extent.

3.1.5. Acquisition software

Most microscope systems come with their own acquisition software and in general this is sufficient to perform basic time-lapse imaging. The software needs to be able to control all aspects of the microscope (objectives, filters, shutters, stage, and camera). Independent proprietary control software may give more power and flexibility to system control, but at a cost premium. Free software also exists (e.g., MicroManager), but comes with a potentially steep learning curve, lack of support and limited hardware support.

3.1.6. Data management

Live imaging potentially generates large amounts of data with multi-day experiments exceeding tens of gigabytes. Usually data is captured to the local computer driving the microscope system, but it must be transferred for storage, backup, and later analysis. For efficient, rapid transfer, central storage requires high speed network connections (1 gigabit or higher). Once the raw data is analyzed, it can usually be backed up and removed from central storage but a detailed record of the backups is required to make future recovery as easy as possible.

3.1.7. Stage movement

During multipoint imaging, the motorized/robotic microscope stage needs to reproducibly return to the point of interest over a prolonged period. Most stages now contain linear encoded motors and high quality control systems that achieve this reliably, providing a resolution of 25–100 nm which is beyond the resolution of most objective lenses and placing any drift that may occur beyond the level of detection. The speed of the stage movement is an important factor in governing the maximum temporal resolution. Moving the stage at its maximum travel speed may have consequences for sample stability and should be minimized when setting up the experiment.

3.1.8. Automatic focus

During long time lapses, samples may drift out of focus due to thermal instability or because the sample alters due to inherent biological processes such as the progressive morphogenesis of a developing embryo. Image-based autofocus systems are usually built into the capture software and rely on taking multiple images at different focal positions and picking the one that is most in focus by use of various algorithms. This method can be very slow and exposes the sample to more potentially damaging excitation light than may be necessary. It is advantageous in situations where the sample moves within its environment for this movement to be in an axial direction, as the microscope will always find the sample if the parameters are set correctly. Laser-based focus systems that constantly check for the cover slip or base of the dish are much faster but are expensive and not able to adapt to samples moving up or down. If the system is thermally stable and the sample is generally flat, then autofocus is generally not necessary.

3.2. Specific considerations for determining changes in H_2O_2 levels using HyPer

Genetically encoded sensors offer several advantages over chemically synthesized sensitive fluorescent dyes. Genetic delivery means that the sensor is generated by the cells themselves, eliminating issues with dye uptake and leakage during long time lapses. Genetic-encoded sensors can have their expression pattern constrained in space and time by using tissue- or developmental stage-specific promoters, or localized to specific subcellular compartments by incorporating appropriate targeting sequences.

HyPer is a genetically engineered fluorophore that can be used to display changes in H_2O_2 concentrations (Belousov *et al.*, 2006). HyPer consists of cpYFP inserted into the H_2O_2-sensitive regulatory domain of the bacterial gene OxyR. HyPer can be used as a tool for studying fluctuating H_2O_2 concentrations because its spectral properties change depending on the concentration of H_2O_2. HyPer has two excitation maxima, 420 and 500 nm, and one emission maximum at 516 nm. At higher H_2O_2 concentrations, the HyPer fluorophore emits more light when excited at 500 nm and slightly less light when excited at 420 nm. The ratio between the emission signal intensities acquired from samples excited at 500 nm ($HyPer_{500}$ image) and 420 nm ($HyPer_{420}$ image) reflects the current ambient H_2O_2 concentration, and changes in the HyPer ratio (i.e., $HyPer_{500}/HyPer_{420}$) can be used to display H_2O_2 concentration changes across both time and space. A significant advantage of HyPer ratiometric sensing of H_2O_2 concentration is that the value is a ratio. As such, the HyPer ratio provides an internally controlled parameter that normalizes artifacts caused by uneven protein concentrations throughout the tissue and variations over time due to cell and tissue thickness, cell movement, and excitation intensity.

The following are some important considerations for successfully and accurately capturing quantitative HyPer fluorescence data. We explain how to process acquired HyPer images and provide a MetaMorph journal to generate the HyPer ratios (Figs. 8.3 and 8.4). The method is demonstrated by an experimental example of the visualization of the H_2O_2 burst following caudal fin injury (Fig. 8.5).

- To accommodate the spectral properties of HyPer, optimal filter blocks need to be selected or custom made. For widefield microscopes, HyPer ratios have been successfully generated using the following filters:

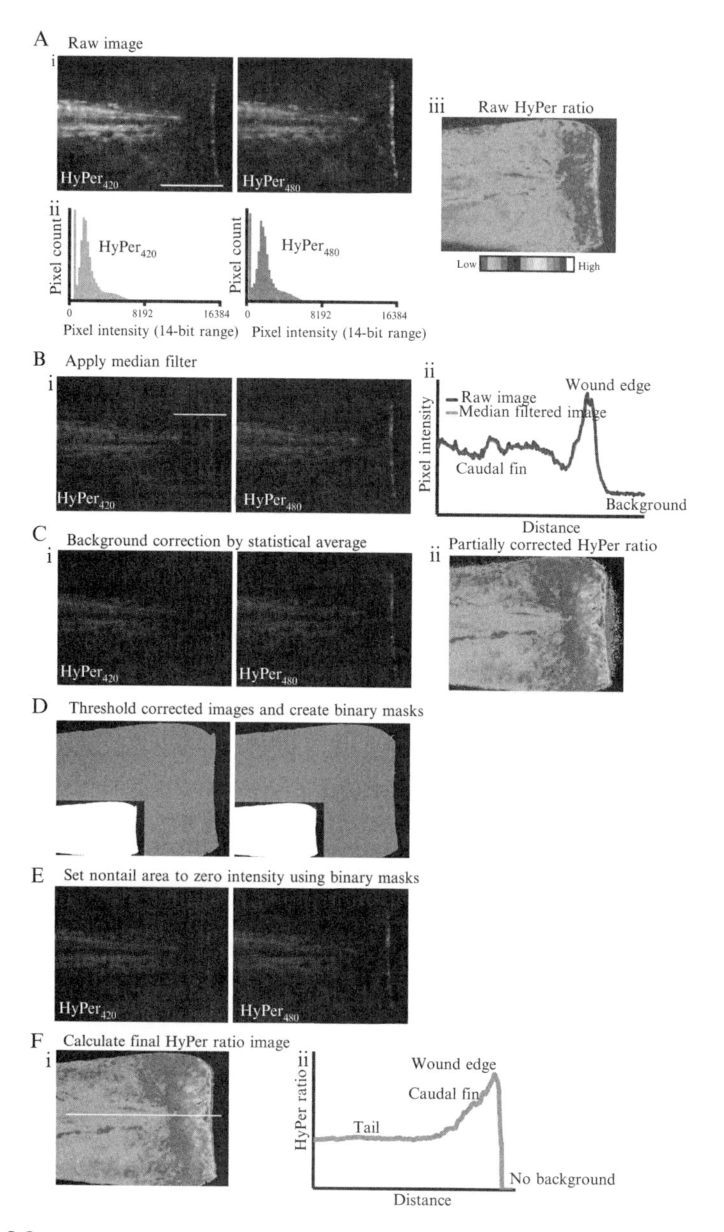

Figure 8.3 A method for processing images to generate specific HyPer ratios reflecting intracellular hydrogen peroxide concentrations. Still images are of a HyPer-expressing 3 days postfertilization zebrafish embryo, 20 min post-caudal fin injury. (A) (i) Acquire a raw HyPer fluorescent images. This example was acquired on a Nikon Ti-E inverted microscope using a (ex420/20, 505LP, em535/20) filter cube ($HyPer_{420}$) and a (ex480/15, 505LP, em535/20) filter cube ($HyPer_{480}$). (ii) Histogram illustrating the distribution of pixel intensities across the 14-bit range for both $HyPer_{420}$ and $HyPer_{480}$ images.

HyPer fluorescence excited with 420/40 and 501/16 band-pass excitation filters and a 535/30 band-pass emission filter (Chroma) (Niethammer *et al.*, 2009) or 420/40 and 480/15 band-pass excitation filters and 535/20 band-pass emission filter (Nikon) (our experience). We have captured changes in HyPer ratios using two different confocal instruments, a Zeiss LSM 5 Live microscope and an Olympus FV1000, exciting specimens with 405 and 488 nm (473 nm on FV1000) lasers. However, the quality of the data generated from multiple laser confocal instruments was inferior to that acquired on a widefield microscope using a single light source and filters. We performed imaging with a Nikon Ti-E microscope using a 20× PlanFluor NA/0.5 objective (Nikon). Acquisition used a SPOT Pursuit Slider CCD camera (Diagnostic Instruments) and MetaMorph software (v7.7.0, Molecular Devices). A Prior ProScan II (Prior, UK) motorized stage was used for imaging multiple embryos within the same dish.

Importantly, the settings result in a signal intensity from the embryo tail that provides distinction from the background without there being any saturated pixels. (iii) Raw HyPer ratio ($HyPer_{480}/HyPer_{420}$) represented by a rainbow color spectrum. High HyPer ratio values (i.e., high H_2O_2 concentrations) are represented by yellow and red colors, while low HyPer ratio values (low H_2O_2 levels) are represented by violet and blue colors. Note that background noise generates a ratio value that will be removed by the end of this process. (B) Apply a median filter (5,5,1) to reduce subtle fluctuations in the pixel values that may be amplified in the ratio image. (i) Median filtered $HyPer_{420}$ and $HyPer_{480}$ images. (ii) Graph of the pixel intensity along the white line drawn on the $HyPer_{420}$ image. Note that the small fluctuations between adjacent pixels in the raw image are smoothed out in the median filtered image without affecting the trend in pixel intensity. (C) Correct for background noise. (i) Background corrected $HyPer_{420}$ and $HyPer_{480}$ images, correction performed using the statistical average algorithm. (ii) HyPer ratio generated from background corrected images. Note that correcting for the background noise slightly increased the HyPer ratio values throughout the tail compared to panel (A, iii) (i.e., there are more red assigned pixels). However, there is still background noise in the HyPer ratio image. (D) Remove all background pixel values using a binary mask. (i) Images depicting how the whole tail could be selected using a simple pixel intensity threshold (orange) and the resulting binary images (insert). The binary mask provides a method to extract pixel values within a defined area (white area). (E) Assign zero intensity to background using the binary mask. $HyPer_{420}$ and $HyPer_{480}$ with zero intensity background. Note there are no changes to tail intensity values. (F) Final HyPer ratio image that is specific to the tissue expressing the HyPer fluorophore. (i) Corrected HyPer ratio image. (ii) A graph showing the average HyPer ratio 25 μm either side of the white line in panel (F, i). Note that the HyPer ratio demonstrates a H_2O_2 concentration gradient extending from the wound edge. All images and data were generated using the MetaMorph journal presented in Fig. 8.4, except of the panel (F, ii) which was generated using the MetaMorph line scan function to show average intensities along the user described line. Scale bar (in panel (A, i) applies to all panels): 200 μm. (See Color Insert.)

Step 1 - Image Selection and Median Filtering

Ask user to select Images representing each channel and store their names in temp variables (HyPer420.store and HyPer480.store)

1: Select Image("Select HyPer420 Image")

HyPer420.store = image.name

2: Select Image("Select HyPer480 Image")

HyPer480.store = image.name

Apply a median filter (5,5,1) to all planes of the HyPer420 and HyPer480 image stacks

3: New "HyPer420 Median" = MedianFilter("%HyPer420.store%", 5, 5, 1)

4: New "HyPer480 Median" = MedianFilter("%HyPer480.store%", 5, 5, 1)

Step 2a - Background Correction of HyPer420 Image

Place a small region on the image and ask the user to move it over an area of backround signal in the HyPer420 image

5: Select Image([3: MedianFilter])

6: Create Region(Position:X/From:X = 10, Position:Y/From:Y = 10, Width/To:X = 100, Height/To:Y = 100)

7: Show Message and Wait("", NO TIMEOUT)

Loop the "Correct HyPer420" Journal for all planes of the median filtered blue image

8: Loop for all Planes([3: MedianFilter], "Correct HyPer420")

Step 2b - Correct HyPer420 sub-Journal

use user asigned area to perform average background correction

1: Add to "Statistical Correction-HyPer420" = Background and Shading Correction("HyPer420 Median", 1, Average)

*** End of Journal ***

Step 3a - Background Correction of HyPer480 Image

Place a small region on the image and ask the user to move it over an area of backround signal in the HyPer480 image

9: Select Image([4: MedianFilter])

10: Create Region(Position:X/From:X = 10, Position:Y/From:Y = 10, Width/To:X = 100, Height/To:Y = 100)

11: Show Message and Wait("", NO TIMEOUT)

Loop the "Correct HyPer480" Journal for all planes of the median filtered green image

12: Loop for all Planes([4: MedianFilter], "Correct HyPer480")

Step 3b - Correct HyPer480 sub-Journal

use user asigned area to perform average background correction

1: Add to "Statistical Correction-HyPer480" = Background and Shading Correction("HyPer480 Median", 1, Average)

*** End of Journal ***

Step 4 - Close Unwanted Median FIlter Images

Close the median filtered images

13: Close([3: MedianFilter])

14: Close([4: MedianFilter])

Step 5 - Create Binary Mask of the HyPer420 Image

Select the corrected HyPer420 image and ask the user to set a threshold that selects the whole tail

15: Select Image("Statistical Correction-HyPer420")

16: Auto Threshold for Light Objects(Legacy heuristic algorithm)

17: Show Message and Wait("", NO TIMEOUT)

Create a binary image of the selected HyPer420 tail

18: New "HyPer420 Binary" = Binarize("Statistical Correction-HyPer420"), high = current value, low = current value

Step 6 - Create Binary Mask of the HyPer480 Image

Select the corrected HyPer480 image and ask the user to set a threshold that selects the whole tail

19: Select Image("Statistical Correction-HyPer480")

20: Auto Threshold for Light Objects(Legacy heuristic algorithm)

21: Show Message and Wait("", NO TIMEOUT)

Create a binary image of the selected HyPer480 tail

22: New "HyPer480 Binary" = Binarize("Statistical Correction-HyPer480"), high = current value, low = current value

Step 7 - Create Final Correction Image using Boolean AND Function

Perform logical AND functions between the respective binary and corrected images to remove signal unasociated with the fish tail

23: New "HyPer420 Corrected" = "Statistical Correction-HyPer420" AND [18: Binary Operations]

24: New "HyPer480 Corrected" = "Statistical Correction-HyPer480" AND [22: Binary Operations]

Step 8 - Turn Off Thresholds and Close Unwanted Images

Turn thresholds off on the corrected images

25: Threshold Image([23: Arithmetic], 0, 0, Off)

26: Threshold Image([24: Arithmetic], 0, 0, Off)

close unwanted images

27: Close("Statistical Correction-HyPer420")

28: Close("Statistical Correction-HyPer480")

29: Close([18: Binary Operations])

30: Close([22: Binary Operations])

Step 9 - Generate Ratio Image and Apply Heat Map

Divide the HyPer480 corrected image by the HyPer420 corected image to obtain the ratio image and set the LUT colour to heat map

31: New "Divide" = ([24: Arithmetic] * 1000) / ([23: Arithmetic] * 1)

32: Select Image([31: Arithmetic])

33: Set LUT Model([31: Arithmetic], 1)

*** End of Journal ***

Figure 8.4 MetaMorph journal script to generate specific HyPer ratios of tissues. Narrated journal commands written in MetaMorph (v7.7.0, Molecular Devices, USA).

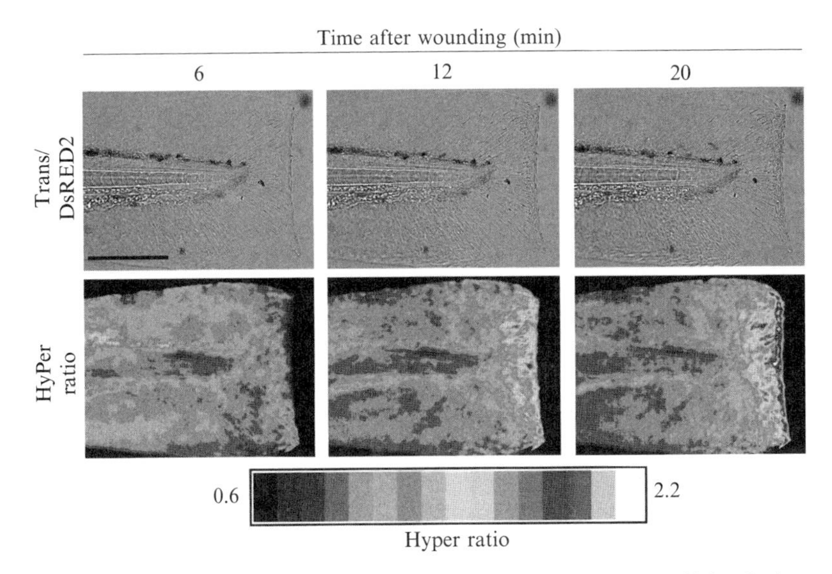

Figure 8.5 *In vivo* visualization of leukocyte behavior and intracellular hydrogen peroxide concentration over time in a single zebrafish embryo. Representative HyPer imaging data at specific times after wounding. Overlayed transmission (Trans) and *lyz*: DsRED2 positive neutrophil images are displayed to demonstrate the proximity of neutrophils to the wound margin over time. HyPer ratio data represented using a 16-color spectrum, reflecting a local increase over time in intracellular H_2O_2 concentrations across the wounded zebrafish caudal fin. Scale bar: 200 μm, applies to all panels. (For interpretation of the references to color in this figure legend, the reader is referred to the Web version of this chapter.)

- HyPer can be expressed ubiquitously throughout the embryo by transient overexpression from *in vitro* transcribed mRNA microinjected into 1–2-cell embryos using established techniques (Pase and Lieschke, 2009). We find that embryos microinjected with 1–2 nL HyPer mRNA (250 μg/mL) display sufficient fluorescence levels up to 4 dpf. For this the HyPer cDNA available commercially (Evrogen) requires subcloned into an appropriate zebrafish expression vector (e.g., pCS2+).
- Strongly fluorescent embryos are preferred, as strong fluorescence reduces the acquisition time and produces images with better signal-to-noise ratios.
- If either of the $HyPer_{420}$ or $HyPer_{500}$ images is saturated, the HyPer ratio cannot be accurately calculated. To avoid saturation of the HyPer signal during the injury-induced H_2O_2 burst, set the initial exposure times to a level that the majority of tail tissue signal falls in the bottom quarter of the digital image dynamic range (e.g., <1000 intensity unity on a 12-bit detection system) (Fig. 8.3). This permits the H_2O_2-dependent increase in $HyPer_{500}$ signal to be captured within the dynamic range. The tissue at the wound edge usually thickens, further increasing the intensity of both $HyPer_{420}$ and $HyPer_{500}$ signal.

- Prolonged exposure to high intensity light can cause tissue photodamage and generate reactive oxygen species, including H_2O_2. It is therefore important, for the microscope used, to find a balance between the exposure time and time-lapse interval that best minimizes photodamage yet generates adequate HyPer images. An uninjured HyPer fluorescent embryo should be used as a control for embryo preparation and microscope generated H_2O_2.
- Leukocyte behaviors and intracellular H_2O_2 dynamics can be visualized in a single embryo by microinjecting HyPer mRNA into embryos of transgenic lines with fluorophore-marked leukocytes. However, it is important to note that if the leukocyte-marking fluorophore can be detected with $HyPer_{420}$ or $HyPer_{500}$ filter sets (e.g., EGFP), the signal generated by the leukocyte fluorophore will distort the HyPer ratio in the area it is expressed. Therefore, a misleading HyPer ratio for the leukocyte will be calculated. If it is necessary to avoid this leukocyte fluorophore corruption of the HyPer signal, leukocytes marked with transgenes expressing red or far-red fluorophore should be used (e.g., mCherry and mKate, but not DsRed2 which is detected with the $HyPer_{500}$ filter block).
- The ratiometric properties of the genetic sensor HyPer provide an excellent internal control; however, it is critical that the only variable contributing to the $HyPer_{420}$ and the $HyPer_{500}$ images is the change in H_2O_2 concentration. The microscope itself is a potential source of artifact, and therefore it is critical to perform a full time-lapse control experiment with a uniformly fluorescent slide (Chroma). The calculated HyPer ratio for this consistent source of fluorescence should be uniform and consistent over the entire image and time period. Variability in the HyPer ratio across the image is due to variation in the light intensity from the source to the camera, possibly due to variable vignetting of each excitation wavelength or impurities in the different filters. If this is the case, the uniform HyPer ratio image can be used to correct for this consistent artifact influencing the experimental sample. Variability in the HyPer ratio across time would be due to an unstable light in only one channel. This is a possibility if the light source exciting $HyPer_{420}$ and $HyPer_{500}$ is independent (e.g., it would result from a fluctuating light intensity from a 405 nm laser but not the 488 nm laser). Unless appropriate controls can be introduced to experimental samples, the use of such a microscope should be avoided.

The zebrafish model provides the opportunity to visualize multiple aspects of the inflammatory response *in vivo* and in real time. Overall, combining the above protocols and considerations will facilitate successful *in vivo* time-lapse imaging during the acute inflammatory response of leukocyte behaviors simultaneously with dynamic changes in H_2O_2 concentrations (Fig. 8.5).

ACKNOWLEDGMENTS

We thank Kelly Rogers for her assistance with time-lapse imaging using the LSM 5 Live microscope; Stephen Renshaw and Constantino Carlos Reyes-Aldasoro for great discussions regarding time-lapse imaging of leukocytes and Felix Ellett for input into Table 8.1. Work contributing to these methods has been undertaken in projects supported by grants to G. L. from NIH (R01 HL079545), the NHMRC (234708, 461208, 637394), and ARC (DP0346823). WEHI receives infrastructure support from the Commonwealth NHMRC Independent Research Institutes Infrastructure Support Scheme (361646) and a Victorian State Government Operational Infrastructure Support Scheme grant. The Australian Regenerative Medicine Institute is supported by grants from the State Government of Victoria and the Australian Government.

REFERENCES

Balla, K. M., Lugo-Villarino, G., Spitsbergen, J. M., Stachura, D. L., Hu, Y., Banuelos, K., Romo-Fewell, O., Aroian, R. V., and Traver, D. (2010). Eosinophils in the zebrafish: Prospective isolation, characterization, and eosinophilia induction by helminth determinants. *Blood* **116,** 3944–3954.

Belousov, V. V., Fradkov, A. F., Lukyanov, K. A., Staroverov, D. B., Shakhbazov, K. S., Terskikh, A. V., and Lukyanov, S. (2006). Genetically encoded fluorescent indicator for intracellular hydrogen peroxide. *Nat. Methods* **3,** 281–286.

d'Alencon, C. A., Pena, O. A., Wittmann, C., Gallardo, V. E., Jones, R. A., Loosli, F., Liebel, U., Grabher, C., and Allende, M. L. (2010). A high-throughput chemically induced inflammation assay in zebrafish. *BMC Biol.* **8,** 151.

Ellett, F., and Lieschke, G. J. (2010). Zebrafish as a model for vertebrate hematopoiesis. *Curr. Opin. Pharmacol.* **10,** 563–570.

Ellett, F., Pase, L., Hayman, J. W., Andrianopoulos, A., and Lieschke, G. J. (2011). mpeg1 promoter transgenes direct macrophage-lineage expression in zebrafish. *Blood* **117,** e49–e56.

Feng, Y., Santoriello, C., Mione, M., Hurlstone, A., and Martin, P. (2010). Live imaging of innate immune cell sensing of transformed cells in zebrafish larvae: Parallels between tumor initiation and wound inflammation. *PLoS Biol.* **8,** e1000562.

Gray, C., Loynes, C. A., Whyte, M. K., Crossman, D. C., Renshaw, S. A., and Chico, T. J. (2011). Simultaneous intravital imaging of macrophage and neutrophil behaviour during inflammation using a novel transgenic zebrafish. *Thromb. Haemost.* **105,** 811–819.

Hall, C., Flores, M. V., Storm, T., Crosier, K., and Crosier, P. (2007). The zebrafish lysozyme C promoter drives myeloid-specific expression in transgenic fish. *BMC Dev. Biol.* **7,** 42.

Hall, C., Flores, M. V., Chien, A., Davidson, A., Crosier, K., and Crosier, P. (2009). Transgenic zebrafish reporter lines reveal conserved Toll-like receptor signaling potential in embryonic myeloid leukocytes and adult immune cell lineages. *J. Leukoc. Biol.* **85,** 751–765.

Herbomel, P., Thisse, B., and Thisse, C. (1999). Ontogeny and behaviour of early macrophages in the zebrafish embryo. *Development* **126,** 3735–3745.

Hsu, K., Traver, D., Kutok, J. L., Hagen, A., Liu, T. X., Paw, B. H., Rhodes, J., Berman, J. N., Zon, L. I., Kanki, J. P., and Look, A. T. (2004). The pu.1 promoter drives myeloid gene expression in zebrafish. *Blood* **104,** 1291–1297.

Kitaguchi, T., Kawakami, K., and Kawahara, A. (2009). Transcriptional regulation of a myeloid-lineage specific gene lysozyme C during zebrafish myelopoiesis. *Mech. Dev.* **126,** 314–323.

Lawson, N. D., and Weinstein, B. M. (2002). In vivo imaging of embryonic vascular development using transgenic zebrafish. *Dev. Biol.* **248,** 307–318.

Lieschke, G. J., Oates, A. C., Crowhurst, M. O., Ward, A. C., and Layton, J. E. (2001). Morphologic and functional characterization of granulocytes and macrophages in embryonic and adult zebrafish. *Blood* **98,** 3087–3096.

Lin, H. F., Traver, D., Zhu, H., Dooley, K., Paw, B. H., Zon, L. I., and Handin, R. I. (2005). Analysis of thrombocyte development in CD41-GFP transgenic zebrafish. *Blood* **106,** 3803–3810.

Long, Q., Meng, A., Wang, H., Jessen, J. R., Farrell, M. J., and Lin, S. (1997). GATA-1 expression pattern can be recapitulated in living transgenic zebrafish using GFP reporter gene. *Development* **124,** 4105–4111.

Martin, P., and Leibovich, S. J. (2005). Inflammatory cells during wound repair: The good, the bad and the ugly. *Trends Cell Biol.* **15,** 599–607.

Mathias, J. R., Perrin, B. J., Liu, T. X., Kanki, J., Look, A. T., and Huttenlocher, A. (2006). Resolution of inflammation by retrograde chemotaxis of neutrophils in transgenic zebrafish. *J. Leukoc. Biol.* **80,** 1281–1288.

Medzhitov, R. (2008). Origin and physiological roles of inflammation. *Nature* **454,** 428–435.

Meijer, A. H., van der Sar, A. M., Cunha, C., Lamers, G. E., Laplante, M. A., Kikuta, H., Bitter, W., Becker, T. S., and Spaink, H. P. (2008). Identification and real-time imaging of a myc-expressing neutrophil population involved in inflammation and mycobacterial granuloma formation in zebrafish. *Dev. Comp. Immunol.* **32,** 36–49.

Niethammer, P., Grabher, C., Look, A. T., and Mitchison, T. J. (2009). A tissue-scale gradient of hydrogen peroxide mediates rapid wound detection in zebrafish. *Nature* **459,** 996–999.

Olivari, F. A., Hernandez, P. P., and Allende, M. L. (2008). Acute copper exposure induces oxidative stress and cell death in lateral line hair cells of zebrafish larvae. *Brain Res.* **1244,** 1–12.

Pase, L., and Lieschke, G. J. (2009). Validating microRNA target transcripts using zebrafish assays. *Methods Mol. Biol.* **546,** 227–240.

Peri, F., and Nusslein-Volhard, C. (2008). Live imaging of neuronal degradation by microglia reveals a role for v0-ATPase a1 in phagosomal fusion in vivo. *Cell* **133,** 916–927.

Redd, M. J., Kelly, G., Dunn, G., Way, M., and Martin, P. (2006). Imaging macrophage chemotaxis in vivo: Studies of microtubule function in zebrafish wound inflammation. *Cell Motil. Cytoskeleton* **63,** 415–422.

Renshaw, S. A., Loynes, C. A., Trushell, D. M., Elworthy, S., Ingham, P. W., and Whyte, M. K. (2006). A transgenic zebrafish model of neutrophilic inflammation. *Blood* **108,** 3976–3978.

Thelen, M., and Stein, J. V. (2008). How chemokines invite leukocytes to dance. *Nat. Immunol.* **9,** 953–959.

Traver, D., Paw, B. H., Poss, K. D., Penberthy, W. T., Lin, S., and Zon, L. I. (2003). Transplantation and in vivo imaging of multilineage engraftment in zebrafish bloodless mutants. *Nat. Immunol.* **4,** 1238–1246.

Walters, K. B., Green, J. M., Surfus, J. C., Yoo, S. K., and Huttenlocher, A. (2010). Live imaging of neutrophil motility in a zebrafish model of WHIM syndrome. *Blood* **116,** 2803–2811.

Ward, A. C., McPhee, D. O., Condron, M. M., Varma, S., Cody, S. H., Onnebo, S. M., Paw, B. H., Zon, L. I., and Lieschke, G. J. (2003). The zebrafish spi1 promoter drives myeloid-specific expression in stable transgenic fish. *Blood* **102,** 3238–3240.

Westerfield, M., Liu, D. W., Kimmel, C. B., and Walker, C. (1990). Pathfinding and synapse formation in a zebrafish mutant lacking functional acetylcholine receptors. *Neuron* **4,** 867–874.

Wittamer, V., Bertrand, J. Y., Gutschow, P. W., and Traver, D. (2011). Characterization of the mononuclear phagocyte system in zebrafish. *Blood* **117,** 7126–7135.

Yoo, S. K., and Huttenlocher, A. (2011). Spatiotemporal photolabeling of neutrophil trafficking during inflammation in live zebrafish. *J. Leukoc. Biol.* **89,** 661–667.

Zhang, Y., Bai, X. T., Zhu, K. Y., Jin, Y., Deng, M., Le, H. Y., Fu, Y. F., Chen, Y., Zhu, J., Look, A. T., Kanki, J., Chen, Z., *et al.* (2008). In vivo interstitial migration of primitive macrophages mediated by JNK-matrix metalloproteinase 13 signaling in response to acute injury. *J. Immunol.* **181,** 2155–2164.

CHAPTER NINE

Imaging Protein Oligomerization in Neurodegeneration Using Bimolecular Fluorescence Complementation

Federico Herrera,* Susana Gonçalves,* *and* Tiago Fleming Outeiro*,†,‡

Contents

1. Introduction 158
2. Required Equipment and Materials 161
3. Protocol 162
   3.1. Choosing the reporter protein 162
   3.2. Creating a BiFC platform 165
   3.3. Testing the BiFC system 167
   3.4. Optimizing the BiFC system 170
Acknowledgements 172
References 172

Abstract

Neurodegenerative disorders such as Alzheimer's, Parkinson's, Huntington's, or Prion diseases belong to a superfamily of pathologies known as protein misfolding disorders. The hallmark of these pathologies is the aberrant accumulation of specific proteins in beta sheet-rich amyloid aggregates either inside or outside cells. Current evidence suggests that oligomeric species, rather than mature protein aggregates, are the most toxic forms of the pathogenic proteins. This is due, at least in part, to their greater solubility and ability to diffuse between intracellular and extracellular compartments. Understanding how oligomerization occurs is essential for the development of new treatments for this group of diseases. Bimolecular fluorescence complementation assays (BiFC) have proved to be excellent systems to study aberrant protein–protein

* Cell and Molecular Neuroscience Unit, Instituto de Medicina Molecular, Lisboa, Portugal
† Instituto de Fisiologia, Faculdade de Medicina da Universidade de Lisboa, Av. Professor Egas Moniz, Lisboa, Portugal
‡ Department of Neurodegeneration and Restaurative Research, Center for Molecular Physiology of the Brain, University of Göttingen, Waldweg 33, Göttingen, Germany

Methods in Enzymology, Volume 506
ISSN 0076-6879, DOI: 10.1016/B978-0-12-391856-7.00033-0

interactions, including those involved in neurodegenerative diseases. Here, we provide a detailed description of the rationale to develop and validate BiFC assays for the visualization of oligomeric species in living cells in the context of neurodegeneration. These systems could constitute powerful tools for the identification of genetic and pharmacological modifiers of protein misfolding and aggregation.

1. Introduction

The excessive production, aberrant folding or failure to process particular proteins can cause them to aggregate into toxic beta-sheet-rich amyloid structures. The presence of these amyloid proteinaceous inclusions is a common histopathological hallmark of a large group of human pathologies known as protein misfolding disorders. Neurodegenerative disorders such as Alzheimer's, Parkinson's, or Huntington's disease belong to this group of pathologies. The specific localization and composition of protein aggregates are characteristic of each disorder (Table 9.1).

Aggregates of misfolded proteins may range in size and degree of solubility (Lindgren and Hammarstrom, 2010). Growing evidence suggests that the most toxic species are the small-order oligomers due, at least in part, to their higher degree of solubility and ability to diffuse between different cell and tissue compartments (Hansen *et al.*, 2011; Herrera *et al.*, 2011; Lee *et al.*, 2010; Outeiro *et al.*, 2008). According to this hypothesis, larger and more insoluble aggregates such as protofibrils, fibrils, and inclusion bodies would be rather neuroprotective by recruiting toxic oligomers to their core and preventing them from interacting with more sensitive cell structures (Arrasate *et al.*, 2004; Bodner *et al.*, 2006). Understanding how the aggregation process begins, that is, how dimers and oligomers are generated, is essential to identify possible molecular targets for the treatment of neurodegenerative disorders involving protein misfolding and aggregation.

Diverse biochemical and imaging methods have been developed to visualize and study aberrant protein–protein interactions (PPIs), and to identify modulators of the aggregation process in protein misfolding disorders. Biochemical assays, such as immunoblotting, provide relevant information about the physical–chemical properties of oligomers and aggregates, but they require cells to be lysed and are, therefore, more static and less physiological than studies on living cells and organisms. Imaging methods include various fluorescence- or bioluminescence-based approaches and protein-fragment complementation assays (PCAs). (Funke *et al.*, 2007; Galarneau *et al.*, 2002; Ghosh, 2000; Goo and Park, 2004; Hu *et al.*, 2002; Johnsson and Varshavsky, 1994; Outeiro *et al.*, 2009; Rossi *et al.*, 1997). These methods are less informative in terms of the precise oligomeric structure, but they allow the

Table 9.1 Localization and composition of protein aggregates in protein misfolding disorders

Disease	Location	Aggregate types	Composition	References
Parkinson's disease	Substantia nigra Cerebral cortex (1)	Cytoplasmic Lewy Bodies (1)	asyn (1) Ubiquitin (1) LRRK2 (2)	(1) Savitt *et al.* (2006) (2) Zhu *et al.* (2006)
Alzheimer's disease	Neocortex Limbic region (3)	Extracellular amyloid plaques and intracellular tau tangles (3)	β-amyloid (3) HSP90, Tau, Ubiquitin (4)	(3) Bekris *et al.* (2010) (4) Liao *et al.* (2004)
Huntington's disease	Striatal and cortical areas (5)	Nuclear and cytoplasmic aggregates (6)	Huntingtin (6) Ubiquitin (6)	(5) Gourfinkel-An *et al.* (1998) (6) Davies *et al.* (1997)
ALS	Motor neurons (7)	Cytoplasmatic aggregates in neurons and astrocytes (7)	SOD I (7) TDP-43 (8) FUS/TLS (9) Ubiquitin (10)	(7) Bruijn *et al.* (1998) (8) Arai *et al.* (2006) (9) Kwiatkowski *et al.* (2009) (10) Niwa *et al.* (2002)

visualization and study of the dynamics and subcellular localization of oligomers in living cells.

The fluorescent variant of PCA methods is called bimolecular fluorescence complementation (BiFC) assay (Ghosh, 2000; Hu *et al.*, 2002). In the BiFC assay, a protein of interest is fused to a half of a fluorescent reporter protein, such as members of the green fluorescent protein (GFP) family. A second protein of interest is fused to the other half of the reporter protein. If the proteins of interest interact, they get close enough for the two halves of the reporter protein to come together and reconstitute the functional reporter protein (Fig. 9.1). The signal from the reconstituted reporter protein can be quantified by conventional methods, and is, in principle, proportional to the levels of interaction between the two proteins of interest. These systems can also be used for the study of homo-dimerization and -oligomerization of a single protein in normal or pathological conditions, by fusing the same protein

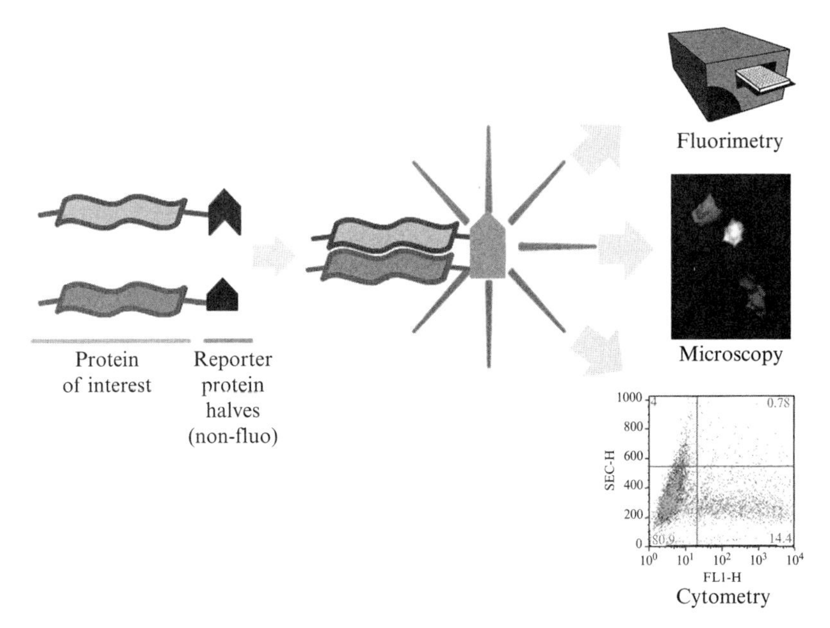

Figure 9.1 Fundamentals of BiFC. Two proteins of interest (or the same protein of interest) are fused to two nonfluorescent halves of a fluorescent reporter protein. If the proteins interact, they bring together the halves of the reporter protein and reconstitute the functional fluorophore. Fluorescence can be then visualized and measured by conventional methods, such as microscopy, flow cytometry, or fluorimetry. (For color version of this figure, the reader is referred to the Web version of this chapter.)

of interest to both halves of the reporter protein. This is the case of the BiFC systems applied to models of protein misfolding disorders (Goncalves *et al.*, 2010; Herrera *et al.*, 2011; Lajoie and Snapp, 2010; Outeiro *et al.*, 2008).

Although the characterization of individual PPI pairs is possible with BiFC, cell processes are governed by multiple and highly synchronized PPIs. Thus, a multicolour BiFC approach has been developed to identify alternative interaction partners in living cells (Hu and Kerppola, 2003; Kodama and Wada, 2009). The sequences of the C-terminal fragments of GFP-derived fluorescent proteins are basically identical, but they have different excitation and emission spectra when they are combined with the N-terminal fragments characteristic of each fluorescent protein. It has been described, that up to three pairs of PPIs can be visualized in the same cells, using the N-terminal fragments of mLumin, mCerulean, and Venus and the common C-terminal fragment (Chu *et al.*, 2009).

In the present report, we describe in detail how to develop and validate a BiFC system for the visualization and study of toxic oligomeric species in living cells. Although BiFC is the main focus of this report, split-protein systems can be developed with virtually any reporter protein whose function affords the possibility to be reconstituted from independent fragments, such as luciferase (Kim *et al.*, 2009). We focus on proteins involved in neurodegeneration as an example, but this type of model can be readily used for the study of protein oligomerization in general. First, a fluorescent reporter protein has to be chosen considering its particular properties. For example, the third generation of fluorescent proteins is usually much brighter than the first generation, but they may produce more background and lower signal-to-noise ratios. Particular mutants can overcome these problems. Also, excitation and emission wavelengths should be taken in consideration if the assay will be used in combination with other fluorescent constructs. Second, a series of constructs have to be built, where the fragments of the reporter protein are fused to either the N- or the C-terminus of the protein of interest. Unless detailed information on the 3D structure of the protein is available, there is no way to predict *a priori* the combination of constructs that will allow the dimerization of the protein of interest. Furthermore, control constructs should also be generated to rule out the spontaneous binding of the reporter fragments. Third, constructs have to be tested for their functionality in living cells and their physiological relevance.

2. Required Equipment and Materials

Equipment

- Basic molecular biology/cloning equipment (PCR machine, microcentrifuge, horizontal electrophoresis systems, baths, UV transilluminator, and Gel documentation system).
- Basic cell culture equipment (cell culture hood, incubator, and electroporator if this is the method of choice for transfection).
- Basic immunoblot equipment (vertical electrophoresis system, film cassettes, and developer).
- Inverted widefield or confocal fluorescence microscopes, equipped with filters, objectives, charge-coupled device camera and software suitable for the visualization of the cells, and fluorophores of choice.
- Flow cytometer and/or fluorimeter/microtiter plate reader, equipped with filters and mirrors suitable for the detection of the fluorophores of choice.

Materials

- Transfectable, monolayered and adherent cells.
- Plasmid vectors for expression of fusion proteins.
- DNA sequence of the reporter protein.
- DNA sequence of the proteins of interest.
- PCR primers.

- Basic cloning materials [cloning enzymes (polymerase, phosphatase, restriction enzymes, ligase), dNTPs, competent bacteria, bacteria media and plates, antibiotics, agarose, TAE buffer, DNA extraction and purification kits, DNAse- and RNAse-free miliQ water. . .].
- Basic cell culture materials (cell culture media, serum, dishes, trypsin, antibiotics, transfection reagents. . .).
- Basic immunoblotting materials [acrylamide solution, ammonium persulfate (APS), TEMED, sodium dodecyl sulfate (SDS), Tris–Glycine buffer, methanol, primary and secondary antibodies, nonfat dry milk, bovine serum albumin (BSA), Tween-20].

3. Protocol

3.1. Choosing the reporter protein

A fluorescent reporter protein must be chosen according to the purpose of the experiment and the tools available. First- and second-generation fluorescent proteins, such as GFP, YFP, EGFP, and EYFP, are less bright and require a conformational maturation to efficiently reconstitute the functional fluorophore. Most particularly, BiFC systems based on these proteins require the incubation of cells at 30 °C in order to improve folding efficiency and reach higher levels of fluorescence. Many cell types are particularly sensitive to low temperatures, which trigger many undesired intracellular pathways and may mask biologically relevant events. Third-generation fluorescent proteins, such as Venus, are much brighter and have intrinsically improved folding efficiency, being able to reconstitute the fluorophore very efficiently even at 37 °C. Also, they enable the use of weaker promoters and hence, to approximate the expression of the interactors at physiological levels. However, they also have several drawbacks. They frequently produce higher background and lower signal-to-noise ratios than the original proteins, and their signal is closer to the maximum range of intensity detectable by the instruments. As a consequence, the identification of enhancers of oligomerization may not be as efficient as with the less bright fluorophores.

Besides the brightness and folding efficiency, probably the most limiting aspect is the color of the reporter protein, especially if the BiFC system will be combined with other fluorescent markers. This is also limiting in multicolor BiFC systems, where overlapping of excitation and emission spectra of the fluorophore pair can jeopardize the usefulness of the model. There are many fluorescent protein reporters described to work for BiFC assays covering the whole fluorescent spectrum (Table 9.2). The most widely used pair for multicolor BiFC is Venus (yellow) and Cerulean (blue). BiFC plasmids can be obtained from various sources but, as we will show

Table 9.2 Fluorescent protein reporters used in BiFC systems

	Protein (Acronym)	Max. excitation (nm)	Max. emission (nm)	Molar extinction coefficient (M/cm)	Quantum yield	Relative brightness (% of EGFP)	Splited aminoacid position	Temperature (°C)	Improved point mutants	Species[a]	References
Green fluorescent proteins	GFP (wt)	395/475	509	21,000	0.77	48	155/156	30, 25		P	Kodama and Wada (2009)
	EGFP	484	507	56,000	0.6	100	157/158	30	F223R	M, Y	Barnard *et al.* (2008)
Cyan fluorescent proteins	BFP	380	440	29,000	0.31	27	172/173	30		M	Kodama and Wada (2009),
	ECFP	439, 452	476, 478	32,500	0.4	39	154/155, 172/173	30		P, M	Hu and Kerppola (2003),
	Cerulean	433	475	43,000	0.62	79	154/155, 172/173	30, 37		M	Shyu *et al.* (2006)
Yellow fluorescent proteins	EYFP	514	527	83,400	0.61	151	154/155, 172/173	30	F46L, F64L, M153T, T203Y (Venus), Q69M (Citrine)	P, M	Hu and Kerppola (2003), Kodama *et al.* (2008)
	Venus	515	528	92,200	0.57	156	144/145, 154/155, 158/159, 172/173	30, 37	V150L, I152L, M153T, F46L, F64L, Y203H, V163A S175G	M, X	Shyu *et al.* (2006), Saka *et al.* (2007), Kodama and Hu (2010), Rose *et al.* (2010)
	mCitrine	516	529	77,000	0.76	174	154/155, 172/173	37		M	Shyu *et al.* (2006)
Orange fluorescent proteins	Kusabira Orange	548	559	51,600	0.6	92	–	30		C	Su *et al.* (2006, oral communication)[b]

(Continued)

Table 9.2 (*Continued*)

	Protein (Acronym)	Max. excitation (nm)	Max. emission (nm)	Molar extinction coefficient (M/cm)	Quantum yield	Relative brightness (% of EGFP)	Splited aminoacid position	Temperature (°C)	Improved point mutants	Species[a]	References
Red fluorescent proteins	mRFP1	549	570	38,100	0.43	41	154/155, 168/169	26–28	Q56T	P, T	Jach *et al.* (2006), Kodama and Wada (2009), Hodgson *et al.* (2011)
	mCherry	587	610	72,000	0.22	47	159/160	25		M	Fan *et al.* (2008)
Far red fluorescent proteins	mKate	588	625/635	51,000/ 45,000	0.33	53	151/152, 165/166	37	S158A (mLumin), S158C	M	Chu *et al.* (2009)
	mLumin	587	621	70,000	0.46	100	151/152, 165/167	37		M	Chu *et al.* (2009)

[a] C, *C. elegans;* M, mammals; P, plants; T, *Trichoplusia* sp.; X *Xenopus* sp.; Y, *S. cerevisiae.*
[b] http://www.dmphotonics.com/PW2011/PW2011BiFC.htm.

below, it is not difficult to create home-made BiFC platforms with virtually any protein derived from GFP.

GFP-related proteins have the same length and a very similar sequence, differing only in a small number of amino acids. Since the GFP protein is organized in a β-barrel structure composed of antiparallel-arranged β-strands, the sites of fragmentation are chosen in β-strands surrounding areas that do not interfere with the structural reconstitution of the fluorophore (Hu and Kerppola, 2003). Thus, the partition of these proteins for BiFC systems is usually done in the same sites, typically the amino acids 144, 155, 158, or 172 (Dupre *et al.*, 2007; Rose *et al.*, 2010) (Table 9.2). Since GFP-derived proteins are 238 amino acids in length, the N-terminal halves are always longer than the C-terminal halves.

The combination of the BiFC halves is not necessarily complementary. A half of a GFP-derived protein can be combined with the complementary half of any other GFP-derived protein. Also, a 1–172 N-terminal fragment can reconstitute a functional fluorophore when it is combined with a 155–238 C-terminal fragment. For example, the 1–172 N-terminal halves of Venus and Cerulean, two third-generation GFP-derived proteins, can be combined with a 155–238 C-terminal half of the enhanced cyan fluorescent protein (ECFP), resulting in proteins with two different colors (Shyu *et al.*, 2006). The color of the combination will be mostly determined by the N-terminal halves, because the C-terminal halves of all GFP-derived proteins are almost identical.

In our experience, the length and origin of the split proteins are not a limiting aspect, but rather a matter of choice and availability of tools. Testing pairs of constructs with different lengths and combinations of colors does not improve the chances to get a good BiFC system. Instead, we recommend choosing a color that suits the desired experimental design and a partition that is readily available or more common for that particular protein.

3.2. Creating a BiFC platform

Below, we describe how to make a hypothetical BiFC platform with Venus 1–158 and 159–238 halves inserted in a pcDNA^{TM} 3.1+ plasmid (Invitrogen, Life Technologies Corp., Carslbad, CA, USA) backbone. The relative position of the protein of interest versus the reporter protein is extremely limiting. It is impossible to predict *a priori* the relative position that will allow the interaction between the proteins of interest. While we found huntingtin (Htt) had to be located at the N-termini of both halves of the fluorescent protein, in the case of alpha-synuclein (asyn) we found that complementation worked better if it was located at the C-terminus of the first half and the N-terminus of the second half (Herrera *et al.*, 2011; Outeiro *et al.*, 2008). Additionally, a molecular linker between the protein of interest and the reporter protein might be necessary for the interaction to occur, because it gives more flexibility to the fusion proteins (Kerppola,

2006). This is the case for asyn, but it did not work with Htt (Herrera *et al.*, 2011; Outeiro *et al.*, 2008). Therefore, a series of different constructs should be tested, carrying the protein of interest before or after the reporter protein. Constructs should be ideally tested for the intracellular localization, function, and stability of the protein of interest. Special attention should be paid to possible posttranslational processing of the protein of interest when the constructs are designed. For example, the amyloid beta peptide (APP) suffers posttranslational cleavage and gives rise to the Aβ peptide, which is the toxic fragment that aggregates in AD and other neurodegenerative disorders. A successful APP BiFC system should locate the fluorophore fragment in a position that ensures it remains linked to the Aβ peptide after cleavage.

A full BiFC platform should consist of at least four plasmids, two per Venus half, with or without a stop codon (Fig. 9.2). This will allow the

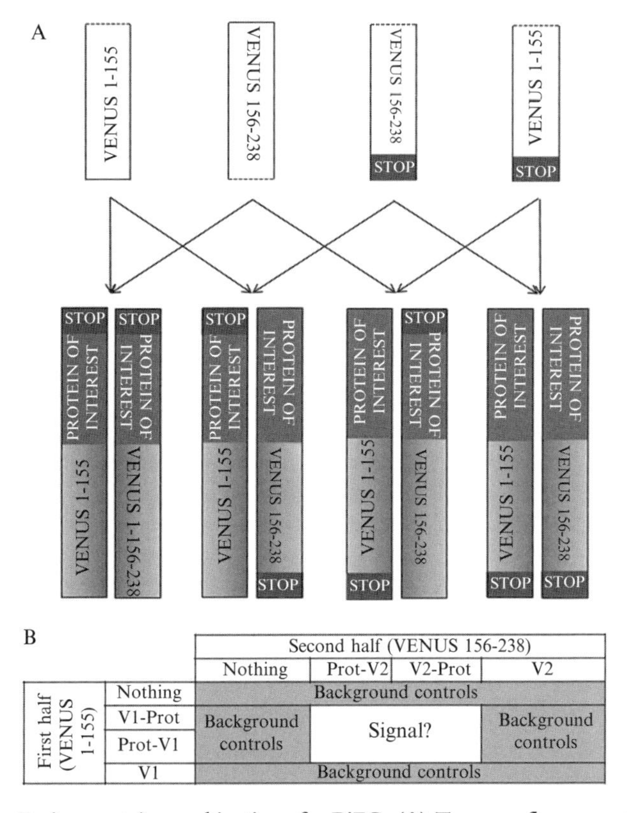

Figure 9.2 Fusion protein combinations for BiFC. (A) Two nonfluorescent fragments of the reporter protein, with or without stop codon, are fused to the interaction partners in different relative positions. All the combinations should be tested for their ability to reconstitute the functional fluorophore and emit fluorescence. (B) Besides these combinations, a series of background controls should be included in the experimental design. (For color version of this figure, the reader is referred to the Web version of this chapter.)

insertion of the cDNAs encoding for the proteins of interest before or after the cDNA of the reporter protein. It is also recommended to have four additional plasmids with a molecular linker of glycine–serine before or after the Venus halves, in case the combinations without the linker fail. Sets of PCR primers can be designed to amplify the desired partitions from the full-length Venus and insert them in the vector of choice. Special attention should be paid to the location of start and stop codons. They have to be removed when they are between the protein of interest and the reporter protein, and this has to be considered when primers are designed. In our example, we included *Bam*HI and *Eco*RI restriction sites in the PCR products, in order to clone them in the middle of the multiple cloning site of the pcDNA™ 3.1+ (Invitrogen, Life Technologies Corp.). Using a proofreading polymerase, we amplified these fragments following manufacturer's instructions. PCR products were purified in order to remove the polymerase, its buffer, primers, and dNTPs, which would interfere with the restriction reaction. The PCR products and the pcDNA™ 3.1+ plasmid (Invitrogen, Life Technologies Corp.) were digested with *Bam*HI and *Eco*RI following manufacturer's instructions. Alkaline phosphatase (0.5 μl, 30 min, two consecutive times) was added to the digested plasmid, but not the PCR products, in order to avoid unspecific religation of the plasmid. Digested plasmid and PCR products were purified again, and ligated with T4 DNA ligase following manufacturer's instructions. Bacteria were transformed with the ligation products and seeded on Agar/LB 1× plates containing ampicillin (100 μg/ml). Individual clones were grown in 3 ml of LB 1× overnight at 37 °C, and minipreps were prepared from these cultures in order to confirm insertion. Insertion was confirmed by a restriction reaction with *Bam*HI and *Eco*RI, and by sequencing using the same primers that were used for amplification of the Venus halves. These procedures lead to a series of plasmids containing two halves of the Venus protein with or without a stop codon or a linker. The DNA sequence for the protein of interest can then be inserted in these plasmids by general cloning procedures.

To set up a multicolour BiFC, the protein that interacts with the other two proteins of interest should be fused to the common C-terminal fragment of the GFP family. The alternative interactors should be fused to N-terminal fragments of fluorophores with different spectral characteristics, such as Venus and Cerulean. Besides this basic difference, the rules and limiting steps for multicolour and single color BiFC are essentially the same.

3.3. Testing the BiFC system

Once the BiFC constructs are made, they have to be tested for their functionality in living cells. A simple test or battery of tests should be defined *a priori*, based on what it is expected to happen, in order to determine which pair of constructs is the best one. For some proteins of interest, such as asyn, changes in fluorescence intensity are enough to determine which combination is the best one. However, other proteins require further tests. For example, the Htt BiFC systems showed fluorescence in several combinations, but aggregates in only one of them. Since we expected aggregate formation, we considered that this combination was the most specific or physiologically relevant.

Cells should be transfected with all possible combinations of plasmids, including a number of controls for background fluorescence and spontaneous binding of the reporter protein halves (Fig. 9.2), in order to prove the specificity of the BiFC system. Besides the combinations and controls shown in Fig. 9.2, additional controls can be designed. For example, a mutant, noninteracting variant of the protein of interest, if available, could be a good negative control; a positive control could be an already described functional BiFC platform with the same fluorophore, or a construct containing the protein of interest and the full-length fluorophore. The result of transfection with the different BiFC constructs should be checked by means of complementary quantitative and imaging methods of choice. A quantitative method, such as cytometry or fluorimetry, will enable the comparison of different groups in terms of fluorescence intensity. Specific binding is expected to be more efficient and, therefore, it should give higher levels of fluorescence.

Microscopy will enable the determination of qualitative changes, such as generation, intracellular localization, and distribution of aggregates and cell morphology. For example, the asyn-venus BiFC system displays homogeneous fluorescence in all cell compartments, including the nucleus (Fig. 9.3A). On the other hand, the Htt-Venus system shows aggregates and the fluorescence is cytosolic, with some exceptions (Fig. 9.3A and B). Some qualitative changes can also be quantified. We were able to quantify number, size, and location of Htt aggregates, and the percentage of cells with diffuse fluorescence (oligomeric species) versus cells with Htt aggregates (Herrera *et al.*, 2011). For example, less than 1% H4 or HEK cells transfected with Htt BiFC plasmids showed fluorescence in their nuclei, but this percentage was highly increased in HT22 cells (6–10%) (Fig. 9.3B and C). This type of data can provide relevant information about the behavior of oligomeric species in normal and pathological conditions.

It is highly recommended to define further the physical–chemical nature of oligomeric species detected by BiFC. Immunoblotting of denaturalized protein samples (SDS-PAGE, where SDS is present in all buffers and gels) should show two bands, corresponding to the two fusion proteins that constitute the BiFC system. This would indicate the relative levels of the fusion proteins present within cells. Immunoblotting of native samples (in the absence of SDS), on the other hand, would give an indication of the relative size and amount of oligomeric species. Sucrose gradients can be used to determine more accurately the size of oligomeric species and larger aggregates. The levels of large or insoluble aggregates in the samples can be

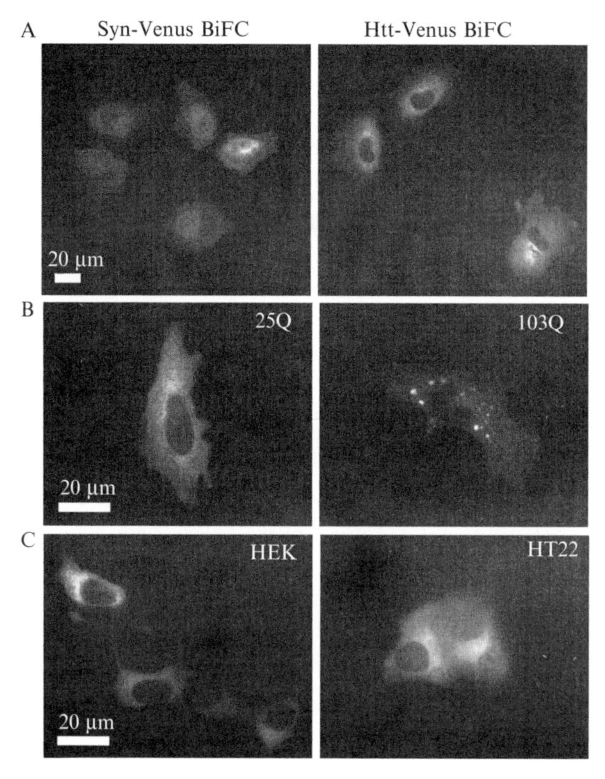

Figure 9.3 Qualitative changes in different BiFC systems. (A) Comparison of the distribution of fluorescence in asyn- and Htt-Venus BiFC systems. While asyn oligomers are present in cytosol and nucleus of H4 cells, Htt oligomers are only present in the cytosol. (B) Cells transfected with wild type Htt (25Q) BiFC constructs show significantly less aggregates then cells transfected with mutant Htt (103Q). The presence of large aggregates is correlated with a decrease in the fluorescence of the rest of the cell, indicating that large aggregates are recruiting Htt oligomers. (C) HT22 cells have significantly more nuclear fluorescence than HEK and H4 cells, suggesting that these cells lack some of the systems that keep Htt in the cytosolic compartment. (For color version of this figure, the reader is referred to the Web version of this chapter.)

determined by filter trap assays. In these assays, protein extracts are passed through a membrane of a particular pore size. The membrane will trap large, insoluble aggregates, but it will allow small-order aggregates, such as dimers and oligomers, to pass through. These large aggregates can be then detected and quantified by immunoblotting. It is noticeable that the degree of solubility of any given protein or aggregate depends strongly on the components of the lysis buffer, especially the detergents (Triton X-100^{TM}, NP-40, Tween-20^{TM}, etc). Each type and concentration of detergent might give a different pattern of aggregation, and this should be considered when choosing the lysis buffers.

Finally, if the oligomeric species of a particular protein are expected to be toxic, it could be interesting to check the toxicity of the BiFC fusion proteins. There are many toxicity assays commercially available, that could be used successfully to determine the toxicity of oligomeric species.

3.4. Optimizing the BiFC system

Absence of signal and background fluorescence are the most relevant issues that can complicate the use of a BiFC system and, sometimes, the interpretation of the results.

Absence of signal may be due to lack of interaction between the proteins of interest, low transfection efficiency, or poor reconstitution of the fluorophore. The choice of transfection reagent depends on their efficiency in each particular cell model. Transfection of H4 cells with Htt exon 1-Venus BiFC plasmids in a proportion 1:3 [DNA(μg): FuGENE® 6 (μl)] allowed the detection of oligomeric species by cytometry and microscopy, but only sporadically by immunoblotting techniques. A proportion 1:5 worked better for detection of Htt by immunoblotting. Therefore, the optimization of transfection conditions is highly recommended.

Transfection efficiency may vary from one experimental group to another or between experiments. Some authors recommend co-transfection of cells with a second full-length fluorophore as a control of transfection efficiency. For example, a plasmid containing full-length DsRed can be co-transfected with an EGFP BiFC system. The meaningful signal would then be the ratio between the average fluorescence of EGFP and DsRed. This procedure is conceptually correct, and we recommend it if there is high variability of transfection efficiency between samples and experiments. However, this procedure could jeopardize genetic screenings where cells are transfected with a fourth plasmid. It is frequently difficult to transfect cells with three plasmids or more, and when it is possible it also increases the variability between samples and experiments. Furthermore, in our experience, transfection efficiency does not vary significantly between samples of the same experiment and reproducibility between experiments can be maintained within reasonable limits, simply by following always exactly the same transfection protocol.

If BiFC systems with two versions of a protein of interest are compared, one version of the protein can be present within cells at higher levels than the other, even when the transfection protocol works perfectly. When cells were transfected with wild type (25Q) or mutant (103Q) Htt BiFC constructs, we and others observed that the levels of 103Q Htt fusion proteins were much lower than the levels of 25Q Htt fusion proteins (Herrera et al., 2011; Lajoie and Snapp, 2010). As a consequence, cells transfected with 103Q constructs showed a wild type phenotype. This was reverted by matching the levels of expression of 25Q and 103Q BiFC systems. This may be due to many factors, such as differences in the length of the fusion proteins, a higher expression efficiency of shorter polyQ tracts, or an increased turnover rate of mutant forms. In any case, efforts should be made to match the levels of expression of the fusion proteins that are going to be compared.

Absence of signal even when transfection efficiency is good and equal between experimental groups means, most likely, that (a) the proteins of interest do not interact or (b) the fluorophore is failing to reconstitute. As stated above, in BiFC models of protein misfolding disorders, it is known *a priori* that the protein of interest dimerizes, oligomerizes, and aggregates. Therefore, at least one of the possible BiFC combinations should give some signal. If there is no signal in any combination, it should be checked whether there is signal at 30 °C, the optimal temperature for the reconstitution of the fluorophore. In all cases, it is recommended to have some positive and negative controls for reconstitution of the fluorophore.

Background fluorescence is another major issue in BiFC systems. Proper controls for background fluorescence and spontaneous binding of the reporter protein halves should be considered and included in the experimental design (Fig. 9.2). Some cells have some degree of auto-fluorescence (e.g., some yeast strains or dying cells) that should be determined by analyzing cells without any construct. In some cases, fusion proteins containing one half of the reporter protein show low levels of fluorescence by themselves, even in the absence of the other half. This should be assessed by testing cells transfected with single constructs for background fluorescence. Finally, there can be some spontaneous binding of reporter protein halves, independent of the interaction of the proteins of interest (Herrera *et al.*, 2011; Robida and Kerppola, 2009; Saka *et al.*, 2007). This type of background is especially present in the brightest members of the GFP family, and it should be assessed by combining constructs with and without the protein of interest. Alternatively, a construct containing the protein of interest could be combined with a complementary construct containing a protein that does not interact with the protein of interest or with a mutant, noninteracting variant of the protein, if it is known.

Spontaneous binding can be strongly reduced by specific mutations in the halves of the reporter protein (Table 9.2). The Venus protein is probably the most studied in this sense. Venus is a third-generation fluorescent protein, derived from EYFP by the insertion of three point mutations. It is very bright, gives significant levels of fluorescence at 37 °C, and is the one recommended for multicolor BiFC in combination with Cerulean (Shyu *et al.*, 2006). However, it has higher levels of spontaneous binding and lower signal-to-noise ratios than its antecessors YFP and EYFP. Because of Venus appealing properties for BiFC, many efforts have been made to find mutants that improve its signal-to-noise ratio. This problem was described for the first time in studies in *Xenopus* (Saka *et al.*, 2007). Their approach was to mutate part of the key residues that produced Venus back to the original

EYFP residues. They found that the mutation T153M reduced significantly the background and increased signal-to-noise ratios in *Xenopus*, and this was confirmed later in mammalian cells (Robida and Kerppola, 2009). However, recent results indicate that T153M mutants do not increase signal-to-noise ratio in all experimental paradigms, and that it should be confirmed for each particular system (Kodama and Hu, 2010). They also identified mutants that increase the signal-to-noise ratio in COS cells, V150L and I152L, the V150L mutant showing weaker fluorescence and lower complementation efficiency.

In summary, BiFC assays should be set up and optimized empirically for each particular protein of interest or experimental condition. Parameters such as the relative location of the protein of interest versus the reporter protein, physical–chemical properties of the reporter protein of choice, temperature of incubation, or point mutations to decrease spontaneous binding and improve signal-to-noise ratios should be considered.

ACKNOWLEDGEMENTS

S. G. is supported by a Ph.D. Fellowship from AXA Research Fund, France. F. H. is supported by Fundação para a Ciência e Tecnologia (SFRH/BPD/63530/2009). T. F. O. is supported by a Marie Curie International Reintegration Grant, an EMBO Installation Grant, and Fundação para a Ciência e Tecnologia.

REFERENCES

Arai, T., Hasegawa, M., Akiyama, H., *et al.* (2006). TDP-43 is a component of ubiquitin-positive tau-negative inclusions in frontotemporal lobar degeneration and amyotrophic lateral sclerosis. *Biochem. Biophys. Res. Commun.* **351,** 602–611.

Arrasate, M., Mitra, S., Schweitzer, E. S., Segal, M. R., and Finkbeiner, S. (2004). Inclusion body formation reduces levels of mutant huntingtin and the risk of neuronal death. *Nature* **431,** 805–810.

Barnard, E., McFerran, N. V., Trudgett, A., Nelson, J., and Timson, D. J. (2008). Detection and localisation of protein-protein interactions in *Saccharomyces cerevisiae* using a split-GFP method. *Fungal Genet. Biol.* **45,** 597–604.

Bekris, L. M., Yu, C. E., Bird, T. D., and Tsuang, D. W. (2010). Genetics of Alzheimer disease. *J. Geriatr. Psychiatry Neurol.* **23,** 213–227.

Bodner, R. A., Outeiro, T. F., Altmann, S., *et al.* (2006). Pharmacological promotion of inclusion formation: A therapeutic approach for Huntington's and Parkinson's diseases. *Proc. Natl. Acad. Sci. USA* **103,** 4246–4251.

Bruijn, L. I., Houseweart, M. K., Kato, S., *et al.* (1998). Aggregation and motor neuron toxicity of an ALS-linked SOD1 mutant independent from wild-type SOD1. *Science* **281,** 1851–1854.

Chu, J., Zhang, Z., Zheng, Y., *et al.* (2009). A novel far-red bimolecular fluorescence complementation system that allows for efficient visualization of protein interactions under physiological conditions. *Biosens. Bioelectron.* **25,** 234–239.

Davies, S. W., Turmaine, M., Cozens, B. A., *et al.* (1997). Formation of neuronal intranuclear inclusions underlies the neurological dysfunction in mice transgenic for the HD mutation. *Cell* **90,** 537–548.

Dupre, D. J., Robitaille, M., Richer, M., Ethier, N., Mamarbachi, A. M., and Hebert, T. E. (2007). Dopamine receptor-interacting protein 78 acts as a molecular chaperone for Ggamma subunits before assembly with Gbeta. *J. Biol. Chem.* **282,** 13703–13715.

Fan, J. Y., Cui, Z. Q., Wei, H. P., Zhang, Z. P., Zhou, Y. F., Wang, Y. P., and Zhang, X. E. (2008). Split mCherry as a new red bimolecular fluorescence complementation system for visualizing protein–protein interactions in living cells. *Biochem. Biophys. Res. Commun.* **367,** 47–53.

Funke, S. A., Birkmann, E., Henke, F., Gortz, P., Lange-Asschenfeldt, C., Riesner, D., and Willbold, D. (2007). Single particle detection of Abeta aggregates associated with Alzheimer's disease. *Biochem. Biophys. Res. Commun.* **364,** 902–907.

Galarneau, A., Primeau, M., Trudeau, L. E., and Michnick, S. W. (2002). Beta-lactamase protein fragment complementation assays as in vivo and in vitro sensors of protein–protein interactions. *Nat. Biotechnol.* **20,** 619–622.

Ghosh, I. (2000). Antiparallel leucine zipper-directed protein reassembly: Application to the green fluorescent protein. *J. Am. Chem. Soc.* **122,** 5658–5659.

Goncalves, S. A., Matos, J. E., and Outeiro, T. F. (2010). Zooming into protein oligomerization in neurodegeneration using BiFC. *Trends Biochem. Sci.* **35,** 643–651.

Goo, J. H., and Park, W. J. (2004). Elucidation of the interactions between C99, presenilin, and nicastrin by the split-ubiquitin assay. *DNA Cell Biol.* **23,** 59–65.

Gourfinkel-An, I., Cancel, G., Duyckaerts, C., *et al.* (1998). Neuronal distribution of intranuclear inclusions in Huntington's disease with adult onset. *Neuroreport* **9,** 1823–1826.

Hansen, C., Angot, E., Bergstrom, A. L., *et al.* (2011). Alpha-Synuclein propagates from mouse brain to grafted dopaminergic neurons and seeds aggregation in cultured human cells. *J. Clin. Invest.* **121,** 715–725.

Herrera, F., Tenreiro, S., Miller-Fleming, L., and Outeiro, T. F. (2011). Visualization of cell-to-cell transmission of mutant huntingtin oligomers. *PLoS Curr.* **3,** RRN1210.

Hodgson, J. J., Arif, B. M., and Krell, P. J. (2011). Interaction of Autographa californica multiple nucleopolyhedrovirus cathepsin protease progenitor (proV-CATH) with insect Baculovirus Chitinase as a mechanism for proV-CATH cellular retention. *J. Virol.* **85,** 3918–3929.

Hu, C. D., and Kerppola, T. K. (2003). Simultaneous visualization of multiple protein interactions in living cells using multicolor fluorescence complementation analysis. *Nat. Biotechnol.* **21,** 539–545.

Hu, C. D., Chinenov, Y., and Kerppola, T. K. (2002). Visualization of interactions among bZIP and Rel family proteins in living cells using bimolecular fluorescence complementation. *Mol. Cell* **9,** 789–798.

Jach, G., Pesch, M., Richter, K., Frings, S., and Uhrig, J. F. (2006). An improved mRFP1 adds red to bimolecular fluorescence complementation. *Nat. Methods* **3,** 597–600.

Johnsson, N., and Varshavsky, A. (1994). Split ubiquitin as a sensor of protein interactions in vivo. *Proc. Natl. Acad. Sci. USA* **91,** 10340–10344.

Kerppola, T. K. (2006). Design and implementation of bimolecular fluorescence complementation (BiFC) assays for the visualization of protein interactions in living cells. *Nat. Protoc.* **1,** 1278–1286.

Kim, S. B., Sato, M., and Tao, H. (2009). Split Gaussia luciferase-based bioluminescence template for tracing protein dynamics in living cells. *Anal. Chem.* **81,** 67–74.

Kodama, Y., and Hu, C. D. (2010). An improved bimolecular fluorescence complementation assay with a high signal-to-noise ratio. *Biotechniques* **49,** 793–805.

Kodama, Y., and Wada, M. (2009). Simultaneous visualization of two protein complexes in a single plant cell using multicolor fluorescence complementation analysis. *Plant Mol. Biol.* **70,** 211–217.

Kodama, Y., Shinya, T., and Sano, H. (2008). Dimerization of N-methyltransferases involved in caffeine biosynthesis. *Biochimie* **90,** 547–551.

Kwiatkowski, T. J., Jr., Bosco, D. A., Leclerc, A. L., *et al.* (2009). Mutations in the FUS/TLS gene on chromosome 16 cause familial amyotrophic lateral sclerosis. *Science* **323,** 1205–1208.

Lajoie, P., and Snapp, E. L. (2010). Formation and toxicity of soluble polyglutamine oligomers in living cells. *PLoS One* **5,** e15245.

Lee, S. J., Desplats, P., Sigurdson, C., Tsigelny, I., and Masliah, E. (2010). Cell-to-cell transmission of non-prion protein aggregates. *Nat. Rev. Neurol.* **6,** 702–706.

Liao, L., Cheng, D., Wang, J., *et al.* (2004). Proteomic characterization of postmortem amyloid plaques isolated by laser capture microdissection. *J. Biol. Chem.* **279,** 37061–37068.

Lindgren, M., and Hammarstrom, P. (2010). Amyloid oligomers: Spectroscopic characterization of amyloidogenic protein states. *FEBS J.* **277,** 1380–1388.

Niwa, J., Ishigaki, S., Hishikawa, N., *et al.* (2002). Dorfin ubiquitylates mutant SOD1 and prevents mutant SOD1-mediated neurotoxicity. *J. Biol. Chem.* **277,** 36793–36798.

Outeiro, T. F., Putcha, P., Tetzlaff, J. E., *et al.* (2008). Formation of toxic oligomeric alpha-synuclein species in living cells. *PLoS One* **3,** e1867.

Outeiro, T. F., Klucken, J., Bercury, K., *et al.* (2009). Dopamine-induced conformational changes in alpha-synuclein. *PLoS One* **4,** e6906.

Robida, A. M., and Kerppola, T. K. (2009). Bimolecular fluorescence complementation analysis of inducible protein interactions: Effects of factors affecting protein folding on fluorescent protein fragment association. *J. Mol. Biol.* **394,** 391–409.

Rose, R. H., Briddon, S. J., and Holliday, N. D. (2010). Bimolecular fluorescence complementation: Lighting up seven transmembrane domain receptor signalling networks. *Br. J. Pharmacol.* **159,** 738–750.

Rossi, F., Charlton, C. A., and Blau, H. M. (1997). Monitoring protein-protein interactions in intact eukaryotic cells by beta-galactosidase complementation. *Proc. Natl. Acad. Sci. USA.* **94,** 8405–8410.

Saka, Y., Hagemann, A. I., Piepenburg, O., and Smith, J. C. (2007). Nuclear accumulation of Smad complexes occurs only after the midblastula transition in Xenopus. *Development* **134,** 4209–4218.

Savitt, J. M., Dawson, V. L., and Dawson, T. M. (2006). Diagnosis and treatment of Parkinson disease: Molecules to medicine. *J. Clin. Invest.* **116,** 1744–1754.

Shyu, Y. J., Liu, H., Deng, X., and Hu, C. D. (2006). Identification of new fluorescent protein fragments for bimolecular fluorescence complementation analysis under physiological conditions. *Biotechniques* **40,** 61–66.

Zhu, X., Siedlak, S. L., Smith, M. A., Perry, G., and Chen, S. G. (2006). LRRK2 protein is a component of lewy bodies. *Ann. Neurol.* **60,** 617–618.

CHAPTER TEN

3D/4D Functional Imaging of Tumor-Associated Proteolysis: Impact of Microenvironment

Kamiar Moin,[*,†] Mansoureh Sameni,[*] Bernadette C. Victor,[†] Jennifer M. Rothberg,[†] Raymond R. Mattingly,[*,†] *and* Bonnie F. Sloane[*,†]

Contents

1. Introduction 176
   1.1. Why image proteolysis rather than protease activity? 176
   1.2. Why image tumor proteolysis? 177
2. Assays for Functional Imaging of Proteolysis 178
   2.1. Probes 178
   2.2. Analysis 181
3. 3D/4D Models for Analysis of Biological Processes Linked to Proteolysis 184
4. Live-Cell Imaging of MAME Models: A Screening Tool for Drug Discovery 186
Acknowledgments 189
References 189

Abstract

Proteases play causal roles in many aspects of the aggressive phenotype of tumors, yet many of the implicated proteases originate from tumor-associated cells or from responses of tumor cells to interactions with other cells. Therefore, to obtain a comprehensive view of tumor proteases, we need to be able to assess proteolysis in tumors that are interacting with their microenvironment. As this is difficult to do *in vivo*, we have developed functional live-cell optical imaging assays and 3D and 4D (i.e., 3D over time) coculture models. We present here a description of the probes used to measure proteolysis and protease activities, the methods used for imaging and analysis of proteolysis and the 3D

[*] Department of Pharmacology, Wayne State University School of Medicine, Detroit, Michigan, USA
[†] Barbara Ann Karmanos Cancer Institute, Wayne State University School of Medicine, Detroit, Michigan, USA

Methods in Enzymology, Volume 506
ISSN 0076-6879, DOI: 10.1016/B978-0-12-391856-7.00034-2

and 4D models used in our laboratory. Of course, all assays have limitations; however, we suggest that the techniques discussed here will, with attention to their limitations, be useful as a screen for drugs to target the invasive phenotype of tumors.

1. Introduction

Recent studies in our laboratory have focused on establishing live-cell assays to image the proteolysis that is associated with the progression of premalignant breast lesions to malignant carcinomas. This chapter will use our studies to illustrate how the interactions of tumor cells with their microenvironment contribute to this proteolysis and how functional imaging assays and 3D/4D coculture models might be used to identify druggable targets and screen therapeutic agents.

1.1. Why image proteolysis rather than protease activity?

Proteolysis or the hydrolytic degradation of proteins occurring as the result of interactions among proteases of more than one catalytic type, that is, proteolytic pathways or networks, has been shown to be critical to malignant progression (e.g., see DeClerck and Laug, 1996; Ellerbroek *et al.*, 1998; Kim *et al.*, 1998; Krol *et al.*, 2003; Muehlenweg *et al.*, 2000; Ramos-DeSimone *et al.*, 1999). Prior research in our laboratory concentrated on one cysteine cathepsin, that is, cathepsin B (for reviews, see Cavallo-Medved and Sloane, 2003; Podgorski and Sloane, 2003; Roshy *et al.*, 2003; Yan and Sloane, 2003). Analysis of any one protease or protease class, however, does not define the "tumor degradome" (Balbin *et al.*, 2003). Therefore, we established an assay to study proteolysis of extracellular matrix substrates encountered by tumor cells as they invade into normal tissues surrounding tumors. This protein-based assay contrasts with those measuring degradation of synthetic substrates that are selective for one protease or one catalytic type of protease. Data generated in our laboratory (Cavallo-Medved *et al.*, 2009; Jedeszko *et al.*, 2009; Li *et al.*, 2008; Podgorski *et al.*, 2009; Sameni *et al.*, 2000, 2001, 2003, 2008, 2009) and other laboratories (Kjoller *et al.*, 2004; Madsen *et al.*, 2011; Wolf *et al.*, 2009) by means of functional proteolysis assays that employ protein substrates support our contention that meaningful analyses of tumor proteolysis require assessment of the roles played by multiple proteases as well as assessment of how those proteases interact to modulate the activities of other proteases.

1.2. Why image tumor proteolysis?

More than one catalytic type of protease has been implicated in the progression of human tumors (for reviews, see Choong and Nadesapillai, 2003; Fuchs, 2002; Hojilla *et al.*, 2003; Podgorski and Sloane, 2003). Although the preclinical data implicating matrix metalloproteinases (MMPs) in malignant progression were particularly compelling, the clinical trials on MMP inhibitors (MMPIs) did not fulfill the promise of MMPs as therapeutic targets in cancer. There are several possible explanations for this apparent "disconnect" between the preclinical and clinical data; please see Chau *et al.* (2003), Coussens *et al.* (2002), Egeblad and Werb (2002) for insightful and thought-provoking reviews on this topic.

The failures of MMPIs in clinical trials have resulted in allegations that MMPs and, by extrapolation, other proteases are not appropriate therapeutic targets in cancer. Is this true or rather might the failure of the MMPI trials reflect problems in clinical trial design for cytostatic agents and in particular the need to use imaging (Adjei *et al.*, 2009; Ang *et al.*, 2010)? This is the case as among the critical questions is whether the MMPIs actually reached and reduced the activity of their target MMPs *in vivo*. The MMPI trials did not include surrogate endpoints so it is not known whether MMPs were actually inhibited in the patients enrolled in the trials (Chau *et al.*, 2003; Li and Anderson, 2003; McIntyre and Matrisian, 2003). Clinical trials without surrogate endpoints to monitor and confirm the efficacy of the therapeutic strategies being tested should not be viewed as definitive (Adjei *et al.*, 2009; Li and Anderson, 2003; McIntyre and Matrisian, 2003; Seymour *et al.*, 2010). There are other concerns about the MMPI trials. Coussens *et al.* (2002) cite data revealing that the tumors studied in the clinical trials did not necessarily express the particular proteases targeted by the MMPIs. None of the clinical trials with BAY 12-9566 or other MMPIs included patients with breast cancers although this is a cancer for which there had been strong preclinical evidence that MMPs impact progression. For example, Sledge and colleagues (Nozaki *et al.*, 2003) demonstrated efficacy for BAY 12-9566 in an orthotopic model in which human breast tumor cells were implanted in the mammary fat pads of mice. Furthermore, not all MMPs should be inhibited; this is clearly true for MMP-8 (collagenase-2), an MMP expressed by inflammatory neutrophils, which plays a protective role in skin cancer (Balbin *et al.*, 2003). The data on MMP-8 and more recent data on a variety of other MMPs (for reviews, see Lopez-Otin and Matrisian, 2007; Lopez-Otin *et al.*, 2009) support the concept that broad-spectrum MMPIs would have unanticipated side effects and indicate how essential it is "to define precisely the tumor degradome" (Balbin *et al.*, 2003) before using MMPIs or for that matter other protease inhibitors for cancer therapy. In fact, unanticipated side effects did occur in the MMPI clinical trials leading to limitations in the amount of MMPIs that the patients could take and in some patients necessitating MMPI-free holidays (Coussens *et al.*, 2002). How endogenous protease inhibitors factor into the equation is also of relevance. As just one example, TIMP-1 has been shown to promote carcinogenesis of squamous cell carcinoma of the skin, exerting "differential regulation on tissues in a stage-dependent manner" (Rhee *et al.*, 2004). We are still far from having a thorough understanding of the roles of proteases in cancer: this includes understanding the roles of MMPs; the roles of other classes of proteases; the roles of endogenous protease inhibitors, activators and receptors, or binding proteins; that those roles are dynamic and may change during the course of malignant progression; whether the proteases playing critical roles in malignant progression come from tumor, stromal, or inflammatory cells; whether the critical proteases are affected by interactions of the tumor with its microenvironment; what the relevant substrates are for proteases that play causal roles in malignant progression; etc. Many of these issues are discussed in a large volume on what constitutes the cancer degradome that was published in 2008 (Edwards *et al.*, 2008). Nonetheless, we do not yet know which protease(s) or proteolytic pathway is the most appropriate target for antiprotease therapies or when antiprotease therapies might prove most effective.

2. Assays for Functional Imaging of Proteolysis

The terminology "functional imaging" was originally used to describe imaging methods that assessed changes in physiological processes such as metabolism and blood flow, for example, positron emission tomography (PET) and magnetic resonance imaging. Here, we use this terminology because the live-cell assays that we have developed for imaging proteolysis can be used to quantify changes in proteolytic activity as well as to localize sites at which proteolysis is occurring (Jedeszko *et al.*, 2008). Functional imaging often uses agents or probes; in PET, for example, a glucose analogue fluorodeoxyglucose is used as an F^{18} radioisotopic tracer to detect and localize the high levels of metabolic activity associated with tumors (Mankoff *et al.*, 2007).

2.1. Probes

Probes for imaging proteases have been based on either substrates or inhibitors. A substrate probe can be a signal amplifier as, in the presence of an active target protease, the substrate will be cleaved continuously and thus the signal intensity increased. This property also makes substrate probes sensitive reagents for detecting reductions in protease activity due to protease inhibitors, etc. A potential negative with substrate probes is that the cleavage products may not remain at the site where cleavage occurred and thus would not accurately localize an active protease. This is especially true for extracellular or cell surface proteolysis where substrates and/or cleavage products might diffuse away from the site at which they are generated. Protease probes based on inhibitors that bind covalently to the active site of a protease are more effective in localizing protease activity. On the other hand, they cannot amplify the signal and therefore are less sensitive in detecting proteases that are not highly expressed or reductions in protease activity (Fonovic and Bogyo, 2007).

Whether protease probes are based on a substrate or an inhibitor, proteases are ideal targets for selective, activatable contrast agents. In this case, a fluorescent probe is synthesized in a quenched state in which a nonfluorescent quencher is attached to the probe in close proximity to the reporting fluorophore via a protease-selective peptide linker. The proximity of the fluorophore to the quencher causes transfer of energy from the former to the latter, thus preventing emission of fluorescence. When such a probe encounters an active target protease, the peptide will be cleaved and the quencher released, resulting in emission of a fluorescent signal. Alternatively, probes can be developed utilizing a synthetic (e.g., dendrimer or polylysine—see Marten *et al.*, 2002; McIntyre *et al.*, 2004, respectively) or protein (collagen) backbone to which a large number of reporters are attached via peptide linkers in close proximity to each other (for review see, Sloane *et al.*, 2006). The overabundance of the reporter in close proximity causes self-quenching due to a FRET (Forster Resonance Energy Transfer) effect (Brzostowski *et al.*, 2009). When an active target protease cleaves the reporter off the backbone, fluorescence is emitted. Such probes have been used for both *in vitro* and *in vivo* systems (Brzostowski *et al.*, 2009; McIntyre *et al.*, 2004).

2.1.1. Fluorescently tagged proteins

The primary probes that we use in our laboratory to image proteolysis are quenched fluorescent or dye quenched (DQ) extracellular matrix proteins that are commercially available from Invitrogen, that is, DQ-collagen IV and DQ-collagen I. Our use of these substrates was featured in their newsletter (Visualizing tumor metastasis: CellTracker™ dyes, DQ™ collagen, and Geltrex™ matrix. BioProbes 60, pp. 32–33, October 2009). We selected the two collagen substrates because type IV collagen is a major component of the basement membrane (Aumailley and Gayraud, 1998) and dissolution of type IV collagen has been shown to be integral to normal developmental processes and an early step in malignant progression (for review, see Liotta and Kohn, 2001). Fibrillar collagen I in the connective tissue through which tumor cells invade is an impediment to cell growth and invasion (Henriet *et al.*, 2000; Sabeh *et al.*, 2009).

By using quenched fluorescent derivatives of type IV and I collagen, we have been able to image and localize proteolysis, that is, a gain-of-function/fluorescence, by live cells (Sameni *et al.*, 2000, 2001, 2003, 2009). This is in contrast to using nonquenched FITC-labeled proteins where one is imaging a loss of fluorescence and has to fix the cells and substrate before imaging (Demchik *et al.*, 1999; Sloane, 1996).

A critical point is that, by virtue of their fluorescent labeling, the DQ-collagens are no longer native proteins. Therefore the ability to cleave these substrates may not be representative of an ability to cleave native forms of these proteins. This is likely more the case for DQ-collagen I as gelatin or denatured collagen I is readily degraded by many proteases. There are collagenase-resistant forms of collagen I in which the cleavage sites in the helical region have been mutated so that they cannot be degraded by true "collagenases" such as MMP-1 (Wu *et al.*, 1990). Studies using collagenase-resistant collagen I have demonstrated that proteolysis of collagen I is required for motility of vascular smooth muscle cells on collagen I (Li *et al.*, 2000) and for invasion of ovarian cancer cells into collagen I (Ellerbroek *et al.*, 2001). Unfortunately, DQ-collagenase-resistant collagen I is not commercially available.

2.1.2. Activity-based probes (ABPs)

We also employ ABPs developed by Bogyo and colleagues for imaging cysteine proteases of the papain family (Greenbaum *et al.*, 2002a,b). These probes are based on the broad-spectrum cysteine protease inhibitor E-64 and bind covalently to the active sites of the enzymes. Therefore, with these probes, one can image active cysteine cathepsins *in situ* in cells and also identify what active protease is being detected by visualizing the fluorescently tagged protease(s) in cell lysates/conditioned media on SDS-PAGE gels. These probes have been used to identify and image cysteine cathepsins in live cells *in vitro* (Blum *et al.*, 2005) and *in vivo* in a transgenic mouse model of pancreatic cancer (Joyce *et al.*, 2004) and a transgenic mouse model of mammary cancer (Vasiljeva *et al.*, 2006). These ABPs are cell permeable and thus allow us to image the intracellular activity of cysteine cathepsins in the lysosomes (Blum *et al.*, 2005). Furthermore, the ABPs are available in both unquenched and quenched versions albeit the selectivity of the ABPs for a given cysteine cathepsin seems to be reduced in the quenched versions (Blum *et al.*, 2005).

2.1.3. Other protease probes

The two other types of protease probes that we are using *in vitro* are fluorogen-activating protein (FAP) protease biosensors developed by Berget and colleagues at Carnegie-Mellon and proteolytic beacons that Matrisian and colleagues have designed and developed for real-time analysis of MMP activity, including activity of individual MMPs like MMP-7

(McIntyre and Matrisian, 2009; McIntyre *et al.*, 2004, 2010; Scherer *et al.*, 2008). The FAPs are engineered proteins based on single chain antibodies (Falco *et al.*, 2009). At present, FAPs to assess the activity on the surface of live cells of MMP-14, -2, -9, and cathepsin K are being tested. The proteolytic beacons are protease substrates, which are built on a nanodendron scaffold and use FRET between two fluorophores linked to a selective peptide substrate (see McIntyre and Matrisian, 2003; Scherer *et al.*, 2008 for review). When in close proximity, the sensor fluorophore is quenched by the reference fluorophore, which also serves to monitor substrate concentration. The ability of these beacons to monitor how much of a probe is delivered and demonstrate that the probe is delivered is a crucial one. Proteolytic cleavage of the peptide linker results in increased fluorescence of the sensor so that the ratio of sensor to reference can be used as a quantitative measure of proteolytic activity. The design permits a great deal of flexibility: fluorophores can be in either the visible or near infrared range (NIR), protease selectivity can be modulated by alterations in the peptide sequence, clearance kinetics and route can be determined by the size and composition of the dendron backbone, and the multifunctionality of the dendrons allows for optimization of fluorophore concentration and solubility characteristics. Published proteolytic beacons include a visible one that is selective for MMP-7 (McIntyre and Matrisian, 2003; Wadsworth *et al.*, 2010), NIR MMP-7- and MMP-9-selective beacons (McIntyre *et al.*, 2010; Scherer *et al.*, 2008), and a NIR beacon that detects general MMP activity (McIntyre *et al.*, 2010).

2.2. Analysis

The development of new fluorogenic dyes and of confocal, multiphoton, and structured illumination microscopes has made optical imaging the method of choice for direct observations in living systems (for review, see Andrews *et al.*, 2002; Swedlow and Platani, 2002). These state-of-the-art advanced imaging systems allow one to perform 3D and 4D analyses using multiple probes.

For live-cell imaging there is an absolute requirement to perform experiments under physiological conditions. For this reason all of our imaging systems are equipped with environmental chambers (temperature, CO_2, and humidity controlled). This allows us to image live cultures over extended times without removing the cultures from the microscope stage. In addition, we can image the dynamics of proteolysis in real-time as impacted by tumor–stromal interactions. Another common feature of our systems is their motorized, fully automated stages. With this type of stage, we are able to select several areas of interest and program the system to image these areas sequentially at multiple time points and in 3D. The environmental chamber allows us to establish 3D models, place them on the microscope stage and program the microscope to acquire images as the cells attach, migrate, and invade into the surrounding extracellular matrices; we illustrate the association of proteolysis with cell migration and cell–cell interaction in Fig. 10.1. We now routinely label the various types of cells in our cocultures by prestaining with vital cytoplasmic dyes (Fig. 10.1) or transducing with fluorescent proteins (Fig. 10.2) so that they can be readily distinguished from one another when analyzing cell–cell interactions. We obtain optical sections through the entire volume

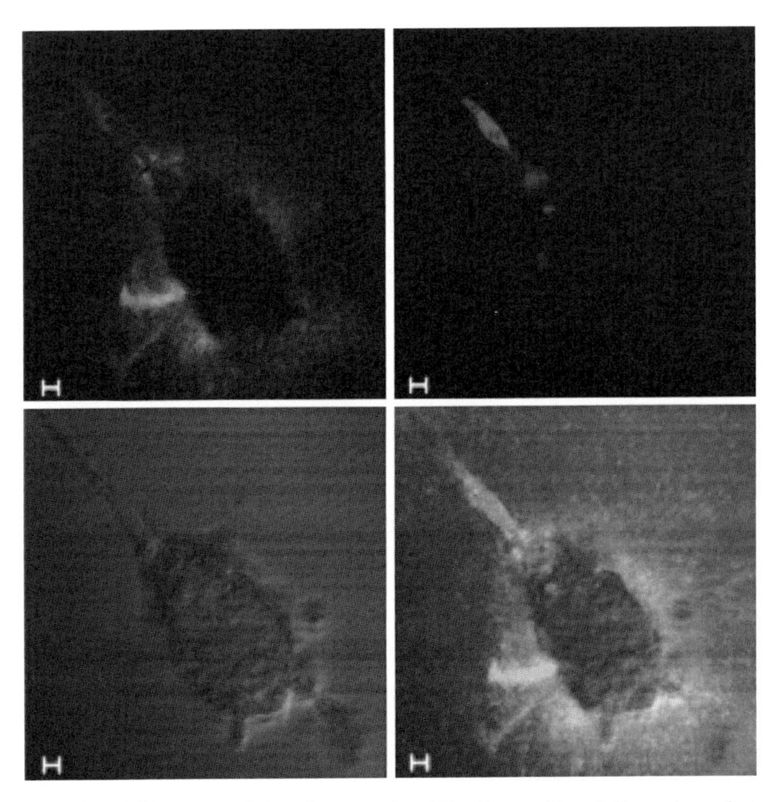

Figure 10.1 Still images of time-lapse series (90 min total) to follow degradation of DQ-collagen IV (green fluorescence) by carcinoma cells in coculture with fibroblasts. CCD-112CoN colon fibroblasts (red due to prestaining with CellTracker Orange) were cocultured with a spheroid of HCT 116 colon carcinoma cells. In this image taken after an overnight period of coculture, the fibroblasts can be seen to be moving toward and entering the HCT 116 spheroid. Note the pericellular fluorescent cleavage products due to degradation of DQ-collagen IV by the fibroblasts as they infiltrate into the HCT 116 spheroid, which is more readily apparent in video format. The four panels of this figure represent degradation products of DQ-collagen IV (green), fibroblasts (red), spheroid of HCT 116 cells (DIC), and a composite of the other three images. Bar, 10 μm. (See Color Insert.)

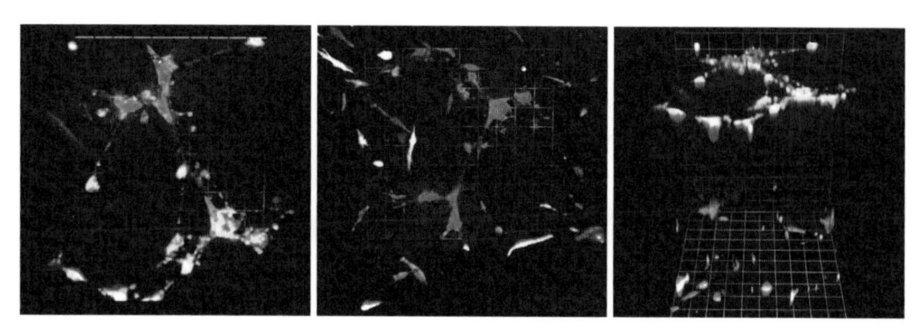

Figure 10.2 MAME tripartite cocultures of human SUM102 breast carcinoma cells (red; transduced with Lenti-RFP) with human umbilicial vein endothelial cells (HUVEC; blue; stained with CellTrace Far Red) in reconstituted basement membrane above a bottom layer of WS-12Ti human breast tumor-associated fibroblasts (magenta; transduced with Lenti-YFP) in interstitial collagen I (see also Fig. 10.3A). At 2 days of coculture as illustrated here, the SUM102 cells cluster around the branching networks formed by the HUVECs. Degradation products of DQ-collagen IV (green) are apparent around the interacting cells. The fibroblasts, which are primarily in the layer of collagen I, are associated with pericellular degradation products of DQ-collagen I (white). A few fibroblasts, also associated with proteolysis, can be seen to have migrated into the upper layer. The three panels from left to right are 3D reconstructions of the MAME tripartite coculture from the top, the bottom and at a 45° angle. Magnification, 20×. (See Color Insert.)

of the fluorescently labeled specimen in real time at various time points. The data are then analyzed with the 3D and 4D image reconstruction software as described below and as illustrated in Fig. 10.2.

The operating software for the imaging systems performs basic quantitative analyses, but is limited in regard to the extensive 3D and 4D quantitation required by the studies described here. Therefore, we use advanced stand-alone image analysis software such as Metamorph™ 7.64 and Volocity™ 5.5 to analyze fluorescence intensities, volumetric areas, surface areas, etc. Volumetric measurements are critical for accurately assessing changes in cell–cell interactions over time in 3D, that is, in 4D. To assess interactions, we label samples for detection of the different cell types, specific target proteins, proteolytic degradation products, etc., as we have described (Jedeszko *et al.*, 2008; Sameni *et al.*, 2003, 2009). Volumetric measurements of the sample components are acquired in multiple channels with each channel representing a single target. Datasets from each time point are then loaded for 4D quantification, analysis, and rendering of the data in 3D for each time point. Importantly, the software allows one to mark specific regions of interests in the 3D image stacks. The software can then quantify each region in 3D as well as quantify changes over time, that is, in 4D. The compiled images can be presented in many formats, including movies of live events.

3. 3D/4D Models for Analysis of Biological Processes Linked to Proteolysis

The working hypothesis for ongoing studies in our laboratory is that 3D mammary cell-based models will recapitulate the proteolytic mechanisms integral to developmental and neoplastic processes. Using such models to image proteolysis over time (i.e., 4D imaging) should allow us to place proteolysis, including the proteases identified and their interactions, within the context of the signaling pathways and other functions already elucidated by Bissell and her many collaborators as essential to malignant progression of mammary cancer. Those studies have convincingly demonstrated the importance of context to developmental and neoplastic processes in the mammary gland (for review, see Lee *et al.*, 2007; Schmeichel and Bissell, 2003). By growing mammary epithelial cells "within 3D basement membrane-like-matrices," the Bissell laboratory has been able to reproduce signaling pathways and functions (e.g., milk protein production) *in vitro* that are not observed when the same cells are cultured in 2D monolayers. Furthermore, the *in vitro* 3D mammary models have shown the need to match cell types with appropriate extracellular matrices, in this case mammary epithelial cells with basement membrane-like-matrices. When Gudjonsson *et al.* (2002) substituted an interstitial connective tissue matrix protein, that is, collagen I, acini formed, but they were inside-out. Adding normal myoepithelial cells to the collagen I gels resulted in correctly polarized acini, an effect dependent on production by the myoepithelial cells of the α-1 chain of the basement membrane protein laminin. There is a wealth of studies by Bissell and her collaborators showing that 3D cell-based assays can be used to study mechanisms for morphogenesis and neoplasia of human breast *in vitro* (for review, see Gudjonsson *et al.*, 2003; Schmeichel and Bissell, 2003). They also have demonstrated that proteases, in particular MMPs, are involved in morphogenesis and neoplasia. Studies by Weiss and his colleagues also implicate MMPs, in this case, MMP-14, or MT1-MMP, which they have shown to degrade collagen I and to be a prerequisite for proliferation of tumor cells in 3D collagen I gels (Hotary *et al.*, 2003). The Brugge laboratory has used 3D monotypic cultures grown in reconstituted basement membrane (rBM) to demonstrate a role for caspase-family cysteine proteases and apoptosis in the formation of lumens in the mammary acini (Debnath *et al.*, 2002; Shaw *et al.*, 2004). They have shown that migration and invasion of the mammary epithelial cells can be induced by coexpression of activated ErbB2 and TGF-ß (58), but have not analyzed the proteases responsible for the invasion. Further studies based on the Brugge 3D monotypic cultures by Debnath and colleagues have identified a role for lysosomal proteases. They found that autophagy and proteolysis within lysosomes plays both a suppressive and a promotion role that is context dependent (Chen and Debnath, 2010; Roy and Debnath, 2010).

Liotta and Kohn (2001) have suggested that cancer therapies should target the stroma or the tumor–stroma interface and hypothesized that stromal therapy could require lower doses than therapies that target the tumor. The need to target stroma would certainly appear to be true for protease inhibitors, as cells present in the tumor-associated stroma (e.g., fibroblasts, endothelial cells, inflammatory cells, myofibroblasts) are all important sources of proteases (for reviews, see Almholt and Johnsen, 2003; Bogenrieder and Herlyn, 2003; Coussens and Werb, 2001; DeClerck, 2000; Johnsen *et al.*, 1998; van Kempen and Coussens, 2002). For example, MMP-8 is expressed by neutrophils (Balbin *et al.*, 2003); urokinase plasminogen activator (uPA) by myofibroblasts, macrophages, and endothelial cells in ductal breast cancer (Nielsen *et al.*, 2007); MMP-9 by neutrophils, macrophages, and mast cells in a mouse model of squamous cell carcinoma of the skin (Coussens *et al.*, 2000); MMP-3 by subepithelial myofibroblasts in human colon (Bamba *et al.*, 2003); the cysteine proteases cathepsins B, K, L, and S and MMP-7, -9, and -12 by macrophages (Filippov *et al.*, 2003; Punturieri *et al.*, 2000; Reddy *et al.*, 1995); cathepsin B by endothelial cells from breast, glioma, and prostate (for review, see Keppler *et al.*, 1996); and MMP-2 by endothelial cells (Han *et al.*, 2003). Tumor-associated macrophages in human carcinomas also express high levels of cathepsin B (Campo *et al.*, 1994; Fernandez *et al.*, 2001; McKerrow *et al.*, 2000) and this macrophage cathepsin B enhances malignant progression in mouse transgenic models for mammary carcinoma (Vasiljeva *et al.*, 2006) and pancreatic carcinoma (Gocheva *et al.*, 2010). Stromal cells can also affect tumor proteolysis through the expression of endogenous protease inhibitors. Myofibroblasts, for example, are suggested to be the primary source of plasminogen activator inhibitor-1 (PAI-1) in human breast carcinomas (Offersen *et al.*, 2003). Overall such data indicate that we will not be able to define the "tumor degradome" (Balbin *et al.*, 2003) unless we study proteolysis in the context of tumor cells interacting with their microenvironment, interactions that we contend can be modeled *in vitro* in organotypic cocultures.

Using cocultures, we have established that stromal cells significantly impact tumor proteolysis (Sameni *et al.*, 2003). Degradation of DQ-collagen IV is increased as much as 17-fold in live 3D cocultures of stromal cells (fibroblasts or fibroblasts + macrophages) with breast or colon tumor cells (Sameni *et al.*, 2003). Such findings are pertinent to whether protease inhibitors would be efficacious *in vivo*. Also relevant is that fibroblasts isolated from invasive ductal breast carcinomas, but not normal breast fibroblasts, can recruit infiltration of other stromal cells, in this case blood monocytes, into 3D spheroids (Silzle *et al.*, 2003). Analyzing the contribution of inflammatory components to the "tumor degradome" is important as infiltration of macrophages *in vivo* potentiates malignant progression of tumors, for example, in a transgenic mouse model for mammary carcinoma (Gouon-Evans *et al.*, 2002; Lin *et al.*, 2001).

The composition and density of the extracellular matrix appears to be critical to tumor cell invasion as well as to fibril formation by collagen I. Brugge and colleagues (Seton-Rogers *et al.*, 2004) have reported that breast epithelial cells do not invade in 3D cultures plated on undiluted rBM; however, when the basement membrane was diluted with collagen I (in this case pepsin-solubilized collagen I), the cells did invade. Such observations indicate the need to compare various matrix compositions if we are to reach any definitive conclusions about whether proteolysis is or is not required for tumor invasion in such models. Intriguing work by Weaver and colleagues suggests that increased cross-linking of collagen I promotes tumor invasion *in vitro* and *in vivo* (Levental *et al.*, 2009). This finding seems counterintuitive to a role for proteolysis in tumor invasion, as cross-linked collagen is more resistant to proteolysis (Sabeh *et al.*, 2009). It is the dynamics of collagen remodeling that appear to be critical (Egeblad *et al.*, 2010), supporting a need for not only 3D models but also for studying 3D models over time, that is, in 4D.

4. Live-Cell Imaging of MAME Models: A Screening Tool for Drug Discovery

There is substantial evidence that 3D cultures are predictive of the resistance of tumor cells to cytotoxic therapy and can be used to identify targets and validate potential therapeutic agents (Li *et al.*, 2008, 2010; Nam *et al.*, 2010). We hypothesize that functional imaging of proteolysis by live cells in 3D and 4D can be used as an *in vitro* screen for testing alternative strategies to target the malignant phenotype of tumor cells, using as a readout proteolysis of extracellular matrix proteins. Cell-based 3D models have already been proposed for use in high-throughput screening of drugs (Schmeichel and Bissell, 2003) as well as their use for analyzing dynamic interactions between tumor cells and cellular and noncellular constituents of their microenvironment (Ng and Brugge, 2009). Therefore, we contend that the dimension of time needs to be part of high-throughput screening, including screening of therapeutic strategies to reduce protease activity, whether those strategies are ones that directly impact activity such as protease inhibitors or ones that target upstream effectors of protease activity.

We have developed a robust preclinical *in vitro* 3D/4D model to recapitulate paracrine interactions between tumor cells and other cells that comprise the tumor microenvironment. We have named these models MAME for *m*ammary *a*rchitecture and *m*icroenvironment *e*ngineering (Sameni *et al.*, 2009). Our MAME models are designed to closely mimic the architecture of normal breast tissue, a need strongly advocated by Weigelt and Bissell

(2008), as they provide a readily adaptable system through which to determine the contribution of individual cell types of the tumor microenvironment to the aggressive phenotype of breast cancers. In 4D (3D + time), MAME models in conjunction with live-cell imaging techniques are allowing us to determine the timing as well as the respective contributions of various cell types, proteolytic pathways, signaling pathways, etc., to progression from a preinvasive to an invasive phenotype.

Our MAME tripartite cocultures (Figs. 10.2 and 10.3A) consist of a bottom layer of interstitial type I collagen in which we embed breast fibroblasts, a second layer of rBM on which we plate normal breast epithelial cells, premalignant breast epithelial cells or breast carcinoma cells and a top layer of 2% rBM. Thus, the second and top layers are based on the rBM overlay cultures used by the Brugge laboratory (Debnath and Brugge, 2005; Debnath *et al.*, 2003; Shaw *et al.*, 2004). In order to study proteolysis, we incorporate DQ-collagen I in the collagen I layer and DQ-collagen IV in the second layer of rBM. Other iterations of the MAME models consist of only the second and top layers of rBM in which DQ-collagen IV is incorporated (Fig. 10.3B). The tripartite cocultures recapitulate tumor–tumor microenvironment interactions that occur *in vivo* in human breast tumors as a result of indirect interactions. With the tripartite model we have observed over time the migration of fibroblasts toward the tumor cells, eventually infiltrating into the tumor structures over a period of 7 days. The breast tumor cells also migrate towards the lower layer of fibroblasts, but do so more slowly over a period of 3 weeks. Proteolysis is associated with the migrating tumor cells and fibroblasts, in this case pericellular fluorescent degradation products. In contrast, there is extensive diffuse fluorescence associated with the bottom layer of fibroblasts, indicative of the high

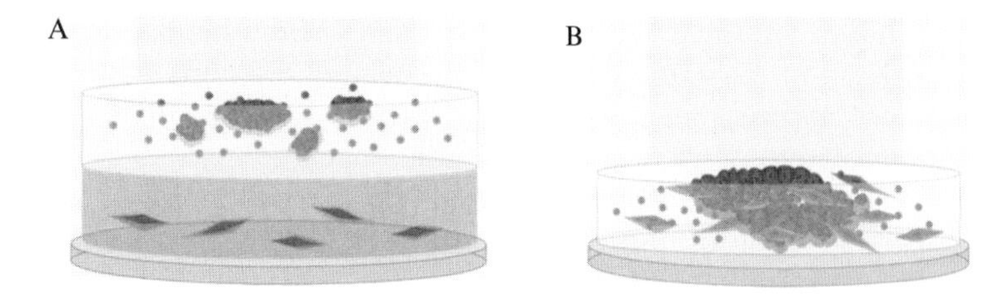

Figure 10.3 Schematic of MAME tripartite and mixed cocultures of tumor cells, fibroblasts, and macrophages. (A) Coverslips are coated with collagen I containing DQ-collagen I^{TM} and fibroblasts (elongated red cells). A second layer of rBM containing DQ-collagen IV^{TM} is added and tumor cells (round red cells) plated on top along with macrophages (blue). The cultures are then overlaid with a third layer of 2% rBM, which also is included in subsequent changes of media. (B) Coverslips are coated with rBM containing DQ-collagen IV^{TM} and a mixture of fibroblasts, tumor cells, and macrophages plated on top. Cocultures are imaged live to follow changes in morphogenesis and collagen degradation, depicted here in green. (See Color Insert.)

protease activity produced by fibroblasts. To date, we have maintained the tripartite MAME models for as long as 24 days and imaged live cultures at intervals over that period. If we use preinvasive breast epithelial cells in the MAME models, we are able to image the progression of those cells to an invasive phenotype, accompanied by an increase in proteolysis. Furthermore, we have demonstrated the ability of a variety of antagonists to reduce the invasive phenotype and the proteolysis (Jedeszko *et al.*, 2009; Sameni *et al.*, 2009). We have used comparable models for analysis of the invasive phenotype of a variety of cancers, including inflammatory breast cancer (Victor *et al.*, in press), and prostate cancer (Hayward and Sloane, unpublished data). Recently, we have used MAME models to demonstrate that pericellular proteolysis is increased by incubating the cultures at a slightly acidic pH comparable to that found in tumors *in vivo* (Rothberg and Sloane, unpublished data).

Through live-cell imaging of MAME models, we are able to both image and quantify the cleavage products of the DQ-collagens (Jedeszko *et al.*, 2008; Sameni *et al.*, 2009). We are of course limited to localizing and quantifying cleavage of the labeled collagens and therefore our findings need to be considered in that context. Despite this limitation, however, we can visualize proteolysis associated with migration of individual cells and cellular structures and invasive protrusions from those structures (Jedeszko *et al.*, 2009) and to do so over long time periods. In the case of individual endothelial cells migrating to form tube-like structures, we have imaged proteolysis over a 20-h period (Cavallo-Medved *et al.*, 2009). That these labeled collagens may be more easily degraded than native collagens is another caveat that we must consider. Nonetheless, the labeled collagens do allow us to monitor protease activity and in combination with other protease probes such as those discussed above should allow us to identify proteases that participate in progression to an invasive phenotype and proteases that may be read-outs for therapeutic strategies to abrogate that progression.

We are now adapting a WaferGen SmartSlide Microincubation System for real-time monitoring of our MAME models. The system was designed for monolayer culture of cells in six-well plates on the stage of an inverted confocal microscope. Each well can be individually perfused and effluents individually collected for immunochemical and biochemical analysis of secreted proteins without disturbing the integrity of the cultures. Furthermore, we can acquire optical sections of the six wells sequentially at multiple time points. In addition, we can harvest media for immunochemical and biochemical analyses at times corresponding to changes in aggressive phenotype. We will be able to image and assay conditioned media from six cultures at once and at multiple time points, thus increasing the throughput of our live-cell imaging assays of MAME models. Although we are primarily interested in using MAME models for studying proteolysis, the tripartite and other modifications of the MAME models along with microincubation systems such as the Wafergen provide an experimental system in which one can test various cellular and noncellular aspects of the tumor microenvironment as it affects the progression of human breast cancer and other cancers.

ACKNOWLEDGMENTS

The research described in this paper and the confocal facility in which the imaging was performed were supported by the National Cancer Institute, the National Center for Research Resources and the National Institute of Child Health and Human Development of the National Institutes of Health, a Department of Defense Breast Cancer Center of Excellence and the Avon Foundation.

REFERENCES

Adjei, A. A., Christian, M., and Ivy, P. (2009). Novel designs and end points for phase II clinical trials. *Clin. Cancer Res.* **15,** 1866–1872.

Almholt, K., and Johnsen, M. (2003). Stromal cell involvement in cancer. *Recent Results Cancer Res.* **162,** 31–42.

Andrews, P. D., Harper, I. S., and Swedlow, J. R. (2002). To 5D and beyond: Quantitative fluorescence microscopy in the postgenomic era. *Traffic* **3,** 29–36.

Ang, M. K., Tan, S. B., and Lim, W. T. (2010). Phase II clinical trials in oncology: Are we hitting the target? *Expert Rev. Anticancer Ther.* **10,** 427–438.

Aumailley, M., and Gayraud, B. (1998). Structure and biological activity of the extracellular matrix. *J. Mol. Med. (Berl)* **76,** 253–265.

Balbin, M., Fueyo, A., Tester, A. M., Pendas, A. M., Pitiot, A. S., Astudillo, A., Overall, C. M., Shapiro, S. D., and Lopez-Otin, C. (2003). Loss of collagenase-2 confers increased skin tumor susceptibility to male mice. *Nat. Genet.* **35,** 252–257.

Bamba, S., Andoh, A., Yasui, H., Araki, Y., Bamba, T., and Fujiyama, Y. (2003). Matrix metalloproteinase-3 secretion from human colonic subepithelial myofibroblasts: Role of interleukin-17. *J. Gastroenterol.* **38,** 548–554.

Blum, G., Mullins, S. R., Keren, K., Fonovic, M., Jedeszko, C., Rice, M. J., Sloane, B. F., and Bogyo, M. (2005). Dynamic imaging of protease activity with fluorescently quenched activity-based probes. *Nat. Chem. Biol.* **1,** 203–209.

Bogenrieder, T., and Herlyn, M. (2003). Axis of evil: Molecular mechanisms of cancer metastasis. *Oncogene* **22,** 6524–6536.

Brzostowski, J. A., Meckel, T., Hong, J., Chen, A., and Jin, T. (2009). Imaging protein-protein interactions by Forster resonance energy transfer (FRET) microscopy in live cells. *Curr. Protoc. Protein Sci.* (Chapter 19, Unit19 15).

Campo, E., Munoz, J., Miquel, R., Palacin, A., Cardesa, A., Sloane, B. F., and Emmert-Buck, M. R. (1994). Cathepsin B expression in colorectal carcinomas correlates with tumor progression and shortened patient survival. *Am. J. Pathol.* **145,** 301–309.

Cavallo-Medved, D., and Sloane, B. F. (2003). Cell-surface cathepsin B: Understanding its functional significance. *Curr. Top. Dev. Biol.* **54,** 313–341.

Cavallo-Medved, D., Rudy, D., Blum, G., Bogyo, M., Caglic, D., and Sloane, B. F. (2009). Live-cell imaging demonstrates extracellular matrix degradation in association with active cathepsin B in caveolae of endothelial cells during tube formation. *Exp. Cell Res.* **315,** 1234–1246.

Chau, I., Rigg, A., and Cunningham, D. (2003). Matrix metalloproteinase inhibitors—an emphasis on gastrointestinal malignancies. *Crit. Rev. Oncol. Hematol.* **45,** 151–176.

Chen, N., and Debnath, J. (2010). Autophagy and tumorigenesis. *FEBS Lett.* **584,** 1427–1435.
Choong, P. F., and Nadesapillai, A. P. (2003). Urokinase plasminogen activator system: A multifunctional role in tumor progression and metastasis. *Clin. Orthop. Relat. Res.* (Suppl. 415), S46–S58.
Coussens, L. M., and Werb, Z. (2001). Inflammatory cells and cancer: Think different!. *J. Exp. Med.* **193,** F23–F26.
Coussens, L. M., Tinkle, C. L., Hanahan, D., and Werb, Z. (2000). MMP-9 supplied by bone marrow-derived cells contributes to skin carcinogenesis. *Cell* **103,** 481–490.
Coussens, L. M., Fingleton, B., and Matrisian, L. M. (2002). Matrix metalloproteinase inhibitors and cancer: Trials and tribulations. *Science* **295,** 2387–2392.
Debnath, J., and Brugge, J. S. (2005). Modelling glandular epithelial cancers in three-dimensional cultures. *Nat. Rev. Cancer* **5,** 675–688.
Debnath, J., Mills, K. R., Collins, N. L., Reginato, M. J., Muthuswamy, S. K., and Brugge, J. S. (2002). The role of apoptosis in creating and maintaining luminal space within normal and oncogene-expressing mammary acini. *Cell* **111,** 29–40.
Debnath, J., Muthuswamy, S. K., and Brugge, J. S. (2003). Morphogenesis and oncogenesis of MCF-10A mammary epithelial acini grown in three-dimensional basement membrane cultures. *Methods* **30,** 256–268.
DeClerck, Y. A. (2000). Interactions between tumour cells and stromal cells and proteolytic modification of the extracellular matrix by metalloproteinases in cancer. *Eur. J. Cancer* **36,** 1258–1268.
DeClerck, Y. A., and Laug, W. E. (1996). Cooperation between matrix metalloproteinases and the plasminogen activator-plasmin system in tumor progression. *Enzyme Protein* **49,** 72–84.
Demchik, L. L., Sameni, M., Nelson, K., Mikkelsen, T., and Sloane, B. F. (1999). Cathepsin B and glioma invasion. *Int. J. Dev. Neurosci.* **17,** 483–494.
Edwards, D. R., Hoyer-Hansen, G., Blasi, F., and Sloane, B. F. (2008). The Cancer Degradome—Proteases and Cancer Biology. Springer, New York(p. 896).
Egeblad, M., and Werb, Z. (2002). New functions for the matrix metalloproteinases in cancer progression. *Nat. Rev. Cancer* **2,** 161–174.
Egeblad, M., Rasch, M. G., and Weaver, V. M. (2010). Dynamic interplay between the collagen scaffold and tumor evolution. *Curr. Opin. Cell Biol.* **22,** 697–706.
Ellerbroek, S. M., Hudson, L. G., and Stack, M. S. (1998). Proteinase requirements of epidermal growth factor-induced ovarian cancer cell invasion. *Int. J. Cancer* **78,** 331–337.
Ellerbroek, S. M., Wu, Y. I., Overall, C. M., and Stack, M. S. (2001). Functional interplay between type I collagen and cell surface matrix metalloproteinase activity. *J. Biol. Chem.* **276,** 24833–24842.
Falco, C. N., Dykstra, K. M., Yates, B. P., and Berget, P. B. (2009). scFv-based fluorogen activating proteins and variable domain inhibitors as fluorescent biosensor platforms. *Biotechnol. J.* **4,** 1328–1336.
Fernandez, P. L., Farre, X., Nadal, A., Fernandez, E., Peiro, N., Sloane, B. F., Shi, G. P., Chapman, H. A., Campo, E., and Cardesa, A. (2001). Expression of cathepsins B and S in the progression of prostate carcinoma. *Int. J. Cancer* **95,** 51–55.
Filippov, S., Caras, I., Murray, R., Matrisian, L. M., Chapman, H. A., Jr., Shapiro, S., and Weiss, S. J. (2003). Matrilysin-dependent elastolysis by human macrophages. *J. Exp. Med.* **198,** 925–935.
Fonovic, M., and Bogyo, M. (2007). Activity based probes for proteases: Applications to biomarker discovery, molecular imaging and drug screening. *Curr. Pharm. Des.* **13,** 253–261.
Fuchs, S. Y. (2002). The role of ubiquitin-proteasome pathway in oncogenic signaling. *Cancer Biol. Ther.* **1,** 337–341.

Gocheva, V., Wang, H. W., Gadea, B. B., Shree, T., Hunter, K. E., Garfall, A. L., Berman, T., and Joyce, J. A. (2010). IL-4 induces cathepsin protease activity in tumor-associated macrophages to promote cancer growth and invasion. *Genes Dev.* **24,** 241–255.

Gouon-Evans, V., Lin, E. Y., and Pollard, J. W. (2002). Requirement of macrophages and eosinophils and their cytokines/chemokines for mammary gland development. *Breast Cancer Res.* **4,** 155–164.

Greenbaum, D., Baruch, A., Hayrapetian, L., Darula, Z., Burlingame, A., Medzihradszky, K. F., and Bogyo, M. (2002a). Chemical approaches for functionally probing the proteome. *Mol. Cell. Proteomics* **1,** 60–68.

Greenbaum, D. C., Arnold, W. D., Lu, F., Hayrapetian, L., Baruch, A., Krumrine, J., Toba, S., Chehade, K., Bromme, D., Kuntz, I. D., and Bogyo, M. (2002b). Small molecule affinity fingerprinting. A tool for enzyme family subclassification, target identification, and inhibitor design. *Chem. Biol.* **9,** 1085–1094.

Gudjonsson, T., Ronnov-Jessen, L., Villadsen, R., Rank, F., Bissell, M. J., and Petersen, O. W. (2002). Normal and tumor-derived myoepithelial cells differ in their ability to interact with luminal breast epithelial cells for polarity and basement membrane deposition. *J. Cell Sci.* **115,** 39–50.

Gudjonsson, T., Ronnov-Jessen, L., Villadsen, R., Bissell, M. J., and Petersen, O. W. (2003). To create the correct microenvironment: Three-dimensional heterotypic collagen assays for human breast epithelial morphogenesis and neoplasia. *Methods* **30,** 247–255.

Han, X., Boyd, P. J., Colgan, S., Madri, J. A., and Haas, T. L. (2003). Transcriptional up-regulation of endothelial cell matrix metalloproteinase-2 in response to extracellular cues involves GATA-2. *J. Biol. Chem.* **278,** 47785–47791.

Henriet, P., Zhong, Z. D., Brooks, P. C., Weinberg, K. I., and DeClerck, Y. A. (2000). Contact with fibrillar collagen inhibits melanoma cell proliferation by up-regulating p27KIP1. *Proc. Natl. Acad. Sci. USA* **97,** 10026–10031.

Hojilla, C. V., Mohammed, F. F., and Khokha, R. (2003). Matrix metalloproteinases and their tissue inhibitors direct cell fate during cancer development. *Br. J. Cancer* **89,** 1817–1821.

Hotary, K. B., Allen, E. D., Brooks, P. C., Datta, N. S., Long, M. W., and Weiss, S. J. (2003). Membrane type I matrix metalloproteinase usurps tumor growth control imposed by the three-dimensional extracellular matrix. *Cell* **114,** 33–45.

Jedeszko, C., Sameni, M., Olive, M. B., Moin, K., and Sloane, B. F. (2008). Visualizing protease activity in living cells: From two dimensions to four dimensions. *Curr. Protoc. Cell Biol.(Chapter 4, Unit 4 20).*

Jedeszko, C., Victor, B. C., Podgorski, I., and Sloane, B. F. (2009). Fibroblast hepatocyte growth factor promotes invasion of human mammary ductal carcinoma in situ. *Cancer Res.* **69,** 9148–9155.

Johnsen, M., Lund, L. R., Romer, J., Almholt, K., and Dano, K. (1998). Cancer invasion and tissue remodeling: Common themes in proteolytic matrix degradation. *Curr. Opin. Cell Biol.* **10,** 667–671.

Joyce, J. A., Baruch, A., Chehade, K., Meyer-Morse, N., Giraudo, E., Tsai, F. Y., Greenbaum, D. C., Hager, J. H., Bogyo, M., and Hanahan, D. (2004). Cathepsin cysteine proteases are effectors of invasive growth and angiogenesis during multistage tumorigenesis. *Cancer Cell* **5,** 443–453.

Keppler, D., Sameni, M., Moin, K., Mikkelsen, T., Diglio, C. A., and Sloane, B. F. (1996). Tumor progression and angiogenesis: Cathepsin B & Co. *Biochem. Cell Biol.* **74,** 799–810.

Kim, J., Yu, W., Kovalski, K., and Ossowski, L. (1998). Requirement for specific proteases in cancer cell intravasation as revealed by a novel semiquantitative PCR-based assay. *Cell* **94,** 353–362.

Kjoller, L., Engelholm, L. H., Hoyer-Hansen, M., Dano, K., Bugge, T. H., and Behrendt, N. (2004). uPARAP/endo180 directs lysosomal delivery and degradation of collagen IV. *Exp. Cell Res.* **293,** 106–116.

Krol, J., Kopitz, C., Kirschenhofer, A., Schmitt, M., Magdolen, U., Kruger, A., and Magdolen, V. (2003). Inhibition of intraperitoneal tumor growth of human ovarian cancer cells by bi- and trifunctional inhibitors of tumor-associated proteolytic systems. *Biol. Chem.* **384,** 1097–1102.

Lee, G. Y., Kenny, P. A., Lee, E. H., and Bissell, M. J. (2007). Three-dimensional culture models of normal and malignant breast epithelial cells. *Nat. Methods* **4,** 359–365.

Levental, K. R., Yu, H., Kass, L., Lakins, J. N., Egeblad, M., Erler, J. T., Fong, S. F., Csiszar, K., Giaccia, A., Weninger, W., Yamauchi, M., Gasser, D. L., *et al.* (2009). Matrix crosslinking forces tumor progression by enhancing integrin signaling. *Cell* **139,** 891–906.

Li, W. P., and Anderson, C. J. (2003). Imaging matrix metalloproteinase expression in tumors. *Q. J. Nucl. Med.* **47,** 201–208.

Li, S., Chow, L. H., and Pickering, J. G. (2000). Cell surface-bound collagenase-1 and focal substrate degradation stimulate the rear release of motile vascular smooth muscle cells. *J. Biol. Chem.* **275,** 35384–35392.

Li, Q., Mullins, S. R., Sloane, B. F., and Mattingly, R. R. (2008). p21-Activated kinase 1 coordinates aberrant cell survival and pericellular proteolysis in a three-dimensional culture model for premalignant progression of human breast cancer. *Neoplasia* **10,** 314–329.

Li, Q., Chow, A. B., and Mattingly, R. R. (2010). Three-dimensional overlay culture models of human breast cancer reveal a critical sensitivity to mitogen-activated protein kinase kinase inhibitors. *J. Pharmacol. Exp. Ther.* **332,** 821–828.

Lin, E. Y., Nguyen, A. V., Russell, R. G., and Pollard, J. W. (2001). Colony-stimulating factor 1 promotes progression of mammary tumors to malignancy. *J. Exp. Med.* **193,** 727–740.

Liotta, L. A., and Kohn, E. C. (2001). The microenvironment of the tumour-host interface. *Nature* **411,** 375–379.

Lopez-Otin, C., and Matrisian, L. M. (2007). Emerging roles of proteases in tumour suppression. *Nat. Rev. Cancer* **7,** 800–808.

Lopez-Otin, C., Palavalli, L. H., and Samuels, Y. (2009). Protective roles of matrix metalloproteinases: From mouse models to human cancer. *Cell Cycle* **8,** 3657–3662.

Madsen, D. H., Ingvarsen, S., Jurgensen, H. J., Melander, M. C., Kjoller, L., Moyer, A., Honore, C., Madsen, C. A., Garred, P., Burgdorf, S., Bugge, T. H., Behrendt, N., *et al.* (2011). The non-phagocytic route for collagen uptake: A distinct degradation pathway. *J. Biol. Chem.* **286**(30), 26996–27010.

Mankoff, D. A., O'Sullivan, F., Barlow, W. E., and Krohn, K. A. (2007). Molecular imaging research in the outcomes era: Measuring outcomes for individualized cancer therapy. *Acad. Radiol.* **14,** 398–405.

Marten, K., Bremer, C., Khazaie, K., Sameni, M., Sloane, B., Tung, C. H., and Weissleder, R. (2002). Detection of dysplastic intestinal adenomas using enzyme sensing molecular beacons. *Gastroenterology* **122**(2), 406–414.

McIntyre, J. O., and Matrisian, L. M. (2003). Molecular imaging of proteolytic activity in cancer. *J. Cell. Biochem.* **90,** 1087–1097.

McIntyre, J. O., and Matrisian, L. M. (2009). Optical proteolytic beacons for in vivo detection of matrix metalloproteinase activity. *Methods Mol. Biol.* **539,** 155–174.

McIntyre, J. O., Fingleton, B., Wells, K. S., Piston, D. W., Lynch, C. C., Gautam, S., and Matrisian, L. M. (2004). Development of a novel fluorogenic proteolytic beacon for in vivo detection and imaging of tumour-associated matrix metalloproteinase-7 activity. *Biochem. J.* **377,** 617–628.

McIntyre, J. O., Scherer, R. L., and Matrisian, L. M. (2010). Near-infrared optical proteolytic beacons for in vivo imaging of matrix metalloproteinase activity. *Methods Mol. Biol.* **622,** 279–304.

McKerrow, J. H., Bhargava, V., Hansell, E., Huling, S., Kuwahara, T., Matley, M., Coussens, L., and Warren, R. (2000). A functional proteomics screen of proteases in colorectal carcinoma. *Mol. Med.* **6,** 450–460.

Muehlenweg, B., Assfalg-Machleidt, I., Parrado, S. G., Burgle, M., Creutzburg, S., Schmitt, M., Auerswald, E. A., Machleidt, W., and Magdolen, V. (2000). A novel type of bifunctional inhibitor directed against proteolytic activity and receptor/ligand interaction. Cystatin with a urokinase receptor binding site. *J. Biol. Chem.* **275,** 33562–33566.

Nam, J. M., Onodera, Y., Bissell, M. J., and Park, C. C. (2010). Breast cancer cells in three-dimensional culture display an enhanced radioresponse after coordinate targeting of integrin alpha5beta1 and fibronectin. *Cancer Res.* **70,** 5238–5248.

Ng, M. R., and Brugge, J. S. (2009). A stiff blow from the stroma: Collagen crosslinking drives tumor progression. *Cancer Cell* **16,** 455–457.

Nielsen, B. S., Rank, F., Illemann, M., Lund, L. R., and Dano, K. (2007). Stromal cells associated with early invasive foci in human mammary ductal carcinoma in situ coexpress urokinase and urokinase receptor. *Int. J. Cancer* **120,** 2086–2095.

Nozaki, S., Sissons, S., Chien, D. S., and Sledge, G. W., Jr. (2003). Activity of biphenyl matrix metalloproteinase inhibitor BAY 12–9566 in a human breast cancer orthotopic model. *Clin. Exp. Metastasis* **20,** 407–412.

Offersen, B. V., Nielsen, B. S., Hoyer-Hansen, G., Rank, F., Hamilton-Dutoit, S., Overgaard, J., and Andreasen, P. A. (2003). The myofibroblast is the predominant plasminogen activator inhibitor-1-expressing cell type in human breast carcinomas. *Am. J. Pathol.* **163,** 1887–1899.

Podgorski, I., and Sloane, B. F. (2003). Cathepsin B and its role(s) in cancer progression. *Biochem. Soc. Symp.* **70,** 263–276.

Podgorski, I., Linebaugh, B. E., Koblinski, J. E., Rudy, D. L., Herroon, M. K., Olive, M. B., and Sloane, B. F. (2009). Bone marrow-derived cathepsin K cleaves SPARC in bone metastasis. *Am. J. Pathol.* **175,** 1255–1269.

Punturieri, A., Filippov, S., Allen, E., Caras, I., Murray, R., Reddy, V., and Weiss, S. J. (2000). Regulation of elastinolytic cysteine proteinase activity in normal and cathepsin K-deficient human macrophages. *J. Exp. Med.* **192,** 789–799.

Ramos-DeSimone, N., Hahn-Dantona, E., Sipley, J., Nagase, H., French, D. L., and Quigley, J. P. (1999). Activation of matrix metalloproteinase-9 (MMP-9) via a converging plasmin/stromelysin-1 cascade enhances tumor cell invasion. *J. Biol. Chem.* **274,** 13066–13076.

Reddy, V. Y., Zhang, Q. Y., and Weiss, S. J. (1995). Pericellular mobilization of the tissue-destructive cysteine proteinases, cathepsins B, L, and S, by human monocyte-derived macrophages. *Proc. Natl. Acad. Sci. USA* **92,** 3849–3853.

Rhee, J. S., Diaz, R., Korets, L., Hodgson, J. G., and Coussens, L. M. (2004). TIMP-1 alters susceptibility to carcinogenesis. *Cancer Res.* **64,** 952–961.

Roshy, S., Sloane, B. F., and Moin, K. (2003). Pericellular cathepsin B and malignant progression. *Cancer Metastasis Rev.* **22,** 271–286.

Roy, S., and Debnath, J. (2010). Autophagy and tumorigenesis. *Semin. Immunopathol.* **32,** 383–396.

Sabeh, F., Shimizu-Hirota, R., and Weiss, S. J. (2009). Protease-dependent versus -independent cancer cell invasion programs: Three-dimensional amoeboid movement revisited. *J. Cell Biol.* **185,** 11–19.

Sameni, M., Moin, K., and Sloane, B. F. (2000). Imaging proteolysis by living human breast cancer cells. *Neoplasia* **2,** 496–504.

Sameni, M., Dosescu, J., and Sloane, B. F. (2001). Imaging proteolysis by living human glioma cells. *Biol. Chem.* **382,** 785–788.

Sameni, M., Dosescu, J., Moin, K., and Sloane, B. F. (2003). Functional imaging of proteolysis: Stromal and inflammatory cells increase tumor proteolysis. *Mol. Imaging* **2,** 159–175.

Sameni, M., Dosescu, J., Yamada, K. M., Sloane, B. F., and Cavallo-Medved, D. (2008). Functional live-cell imaging demonstrates that beta1-integrin promotes type IV collagen degradation by breast and prostate cancer cells. *Mol. Imaging* **7,** 199–213.

Sameni, M., Cavallo-Medved, D., Dosescu, J., Jedeszko, C., Moin, K., Mullins, S. R., Olive, M. B., Rudy, D., and Sloane, B. F. (2009). Imaging and quantifying the dynamics of tumor-associated proteolysis. *Clin. Exp. Metastasis* **26,** 299–309.

Scherer, R. L., VanSaun, M. N., McIntyre, J. O., and Matrisian, L. M. (2008). Optical imaging of matrix metalloproteinase-7 activity in vivo using a proteolytic nanobeacon. *Mol. Imaging* **7,** 118–131.

Schmeichel, K. L., and Bissell, M. J. (2003). Modeling tissue-specific signaling and organ function in three dimensions. *J. Cell Sci.* **116,** 2377–2388.

Seton-Rogers, S. E., Lu, Y., Hines, L. M., Koundinya, M., LaBaer, J., Muthuswamy, S. K., and Brugge, J. S. (2004). Cooperation of the ErbB2 receptor and transforming growth factor beta in induction of migration and invasion in mammary epithelial cells. *Proc. Natl. Acad. Sci. USA* **101,** 1257–1262.

Seymour, L., Ivy, S. P., Sargent, D., Spriggs, D., Baker, L., Rubinstein, L., Ratain, M. J., Le Blanc, M., Stewart, D., Crowley, J., Groshen, S., Humphrey, J. S., *et al.* (2010). The design of phase II clinical trials testing cancer therapeutics: Consensus recommendations from the clinical trial design task force of the national cancer institute investigational drug steering committee. *Clin Cancer Res.* **16,** 1764–1769.

Shaw, K. R., Wrobel, C. N., and Brugge, J. S. (2004). Use of three-dimensional basement membrane cultures to model oncogene-induced changes in mammary epithelial morphogenesis. *J. Mammary Gland Biol. Neoplasia* **9,** 297–310.

Silzle, T., Kreutz, M., Dobler, M. A., Brockhoff, G., Knuechel, R., and Kunz-Schughart, L. A. (2003). Tumor-associated fibroblasts recruit blood monocytes into tumor tissue. *Eur. J. Immunol.* **33,** 1311–1320.

Sloane, B. F. (1996). Suicidal tumor proteases. *Nat. Biotechnol.* **14,** 826–827.

Sloane, B. F., Sameni, M., Podgorski, I., Cavallo-Medved, D., and Moin, K. (2006). Functional imaging of tumor proteolysis. *Annu. Rev. Pharmacol. Toxicol.* **46,** 301–315.

Swedlow, J. R., and Platani, M. (2002). Live cell imaging using wide-field microscopy and deconvolution. *Cell Struct. Funct.* **27,** 335–341.

van Kempen, L. C., and Coussens, L. M. (2002). MMP9 potentiates pulmonary metastasis formation. *Cancer Cell* **2,** 251–252.

Vasiljeva, O., Papazoglou, A., Kruger, A., Brodoefel, H., Korovin, M., Deussing, J., Augustin, N., Nielsen, B. S., Almholt, K., Bogyo, M., Peters, C., and Reinheckel, T. (2006). Tumor cell-derived and macrophage-derived cathepsin B promotes progression and lung metastasis of mammary cancer. *Cancer Res.* **66,** 5242–5250.

Victor, B. C., Anbalagan, A., Mohamed, M. M., Sloane, B. F. and Cavallo-Medved, D. *Breast Cancer Res.* (in press).

Wadsworth, S. J., Atsuta, R., McIntyre, J. O., Hackett, T. L., Singhera, G. K., and Dorscheid, D. R. (2010). IL-13 and TH2 cytokine exposure triggers matrix metalloproteinase 7-mediated Fas ligand cleavage from bronchial epithelial cells. *J. Allergy Clin. Immunol.* **126**(366–374), e361–e368.

Weigelt, B., and Bissell, M. J. (2008). Unraveling the microenvironmental influences on the normal mammary gland and breast cancer. *Semin. Cancer Biol.* **18,** 311–321.

Wolf, K., Alexander, S., Schacht, V., Coussens, L. M., von Andrian, U. H., van Rheenen, J., Deryugina, E., and Friedl, P. (2009). Collagen-based cell migration models in vitro and in vivo. *Semin. Cell Dev. Biol.* **20,** 931–941.

Wu, H., Byrne, M. H., Stacey, A., Goldring, M. B., Birkhead, J. R., Jaenisch, R., and Krane, S. M. (1990). Generation of collagenase-resistant collagen by site-directed mutagenesis of murine pro alpha 1(I) collagen gene. *Proc. Natl. Acad. Sci. USA* **87,** 5888–5892.

Yan, S., and Sloane, B. F. (2003). Molecular regulation of human cathepsin B: Implication in pathologies. *Biol. Chem.* **384,** 845–854.

SECTION THREE

OTHER TECHNIQUES

CHAPTER ELEVEN

Live Cell Imaging in Live Animals with Fluorescent Proteins

Robert M. Hoffman

Contents

1. Introduction 199
    1.1. Pioneering *in vivo* imaging with GFP 199
    1.2. GFP is the method of choice for *in vivo* imaging 199
    1.3. Imaging the cell biology of metastasis *in vivo* 200
2. Imaging Angiogenesis *In Vivo* 201
    2.1. Two-photon imaging with fluorescent proteins 201
    2.2. Red fluorescent protein 202
    2.3. Establishment of dual-color cancer cells 202
3. Imaging Cell Trafficking 203
    3.1. Imaging cancer cell deformation and trafficking in blood vessels 203
    3.2. Imaging cancer cell trafficking in lymphatic vessels 204
    3.3. Color-coded imaging of tumor–host interaction 204
    3.4. Transgenic GFP nude mouse 204
    3.5. Transgenic RFP nude mouse 205
    3.6. Transgenic CFP nude mouse 206
    3.7. Noninvasive color-coded imaging of the tumor microenvironment 206
    3.8. Imaging circulating tumor cells and gene transfer between them 207
    3.9. Color-coded imaging of gene transfer from high- to low-metastatic osteosarcoma cells *in vivo* 207
    3.10. Imaging dormant cancer cells 208
    3.11. Determining clonality of metastasis using color-coded cancer cells 208
    3.12. Telomerase-dependent adenovirus to label tumors *in vivo* for surgical navigation 209

AntiCancer, Inc., San Diego, California, USA
Department of Surgery, University of California San Diego, San Diego, California, USA

Methods in Enzymology, Volume 506
ISSN 0076-6879, DOI: 10.1016/B978-0-12-391856-7.00035-4

4. Methods 209
4.1. Production of GFP retrovirus 209
4.2. Production of RFP retroviral vector 210
4.3. Production of histone H2B-GFP vector 210
4.4. GFP or RFP gene transduction of cancer cells 210
4.5. Imaging apparatus 211
5. Technical Details 213
5.1. RFP retrovirus production 213
5.2. GFP retrovirus production 214
5.3. RFP or GFP gene transduction of tumor cell lines 215
5.4. Cell injection to establish a two-color experimental metastasis model 215
5.5. Surgical orthotopic implantation to establish a spontaneous metastasis model 216
6. Imaging 216
6.1. Fluorescence microscopy 216
6.2. Fluorescence stereomicroscopy 217
6.3. Long-working-distance variable-magnification fluorescence microscope for *in vivo* imaging 217
7. Chamber Imaging Systems 218
7.1. Olympus OV100 218
7.2. INDEC FluorVivo 218
7.3. UVP iBox 218
8. Histological Techniques 219
8.1. Tumor tissue sampling 219
8.2. Measurement of GFP-expressing tumor blood vessel length and evaluation of antiangiogenetic agents 219
8.3. Immunohistochemical staining 220
9. Conclusions 220
References 221

Abstract

The discovery, cloning, and characterization of GFP and related proteins of many colors have enabled live cell imaging to an unprecedented extent and resolution. Essentially, any cellular process can be imaged with a fluorescent protein. These proteins serve as genetic reporters and therefore can be used to follow cellular processes over indefinite periods *in vivo* as well as *in vitro*. The brightness and specific spectra of fluorescent proteins allow them to be imaged *in vivo*, using specific filters, without interference from autofluorescence. This chapter describes the development of live imaging in live animals with subcellular resolution, emphasizing the study of *in vivo* cell biology of cancer growth, spread, and metastasis.

1. Introduction

Green fluorescent protein (GFP) was discovered in the bioluminescent jellyfish *Aequorea victoria* by Shimomura (2009). The GFP gene was cloned from *A. victoria* by Doug Prasher which enabled GFP to become the most powerful tool in cell biology (Betzig *et al.*, 2006; Chalfie, 2009; Condeelis and Segall, 2003; Cheng *et al.*, 1996; Morin and Hastings, 1971; Sakaue-Sawano *et al.*, 2008). The GFP cDNA encodes a 283-amino acid polypeptide with a molecular weight of 27 kDa (Prasher *et al.*, 1992; Yang *et al.*, 1996). The monomeric GFP requires no other *Aequorea* proteins, substrates, or cofactors to fluoresce (Cody *et al.*, 1993; Hoffman, 2005).

In 2008, the Nobel Prize for chemistry was awarded for discovery and modification of GFP (Tsien, 2009). The Nobel announcement cited two uses of GFP (Nobel background), one of which was the use of GFP to track cancer cells *in vivo*, which was pioneered in our laboratory (Chishima *et al.*, 1997; Hoffman, 2005; Hayashi, *et al.*, 2007; Yamauchi *et al.*, 2006; Yang, *et al.*, 2000).

1.1. Pioneering *in vivo* imaging with GFP

Our laboratory pioneered the use of fluorescent protein for *in vivo* imaging (Chishima *et al.*, 1997), including noninvasive whole-body imaging (Yang *et al.*, 2000). These early studies demonstrated the power of fluorescent proteins for imaging tumors and metastasis, including possibilities to observe single cancer cells in the open mouse (Chishima *et al.*, 1997).

1.2. GFP is the method of choice for *in vivo* imaging

The GFP approach has several important advantages over other optical approaches to imaging tumor growth *in vivo*. In comparison with the luciferase reporter, GFP has a much stronger signal, and therefore can be used to image unrestrained animals. Irradiation with nondamaging blue light is the only step needed. Images can be captured with fairly simple apparatus and there is no need for total darkness. The fluorescence intensity of GFP is strong (Cormack *et al.*, 1996; Delagrave *et al.*, 1995; Heim *et al.*, 1995; Morin and Hastings, 1971) and the protein sequence of GFP has also been "humanized," which enables it to be highly expressed in mammalian cells (Zolotukhin *et al.*, 1996). Importantly, unlike luciferase, fluorescent proteins come in a multitude of colors (Shaner *et al.*, 2004), allowing for multiple events to be imaged. In addition, GFP fluorescence is relatively unaffected by the external environment, as the chromophore is protected by the three-dimensional structure of the protein (Cody *et al.*, 1993). A triple fusion reporter vector harboring a *Renilla* luciferase reporter gene, a reporter gene encoding a monomeric red fluorescent protein (RFP), and a mutant herpes simplex virus type thymidine kinase was tested *in vivo*. A highly sensitive cooled CCD camera that is compatible with both luciferase and fluorescence imaging compared these two signals from the fused reporter gene expressed with a lentivirus vector in 293T cells implanted in nude mice. The signal from RFP was found to be approximately 1000 times stronger than that from luciferase (Ray *et al.*, 2004). The weak signal from luciferase necessitates photon counting, with the construction of a pseudo-image *in vivo* rather than true imaging, therefore greatly reducing resolution and precluding the *in vivo* cellular imaging that is an important feature of GFP imaging. In addition, the rapid clearance of the injected luciferase results in an unstable signal that makes comparison of data difficult (Burgos *et al.*, 2003). The stronger signals from fluorescent proteins allow much more cost-efficient instrumentation. To overcome limits on fluorescent protein imaging imposed by the skin, reversible skin-flap window models have been developed that allow single-cell imaging on most organs of the mouse (Yang *et al.*, 2002). The main advantage of luciferase-based imaging is that no excitation light is required (Hoffman and Yang, 2006a).

1.3. Imaging the cell biology of metastasis *in vivo*

High-resolution intravital videomicroscopy has provided a powerful tool for directly observing steps in the metastatic process and for clarifying molecular mechanisms of metastasis and modes of action of antimetastasis therapeutics. Cells previously have been identified *in vivo* using exogenously-added fluorescent labels, limiting observations to a few cell divisions, or by natural markers (e.g., melanin) expressed only by specific cell types. GFP-transfected cells were used for monitoring and quantifying sequential steps in the metastatic process. Using CHO-K1 cells that stably express GFP, intravital videomicroscopy visualized sequential steps in metastasis within mouse liver, from initial arrest of cells in the microvasculature to the growth and angiogenesis of metastases. Individual, nondividing cells, as well as micro- and macrometastases could clearly be detected and quantified. Fine cellular details such as pseudopodial projections were visualized, even after extended periods of *in vivo* growth. The data suggest preferential growth and survival of micrometastases near the liver surface. A small population of single cells was imaged that persisted over the 11-day observation period, which may represent dormant cells (Naumov *et al.*, 1999).

Cellular details, such as pseudopodial projections, could be clearly seen (Naumov *et al.*, 1999). Farina *et al.* (1998) observed tumor cell motility at the single-cell level, including movement in and out of blood vessels, using GFP-expressing cells. Condeelis and Segall (2003) used GFP imaging to view cells in time-lapse images in a single optical section using a confocal microscope. The polarity of tumor cells, along with their response to chemotatic cytokines, has been visualized by intravital imaging (Farina *et al.*, 1998). Wyckoff *et al.* (2007) have shown with multiphoton microscopy that tumor cell intravasation occurs in association with perivascular macrophages in mammary tumors. These techniques enable a greater understanding of tumor cell migration *in vivo* (Hoffman, 2009).

2. Imaging Angiogenesis *In Vivo*

The dynamics of blood vessel recruitment by cancer cells were investigated by Amoh *et al.* (2004). They used transgenic mice expressing GFP under the control of a neural-stem-cell marker nestin. The nestin-promoter-driven GFP (ND-GFP) has shown that nestin is expressed in hair follicle stem cells and the blood vessel network interconnecting hair follicles in the skin. The hair follicles were shown to directly give rise to the nestin-expressing blood vessels. Following transplantation of the RFP-expressing murine melanoma cell line B16F10 into these animals, tumor angiogenesis was visualized using dual-color fluorescence imaging. ND-GFP was expressed in the proliferating endothelial cells and nascent blood vessels in the growing tumor. Immunohistochemical staining showed that the endothelial-cell-specific antigen CD31 was expressed in ND-GFP-expressing nascent blood vessels, showing that the tumor directly recruited the nestin-expressing cells. Doxorubicin inhibited tumor angiogenesis as well as tumor growth in these mice (Amoh *et al.*, 2005a; Hoffman, 2009).

2.1. Two-photon imaging with fluorescent proteins

Two-photon imaging results in simultaneous absorption of $n \geq 2$ photons that together provide the energy needed for excitation. This predicts optimum excitation wavelengths at roughly twice the corresponding single-photon wavelength (Dickinson *et al.*, 2003). The excitation and emission spectra of EGFP and DsRed2 were obtained from living dual-color human fibrosarcoma cells (HT-1080) (Yamamoto *et al.*, 2004) expressing both nuclear EGFP/histone-2B (H2B) and cytoplasmic DsRed2. The most efficient excitation of EGFP was at 930 nm, as reported (Dickinson *et al.*, 2003), but no excitation was observed between 1060 and 1350 nm. By contrast, DsRed2 showed a small excitation peak at 760 nm and 20-fold higher efficiency from 1090 to 1120 nm (Andresen *et al.*, 2009).

Brown *et al.* (2001) showed that multiphoton laser-scanning microscopy (MPLSM) (Denk *et al.*, 1990) could be used to investigate deeper regions of GFP-expressing tumors in dorsal skin-fold chambers. MPLSM offers significant advantages over other visualization techniques, such as improved signal/background ratios and longer sample lifetimes, as well as greater imaging depths. Fukumura *et al.* (1997, 1998) monitored the activity of the vascular endothelial growth factor (VEGF) promoter in transgenic mice that expressed GFP under the control of the VEGF promoter. MPLSM showed that the tumor was able to induce activity of the VEGF promoter and subsequent blood vessel formation.

2.2. Red fluorescent protein

The first red fluorescent proteins were isolated from the coral *Discosoma* by Matz *et al.*, (1999). A red protein has been further developed for expression in mammalian cells and is now known as DsRed2. Monomer variants of the *Discosoma* red protein have been developed with distinguishable colors from yellow-orange to red-orange with fruit names similar to their color, mCherry, mTomato, etc. (Shaner *et al.*, 2004). Fluorescent proteins that can be converted from one color to another or become fluorescent upon light activation have also been developed (Sakaue-Sawano *et al.*, 2008; Verkhusha and Lukyanov, 2004).

mCherry, mRaspberry, mPlum, and mTomato have emission maxima as long as 649 nm. However, these mutants have low quantum yields, thereby reducing their brightness. A very bright, red-shifted variant has subsequently been isolated, called Katushka, with an excitation peak at 588 mm emission peak at 635 nm both of which are relatively nonabsorbed by issues and hemoglobin. Katushka has many favorable properties in addition to its absorption and emission peaks including a rapid maturation time of 20 min. Importantly, an extinction coefficient of 65,000 M^{-1} cm^{-1} and quantum yield of 0.34, make Katushka the brightest fluorescent protein with an emission maximum beyond 620 nm. In cells, Katushka demonstrated no visible aggregates or other toxic effects (Shcherbo *et al.*, 2007).

2.3. Establishment of dual-color cancer cells

For establishing dual-color cells, RFP-expressing cancer cells were incubated with a 1:1 precipitated mixture of retroviral supernatants of PT67H2B-GFP cells and culture medium. To select the double transformants, the cells were incubated with hygromycin 72 h after transfection. The level of hygromycin was increased stepwise up to 400 mg/ml. Clones of dual-color cancer cells were isolated with cloning cylinders under fluorescence microscopy. These clones were amplified by conventional culture methods. These sublines stably expressed GFP in the nucleus and RFP in the cytoplasm (Yang *et al.*, 2007). Nuclear GFP expression enabled visualization of nuclear dynamics, whereas simultaneous cytoplasmic RFP expression enabled visualization of nuclear–cytoplasmic ratios as well as simultaneous cell and nuclear shape changes. Thus, total cellular dynamics can be visualized in the living dual-color cells in real time. The cell-cycle position of individual living cells was readily visualized by the nuclear–cytoplasmic ratio and nuclear morphology. Real-time induction of apoptosis was observed by nuclear size changes and progressive nuclear fragmentation. Thus, the dual-color cells are a useful tool for visualizing living-cell dynamics *in vivo* as well as *in vitro* (Hoffman, 2011).

Miyawaki's group has used oscillating proteins that mark cell-cycle transitions to effectively label individual G1 phase nuclei red and those in S/G2/M phases green. Every cell exhibits either red or green fluorescence (Sakaue-Sawano *et al.*, 2008).

3. Imaging Cell Trafficking

3.1. Imaging cancer cell deformation and trafficking in blood vessels

The mechanism of cancer cell deformation and migration in narrow vessels is incompletely understood. In order to visualize the cytoplasmic and nuclear dynamics of cancer cells migrating in capillaries, dual color cancer-cells with RFP expressed in the cytoplasm, and GFP, linked to histone H2B, expressed in the nucleus, were used. Immediately after the cells were injected in the heart of nude mice, a skin flap on the abdomen was made. With a color CCD camera, highly elongated cancer cells and nuclei in capillaries in the skin flap in living mice were observed. The migration velocities of the cancer cells in the capillaries were measured by capturing images of the dual-color fluorescent cells over time. The cells and nuclei in the capillaries elongated to fit the width of these vessels. The average length of the major axis of the cancer cells in the capillaries increased to approximately four times their normal length. The length of the nuclei in the capillaries increased 1.6 times. Cancer cells in capillaries over 8 mm in diameter could migrate up to 48.3 mm/h. The data suggest that the minimum diameter of capillaries where cancer cells are able to migrate is approximately 8 mm (Yamauchi *et al.*, 2005).

Dual-color cancer cells were also injected by a vascular route in an abdominal skin flap in nude mice. The mice were imaged with an Olympus OV100 small-animal imaging system. Nuclear and cytoplasmic behavior of cancer cells in real time in blood vessels were observed as they moved by various means or adhered to the vessel surface in the abdominal skin flap.

Real-time dual-color imaging showed that during extravasation, cytoplasmic processes of the cancer cells exited the vessels first, with nuclei following along the cytoplasmic projections. Both cytoplasm and nuclei underwent deformation during extravasation. Different cancer cell lines seemed to vary strongly in their ability to extravasate (Yamauchi *et al.*, 2006).

3.2. Imaging cancer cell trafficking in lymphatic vessels

Cancer cells labeled with both GFP in the nucleus and RFP in the cytoplasm, or with GFP only or RFP only were injected into the inguinal lymph node of nude mice. The labeled cancer cells trafficked through lymphatic vessels where they were imaged via a skin flap in real time at the cellular level until they entered the axillary lymph node. The bright fluorescence of the cancer cells and the real-time microscopic imaging capability of the Olympus OV100 small-animal imaging system enabled visualization of the trafficking cancer cells in the lymphatics. The role of pressure on tumor-cell shedding into lymphatic vessels was also investigated. Pressure was generated by placing 25- and 250-g weights for 10 s on the bottom surface of a tumor-bearing footpad. Increasing pressure on the tumor increased the numbers of shed cells, fragments, and embolished into the lymphatic vessel. Pressure also deformed the shed emboli, increasing their maximum major axis (Hayashi *et al.*, 2007).

3.3. Color-coded imaging of tumor–host interaction

Tumor–host models can be made with transgenic nude mice expressing GFP or RFP or cyan fluorescent protein (CFP) in almost all cells and tissues. Colored transgenic nude mice are particularly useful, as they can accept human tumors cells. For example, when tumor cells expressing RFP are implanted in mice expressing GFP in most tissues, various types of tumor–host interactions can be observed, including those involving host blood vessels, lymphocytes, tumor-associated fibroblasts, macrophages, dendritic cells, and others. The "color-coded" tumor–host models enable imaging and, therefore, a deeper understanding of the host cells involved and their function in tumor progression (Hoffman and Yang, 2006b).

3.4. Transgenic GFP nude mouse

We have developed a transgenic GFP nude mouse with ubiquitous GFP expression. The GFP nude mouse was obtained by crossing nontransgenic nude mice with the transgenic C57/B6 mouse in which the beta-actin promoter drives GFP expression in essentially all tissues. In crosses between nu/nu GFP male mice and nu/+ GFP female mice, the embryos fluoresced green. Approximately 50% of the offspring of these mice were GFP nude mice. Newborn mice and adult mice fluoresced very bright green and could be detected with a simple blue-light-emitting diode flashlight with a central peak of 470 nm and a bypass emission filter. In the adult mice, most of the organs, including the heart, lungs, spleen, pancreas, esophagus, stomach, and duodenum, brightly expressed GFP. The following systems were dissected out and shown to have brilliant GFP fluorescence: the entire digestive system from tongue to anus; the male and female reproductive systems; brain and spinal cord; and the circulatory system, including the heart and major arteries and veins. The skinned skeleton highly expressed GFP. Pancreatic islets showed GFP fluorescence. The spleen cells were also GFP positive. RFP-expressing human tumors grew extensively in the transgenic GFP nude mouse (Yang *et al.*, 2004).

For example, when tumor cells expressing RFP are implanted into mice expressing GFP, various types of tumor interactions can be observed (Amoh *et al.*, 2005a,b; Yang *et al.*, 2003, 2004). In fresh tissue specimens, tumor vessels expressing GFP can be visualized vascularizing tumors expressing RFP in primary and metastatic sites. Dendritic cells expressing GFP can be seen in close contact with tumor cells with their dendritic processes. Stromal fibroblasts expressing GFP can be seen in contact with multiple cancer cells through their pseudopodia. Lymphocytes, expressing GFP, can be observed in the process of rejecting tumor cells from growing in immunocompetent mice. Macrophages, expressing GFP, can be observed engulfing tumor cells expressing RFP. The color-coded tumor–host models will enable imaging and, therefore, help elucidate to a much greater extent the function of the stromal host cells involved in tumor progression (Hoffman and Yang, 2006b).

3.5. Transgenic RFP nude mouse

The RFP nude mouse was obtained by crossing nontransgenic nude mice with the transgenic C57/B6 mouse in which the beta-actin promoter drives RFP (DsRed2) expression in essentially all tissues. In crosses between nu/nu RFP male mice and nu/+ RFP female mice, the embryos fluoresced red. Approximately 50% of the offspring of these mice were RFP nude mice. In the RFP nude mouse, all the organs, including the heart, lungs, spleen, pancreas, esophagus, stomach, duodenum, the male and female reproductive systems, brain and spinal cord, and the circulatory system, including the heart, and major arteries and veins, brightly expressed RFP. The skinned skeleton highly expressed RFP. The bone marrow and spleen cells were also RFP positive. GFP-expressing human cancer cell lines, were orthotopically transplanted to the transgenic RFP nude mice. These human tumors grew extensively in the transgenic RFP nude mouse (Yang *et al.*, 2009).

3.6. Transgenic CFP nude mouse

The CFP nude mouse was developed by crossing nontransgenic nude mice with the transgenic CK/ECFP mouse in which the beta-actin promoter drives expression of CFP in almost all tissues. In crosses between nu/nu CFP male mice and nu/+ CFP female mice, approximately 50% of the embryos fluoresced blue. In the CFP nude mice, of all internal organs, the pancreas and reproductive organs displayed the strongest fluorescent signals, which varied in intensity. XPA-1 human pancreatic cancer cells expressing RFP, or GFP in the nucleus and RFP in the cytoplasm, was transplanted in nude CFP mice. Color-coded fluorescence imaging of these human pancreatic cancer cells implanted into the bright blue fluorescent pancreas of the CFP nude mouse demonstrated the interaction of the pancreatic tumor and the normal pancreas, in particular the strong desmoplastic reaction of the tumor. (Tran Cao *et al.*, 2009).

3.7. Noninvasive color-coded imaging of the tumor microenvironment

The Olympus IV100 laser-scanning microscope, with ultra-narrow microscope objectives ("stick objectives"), was used for three-color whole-body imaging of two-color cancer cells interacting with the GFP-expressing stromal cells of the transgenic GFP nude mouse. In this model, drug response of both cancer and stromal cells in the intact live animal was also imaged in real time. Various *in vivo* phenomena of tumor–host interaction and cellular dynamics were imaged, including mitotic and apoptotic tumor cells, stromal cells interacting with the tumor cells, tumor vasculature, and tumor blood flow. This model enabled the first cellular and subcellular images of unperturbed tumors in the live intact animal (Yang *et al.*, 2007).

Our laboratory has previously isolated CTCs from orthotopic nude mouse models of human prostate cancer cells where the PC-3 cancer cells express GFP (Glinskii *et al.*, 2003). It was found that orthotopic tumors produced CTCs and not subcutaneous tumors, which may explain why orthotopic tumors metastasize and subcutaneous tumors do not. Using the GFP-expressing PC-3 orthotopic model and immunomagnetic beads coated with antiepithelial cell adhesion molecule (EpCAM) and antiprostate specific membrane antigen (PSMA), GFP-expressing CTC were isolated within 15 min and were readily visualized by GFP fluorescence (Kolostova *et al.*, 2011). The GFP CTC could be expanded in culture and then transplanted *in vivo* for subsequent analysis.

3.8. Imaging circulating tumor cells and gene transfer between them

Color-coded imaging visualized circulating yellow fluorescent prostate-cancer metastatic cells that were readily isolated from the circulation of tumor-bearing mice after mixtures of RFP- and GFP-expressing PC-3 human prostate carcinoma cells were implanted in the nude mouse prostate. The yellow fluorescent cells were purified from the circulation of nude mice to 99% homogeneity by FACS, expanded in culture, and reimplanted in the prostate of nude mice. The yellow fluorescent phenotype was heritable and stably maintained *in vitro* and *in vivo* by tumor cells for many generations. In the animals implanted with the yellow fluorescing cells, 100% developed aggressive metastatic cancer. Lung metastases were demonstrated in 100% of the animals as early as 4 weeks after injection of the yellow fluorescing cells in the mouse prostate. In contrast, when the GFP- and RFP-expressing parental cells were inoculated into the mouse prostate separately, none of the animals developed lung metastasis. These results are consistent with the idea that spontaneous genetic exchange between tumor cells occurs *in vivo* and contributes to genomic instability and creation of highly metastatic cells (Glinsky *et al.*, 2006).

3.9. Color-coded imaging of gene transfer from high- to low-metastatic osteosarcoma cells *in vivo*

The 143B-GFP human osteosarcoma cell line, with high metastatic potential, and the human MNNG/HOS-RFP human osteosarcoma cell line with low metastatic potential, both derived from the TE85 human osteosarcoma cell line, were either cotransplanted or transplanted alone in the tibia in nude mice. Upon mixed transplantation of the two differently-labeled sublines, resulting metastatic colonies were single colored, either red or green, thereby demonstrating their clonality and enabling facile color-coded quantification. When MNNG/HOS-RFP and 143B-GFP were cotransplanted in the tibia, the number of lung metastases of MNNG/HOS-RFP increased eightfold compared to that when MNNG/HOS-RFP was transplanted alone ($P < 0.01$). In contrast, no enhancement of MNNG/HOS-RFP metastases occurred when MNNG/HOS-RFP and 143B-GFP were transplanted separately in the right and left tibiae, respectively. This result suggests that the presence of 143B-GFP increased the metastatic potential of MNNG/HOS-RFP within the mixed tumor. We observed transfer of the Ki-ras gene from 143B-GFP to MNNG/HOS-RFP after they were co-implanted, suggesting that the Ki-ras played a role in increasing the metastatic potential of MNNG/HOSRFP in the presence of 143B-GFP (Tome *et al.*, 2009).

These experiments further suggest the possible role of *in vivo* gene transfer in enhancing the metastatic potential of cancer cells. The data also further demonstrated the power of color-coded imaging to visualize gene transfer between cancer cells *in vivo*.

3.10. Imaging dormant cancer cells

An isogenic pair of metastatic (M4A4) and nonmetastatic (NM2C5), GFP-labeled human breast cancer cell lines were inoculated into the mammary glands of nude mice to investigate cancer cell dormancy. Single GFP-expressing cancer cells were observed in the lungs, even in animals inoculated with NM2C5, which fail to form secondary tumors in other organs. NM2C5 cells remained in the lung after primary tumor resection, although they formed no metastases by 6 months. When isolated, these cells proliferate *in vitro* and when implanted, formed tumors (Goodison *et al.*, 2003).

3.11. Determining clonality of metastasis using color-coded cancer cells

We used GFP-labeled and RFP-labeled HT-1080 human fibrosarcoma cells to determine clonality of metastatic colonies after mixed implantation of the red and green fluorescent cells. Resulting pure red or pure green colonies were scored as clonal, whereas mixed yellow colonies were scored as nonclonal. In a spontaneous metastasis model originating from footpad injection in severe combined immunodeficient (SCID) mice, 95% of the resulting lung colonies were either pure green or pure red, indicating monoclonal origin, whereas 5% were of mixed color, indicating polyclonal origin. In an experimental lung metastasis model established by tail-vein injection in SCID mice, clonality of lung metastasis was dependent on cell number. With a minimum cell number injected, almost all (96%) colonies were pure red or green and, therefore, monoclonal. When a large number of cells were injected, almost all (87%) colonies were of mixed color and, therefore, heteroclonal. We concluded that spontaneous metastasis may be clonal because they are rare events, thereby supporting the rare-cell clonal origin of metastasis hypothesis. The clonality of the experimental metastasis model depended on the number of input cells (Yamamoto *et al.*, 2003a,b).

Color-coded lung metastases were also visualized by external fluorescence imaging in live animals through skin-flap windows over the chest wall. Lung metastases were observed on the lung surface through the skin flap. Real-time metastatic growth of the color-coded clones in the same lung was externally imaged with resolution and quantification of green, red, or yellow colonies in live animals that signified clonal or non-clonal metastasis, respectively (Yamamoto *et al.*, 2003a).

3.12. Telomerase-dependent adenovirus to label tumors *in vivo* for surgical navigation

Introducing and selectively activating the *GFP* gene in malignant tissue *in vivo* were made possible by the development of OBP-401, a telomerase-dependent, replication-competent adenovirus expressing GFP cancer. HCT-116, a model of intraperitoneal disseminated human colon cancer, was labeled with GFP by OBP-401 virus injection into the peritoneal cavityof nude mice and A549, a model of pleural dissemination of human lung cancer, was labeled with GFP by OBP-401 virus administered into the pleural cavity. Only the malignant tissue fluoresced brightly in both models. In the intraperitoneal model of disseminated cancer, fluorescence-guided surgery enabled resection of all tumor nodules labeled with GFP by OBP-401. These results suggest that adenoviral-GFP labeling of tumors in patients can enable fluorescence-guided surgical navigation (Kishimoto *et al.*, 2009).

4. Methods

4.1. Production of GFP retrovirus

The pLEIN retroviral vector (Clontech Laboratories, Inc., Palo Alto, CA) expressing GFP and the neomycin resistance gene on the same bicistronic message was used as a GFP expression vector. PT67, an NIH3T3-derived packaging cell line, expressing the 10 A1 viral envelope, was purchased from Clontech Laboratories, Inc. PT67 cells were cultured in DMEM (Irvine Scientific, Santa Ana, CA) supplemented with 10% heat-inactivated fetal bovine serum (FBS; Gemini Bio-products, Calabasas, CA). For vector production, packaging cells (PT67), at 70% confluence, were incubated with a precipitated mixture of LipofectAMINE reagent (Life Technologies Inc., Grant Island, NY) and saturating amounts of pLEIN plasmid for 18 h. Fresh medium was replenished at this time. The cells were examined by fluorescence microscopy 48 h after transfection. For selection, the cells were cultured in the presence of 500–2000 mg/ml of G418 (Life Technologies, Inc., Grand Island, NY) for 7 days to select for a clone producing high amounts of a GFP retroviral vector (PT67-GFP) (Yamamoto *et al.*, 2004; Hoffman and Yang, 2006b).

4.2. Production of RFP retroviral vector

For RFP retrovirus production, the *Hin*dIII/*Not*I fragment from pDsRed2 (Clontech Laboratories, Inc.), containing the full-length RFP (DsRed2) cDNA, was inserted into the *Hin*dIII/*Not*I site of pLNCX2 (Clontech Laboratories) that has the neomycin resistance gene to establish the pLNCX2-DsRed2 plasmid. PT67 cells were cultured in DMEM (Irvine Scientific) supplemented with 10% heat-inactivated FBS (Gemini Bio-products). For vector production, PT67 cells, at 70% confluence, were incubated with a precipitated mixture of Lipofectamine reagent (Life Technologies, Inc., Grand Island, NY) and saturating amounts of pLNCX2-DsRed2 plasmid for 18 h. Fresh medium was replenished at this time. The cells were examined by fluorescence microscopy 48 h after transfection. For selection of a clone producing high amounts of a RFP retroviral vector (PT67-DsRed2), the cells were cultured in the presence of 200–1000 mg/ml of G418 which was increased stepwise in order to select for brighter cells (Yang *et al.*, 2007; Hoffman and Yang, 2006c).

4.3. Production of histone H2B-GFP vector

The histone H2B gene has no stop codon, thereby enabling the ligation of the H2B gene to the 5′-coding region of the *A. victoria* EGFP gene (Clontech Laboratories). The histone H2B-GFP fusion gene was then inserted at the *Hin*dIII/*Cla*I site of the pLHCX (Clontech Laboratories) that contains the hygromycin resistance gene. To establish a packaging cell clone producing high amounts of a histone H2B-GFP retroviral vector, the pLHCX histone H2B-GFP plasmid was transfected in PT67 cells using the same methods described above for PT67-DsRed2. The transfected cells were cultured in the presence of 200–400 mg/ml of hygromycin (Life Technologies) for 15 days to establish stable PT67H2B-GFP packaging cells (Yamamoto *et al.*, 2004; Yang *et al.*, 2007; Hoffman and Yang, 2006c).

4.4. GFP or RFP gene transduction of cancer cells

For GFP or RFP gene transduction, 70% confluent cultures of cancer lines are used. Cancer cells were incubated with a 1:1 precipitated mixture of retroviral supernatants of PT67-GFP or PT67-RFP cells and RPMI 1640 (Mediatech, Inc.) containing 10% FBS for 72 h. Fresh medium was replenished at this time. Cells were harvested with trypsin/EDTA 72 h after transduction and subcultured at a ratio of 1:15 in selective medium, which contained 200 mg/ml of G418. The level of G418 was increased stepwise up to 800 mg/ml. GFP- or RFP-expressing cancer cells were isolated with cloning cylinders (Bel-Art Products) using trypsin/EDTA and amplified by conventional culture methods (Yang *et al.*, 2007; Hoffman and Yang, 2006c).

To establish dual-color cancer cells, clones of cancer cells expressing RFP in the cytoplasm were initially established as described above. In brief, cancer cells were incubated with a 1:1 precipitated mixture of retroviral supernatants of PT67-RFP cells and RPMI 1640 (Mediatech, Inc., Herndon, VA) containing 10% FBS for 72 h. Fresh medium was replenished at this time. Cells were harvested with trypsin/EDTA 72 h posttransduction and subcultured at a ratio of 1:15 into selective medium, which contains 200 mg/ml G418. The level of G418 was increased stepwise up to 800 mg/ml. RFP cancer cells were isolated with cloning cylinders (Bel-Art Products, Pequannock, NJ) using trypsin/EDTA and amplified by conventional culture methods. For establishing dual-color cells, RFP cancer cells were then incubated with a 1:1 precipitated mixture of retroviral supernatants of PT67 H2B-GFP cells and culture medium. To select the double transformants, cells were incubated with hygromycin 72 h after transfection. The level of hygromycin was increased stepwise up to 400 mg/ml (Jiang *et al.*, 2006).

4.5. Imaging apparatus

4.5.1. *In vivo* imaging with an LED flashlight and filters

Low-cost instrumentation and standard GFP and RFP biomarkers can be used to visualize tumors completely noninvasively. Utilizing a blue LED flashlight with a 470-nm excitation filter, tumors in—and on—several organs (including liver, pancreas, colon, bone, and brain) could be clearly imaged. The clearest demonstration of the power of this technique is the data showing that when the image of a surgically-exposed colon tumor was analyzed and compared to the same tumor from a whole-body (unopened) image, the intensity of the GFP signal from the unopened mouse was 70% that of the opened. Further, despite the expected distortion caused by light scattering, the image sizes were comparable (Yang *et al.*, 2005).

A blue LED flashlight (LDP LLC, Woodcliff Lake, NJ, USA; www.maxmax.com/OpticalProducts.htm) with an excitation filter (midpoint wavelength peak of 470 nm) and an emission D470/40 filter (Chroma Technology, Brattleboro, VT, USA) for viewing could be used for whole-body imaging of mice with GFP- and RFP-expressing tumors growing in or on internal organs (Yang *et al.*, 2000, 2005).

4.5.2. Simple light-box imaging

Whole-body imaging that visualized the entire animal at lower magnification was carried out in a light box illuminated by fiberoptic lighting (Lightools Research, Inc., Encinitas, CA) and imaged using a thermoelectrically cooled color CCD camera (Yang *et al.*, 2000).

4.5.3. *In vivo* imaging with a fluorescence dissecting microscope

A Leica fluorescence stereo microscope, model LZ12, equipped with a 50-W mercury lamp, was used for high-magnification imaging of GFP-expressing tumors and metastasis *in situ* or for whole-body imaging of animals with GFP-expressing tumors. Selective excitation of GFP was produced through a D425/60 band-pass filter and 470 DCXR dichroic mirror. Emitted fluorescence was collected through a long-pass filter GG475 (Chroma Technology, Brattleboro, VT) on a Hamamatsu C5810 three-chip cooled color CCD camera (Hamamatsu Photonics Systems, Bridgewater, NJ). Images were processed for contrast and brightness and analyzed with the use of Image-Pro Plus 3.1 software (Media Cybernetics, Silver Springs, MD). Images of 1024 × 724 pixels were captured directly on an IBM PC or continuously through video output on a high-resolution Sony VCR model SLVR1000 (Sony Corp., Tokyo, Japan) (Hoffman, 2011).

4.5.4. Long-working-distance variable-magnification fluorescence microscope for *in vivo* imaging

The Olympus MVX10 Macro View (OLYMPUS Corporation, Center Valley, PA, USA) and a FluoCam 1500G color CCD camera were used for macro, cellular and subcellular imaging in live mice (Kimura *et al.*, 2010).

4.5.5. *In vivo* cellular imaging with a variable-magnification imaging chamber

An Olympus OV100 small-animal imaging system with a sensitive CCD camera and four objective lenses, parcentered and parfocal, enabling imaging from macrocellular to subcellular was developed. Nuclear and cytoplasmic behavior of cancer cells expressing GFP in the nucleus and RFP in the cytoplasm was observed in real time in live mice (Yamauchi *et al.*, 2006).

4.5.6. Imaging chambers designed for whole-body imaging

The FluorVivo small-animal imaging system (INDEC Systems, Inc.) can be used for whole-body imaging in live mice. FluorVivo uses extremely bright, solid state, LED illuminators, and a full color CCD camera to provide high speed, multicolor imaging of up to three animals with single exposures. The instrument's high-speed acquisition permits *in vivo* monitoring of both static and dynamic processes, as well as real-time recordings of fluorescence-guided surgeries. FluorVivo's fully integrated software provides complete control of the instrument, ease of use, and powerful analytical tools for extracting quantitative data from acquired images (Yang *et al.*, 2009).

The UVP iBox small-animal imaging system is capable of fluorescent protein imaging with a range of cameras that use front and back illuminated CCDs with sizes up to a 43-mm diagonal, greatly expanding the applications for high-resolution, large-field-of-view, and increased-throughput imaging. The iBox imaging system can be configured with both monochrome and color CCDs, with CCD resolution currently up to 8.3 megapixels and sensitive to a wide range of spectrum (CFP to near infrared). The range of fast lenses includes several interchangeable, fully automated optics: a 50-mm f1.2, a 28-mm f1.8, and a 24- to 70-mm f2.8 zoom lens. These lenses give maximum imaging flexibility, with the field of view ranging from one to several animals. At f1.2, the typical exposures are less than 50 ms, minimizing the effect of animal movement. The camera, optics, sample platform position, and excitation and emission filters are under full software control, permitting reproducible, and rapid imaging with software presets and macros (Tran Cao *et al.*, 2009).

5. Technical Details

5.1. RFP retrovirus production

1. Insert the *Hin*dIII–*Not*I fragment from pDsRed2, containing full-length RFP cDNA, into the *Hin*dIII–*Not*I site of pLNCX2, which contains a neomycin resistance gene, to establish the pLNCX2-DsRed2 plasmid.
2. Use PT67, an NIH3T3-derived packaging cell line expressing the 10 A1 viral envelope, to produce retrovirus. Culture 3×10^5 PT67 cells in a 25-mm^2 flask with DMEM supplemented with 10% heat-inactivated FBS. It takes approximately 3 days for the cells to reach about 70% confluence.
3. For vector production, use PT67 packaging cells at 70% confluence. Plate PT67 cells on a 60-mm culture dish at 60–80% confluence 12 h before transfection. Use 10 μg of pLNCX2-DsRed2 DNA and the Lipofectamine Plus transfection kit. Add 7 μl of pLNCX2-DsRed2 DNA to 87 μl serum-free medium in a tube and then add 6 μl Lipofectamine reagent, mix, and incubate for 15 min at 22–26 °C (room temperature).
4. Dilute 4 μl of Lipofectamine reagent in 96 μl of serum-free medium in a second tube. Mix and incubate for 15 min at room temperature.
5. Combine the DNA prepared in Step 3 and diluted Lipofectamine reagent, then mix and incubate for 15 min at room temperature.
6. While the complexes are forming, replace the medium on the cells with 800 μl of serum-free DMEM. Add the DNA–Lipofectamine complex to

the dish with cells containing fresh DMEM. Mix the complexes into the medium gently and incubate for 4 h at 37 °C in 5% CO_2.

7. After 4 h of incubation, increase the volume of the medium to 5 ml and incubate for 24 h at 37 °C.
8. After 24 h of incubation, clone the packaging cells by limiting dilution in 96-well plates.
9. For selection of a PT67 packaging cell clone producing large amounts of RFP retroviral vector (PT67-DsRed2), culture the cells in the presence of 100–1000 μg/ml of G418. Culture the cells for 1–2 days in each concentration of G418. Clones of PT67-DsRed2 cells with high-viral titer production are identified with 3T3 cells used for virus titering. Clones with a titer higher than 1×10^6 plaque-forming units per ml are used for RFP vector production.

5.2. GFP retrovirus production

1. Use the pLEIN or equivalent retroviral vector expressing enhanced GFP or equivalent GFP and the neomycin resistance gene, on the same bicistronic message, as a GFP expression vector.
2. Use PT67, an NIH3T3-derived packaging cell line expressing the 10 A1 viral envelope, to produce retrovirus. Culture 3×10^5 PT67 cells in a 25-mm^2 flask with DMEM supplemented with 10% heat-inactivated FBS. It takes approximately 3 days for the cells to reach ~70% confluence.
3. For vector production, use PT67 cells at 70% confluence. Plate PT67 cells on a 60-mm dish at 60–80% confluence 12 h before transfection. Use 10 μg pFB-GFP (Clontech) with the Lipofectamine Plus transfection kit. Add 7 ml precomplexed pFB-GFP DNA in 87 μl of serum-free medium and then add 6 μl Lipofectamine reagent in a tube; mix and incubate at room temperature (22–26 °C) for 15 min.
4. Dilute 4 μl Lipofectamine in 96 μl serum-free medium in a second tube. Mix and incubate at RT for 15 min.
5. Combine precomplexed DNA and diluted Lipofectamine reagent; then mix and incubate at RT for 15 min.
6. While the complexes are forming, replace the medium on the cells with 800 μl serum-free DMEM. Add the DNA–Lipofectamine reagent complex to the dish with cells containing fresh DMEM. Mix the complexes into the medium gently; incubate in a humidified incubator at 37 °C and 5% CO_2 for 4 h.
7. After 4 h of incubation, increase the volume of the medium to 5 ml. Incubate in the same conditions for 24 h.

8. After 24 h incubation, clone the packaging cells by limiting dilution in 96-well plates.
9. Examine the cells by fluorescence microscopy 48 h posttransduction.
10. For selection, culture the cells in the presence of 500–2000 μg/ml G418 to select for a clone producing high amounts of a GFP retroviral vector (PT67-GFP). Culture the cells for 1–2 days in each concentration of G418. High-viral titer production clones of GFP PT67 cells are identified with 3T3 cells used for virus titering. Clones with titer higher than 10^6 plaque-forming units per ml are used for GFP vector production.

5.3. RFP or GFP gene transduction of tumor cell lines

1. For RFP or GFP gene transduction, use cancer cells that are 20% confluent. Plate cancer cells at a density of 1×10^5 to 2×10^5 cells per 60-mm plate 12–18 h before infection with RFP retrovirus.
2. For retroviral infection, collect conditioned medium from packaging cells (PT67-DsRed2 or PT67-eGFP) and filter medium through a 0.45-mm polysulfonic filter. Add virus-containing filtered medium to the target cells. Add polybrene to a final concentration of 8 μg/ml. Incubate the cells for 24 h at 37 °C.
3. Replace the medium with DMEM and 10% FBS after 24 h of incubation and check for RFP-expressing cells by fluorescence microscopy.
4. Collect tumor cells with trypsin–EDTA and subculture them at a ratio of 1:15 in selective medium, which contains 50 μg/ml G418.
5. To select brightly fluorescent cells, increase the concentration of G418 to 800 μg/ml in a stepwise way. Culture the cells for 1–2 days in each concentration of G418.
6. Isolate clones expressing RFP with cloning cylinders using trypsin–EDTA and amplify them in DMEM in the absence of the selective agent. Then select cells for brightness and stability.

To establish dual-color cancer cells, use clones of cancer cells expressing RFP in the cytoplasm at 70% confluence. Incubate RFP cancer cells, produced as described above, with the retroviral-containing medium supernatants of PT67 H2B-GFP cells and the above culture medium. To obtain the double transformants, incubate cells with hygromycin 48 h after transfection and select as described above (Hoffman and Yang, 2006c).

5.4. Cell injection to establish a two-color experimental metastasis model

1. Collect RFP expressing cancer cells by trypsinization for 3 min at 37 °C with 0.25% trypsin.

2. Wash the cells three times with cold serum-free medium using a tabletop centrifuge at 500×*g*.
3. Resuspend the cells in approximately 0.2 ml serum-free medium.
4. Within 30 min of collecting cells, inject 1×10^6 tumor cells in a total volume of 0.2 ml into 6-week-old C57BL/6-GFP mice or nude (*nu/nu*) GFP mice in the lateral tail vein, or subcutaneously using a 1-ml 27G2 latex-free syringe.
5. For liver colonization, inject fluorescent protein-expressing cells directly into the portal vein or spleen in anesthetized mice (Bouvet *et al.*, 2006).

5.5. Surgical orthotopic implantation to establish a spontaneous metastasis model

1. Induce anesthesia with a "ketamine mixture" (10 ml ketamine HCl, 7.6 ml xylazine, and 2.4 ml acepromazine maleate), injected sc.
2. Use a microscope (Leica MZ6) with magnification of about 6× to about 40× for all procedures of the operation.
3. Isolate fluorescent protein-expressing tumor fragments (1 mm^3) from subcutaneously growing tumors, formed by injection of RFP-expressing cancer cells by mincing tumor tissue into 1-mm^3 fragments. After proper exposure of the target organ, implant three tumor fragments per transgenic GFP mouse.
4. With an 8–0 surgical suture, penetrate the tumor fragments and suture the fragments onto the target organ.
5. Keep the mice in a barrier facility under high-efficiency particulate air filtration (Hoffman, 1999).

6. Imaging

6.1. Fluorescence microscopy

1. Use an Olympus BH2-RFCA fluorescence microscope equipped with a mercury 100-W lamp power supply or its equivalent.
2. To visualize both GFP and RFP fluorescence at the same time, produce excitation light using a D425/60 band-pass filter and a 470-DCXR dichroic mirror.
3. Collect emitted fluorescence light through a GG475 long-pass filter.
4. Capture high-resolution images of 1024 × 724 pixels with a Hamamatsu C5810 three-chip cooled color CCD camera or its equivalent, and store directly on an IBM PC or its equivalent.
5. Process images for contrast and brightness using Image-Pro Plus 4.0 software or its equivalent.

6.2. Fluorescence stereomicroscopy

1. Use a Leica fluorescence stereomicroscope (model LZ12) equipped with a mercury 50-W lamp power supply or its equivalent.
2. Produce selective excitation of GFP and/or RFP via a D425/60 band-pass filter and 470 DCXR dichroic mirror.
3. Collect emitted fluorescence through a long-pass filter (GG475) on a Hamamatsu C5810 three-chip cooled color CCD camera or its equivalent.
4. Process images for contrast and brightness with Image-Pro Plus 4.0 software or its equivalent.
5. Capture high-resolution images of 1024 × 724 pixels directly on an IBM PC or continuously through video output on a high-resolution Sony VCR model SLVR1000 or its equivalent.
6. For C57BL/6 mice, remove hair with Nair or by shaving before images are obtained.

6.3. Long-working-distance variable-magnification fluorescence microscope for *in vivo* imaging

The Olympus MVX10 Macro View (OLYMPUS Corporation, Center Valley, PA, USA) and a FluoCam 1500G color CCD camera were used for imaging in live mice (Kimura *et al.*, 2010).

The FluoCam 1500G is a digital color CCD camera manufactured to the specifications of INDEC Systems, Inc. (Santa Clara, CA, USA). It is based on a Sony ICX285 sensor and has 1392 × 1040 pixels, each 6.45 mm^2. Its quantum efficiency is over 60%. It supports real-time imaging at rates greater than 20 images/s (Kimura *et al.*, 2010).

The Olympus MVX10 MacroView employs a single-zoom optical path with a large diameter, which is optimized to collect light with unprecedented efficiency and resolution at all magnifications while still using the two-position revolving nosepiece (Kimura *et al.*, 2010).

The 0.63× and 2× objectives expand the usable zoom range up to 31. The objectives are parfocal corrected, making refocusing after objective switching very quick and easy. Only a small amount of fine focusing is necessary to return to the optimal focus position, making macro- to micro-changes seamless. The 2× objective is also equipped with an additional correction collar to adjust the image quality independently of the specimen medium. This microscope enables fluorescence imaging of whole organisms in a range from low magnification to the detailed observation at the cellular level at high magnification (Kimura *et al.*, 2010).

In comparison with currently available microscopes, the MVX10 provides a long-working-distance and a much higher NA. This makes variable-magnification fluorescence imaging especially powerful, improves speed and precision, and eliminates the necessity to switch back and forth between a stereomicroscope and an inverted microscope (Kimura *et al.*, 2010).

7. Chamber Imaging Systems

7.1. Olympus OV100

1. Perform whole-body or intravital imaging with an Olympus OV100 imaging system using 470-nm excitation light originating from an MT-20 light source.
2. Collect emitted fluorescence through appropriate filters configured on a filter wheel using a DP70 CCD camera. Variable-magnification imaging can be done with a series of four objective lenses for macro- or cellular and subcellular imaging *in vivo*.
3. Capture images on a PC (Fujitsu-Siemens) and process images for contrast and brightness with Paint Shop Pro 8 and cellR.

7.2. INDEC FluorVivo

1. Perform whole-body imaging with an INDEC FluorVivo imaging system using 470-nm excitation light.
2. Collect emitted fluorescence with the instrument's full color CCD camera, using the appropriate emission filter.
3. Use the integrated FluorVivo software to adjust acquisition parameters.
4. Capture still and streaming images to the PC's hard disc using the FluorVivo software.
5. Make required spatial and intensity measurements with the FluorVivo software.

7.3. UVP iBox

1. Turn on the power of the imaging system and (optional) turn on the warming pad.
2. Place the animal in the imaging system (optional) with their nose in the anesthesia cone. Illuminate the animal without excitation filter first to capture a wide spectrum (white-light) image. Set the F number of the camera lens to over 10 and the exposure time at 200 ms. Use the preview function of the imaging system while adjusting the height of the platform, supporting the animal to obtain a comfortable field of view. Adjust the focus of the lens until the image is clear. Reduce the intensity of illumination if horizontal strips are seen in the white-light image (blooming). Press the capture button when the preview image is satisfactory.

3. Change to the appropriate excitation and emission filter for fluorescence capturing. Adjust the focus with the aperture wide open (smallest number). The camera exposure time is typically about 1 s. The exposure time can be lengthened to increase the brightness.
4. (Optional) Use the VisionWorks LS software to create an overlay of the white-light and the fluorescent images.

8. Histological Techniques

8.1. Tumor tissue sampling

1. Obtain tumor tissue biopsies from 3 days to 4 weeks after inoculation of tumor cells. Biopsies of tumor tissue can be obtained from anesthetized mice by removal of a small piece of tumor tissue (1 mm^3 or less) with a scalpel. Staunch bleeding by pressing the wound with sterile gauze. Alternatively, the mouse can be killed and the tissue can be collected and processed for analysis.
2. Cut fresh tissue into pieces of about 1 mm^3 and gently press onto slides for fluorescence microscopy. This procedure is done manually on normal slides.
3. To analyze tumor angiogenesis, digest the tissues with trypsin–EDTA for 5 min at 37 °C before examination.
4. After trypsinization, place the tissues on precleaned microscope slides and cover with another microscope slide (Yang *et al.*, 2003).

8.2. Measurement of GFP-expressing tumor blood vessel length and evaluation of antiangiogenetic agents

1. Give GFP transgenic mice or nestin-GFP transgenic mice daily i.p. injections of doxorubicin or other drugs (vehicle controls) on days 0, 1, and 2 after implantation of tumor cells.
2. Anesthetize mice with the ketamine mixture and obtain biopsies on days 10, 14, 21, and 28 after implantation (Step 18 provides biopsy sample details).
3. Gently flatten the tumor tissue between the slide and coverslip.
4. Quantify angiogenesis in the tumor tissue by measuring the length of GFP-expressing blood vessels in all fields using fluorescence microscopy.
5. Obtain measurements in all fields at 40× or 100× magnification to calculate the total length of GFP-expressing blood vessels.
6. Calculate the vessel density by dividing the total length of GFP-expressing blood vessels (in mm) by the tumor volume (in mm^3) (Amoh *et al.*, 2005a).

8.3. Immunohistochemical staining

1. "Snap-freeze" fresh tissue with liquid nitrogen, then orient and embed the frozen tissue in optimum cutting temperature blocks and store at −80 °C. Cut the frozen sections to a thickness of 5 mm with a Leica CM1850 cryostat.
2. Detect colocalization of GFP fluorescence, CD31, and nestin in the frozen skin sections of mice transgenic for nestin GFP expression using the anti-rat immunoglobulin and anti-mouse immunoglobulin horseradish peroxidase detection kits following the manufacturer's instructions.
3. Use monoclonal anti-CD31 (1:50 dilution) and monoclonal antinestin (1:80 dilution) as primary antibodies. To identify GFP-expressing tumor-infiltrating natural killer cells, macrophages, and dendritic cells, detect localization of GFP together with cell surface markers using immunohistochemical staining with monoclonal antibodies to NK1.1, CD11b, and CD11c, respectively.
4. Use staining with substrate-chromogen 3,3′-diaminobenzidine for antigen detection (Amoh *et al.*, 2005a).

9. Conclusions

Whole-body imaging with fluorescent proteins has been shown to be a powerful technology to follow the dynamics of metastatic cancer. Whole-body imaging of fluorescent protein-expressing cancer cells enables the facile determination of efficacy of candidate antitumor and antimetastatic agents in mouse models. Transgenic mice expressing one color fluorescent protein transplanted with the cancer cells expressing another color fluorescent protein enable the distinction of cancer and host cells and the efficacy of drugs on each type of cell. This is particularly useful for imaging tumor angiogenesis. Cancer cell trafficking through the cardiovascular and lymphatic systems is the critical means of spread of cancer. The use of fluorescent proteins to differentially label cancer calls in the nucleus and cytoplasm and high-powered imaging technology are used to visualize the nuclear–cytoplasmic dynamics of cancer cell trafficking in both blood vessels and lymphatic vessels in the live animal. This technology has furthered our understanding of the spread of cancer at the subcellular level in the live mouse. Fluorescent proteins thus enable both macro- and microimaging technology and thereby provide the basis for the new field of *in vivo* cell biology (Hoffman, 2009).

REFERENCES

Amoh, Y., Li, L., Yang, M., Moossa, A. R., Katsuoka, K., Penman, S., and Hoffman, R. M. (2004). Nascent blood vessels in the skin arise from nestin-expressing hair follicle cells. *Proc. Natl. Acad. Sci. USA* **101,** 13291–13295.

Amoh, Y., Li, L., Yang, M., Jiang, P., Moossa, A. R., Katsuoka, K., and Hoffman, R. M. (2005a). Hair follicle-derived blood vessels vascularize tumors in skin and are inhibited by doxorubicin. *Cancer Res.* **65,** 2337–2343.

Amoh, Y., Yang, M., Li, L., Reynoso, J., Bouvet, M., Moossa, A. R., Katsuoka, K., and Hoffman, R. M. (2005b). Nestin-linked green fluorescent protein transgenic nude mouse for imaging human tumor angiogenesis. *Cancer Res.* **65,** 5352–5357.

Andresen, V., Alexander, S., Heupel, W.-M., Hirschberg, M., Hoffman, R. M., and Friedl, P. (2009). Infrared multiphoton microscopy: Subcellular-resolved deep tissue imaging. *Curr. Opin. Biotechnol.* **20,** 54–62.

Betzig, E., Patterson, G. H., Sougrat, R., Lindwasser, O. W., Olenych, S., Bonifacino, J. S., Davidson, M. W., Lippincott-Schwartz, J., and Hess, H. F. (2006). Imaging intracellular fluorescent proteins at nanometer resolution. *Science* **313,** 1642–1645.

Bouvet, M., Tsuji, K., Yang, M., Jiang, P., Moossa, A. R., and Hoffman, R. M. (2006). *In vivo* color-coded imaging of the interaction of colon cancer cells and splenocytes in the formation of liver metastases. *Cancer Res.* **66,** 11293–11297.

Brown, E. B., Campbell, R. B., Tsuzuki, Y., Xu, L., Carmeliet, P., Fukumura, D., and Jain, R. K. (2001). *In vivo* measurement of gene expression, angiogenesis and physiological function in tumors using multiphoton laser scanning microscopy. *Nat. Med.* **7,** 864–868.

Burgos, J. S., Rosol, M., Moats, R. A., Khankaldyyan, V., Kohn, D. B., Nelson, M. D., Jr., and Laug, W. E. (2003). Time course of bioluminescent signal in orthotopic and heterotopic brain tumors in nude mice. *Biotechniques* **34,** 1184–1188.

Chalfie, M. (2009). GFP: Lighting up life (Nobel Lecture). *Angew. Chem. Int. Ed. Engl.* **48,** 5603–5611.

Cheng, L., Fu, J., Tsukamoto, A., and Hawley, R. G. (1996). Use of green fluorescent protein variants to monitor gene transfer and expression in mammalian cells. *Nat. Biotechnol.* **14,** 606–609.

Chishima, T., Miyagi, Y., Wang, X., Yamaoka, H., Shimada, H., Moossa, A. R., and Hoffman, R. M. (1997). Cancer invasion and micrometastasis visualized in live tissue by green fluorescent protein expression. *Cancer Res.* **57,** 2042–2047.

Cody, C. W., Prasher, D. C., Westler, W. M., Prendergast, F. G., and Ward, W. W. (1993). Chemical structure of the hexapeptide chromophore of the Aequorea green fluorescent protein. *Biochemistry* **32,** 1212–1218.

Condeelis, J., and Segall, J. E. (2003). Intravital imaging of cell movement in tumours. *Nat. Rev. Cancer* **3,** 921–930.

Cormack, B., Valdivia, R., and Falkow, S. (1996). FACS-optimized mutants of the green fluorescent protein (GFP). *Gene* **173,** 33–38.

Delagrave, S., Hawtin, R. E., Silva, C. M., Yang, M. M., and Youvan, D. C. (1995). Red-shifted excitation mutants of the green fluorescent protein. *Biotechnology* **13,** 151–154.

Denk, W., Strickler, J. H., and Webb, W. W. (1990). Two-photon laser scanning fluorescence microscopy. *Science* **248,** 73–76.

Dickinson, M. E., Simbuerger, E., Zimmermann, B., Waters, C. W., and Fraser, S. E. (2003). Multiphoton excitation spectra in biological samples. *J. Biomed. Opt.* **8,** 329–338.

Farina, K. L., Wyckoff, J. B., Rivera, J., Lee, H., Segall, J. E., Condeelis, J. S., and Jones, J. G. (1998). Cell motility of tumor cells visualized in living intact primary tumors using green fluorescent protein. *Cancer Res.* **58,** 2528–2532.

Fukumura, D., Yuan, F., Monsky, W. L., Chen, Y., and Jain, R. K. (1997). Effect of host microenvironment on the microcirculation of human colon adenocarcinoma. *Am. J. Pathol.* **151,** 679–688.

Fukumura, D., Xavier, R., Sugiura, T., Chen, Y., Park, E. C., Lu, N., Selig, M., Nielsen, G., Taksir, T., Jain, R. K., and Seed, B. (1998). Tumor induction of VEGF promoter activity in stromal cells. *Cell* **94,** 715–725.

Glinskii, A. B., Smith, B. A., Jiang, P., Li, X.-M., Yang, M., Hoffman, R. M., and Glinsky, G. V. (2003). Viable circulating metastatic cells produced in orthotopic but not ectopic prostate cancer models. *Cancer Res.* **63,** 4239–4243.

Glinsky, G. V., Glinskii, A. B., Berezovskaya, O., Smith, B. A., Jiang, P., Li, X.-M., Yang, M., and Hoffman, R. M. (2006). Dual-color-coded imaging of viable circulating prostate carcinoma cells reveals genetic exchange between tumor cells *in vivo*, contributing to highly metastatic phenotypes. *Cell Cycle* **5,** 191–197.

Goodison, S., Kawai, K., Hihara, J., Jiang, P., Yang, M., Urquidi, V., Hoffman, R. M., and Tarin, D. (2003). Prolonged dormancy and site-specific growth potential of cancer cells spontaneously disseminated from non-metastatic breast tumors revealed by labeling with green fluorescent protein. *Clin. Cancer Res.* **9,** 3808–3814.

Hayashi, K., Jiang, P., Yamauchi, K., Yamamoto, N., Tsuchiya, H., Tomita, K., Moossa, A. R., Bouvet, M., and Hoffman, R. M. (2007). Real-time imaging of tumor-cell shedding and trafficking in lymphatic channels. *Cancer Res.* **67,** 8223–8228.

Heim, R., Cubitt, A. B., and Tsien, R. Y. (1995). Improved green fluorescence. *Nature* **373,** 663–664.

Hoffman, R. M. (1999). Orthotopic metastatic mouse models for anticancer drug discovery and evaluation: A bridge to the clinic. *Invest. New Drugs* **17,** 343–359.

Hoffman, R. M. (2005). The multiple uses of fluorescent proteins to visualize cancer *in vivo*. *Nat. Rev. Cancer* **5,** 796–806.

Hoffman, R. M. (2009). Imaging cancer dynamics *in vivo* at the tumor and cellular level with fluorescent proteins. *Clin. Exp. Metastasis* **26,** 345–355.

Hoffman, R. M. (2011). Imaging the steps of metastasis at the macro and cellular level with fluorescent proteins in real time (Chapter 6). *In* "Tumor Models in Cancer Research," (B. A. Teicher, ed.) vol. 2, pp. 125–166. Humana Press, New York.

Hoffman, R. M., and Yang, M. (2006a). Whole-body imaging with fluorescent proteins. *Nat. Protoc.* **1,** 1429–1438.

Hoffman, R. M., and Yang, M. (2006b). Color-coded fluorescence imaging of tumor-host interactions. *Nat. Protoc.* **1,** 928–935.

Hoffman, R. M., and Yang, M. (2006c). Subcellular imaging in the live mouse. *Nat. Protoc.* **1,** 775–782.

Jiang, P., Yamauchi, K., Yang, M., Tsuji, K., Xu, M., Maitra, A., Bouvet, M., and Hoffman, R. M. (2006). Tumor cells genetically labeled with GFP in the nucleus and RFP in the cytoplasm for imaging cellular dynamics. *Cell Cycle* **5,** 1198–1201.

Kimura, H., Momiyama, M., Tomita, K., Tsuchiya, H., and Hoffman, R. M. (2010). Long-working-distance fluorescence microscope with high-numerical-aperture objectives for variable-magnification imaging in live mice from macro- to subcellular. *J. Biomed. Opt.* **15**(6), 066029.

Kishimoto, H., Zhao, M., Hayashi, K., Urata, Y., Tanaka, N., Fujiwara, T., Penman, S., and Hoffman, R. M. (2009). *In vivo* internal tumor illumination by telomerase-dependent adenoviral GFP for precise surgical navigation. *Proc. Natl. Acad. Sci. USA* **106,** 14514–14517.

Kolostova, K., Pinterova, D., Hoffman, R. M., and Bobek, V. (2011). Circulating human prostate cancer cells from an orthotopic mouse model rapidly captured by immunomagnetic beads and imaged by GFP expression. *Anticancer Res.* **31,** 1535–1539.

Matz, M. V., Fradkov, A. F., Labas, Y. A., Savitsky, A. P., Zaraisky, A. G., Markelov, M. L., and Lukyanov, S. A. (1999). Fluorescent proteins from nonbioluminescent Anthozoa species. *Nat. Biotechnol.* **17,** 969–973.

Morin, J., and Hastings, J. (1971). Energy transfer in a bioluminescent system. *J. Cell. Physiol.* **77,** 313–318.

Naumov, G. N., Wilson, S. M., MacDonald, I. C., Schmidt, E. E., Morris, V. L., Groom, A. C., Hoffman, R. M., and Chambers, A. F. (1999). Cellular expression of green fluorescent protein, coupled with high-resolution *in vivo* videomicroscopy, to monitor steps in tumor metastasis. *J. Cell Sci.* **112,** 1835–1842.

Prasher, D. C., Eckenrode, V. K., Ward, W. W., Prendergast, F. G., and Cormier, M. J. (1992). Primary structure of the *Aequorea victoria* green-fluorescent protein. *Gene* **111,** 229–233.

Ray, P., De, A., Min, J. J., Tsien, R. Y., and Gambhir, S. S. (2004). Imaging tri-fusion multimodality reporter gene expression in living subjects. *Cancer Res.* **64,** 1323–1330.

Sakaue-Sawano, A., Kurokawa, H., Morimura, T., Hanyu, A., Hama, H., Osawa, H., Kashiwagi, S., Fukami, K., Miyata, T., Miyoshi, H., Imamura, T., Ogawa, M., *et al.* (2008). Visualizing spatiotemporal dynamics of multicellular cell-cycle progression. *Cell* **132,** 487–498.

Shaner, N. C., Campbell, R. E., Steinbach, P. A., Giepmans, B. N., Palmer, A. E., and Tsien, R. Y. (2004). Improved monomeric red, orange and yellow fluorescent proteins derived from *Discosoma* sp. red fluorescent protein. *Nat. Biotechnol.* **22,** 1567–1572.

Shcherbo, D., Merzlyak, E. M., Chepurnykh, T. V., Fradkov, A. F., Ermakova, G. V., Solovieva, E. A., Lukyanov, K. A., Bogdanova, E. A., Zaraisky, A. G., Lukyanov, S., and Chudakov, D. M. (2007). Bright far-red fluorescent protein for whole-body imaging. *Nat. Methods* **4,** 741–746.

Shimomura, O. (2009). Discovery of green fluorescent protein (GFP) (Nobel Lecture). *Angew. Chem. Int. Ed. Engl.* **48,** 5590–5602.

Tome, Y., Tsuchiya, H., Hayashi, K., Yamauchi, K., Sugimoto, N., Kanaya, F., Tomita, K., and Hoffman, R. M. (2009). *In vivo* gene transfer between interacting human osteosarcoma cell lines is associated with acquisition of enhanced metastatic potential. *J. Cell. Biochem.* **108,** 362–367.

Tran Cao, H. S., Reynoso, J., Yang, M., Kimura, H., Kaushal, S., Snyder, C. S., Hoffman, R. M., and Bouvet, M. (2009). Development of the transgenic cyan fluorescent protein (CFP)-expressing nude mouse for "Technicolor" cancer imaging. *J. Cell. Biochem.* **107,** 328–334.

Tsien, R. Y. (2009). Constructing and exploiting the fluorescent protein paintbox (Nobel Lecture). *Angew. Chem. Int. Ed. Engl.* **48,** 5612–5626.

Verkhusha, V. V., and Lukyanov, K. A. (2004). The molecular properties and applications of Anthozoa fluorescent proteins and chromoproteins. *Nat. Biotechnol.* **22,** 289–296.

Wyckoff, J. B., Wang, Y., Lin, E. Y., Li, J. F., Goswami, S., Stanley, E. R., Segall, J. E., Pollard, J. W., and Condeelis, J. (2007). Direct visualization of macrophage-assisted tumor cell intravasation in mammary tumors. *Cancer Res.* **67,** 2649–2656.

Yamamoto, N., Yang, M., Jiang, P., Xu, M., Tsuchiya, H., Tomita, K., Moossa, A. R., and Hoffman, R. M. (2003a). Real-time imaging of individual fluorescent protein color-coded metastatic colonies in vivo. *Clin. Exp. Metastasis* **20**(7), 633–638.

Yamamoto, N., Yang, M., Jiang, P., Xu, M., Tsuchiya, H., Tomita, K., Moossa, A. R., and Hoffman, R. M. (2003b). Determination of clonality of metastasis by cell-specific color-coded fluorescent-protein imaging. *Cancer Res.* **63,** 7785–7790.

Yamamoto, N., Jiang, P., Yang, M., Xu, M., Yamauchi, K., Tsuchiya, H., Tomita, K., Wahl, G. M., Moossa, A. R., and Hoffman, R. M. (2004). Cellular dynamics visualized in live cells *in vitro* and *in vivo* by differential dual-color nuclear-cytoplasmic fluorescent-protein expression. *Cancer Res.* **64,** 4251–4256.

Yamauchi, K., Yang, M., Jiang, P., Yamamoto, N., Xu, M., Amoh, Y., Tsuji, K., Bouvet, M., Tsuchiya, H., Tomita, K., Moossa, A. R., and Hoffman, R. M. (2005). Real-time *in vivo* dual-color imaging of intracapillary cancer cell and nucleus deformation and migration. *Cancer Res.* **65,** 4246–4252.

Yamauchi, K., Yang, M., Jiang, P., Xu, M., Yamamoto, N., Tsuchiya, H., Tomita, K., Moossa, A. R., Bouvet, M., and Hoffman, R. M. (2006). Development of real-time subcellular dynamic multicolor imaging of cancer cell-trafficking in live mice with a variable-magnification whole-mouse imaging system. *Cancer Res.* **66,** 4208–4214.

Yang, F., Moss, L. G., and Phillips, G. N., Jr. (1996). The molecular structure of green fluorescent protein. *Nat. Biotechnol.* **14,** 1246–1251.

Yang, M., Baranov, E., Jiang, P., Sun, F.-X., Li, X.-M., Li, L., Hasegawa, S., Bouvet, M., Al-Tuwaijri, M., Chishima, T., Shimada, H., Moossa, A. R., *et al.* (2000). Whole-body optical imaging of green fluorescent protein expressing tumors and metastases. *Proc. Natl. Acad. Sci. USA* **97,** 1206–1211.

Yang, M., Baranov, E., Wang, J.-W., Jiang, P., Wang, X., Sun, F.-X., Bouvet, M., Moossa, A. R., Penman, S., and Hoffman, R. M. (2002). Direct external imaging of nascent cancer, tumor progression, angiogenesis, and metastasis on internal organs in the fluorescent orthotopic model. *Proc. Natl. Acad. Sci. USA* **99,** 3824–3829.

Yang, M., Li, L., Jiang, P., Moossa, A. R., Penman, S., and Hoffman, R. M. (2003). Dual-color fluorescence imaging distinguishes tumor cells from induced host angiogenic vessels and stromal cells. *Proc. Natl. Acad. Sci. USA* **100,** 14259–14262.

Yang, M., Reynoso, J., Jiang, P., Li, L., Moossa, A. R., and Hoffman, R. M. (2004). Transgenic nude mouse with ubiquitous green fluorescent protein expression as a host for human tumors. *Cancer Res.* **64,** 8651–8656.

Yang, M., Luiken, G., Baranov, E., and Hoffman, R. M. (2005). Facile whole-body imaging of internal fluorescent tumors in mice with an LED flashlight. *Biotechniques* **39,** 170–172.

Yang, M., Jiang, P., and Hoffman, R. M. (2007). Whole-body subcellular multicolor imaging of tumor-host interaction and drug response in real time. *Cancer Res.* **67,** 5195–5200.

Yang, M., Reynoso, J., Bouvet, M., and Hoffman, R. M. (2009). A transgenic red fluorescent protein-expressing nude mouse for color-coded imaging of the tumor microenvironment. *J. Cell. Biochem.* **106,** 279–284.

Zolotukhin, S., Potter, M., Hauswirth, W. W., Guy, J., and Muzycka, N. (1996). "Humanized" green fluorescent protein cDNA adapted for high-level expression in mammalian cells. *J. Virol.* **70,** 4646–4654.

CHAPTER TWELVE

Protein Activation Dynamics in Cells and Tumor Micro Arrays Assessed by Time Resolved Förster Resonance Energy Transfer

Véronique Calleja,* Pierre Leboucher,*,† *and* Banafshé Larijani*

Contents

Abstract

Analytical time resolved Förster resonance energy transfer (FRET) can be exploited for assessing, in cells and tumor micro arrays, the activation status and dynamics of oncoproteins such as epidermal growth factor receptor (EGFR1) and their downstream effectors such as protein kinase B (PKB) and 3-phosphoinositide-dependent protein kinase 1 (PDK1). The outcome of our research involving the

* Cell Biophysics Laboratory, Cancer Research UK, London Research Institute, London, United Kingdom
† Centre Emotion, Hôpital de la Pitié Salpétrière, Paris, France

Methods in Enzymology, Volume 506
ISSN 0076-6879, DOI: 10.1016/B978-0-12-391856-7.00036-6

application of quantitative imaging for investigating molecular mechanisms of phosphoinositide-dependant enzymes, such as PKB and PDK1, has resulted in a refined model describing the dynamics and regulation of these two oncoproteins in live cells. Our translational research exploits a quantitative FRET method for establishing the activation status of predictive biomarkers in tumor micro arrays. We developed a two-site FRET assay monitored by automated frequency domain Fluorescence lifetime imaging microscopy (FLIM). As a proof of principle, we tested our methodology by assessing EGFR1 activation status in tumor micro arrays from head and neck patients. Our two-site FRET assay, by high-throughput frequency domain FLIM, has great potential to provide prognostic and perhaps predictive biomarkers.

Abbreviations

DABCO 1,4-diazabicyclo [2.2.2] octane
EGF epidermal growth factor
EGFP enhanced green fluorescent protein
EGFR epidermal growth factor receptor
FLIM fluorescence lifetime imaging microscopy
FRET Förster resonance energy transfer
HNSCC head and neck squamous cell carcinoma
IHC immunohistochemistry
mCherry monomeric cherry fluorescent protein
mRFP monomeric red fluorescent protein
PBS phosphate buffered saline
PDGF platelet-derived growth factor
PDGFR platelet-derived growth factor receptor
PDK1 3-phosphoinositide-dependent protein kinase 1
PFA paraformaldehyde
PH domain pleckstrin homology domain
PI3-kinase phosphoinositide 3-kinase
PKB protein kinase B
PtdIns phosphoinositides
RTK receptor tyrosine kinase
TCSPC time correlated single photon counting

1. Introduction

It has been imperative to monitor molecular interactions within the cellular environment without major perturbations; as a result Förster resonance energy transfer (FRET) detected by Fluorescence lifetime imaging microscopy (FLIM) has been exploited to quantitatively measure *in situ* and *in vivo* molecular interactions.

To investigate the *in vivo* mechanism of phosphoinositide-dependent kinase association to the plasma membrane, we have monitored the dynamics and change in conformation of AGC kinases, such as protein kinase B (PKB/Akt) and 3-phosphoinositide-dependent protein kinase 1 (PDK1), in a cellular environment. The outcome of these studies is a new mechanism depicting the allosteric regulation of the pleckstrin homology domain (PH) and kinase domain of PKB which distinguish it from the rest of the AGC kinases. Furthermore, the previously uncharacterised molecular mechanism of PDK1 regulation and its association with PKB in the cell was also determined.

Current methods for assessing a patient's suitability for treatment with signal transduction inhibitors are limited. A very low percentage of cancer patients, those with a high overexpression of growth factor receptors, do not respond to molecular therapy. The capacity of immunohistochemistry (IHC) alone in predicting treatment success is limited, as the overexpression of growth factor receptors and their downstream effectors may not correlate with their functional status. A methodology is therefore required to improve assessment of not just the concentration of oncoproteins but their activation status. We describe how to measure the functional status of signal transduction markers by using a two-site FRET assay using high-throughput frequency domain FLIM (Kong *et al.*, 2006). We depict the application of our methodology on the activation status of EGFR1 in head and neck tumors.

In this chapter, the first section illustrates in detail our methods for monitoring the dynamics of phosphoinositide-dependent kinases and the second section depicts the two-site FRET assay used for *in situ* analyzes of the activation status of oncoproteins, such as epidermal growth factor receptor (EGFR), in tissue micro arrays.

2. Protein–Protein Interactions and Conformation Dynamics in Fixed and Live Cells

2.1. Materials

2.1.1. DNA constructs

- EGFP (enhanced green fluorescent protein)-PDK1 construct: EGFP was fused to the N-terminus of human PDK1 in a pCMV5 vector. This construct was provided by Park *et al.* (2001).
- PDK1-mCherry (monomeric cherry fluorescent protein) construct: mCherry was fused to the C-terminus of human PDK1 in the pCMV5 vector as described in Masters *et al.* (2010).

- mRFP (monomeric red fluorescent protein)-PKB construct: mRFP was fused to the N-terminus of human PKB in the construct pCMV5-HA-PKB kindly provided by Park *et al.* (2001).
- EGFP-PKB-mRFP construct: EGFP and mRFP were fused to the N- and C-termini, respectively, of mouse PKB in a pCDNA3 vector (Calleja *et al.*, 2007).

2.1.2. Reagents for cell line maintenance, transfection, and stimulation

- Mouse embryonic fibroblast NIH3T3 cells (# CRL-1658 American Type Culture Collection, ATCC).
- 35 mm glass bottom microwell dishes (# P35G-1.5-14-C MatTek Corporation).
- Dulbecco's modified eagle's medium (DMEM) containing GlutaMAX (# 31966 Invitrogen).
- Donor bovine serum (# 16030074 Invitrogen).
- LipofectAMINE LTX and PLUS reagent (# 15338-100 Invitrogen).
- OPTIMEM transfection medium with GlutaMAX (# 51985 Invitrogen).
- Human PDGF (platelet-derived growth factor) purified from platelets (# 120-HD R&D Systems).
- Akt inhibitor VIII, isozyme-selective Akti-1/2 (# 124018 Calbiochem).
- LY294002 (# 440204 Calbiochem).
- Bovine serum albumin (BSA) fatty acid free, low endotoxin (# A8806 Sigma).

2.1.3. Reagents for imaging and preparation

- Mowiol 4–88 (# 475904 Calbiochem)

In a 50-ml Falcon tube mix 6 ml water with 2.4 g Mowiol 4–88 and 6 ml glycerol. Vortex and add 12 ml 200 m*M* Tris–HCl, pH 8.5. Incubate overnight (14–16 h) at 50 °C in a beaker containing water on a hot plate with stirring. Filter through a 0.45-μm syringe filter. Store at 4 °C. Before use add 2.5% (w/v) DABCO.

- Paraformaldehyde (PFA) (# P6148 Sigma)

Heat 400 ml PBS (phosphate buffered saline). When the buffer is boiling, add 16 g PFA and stir until dissolution. Allow to cool to room temperature, filter through a 0.22-μm bottle filter (0.22 μm) and store in aliquots at 4 °C.

- 1,4-diazabicyclo[2.2.2] octane (DABCO) (# D-27802 Sigma)
- Sodium borohydride ($NaBH_4$) (# 452882 Sigma)

- 0.2% (v/v) Triton X-100: 50 μl Triton X-100 in 25 ml PBS
- *N*,*N*-dimethylformamide (DMF)
- Bicine-*N*,*N*-bis (hydroxyethyl) aminoacetic acid (# B8660 Sigma)
- Mono-functional fluorescent sulphoindocyanine dyes Cy™3B and Cy5 (GE Healthcare)

2.1.4. Description and utility of the reagents

- Mowiol mount

This is a polyvinyl alcohol (PVA)-based mount that sets as a hard resin, providing the advantage of maintaining cells hydrated under the coverslip. For our experiments this mounting medium is the most reliable for determining donor fluorescence lifetimes of EGFP.

- 1,4-diazabicyclo [2.2.2] octane (DABCO)

Photobleaching is frequently a problem for measuring lifetimes; therefore it is critical to prevent this as much as possible during image acquisition. An antifade, such as DABCO, must be added to Mowiol. DABCO is a free radical "scavenger" that is used to reduce the bleaching of the chromophores. The samples tend to bleach more readily without DABCO.

- Sodium borohydride ($NaBH_4$)

$NaBH_4$ is an effective and very selective reducing agent. It quenches background fluorescence from any remaining PFA and also removes cellular autofluorescence by quenching unreacted glutaraldehyde. It reacts with double bonds, thereby reducing the pi-bond interactions that are partly responsible for auto fluorescence.

- Donor bovine serum

NIH3T3 cells can easily transform in cell culture upon passaging and at least two elements are critical for maintenance of the original cell morphology. First, to avoid the cells becoming over confluent before passaging, it is best to split the cells at 70–80% confluency. Second, the cells should be cultured in donor bovine serum instead of fetal bovine serum. In our experience, fetal bovine serum modifies cell morphology, which in turn influences the dynamics of PKB and PDK1, and in particular creates difficulties for visualizing protein translocation.

2.2. Elemental aspects of Förster resonance energy transfer

- Förster resonance energy transfer is a precise measurement for determining protein–protein, protein–lipid interactions or change in protein conformation in cells or tissue. For measurements of molecular interactions and dynamics in live cells the proteins of choice are labeled with a donor and acceptor chromophore. In this chapter, we have used EGFP as the donor and mRFP or mCherry, as the acceptor, these are referred to as the FRET pair.

- The FRET pairs need to be chosen carefully. One important feature to take into account is the spectral overlap, that is the emission spectrum of the donor and the absorption spectrum of the acceptor, which need to overlap adequately so that transfer of energy between the donor and the acceptor can occur. For this particular parameter the absorption extinction coefficient of the acceptor has to be taken into account. In order to measure accurately the change in the fluorescence lifetime of the donor in presence of an acceptor, the acceptor must not be excited directly. There are many other important parameters that need to be considered—for specific details readers can refer to Alcor *et al.* (2009)
- FRET can be detected by steady state or time-resolved methods. In steady state FRET the fluorescence intensity of the acceptor is monitored. Upon the excitation of the donor the intensity of the acceptor increases when energy transfer occurs, and the donor emission intensity decreases by the same amount. The ratio of the acceptor intensity to donor intensity can be taken as the measurement of steady state FRET.
- In time-resolved FRET, the fluorescence intensity decay of the donor after an excitation pulse is measured. The decay is expressed as follows:

$$I_t = I_0 e^{-t/\tau},$$

where I_0 is the maximum emission intensity of the fluorescence at time $t = 0$, I_t is the intensity of the fluorescence at a time t after excitation, and τ is the lifetime. This fluorescence intensity decay decreases exponentially with a *characteristic* time. This is the specific lifetime that reflects the average time molecules have spent in an excited energy state. The decrease of this lifetime (τ), upon the occurrence of resonance energy transfer, is the measurement of change in the lifetime of the donor.

The efficiency of the energy transfer is calculated by:

$$E = 1 - \left(\tau_{d/a}/\tau_d\right),$$

where τ is the lifetime, d is the donor, and d/a is the donor in presence of the acceptor.

The donor–acceptor distance can be calculated from:

$$r = R_0/\left[\left(\tau_d/\tau_{d/a}\right) - 1\right]^{1/6}.$$

The distance between the donor and the acceptor is in the following range: $0.5R_0 < r < 1.5R_0$. It is in this range that the transfer of energy is the most sensitive. For example, in experiments where EGFP is the donor and mRFP or mCherry is the acceptor the distance between the chromophores is less than 100Å. In this case, R_0 is calculated to be about 40 Å (Calleja *et al.*, 2007).

- One of the main advantages of time-resolved FRET is that, the lifetime measurements are independent of concentration variations. Nevertheless, in all time-resolved FRET experiments, care has to be taken to ensure a high signal-to-noise ratio. This is important as the changes in lifetime distribution should not have a broad distribution and should be as narrow as possible with low standard deviations. A high signal-to-noise ratio is achieved by ensuring low background samples, with the donor intensity being at least four times the background signal, and the acceptor intensity being twice that of the donor intensity. Of course, optimisation of signal variations are dependent on the detection limits of FLIM instruments, which need to be determined for each set of experiments. The FRET signal in each experiment can only be compared in samples with similar donor intensities.
- Lifetime measurements, although concentration independent, are still sensitive to their local environment. For example, variations in local pH and viscosity affect lifetime measurements. In addition the fusion of the protein of interest to the EGFP can cause variations in lifetime measurements. However, since we calculate relative changes in lifetime this effect can be always monitored by taking into account the lifetime measurements of the mono-transfected donor EGFP-PKB or EGFP-PDK1 in each experiment. These values will be regarded as the control lifetime measurements for each experiment.

2.2.1. Time domain and frequency domain FLIM

- In this study, we used two types of time-resolved instruments: a two-photon time-domain and a frequency-domain FLIM. Below is only a brief description of the instruments employed, for further details readers should refer to Alcor *et al.* (2009).

2.2.1.1. Two-photon time domain FLIM

- The two-photon time-domain FLIM is a time correlated single photon counting (TCSPC) system. A TCSPC system records the time of arrival of photons, measures their time in a single period and builds a histogram of photon times. The advantage of such a system is that it records signals with a very high time resolution and a close to perfect efficiency. The two-photon FLIM is equipped with a Ti:Sapphire laser and is tuned at 890 nm for EGFP experiments. The laser generates an ultrafast pulse of 130–170 fs with a 76.46-MHz repetition rate, with a spectral width set to 13–15 nm. All the images described in Section 2.3, 2.4 and 2.5 are acquired on a modified TE 2000-E inverted Nikon microscope with a 40× oil immersion objective lens, 1.3 NA, Plan-Fluor from Nikon.

- The ranges of lifetimes that can be detected can be calculated from: $1/(4f)$ where f is the frequency of the pulsed excitation. The frequency of 76.46 MHz allows us to measure lifetimes up to 3 ns, which is the range of lifetimes for EGFP-labeled PKB and PDK1 in our experiments.
- Samples were routinely excited with 25–30 mW laser power (measured at the microscope inlet aperture). This low power reduces photobleaching of the sample.

2.2.1.2. FRET analysis by two-photon time domain FLIM

- In the experiments depicted in Section 2.3, 2.4 and 2.5 the exponential decays are fitted to a monoexponential Marquardt fit. The number of scans per cell was determined in such a manner that the lowest intensity pixels would have enough photon counts in order for the exponential decay to be fitted accurately. The acquisition time is dependent on the final photon counts that are required. We have tried to reach 1000 photon counts for the maximum of the intensity decay. Care must be taken when varying the photon counts, as although the value of the average lifetime does not change, the lifetime dispersion will increase. This creates a problem if the lifetime variations within the experiment are smaller than the distribution width.
- Upon FRET, an increase in the population of shorter lifetime pixels is detected. However, in cells coexpressing donor and acceptor-tagged molecules, not all of the donor molecules undergo FRET. The fluorescence lifetime measured in each pixel is an average of a mixed population of donor molecules that have undergone FRET (interacting species) and those that have not undergone FRET (noninteracting species). In fixed cells we compare the average lifetime of the donor in presence of the acceptor (lifetime cell map of the donor with acceptor) with average lifetime of the donor alone (lifetime cell map of cells with donor alone). In live experiments we compare the lifetime map of the cell at $t = 0$ with donor and acceptor, to the same stimulated cell over a specific timeframe.
- For each experimental condition a minimum of 10 cells need to be acquired and analyzed. The experiments need to be repeated at least three times for reproducibility. The intensity images are converted to fluorescence lifetime maps by fitting the fluorescence intensity decay at each pixel. The fluorescence lifetime is represented in pseudo-color with the same scale for each cell so that colors are comparable in each experiment. Fluorescence lifetime distributions are represented by histograms. The histograms need to be normalized to the total number of pixels of the cell (Calleja *et al.*, 2007).
- Statistical analyzes are critical for quantitative FRET experiments. To compare the medians of the two data sets a nonparametric Mann–Whitney

test must be performed (GraphPad InStat software version 3.0). To interpret the distribution of data, box and whiskers plots are used. The box and whiskers plot displays upper and lower quartiles, and maximum and minimum values in addition to the median. It is a useful method to use in order to observe the distribution of lifetimes in each experimental condition (Calleja *et al.*, 2007).

2.2.1.3. Frequency-domain FLIM

The advantage of frequency-domain FLIM is its efficiency and speed in acquiring images compared to time-domain FLIM. In the frequency domain the sample is excited by a sinusoidally modulated light source. Both the emission and excitation are modulated at the same frequency (van Munster and Gadella, 2005). However, the fluorescence emission has a reduced modulation depth and its phase is shifted. These two lifetime parameters, the phase (τ_ϕ), and modulation (τ_M), are used for calculating variations in lifetime. They are expressed as follows:

$$\tau_\phi = \tan(\Delta\Phi)/\omega \quad \text{and} \quad \tau_M = \left[\left(\left(1/M^2\right) - 1\right)^{1/2}\right]/\omega,$$

where ω is the angular velocity ($2\pi f$); $\Delta\Phi$ and M are the phase shift and demodulation at each pixel, respectively.

- There are two main features of a frequency domain FLIM imaging system: a standing wave acousto-optic modulator (AOM) modulates the excitation light source at a high frequency, and an image intensifier detects phase-sensitive fluorescence emission. Our instrument is equipped with an Argon/Krypton laser which produces discrete lines at 457, 488, and 514 nm. These excitation sources can be used to excite Cy3b as well as EGFP. The system is set up in such a manner that the output of these excitation sources are modulated by the AOM at 80 MHz. This modulation is optimal for EGFP and Cy3b, which display lifetimes of 1.5–2.5 ns.
- Sixteen phase-dependent images are acquired in a FLIM sequence, each phase-dependent image can be read in less than 20 ms. This entire cycle is acquired from 0° to 360°. Photobleaching is reduced by only acquiring during the image acquisition period. The Fourier transformation of each pixel in the image sequence obtains the phase and modulation images, from which the fluorescence lifetimes are calculated.

2.2.1.4. FRET analysis by frequency-domain FLIM

- A reference image (a cycle of 16 phase-dependent images, each separated by 22.5°) using a scattering foil is acquired. The reference measurement is used to determine the phase angle setting that the maximum intensity has reached in the image series. It is also used for calibrating the system so that image acquisitions always start at 0°.

- The reference measurement from the foil also determines the modulation value which indicates the accuracy of the excitation modulation. The modulation value needs to be close to 1 in order to obtain accurate lifetime measurements.
- To obtain the lifetime of the donor, Cy3b the intensity of the excitation source needs to be restored to a maximum using a variable density filter wheel and the Cy3 filter set (excitation, HQ 545/30 band pass filter, dichroic mirror, Q565 long pass; emission, HQ610/751 band pass filter; Chroma). The exposure time of the acquisition is chosen to obtain the maximum signal-to-noise ratio without causing photobleaching.
- The acceptor images are acquired by a mercury source and a Cy5 filter set (excitation, HQ620/60 band pass filter; dichroic mirror, Q660 long pass; emission, HQ 700/75 band pass filter; Chroma). Care needs to be taken to avoid photobleaching.
- Using the images from the phase series acquisition and the zero lifetime reference from the foil, the donor lifetimes (τ_{ϕ}) and (τ_{M}) are obtained. If these two values are close to each other (± 0.50 ns) then the average lifetime can be calculated. If not, these values should not be averaged and have to be presented individually.
- For each pixel the average lifetime is calculated as: ($[\tau_{\phi} + \tau_{M}]/2$) (Alcor *et al.*, 2009). For visualizing, the average lifetime map can be shown on a pseudo-color scale. It is essential to keep the range of the pseudo-color scale the same from one lifetime map to another.

2.3. Protein Kinase B (PKB/Akt) conformation dynamics

The activation of receptor tyrosine kinases (RTKs) upon growth factor stimulation activates a cascade of proteins, in particular phosphoinositide 3-kinase (PI3-kinase). The studies below focus on the PDGFR (platelet-derived growth factor receptor) pathway. The specific binding of PI3-kinase to phosphotyrosines in the intracellular domain of PDGFR brings it into close proximity to its phosphoinositide (PtdIns) substrate at the plasma membrane. PI3-kinase catalyzes the phosphorylation of PtdIns (4,5) P_2 in the third position of the inositol ring to produce PtdIns (3,4,5) P_3. PKB/Akt, has been shown to translocate to the plasma membrane due to binding of its PH domain to PtdIns (3,4,5) P_3 and PtdIns (3,4) P_2. In this study, we used our EGFP-PKB-mRFP as a sensor and determined that upon PDGF stimulation PKB changes its conformation from a “PH-in” to a “PH-out” at the plasma membrane due to 3-PtdIns binding (Calleja *et al.*, 2007).

2.3.1. Cell seeding and transfection

- NIH3T3 cells are plated at 150,000 cells cells per 35 mm round glass-bottom dish (MatTek) in 2 ml DMEM containing 10% donor bovine serum. The next day (after 14–16 h), the cells are transfected with the mammalian expression vector coding for EGFP-PKB-mRFP sensor.

- 1 μg of EGFP-PKB-mRFP DNA construct is mixed with 4 μl of Plus reagent in 100 μl of OPTIMEM medium (without serum), this mixture is thoroughly mixed by vortexing. Subsequently, 100 μl of OPTIMEM containing 4 μl of Lipofectamine LTX is added to the 100 μl of DNA/ Plus reagent mix. The 200 μl are vortexed for 30 s and incubated at room temperature.
- After 15 min the cell culture medium is removed from the 35-mm dish and replaced by 800 μl of OPTIMEM medium. The 200 μl of transfection mix containing DNA/Plus reagent/Lipofectamine LTX is added to the OPTIMEM and gently mixed by shaking the dish. The dishes are placed at 37 °C in a cell culture incubator for 3 h.

Note: NIH3T3 cells fresh from ATCC do not tolerate long periods of serum deprivation, thus longer transfection times can be harmful for the cells. After 3 h, the transfection mix is aspirated and 2 ml of DMEM containing 10% donor bovine serum is added to the dishes overnight. The following day (14–16 h) the cells are stimulated, fixed with 4% PFA and mounted.

2.3.2. Stimulation, fixation, and mounting

- NIH3T3 cells are stimulated with 30 ng/ml PDGF for 5 min or pretreated for 30 min with 50 μM Akt inhibitor VIII before PDGF stimulation at 37 °C in DMEM containing 10% donor bovine serum.
- The cells are washed twice with PBS on ice and fixed by adding 1 ml 4% PFA for 10 min at room temperature (20–22 °C). After two washes with PBS the cells are treated with 1 mg/ml *freshly prepared* sodium borohydride ($NaBH_4$) in PBS for 2–5 min to quench the autofluorescence. The $NaBH_4$ is washed with PBS and a coverslip is mounted on top of the glass-bottom on the 35-mm dish with Mowiol containing 2.5% (w/v) of DABCO. The coverslips are left to dry overnight at room temperature and stored at 4 °C.

Note: the NIH3T3 cells are not starved as they are very sensitive to serum deprivation; however the basal activity of PKB in these cells is sufficiently low to allow for an adequate response upon PDGF stimulation. The read-out for proper responsiveness to stimulation is the phosphorylation of PKB Thr308 and Ser473 activation sites, as well as translocation of PKB to the plasma membrane (Calleja *et al.*, 2007).

2.3.3. Data analysis

- Figure 12.1 shows the lifetime maps of NIH3T3 cells acquired before and after stimulation with PDGF, or together with Akt inhibitor VIII as indicated.
- For each condition the cells are acquired at the mid-section. The control image of a cell that only expresses the donor EGFP-PKB shows the

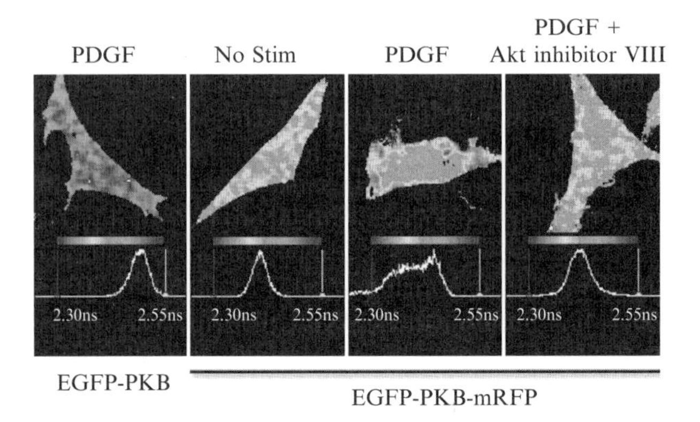

Figure 12.1 PKB conformation dynamics are prevented by an allosteric inhibitor Lifetime maps of fixed NIH3T3 cells expressing either EGFP-PKB or our EGFP-PKB-mRFP conformation sensor. The cells are treated as indicated. Akt inhibitor VIII prevents the change in conformation of the PKB sensor upon PDGF stimulation. The lifetime scale ranges from 2.30 (red) to 2.55 ns (blue). (See Color Insert.)

distribution of the donor lifetime. Here FRET does not occur, as the acceptor is not present. When the sensor EGFP-PKB-mRFP is expressed, a decrease in the EGFP lifetime due to the presence of the acceptor mRFP is detected. This basal FRET may be due to the short distance between the two chromophores in the sensor.

- Upon 5 min PDGF stimulation the sensor translocates to the plasma membrane through the interaction of the PKB PH domain with PtdIns(3,4,5)P_3. Concomitantly, an increase in the population of shorter lifetime pixels (indicative of an increase in FRET) is measured at the plasma membrane. This indicates that the distance between EGFP and mRFP in the sensor, has varied and hence a change in conformation of the sensor at the plasma membrane has occurred.
- When cells are pretreated with Akt inhibitor VIII (an allosteric inhibitor) the translocation of the sensor and its change in conformation are prevented. In the presence of the inhibitor the lifetime distribution is very similar to the basal level.
- We conclude that Akt inhibitor VIII locks PKB in its PH-in conformation thereby preventing its change in conformation.

2.4. Heterodimerisation of PKB with PDK1

PDK1 is a direct PKB activator. Like PKB, PDK1 possess a PH domain that binds to PtdIns(3,4,5)P_3 upon growth factor stimulation, which triggers its translocation to the plasma membrane. At the plasma membrane PDK1 phosphorylates PKB on the activation loop Thr308. The dual phosphorylation of PKB at Thr308, and also at Ser473 by another kinase, is critical for its full activation. In our studies, we demonstrate using EGFP-PDK1 and mRFP-PKB that the proteins do not only co-localize at the plasma membrane, they directly *interact*. We show that PKB/PDK1 heterodimer formation is independent of PtdIns(3,4,5)P_3 binding and occurs in the cytoplasm prior to stimulation. However, the formation of PtdIns(3,4,5)P_3 upon PDGF stimulation allows the translocation of the PKB/PDK1 heterodimer to the plasma membrane where PDK1 phosphorylates and activates the PH-out conformer of PKB.

2.4.1. Cell seeding and transfection

- NIH3T3 cells are plated at 150,000 cells per 35 mm round glass-bottom dish (MatTek) in 2 ml DMEM containing 10% donor bovine serum. 24 h later, the cells are transfected with the mammalian expression vectors coding for EGFP-PDK1 and mRFP-PKB.
- 1 μg of EGFP-PDK1 plus 1 μg of mRFP-PKB DNA constructs are mixed with 4 μl of Plus reagent in 100 μl of OPTIMEM medium (without serum). The mixture is thoroughly mixed by vortexing. Subsequently, 100 μl of OPTIMEM containing 4 μl of Lipofectamin LTX is added to the 100 μl DNA/Plus reagent mix. The 200 μl are vortexed for 30 s and allowed to incubate for 15 min.
- The cell culture medium is removed from the 35 mm dish and replaced by 800 μl of OPTIMEM medium. The 200 μl transfection mix containing DNA/Plus reagent/Lipofectamin LTX is added to the OPTIMEM and gently mixed by shaking the dish. The dishes are placed at 37 °C in a cell culture incubator for 3 h. The medium is replaced after 3 h incubation by 2 ml of DMEM containing 10% donor bovine serum. The following day (14–16 h) the cells are stimulated, fixed with 4% PFA and mounted.

2.4.2. Stimulation, fixation, and mounting

- NIH3T3 cells are stimulated with 30 ng/ml of PDGF for 5 min or pretreated 30 min first with 50 μM of Akt inhibitor VIII before PDGF stimulation at 37 °C in DMEM 10% donor bovine serum.
- Since the cells are sensitive to serum deprivation they are not starved. They are washed twice with PBS on ice and fixed by adding 1 ml of 4% PFA for 10 min at room temperature (20–22 °C). After two washes with PBS the cells are treated with 1 mg/ml of *fresh* sodium borohydride ($NaBH_4$) in PBS for 2–5 min to quench the autofluorescence. The $NaBH_4$ is washed with PBS and a coverslip is mounted on top of the glass-bottom on the 35 mm dish with Mowiol containing 2.5% (w/v)

of DABCO. The coverslips are left to dry overnight at room temperature and stored at 4 °C.

2.4.3. Data analysis

- Figure 12.2 shows the lifetime maps of NIH3T3 cells acquired prior to stimulation and after being treated with PDGF alone or after pretreatment with Akt inhibitor VIII as indicated. For each condition the cells are acquired at the mid-section.
- The control (EGFP-PDK1 alone) shows the distribution of the donor lifetime. In unstimulated cells with mRFP-PKB, a decrease in the EGFP lifetime is observed, indicating that a basal interaction between PKB and PDK1 occurs prior to stimulation. Upon 5 min treatment with PDGF an increase in the population of shorter lifetime pixels is detected at the plasma membrane. This indicates, that the PKB/PDK1 complex cotranslocates to the plasma membrane through the interaction of PKB and PDK1 PH domains with PtdIns(3,4,5)P_3.
- Pre-treatment of cells with Akt Inhibitor VIII prior to PDGF stimulation results in the loss of FRET. In the presence of the inhibitor the lifetime distribution is very similar to the basal condition. This may suggest that maintaining PKB in the PH-in conformation by the inhibitor does not preclude the formation of the complex in the cytoplasm.

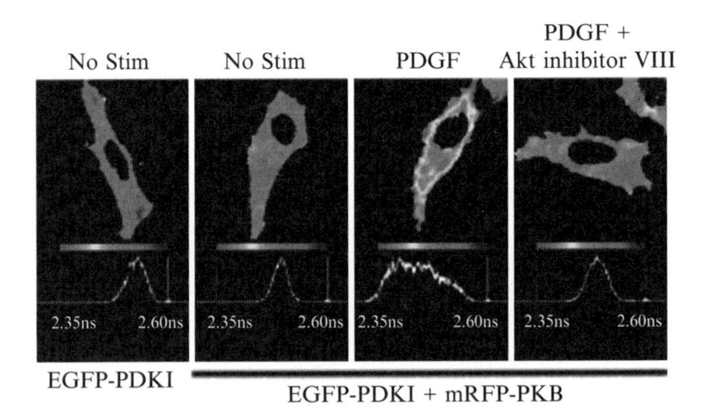

Figure 12.2 Inhibition of PKB and PDK1 interaction by an allosteric inhibitor. Lifetime maps of fixed NIH3T3 cells expressing EGFP-PDK1 alone or with mRFP-PKB. The cells are treated as indicated. Akt inhibitor VIII prevents the heterodimerisation of PKB and PDK1 at the plasma membrane upon PDGF stimulation. The basal interaction is not affected. The scale is 2.35 (red) to 2.60 ns (blue). (See Color Insert.)

2.5. Homodimerisation of PDK1 in live cells

2.5.1. Cell seeding and transfection

- NIH3T3 cells are plated at 150,000 cells per 35 mm round glass-bottom dish (MatTek) in 2 ml of DMEM containing 10% Donor bovine Serum. The following day the cells are cotransfected with the mammalian expression vectors containing the sequences for EGFP-PDK1 and PDK1-mCherry fusion proteins.
- For one dish, 1 μg of EGFP-PDK1 DNA construct + 1 μg of PDK1-mCherry DNA construct are mixed with 4 μl of Plus reagent in 100 μl of OPTIMEM medium (without serum). The mixture is thoroughly mixed by vortexing. Subsequently, 100 μl of OPTIMEM containing 4 μl of Lipofectamin LTX is added to the 100 μl of DNA/Plus reagent mix. The 200 μl are vortexed for 30 s and are incubated for 15 min. After 15 min the cell culture medium is removed from the 35 mm dish and replaced by 800 μl of OPTIMEM medium.
- The 200 μl transfection mix containing DNA/Plus reagent/Lipofectamin LTX is added to OPTIMEM on the dish and gently mixed by shaking the dish. The dishes are placed at 37 °C in a cell culture incubator for 3 h. After 3 h, the transfection mix is aspirated and 2 ml of DMEM containing 10% Donor bovine Serum is added to the dishes overnight (14–16 h). The following day the images are acquired.

2.5.2. Data analysis

- PDK1 homodimerisation dynamics are monitored with time in live NIH3T3 cells. Lifetime maps of a cell are acquired prior to stimulation ($t = 0$) and up to 25 min after being treated with 60 ng/ml of PDGF.
- For each time point the cell image is acquired at the mid-section. The change in EGFP-PDK1 lifetime in the presence of the acceptor PDK1-mCherry in one cell is shown in Fig. 12.3.
- An increase in the population of short lifetime pixels corresponding to the homodimerisation of PDK1 is observed. The control image which only expresses the donor EGFP-PDK1 shows the distribution of the donor lifetime. In the presence of the acceptor at $t = 0$ the lifetime distribution is shifted towards shorter lifetimes compared to the control, indicating that a basal level homodimerisation occurs in the cell prior to stimulation. The shorter lifetime pixels are not restricted to a specific location in the cell, suggesting that the maintenance of the PDK1 homodimer is independent of its binding at the plasma membrane (Masters *et al.*, 2010).

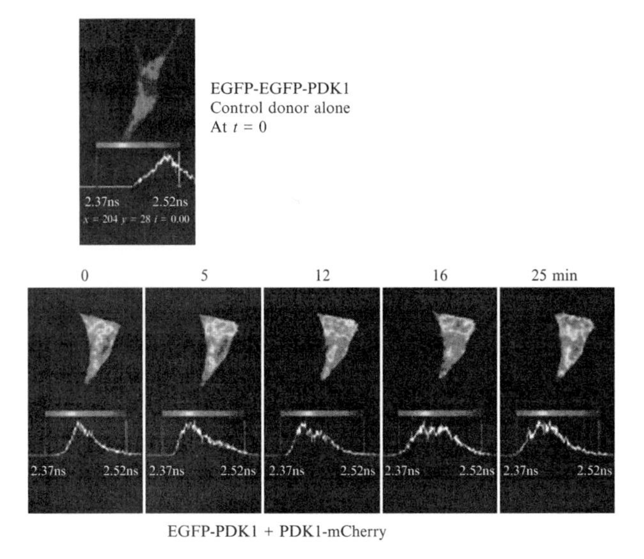

Figure 12.3 PDK1 homodimerisation monitored by FRET-two-photon FLIM lifetime maps of live NIH 3T3 cells expressing EGFP-PDK1 alone or with PDK1-mCherry. The time series shows the increase in PDK1 homodimerisation after stimulation with PDGF. The lifetime map of the donor alone is used as a control to show the lifetime distribution when FRET does not occur. The scale is 2.37 (red) to 2.52 ns (blue). (See Color Insert.)

3. Automated High-Throughput Analysis of Protein Activation in Tumor Micro Arrays

The current methods of measuring EGFR level, including IHC, cannot be endorsed as predictive of patient prognosis or response to treatment. Therefore, there is a need for a quantitative method to measure EGFR and its activation status. To achieve this, FRET monitored by FLIM was exploited as a reporter for the phosphorylation status of the receptor and as a molecular prognostic tool to identify head and neck squamous cell carcinoma (HNSCC) patients who showed overexpression and/or phosphorylation of EGFR. For this purpose, the frequency domain FLIM has been developed as an automated high-throughput instrument. FRET efficiency, indicative of EGFR phosphorylation levels, was shown to correlate with disease recurrence and prognosis of the patients.

Molecular targets in cancer therapy are identified by the relationship between their expression, pathogenesis and clinical outcome. Therefore, in a clinical environment the quantitative assessment of the degree of target molecule expression may provide an indication of both prognosis and whether or not a patient is expected to benefit from the targeted treatment (Dei Tos and Ellis, 2005). In addition, the activation status of oncoproteins, rather than their level of overexpression, would enable a more precise predictive parameter for determining the status of the target.

One critical protein is the EGFR, which takes a central role in regulating cellular processes such as proliferation, differentiation, and survival. When its function is dysregulated it contributes to the growth and survival of cancer cells. Thus, EGFR is recognized as one of the prominent targets for cancer therapy.

For the EGFR pathway, receptor expression as determined by IHC has been the standard. However, not all studies have found a correlation between EGFR overexpression and patient prognosis (Ulanovski *et al.*, 2004; Wen *et al.*, 1996). Part of the variation may derive from the fact that IHC does not provide a truly quantitative analysis of EGFR expression. The results tend to vary between laboratories because of the lack of a standardized scoring system and the subjective process of the interpretation of stained samples (Dei Tos and Ellis, 2005). Moreover, there is variation between laboratories in the fixatives used and the storage conditions of unstained tissue sections (Atkins *et al.*, 2004). The current methods of measuring EGFR levels, including IHC, are unable to detect the functional status of the EGFR pathway (Arteaga, 2002) and thus cannot be a universal parameter to predict patient outcomes. Therefore, there is a need for a quantitative method to measure the expression level of EGFR as well as its activation status.

3.1. Sample preparation

3.1.1. Labelling antibodies with the Cy dyes

- Once the appropriate donor and acceptor pairs are selected (Section 2.2) the monoclonal antibodies are conjugated in the following manner:
- It is essential for the antibodies to be in PBS
- 100 μl of *N,N*-dimethylformamide (DMF) is added to 1 mg Cy3b to make a 10 mg/ml stock solution (15 mM).
- The stock 10 mg/ml Cy3b is diluted in DMF 10-fold to 1 mg/ml (1.5 mM).
- 50 μl of Cy3b/DMF from a stock of 1 mg/ml is added drop by drop into 450 μl anti HER receptor antibody/50 μl Bicine (1 *M*, pH 8) with continuous stirring.
- The final concentration of conjugated anti-HER receptor antibodies (F4) with Cy3b needs to be approximately of 100 μg/ml (150 μM). The solution is stirred in the dark for 2 h.
- To conjugate FB2, an anti-phospho tyrosine antibody, with Cy5, 20 μl of DMF is added to a Cy5 vial. Cy5 dye in DMF is then added drop by drop to 450 μl FB2 antibody, 50 μl Bicine (1 *M*, pH 8) while stirring. The solution is stirred in the dark for 2 h. The conjugated antibodies are separated from free dyes by column chromatography.

- The column is first washed and equilibrated with 3 × 5 ml PBS before loading, the labeled antibody and unconjugated Cy dyes are eluted with PBS. The labeled antibody is collected and the dye/protein ratio (D/P) measured by UV/visible spectroscopy: detection of the antibody concentration is at 280 nm; F4-Cy3b at 561 nm; and FB2-Cy5 at 650 nm.
- The D/P ratio for mono-functional Cy3b or Cy5 dyes is calculated using the formula below:

$$D/P = \frac{[(\text{Absorption}\,A_{\max}) \times (\text{Antibody Extinction Coefficient})]}{[(A_{280} - \text{correction factor} \times A_{\max}) \times (\text{Cy Dye Extinction Coefficient})]}.$$

The ideal D/P is 1–2.

3.2. Automation of frequency-domain FLIM and high-throughput processing for tumor micro arrays (TMAs)

- The frequency-domain FLIM is interfaced to the microscope to process the TMAs and to analyze the data in a high-throughput manner.
- This is achieved by a specific programme which interfaces the motorized stage driver unit with the inverted microscope. Before processing the tumor arrays, the tumor cores are first mapped automatically. The stage is first initialised and the first tumor core is selected. The motorized stage driver unit is set to zero value for this first tumor. The other tumor cores are then mapped and the mapping of the tumor cores on the slide named and saved in a database.
- The determination of the accurate thresholding to obtain an adequate signal-to-noise ratio for each image is performed automatically in a unbiased manner.
- To analyze the data automatically a copy of the reference image (Foil-Section 2.2.1.4) from the experiment is saved on the desktop. The automation of the data analysis lists all the parameters of lifetimes and the intensity of the donor as well as number of pixels for each image in an excel sheet.
- The automation of frequency-domain FLIM is essential in order to implement the two-site FRET assay for wide clinical use.
- As an example, the activation status of EGFR has been assessed in HNSCC tumor arrays using this methodology (Kong *et al.*, 2006).

3.3. Assessing EGFR phosphorylation by two-site FRET assay

- The two-site FRET assay uses two antibodies for the same target to increase signal specificity: an anti-non-phosphoEGFR and an anti-phosphoEGFR. F4 (anti-EGFR antibody, which recognizes the intracellular tyrosine kinase domain of EGFR) and FB2 (anti-phosphotyrosine antibody) are used.

- F4 is conjugated to Cy3b (F4-Cy3b) and FB2 is conjugated to Cy5 (FB2-Cy5).
- The cells need to be permeabilized with 0.2% Triton X-100, labeled with donor F4-Cy3b and acceptor FB2-Cy5.
- When EGFR is not phosphorylated, only F4-Cy3b will bind to the tyrosine kinase domain. Autophosphorylation of the C-terminus of EGFR results in the binding of FB2-Cy5 (pEGFR-Cy5) to the phosphotyrosine sites. The two conjugated antibodies are brought into close proximity only when the receptor is activated, resulting in the transfer of energy from the donor F4-Cy3b (EGFR-Cy3b) to the acceptor FB2-Cy5.
- This transfer results in a decrease in the lifetime of the donor F4-Cy3b which is used as the read out.

3.4. Tumor micro array preparation for two-site FRET assay

- Tissue arrays are prepared from formalin-fixed, paraffin-embedded tumor blocks derived by surgical resection.
- Quality control for tumor specimens needs to be undertaken and all slides need to be examined by at least one specialist consultant pathologist. Usually, it is not possible for each core to be uniform and there will be variations between the individual cores with only a certain percentage of tumors present, therefore a non-tumorous section needs to be provided as a control.
- Typically, 500 μm cores are taken and transferred to an 8 × 15 core recipient array block using a Beecher Instruments Manual Tissue Arrayer-1 (Beecher Instruments Incorporated, Sun Prairie, WI). 5 μm sections are cut from each array block and made into slides which are stored at 4 °C. The approval of an ethics committee is required prior to this type of work.
- To label the tumor arrays with conjugated donor antibodies, the arrays are immersed in 0.2% (v/v) Triton X-100 for 5 min in order to permeabilise the cell membrane; this is followed by incubation with 1 mg/ml *freshly prepared* sodium borohydride in PBS for 10 min to quench the background fluorescence.
- 1% (w/v) BSA in PBS is used for blocking. The tumor arrays are incubated with the donor and acceptor (either F4-Cy3b and FB2-Cy5) for 2 h. The arrays are mounted on a glass coverslip with Mowiol mounting medium containing 2.5% (w/v) DABCO.

3.4.1. Head and neck tumor micro arrays

- HNSCC tumor arrays are labeled with donor F4-Cy3b and acceptor FB2-Cy5 to assess EGFR phosphorylation of the tumor cores.

- Using the automated FLIM system, each tumor core is mapped according to the position on the arrays. All images of the tumor cores are taken using a Zeiss Plan-APOCHROMAT $\times$ 10/0.5 NA phase 1.
- Images from each tumor core are acquired automatically from the arrays according to their positioning. The τ_{ϕ}, τ_M and average lifetimes of each tumor core are calculated, automatically and as well as the average FRET efficiency (Section 2.2).
- Average FRET efficiency (*E*) (Fig. 12.4) of each tumor core is correlated with survival data of the patients. A pilot study, needs to be first carried out to assess the first set of arrays with the automated system (Kong *et al.*, 2006).
- To validate a pilot study, a new set of arrays needs to be prepared with freshly conjugated antibodies and the arrays to be processed with automated FLIM.
- In the validation study, the system is programmed to run two loops so that two lifetime measurements are taken from each tumor core (574 tumor cores in total). However, in any one array a patient would have a duplicate sample, so for each patient the final lifetime represents an average of four measurements.
- In this particular pilot study, a total of 1114 tumor core recordings were analyzed.
- To obtain the EGFR expression status by FLIM, the donor intensity is calculated. For each tumor core, images are acquired, an intensity distribution histogram is plotted and the median of the intensity distribution is calculated. The median is then normalized to the background intensity.

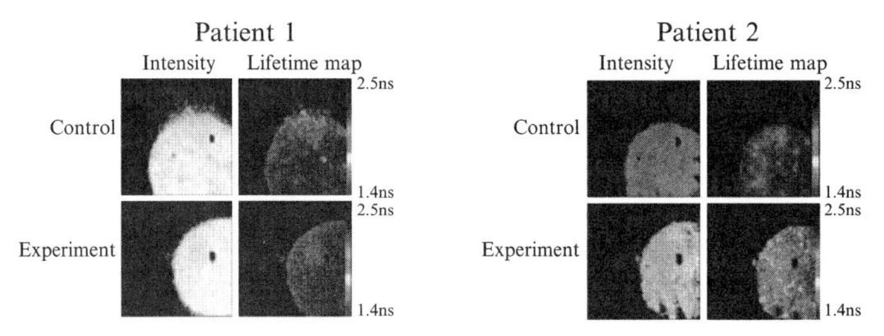

Figure 12.4 Activation status of EGFR1 in head and neck tumors detected by FRET efficiency. The left half of each panel (gray scale) shows the intensity images of tumor cores from the array. The right half images (pseudocolor) are lifetime maps. The upper tumor core of Patient 1 was from an array labeled with donor alone (F4-Cy3b). The average donor lifetime of the tumor core was 2.00 ns. There was no change in the lifetime with a FRET efficiency of 0%. The second set of donor and acceptor cores from Patient 2 indicates a decrease in the average lifetime (2.20–1.90 ns) with a FRET efficiency of 11%. The decrease in donor lifetime induced by FRET is used to calculate the FRET efficiency. Figure modified from Kong *et al.* (2006). (For color version of this figure, the reader is referred to the Web version of this chapter.)

- The average of two median intensity values from the two tumor cores of the same patient is used in each array. Thereafter, these values are used to correlate with IHC stains and average FRET efficiency.

3.5. Statistical analysis of two-site FRET assay in TMAs

- Below are the types of statistical analyzes that need to be performed in order to be able to compare the FRET data with disease-free survival and overall survival. Disease-free survival is defined as the length of time after treatment during which cancer is not found. Overall, survival is a defined period of time that patients in a study have survived since diagnosis or treatment.
- All statistical analyzes are performed using Graphpad Prism. The Kaplan-Meier survival curves are used to compare between the groups, and a log-rank test is used to assess the hazard ratios. For detailed analyzes of this pilot study readers should refer to Kong *et al.*, (2006).

ACKNOWLEDGMENTS

We are grateful to Peter J. Parker for his critical insights and scientific discussions in these projects. We would like to thank Paul Barber from the Gray Institute for Radiation, Oncology & Biology and Medical Sciences Division (University of Oxford) for the analysis software (TRI2) of the two-photon FLIM. We thank Christopher Applebee for the setup of live imaging on the two-photon FLIM. We also thank Nirmal Jethwa and Christopher Applebee for commenting on the manuscript. This work was supported by Cancer Research UK core funding to London Research Institute and by EU grant-QLK3-CT-2000.

REFERENCES

Alcor, D., Calleja, V., and Larijani, B. (2009). Revealing signaling in single cells by single- and two-photon fluorescence lifetime imaging microscopy. *Methods Mol. Biol.* **462,** 307–343.

Arteaga, C. L. (2002). Epidermal growth factor receptor dependence in human tumors: More than just expression? *Oncologist* **7**(Suppl. 4), 31–39.

Atkins, D., Reiffen, K. A., Tegtmeier, C. L., Winther, H., Bonato, M. S., *et al.* (2004). Immunohistochemical detection of EGFR in paraffin-embedded tumor tissues: Variation in staining intensity due to choice of fixative and storage time of tissue sections. *J. Histochem. Cytochem.* **52,** 893–901.

Calleja, V., Alcor, D., Laguerre, M., Park, J., Vojnovic, B., *et al.* (2007). Intramolecular and intermolecular interactions of protein kinase B define its activation *in vivo*. *PLoS Biol.* **5,** e95.

Dei Tos, A. P., and Ellis, I. (2005). Assessing epidermal growth factor receptor expression in tumours: What is the value of current test methods? *Eur. J. Cancer* **41,** 1383–1392.

Kong, A., Leboucher, P., Leek, R., Calleja, V., Winter, S., *et al.* (2006). Prognostic value of an activation state marker for epidermal growth factor receptor in tissue microarrays of head and neck cancer. *Cancer Res.* **66,** 2834–2843.

Masters, T. A., Calleja, V., Armoogum, D. A., Marsh, R. J., Applebee, C. J., *et al.* (2010). Regulation of 3-phosphoinositide-dependent protein kinase 1 activity by homodimerization in live cells. *Sci. Signal.* **3,** ra78.

Park, J., Hill, M. M., Hess, D., Brazil, D. P., Hofsteenge, J., *et al.* (2001). Identification of tyrosine phosphorylation sites on 3-phosphoinositide-dependent protein kinase-1 and their role in regulating kinase activity. *J. Biol. Chem.* **276,** 37459–37471.

Ulanovski, D., Stern, Y., Roizman, P., Shpitzer, T., Popovtzer, A., *et al.* (2004). Expression of EGFR and Cerb-B2 as prognostic factors in cancer of the tongue. *Oral Oncol.* **40,** 532–537.

van Munster, E. B., and Gadella, T. W. (2005). Fluorescence lifetime imaging microscopy (FLIM). *Adv. Biochem. Eng. Biotechnol.* **95,** 143–175.

Wen, Q. H., Miwa, T., Yoshizaki, T., Nagayama, I., Furukawa, M., *et al.* (1996). Prognostic value of EGFR and TGF-alpha in early laryngeal cancer treated with radiotherapy. *Laryngoscope* **106,** 884–888.

CHAPTER THIRTEEN

Imaging Stem Cell Differentiation for Cell-Based Tissue Repair

Zhenghong Lee,* James Dennis,† Jean Welter,‡ *and* Arnold Caplan‡

Contents

1. Introduction 248
2. Identifying Marker Genes Associated with Stem Cell Differentiation 250
3. Lenti-Viral Constructs for Event-Specific Reporter Gene Expression 251
   3.1. Dual promoter reporter gene vector construction for imaging osteogenic differentiation 251
   3.2. Dual promoter reporter gene vector construction for imaging chondrogenic differentiation 253
4. Applications of the Reporter System in Tissue Repair and Regeneration 254
   4.1. *In vitro* osteogenic induction and col1α1 promoter activity of transduced hMSCs 254
   4.2. *In vivo* osteogenesis assays 256
   4.3. *In vitro* chondrogenesis assays 258
5. Concluding Remarks 261
Acknowledgments 261
References 261

Abstract

Mesenchymal stem cells (MSCs) can differentiate into a number of tissue lineages and possess great potential in tissue regeneration and cell-based therapy. For bone fracture or cartilage wear and tear, stem cells need to be delivered to the injury site for repair. Assessing engraftment of the delivered cells and their differentiation status is crucial for the optimization of novel cell-based therapy. A longitudinal and quantitative method is needed to track stem cells transplanted/implanted to advance our understanding of their therapeutic effects and facilitate improvements in cell-based therapy. Currently, there are very few effective noninvasive ways to track the differentiation of infused stem

* Department of Radiology, Case Western Reserve University, Cleveland, Ohio, USA
† Hope Heart Matrix Biology Program, Benaroya Research Institute, Seattle, Washington, USA
‡ Department of Biology, Case Western Reserve University, Cleveland, Ohio, USA

Methods in Enzymology, Volume 506
ISSN 0076-6879, DOI: 10.1016/B978-0-12-391856-7.00037-8

cells. A brief review of a few existing approaches, mostly using transgenic animals, is given first, followed by newly developed *in vivo* imaging strategies that are intended to track implanted MSCs using a reporter gene system. Specifically, marker genes are selected to track whether MSCs differentiate along the osteogenic lineage for bone regeneration or the chondrogenic lineage for cartilage repair. The general strategy is to use the promoter of a differentiation-specific marker gene to drive the expression of an established reporter gene for noninvasive and repeated imaging of stem cell differentiation. The reporter gene system is introduced into MSCs by way of a lenti-viral vector, which allows the use of human cells and thus offers more flexibility than the transgenic animal approach. Imaging osteogenic differentiation of implanted MSCs is used as a demonstration of the proof-of-principle of this differentiation-specific reporter gene approach. This framework can be easily extended to other cell types and for differentiation into any other cell lineage for which a specific marker gene (promoter) can be identified.

1. Introduction

Stem cells are self-renewing, unspecialized cells that can give rise to multiple types of differentiated cells in the body. A subset of stem cells discussed here, the MSCs, are multipotent adult progenitor cells that can differentiate into multiple end-stage phenotypes including osteoblasts, adipocytes, and chondrocytes (Jiang *et al.*, 2002). MSCs have been shown, in animal models, to improve the therapeutic outcomes of myocardial infarction (Tang *et al.*, 2004; Toma *et al.*, 2002), bone marrow transplantation (Maitra *et al.*, 2004), and brain injury (Chen *et al.*, 2003; Honma *et al.*, 2006) although it is disputed as to whether MSCs directly differentiate into these phenotypes or contribute to repair indirectly (Caplan and Dennis, 2006). MSCs are a natural source of progenitor cells during bone turnover or remodeling, and during fracture repair (Caplan, 1991) or during cartilage regeneration for cartilage repair (Solchaga *et al.*, 2001). A key issue in assessing the utility of MSCs for these tissue repair functions is the ability to track their function and interaction with their local microenvironment, such as differentiation into end-stage phenotypes when transplanted, recruited, or homed to injury sites. Currently, the status of differentiation is determined by the assessment of the tissue after sacrificing the animals, which is slow, laborious, and expensive. An alternative method would be to track the differentiation of stem cells in living animals, repeatedly, over the course of the repair.

Since an ultimate goal of cell transplant/implant is to repair or replace damaged tissues, the delivered cells need to engraft into the host tissue, differentiate into the appropriate end-stage phenotype, and integrate into the host tissue or organ. Osteoblasts derived from MSCs or osteoprogenitor cells are the skeletal cells responsible for the synthesis, deposition, and mineralization of the extracellular matrix of bone (Marijanovic *et al.*, 2003).

Adequate numbers of MSCs that retain osteogenic differentiation potential are essential for the acquisition and maintenance of normal bone mass and function. Osteogenic differentiation is associated with matrix secretion, extracellular matrix maturation, and mineralization (Owen *et al.*, 1990). Osteoblasts are postproliferative, cuboidal, strongly alkaline phosphatase (ALP)-positive cells lining the bone matrix at sites of active matrix production. In parallel, specifics for chondroblasts derived from MSCs or alike can be described along the line of tissue engineering (Chen *et al.*, 2006).

Before recent advances in imaging technology, the commonly used methods to track transplanted stem cells' migration, engraftment, and fate relied extensively on the analysis of specific markers that distinguish donor cells from host cells in tissues. These methods required the sacrifice of multiple animals at specific time points post-implantation. Markers for the identification of donor cells during bone formation or cell differentiation include the Y-chromosome (Koc *et al.*, 1999; Pereira *et al.*, 1998), tissue-specific DNA or protein markers, and a variety of reporter genes such as β-galactosidase, luciferase, and green fluorescent protein (GFP). These markers are assayed using histochemical procedures, RT-PCR, or *in situ* hybridization. Major drawbacks for all these methods are that the animals need to be sacrificed and some of the markers are difficult to quantify, for example, *in situ* hybridization or histochemical and fluorescent markers. None of these methods can quantitatively, longitudinally, and noninvasively track stem cells in a living animal in order to reveal the spatial, temporal, and intensity of expression patterns of the marker gene(s).

There have been successful developments of *in vivo* imaging methods or strategies to track stem cells using either direct labeling with a physical label such as radionuclides or other imaging agents (Gleave *et al.*, 2011) or indirect labeling with a reporter driven by a constitutive promoter, rather than a differentiation-related promoter (Lee *et al.*, 2008; Love *et al.*, 2007). However, the timeline for lineage differentiation, where early assessment is crucial for treatment strategies, cannot be revealed by these "beacon"-type imaging strategies.

Molecular imaging of osteogenic differentiation and bone development was implemented via transgenic mouse models that expressed GFP controlled by an osteoblast lineage-specific promoter (Kalajzic *et al.*, 2004; Marijanovic *et al.*, 2003). Although the selected promoter's activity correlated with the developmental and functional state of the osteoblast and revealed topography and a range of osteoblast activity (Shichinohe *et al.*, 2004), the existing technique still required animal sacrifice for histological analysis to achieve quantitative assessment. Similarly, a transgenic mouse model was developed to express the firefly luciferase reporter gene, under the human osteocalcin (hOC) promoter. In this case, luciferase expression was repetitively and noninvasively detected by a cooled charge coupled device (CCD) camera. MSCs isolated from bone marrow of transgenic mice exhibited hOC promoter regulation as detected by luciferase expression that corresponded to osteogenic differentiation (Honigman *et al.*, 2001; Iris *et al.*, 2003).

Since many studies focus on the use of non-murine cells, the transgenic approach may not be suitable for imaging stem cell therapy. Genetic modification of stem cells by inserting a reporter gene into the genome, however, offers a larger degree of flexibility in the design of imaging studies. Before the imaging studies, similar genetic modifications, but with therapeutic gene transfer, have been made to create genetically engineered cells for therapy and tissue regeneration (Sheyn *et al.*, 2010). Currently, the development of molecular imaging techniques for chondrogenic differentiation lags behind that for osteogenic differentiation.

While differentiation-specific imaging techniques allow the tracking of stem cell differentiation, tracking reporter-transduced stem cells by using a "beacon" reporter independent of their differentiation status remains useful for cell purification and histological evaluation. Therefore, we developed a single construct with two promoters to control two reporter systems independently: one beacon system using a constitutively active promoter, the murine leukemia virus-derived MND promoter (*m*yeloproliferative sarcoma virus enhancer, *n*egative control region deleted, *d*l587rev primer-binding site substituted) to drive the expression of GFP for cell sorting, and a second reporter system with a differentiation-specific promoter to drive a luciferase (luc) reporter for imaging differentiation by BLI and other imaging modalities. Our previous study showed that the beacon signals stayed on for several months when the MND promoter was used (Love *et al.*, 2007). Other strong and constitutively active viral promoters such as CMV have been used, but some of these were turned off, possibly due to epigenetic modification (Kim *et al.*, 2005).

2. Identifying Marker Genes Associated with Stem Cell Differentiation

There are several markers associated with different stages of osteogenic differentiation. Early on, when MSCs turn into proliferative fibroblastic cells, STRO-1 (Gronthos *et al.*, 1994), SB-10, and HOP-26 (Stewart *et al.*, 2003) are expressed and can be used as the associated marker genes. Later, the preosteoblast state is accompanied by the expression of osteopontin (OP), collagen type I (col1α1), and alkaline phosphatase (Aubin, 1998). Finally, maturation of the osteoblast and the formation of nonproliferative cuboidal cells are accompanied by the expression of osteocalcin (OC), bone sialoprotein (BSP) that are the markers associated with matrix maturation at different stages of MSC differentiation into the bone (Alberius and Gordh, 1998; Visnjic *et al.*, 2001).

The OC promoter and the col1α1 promoter have been tested *in vivo* for their specificity and shown association with osteogenic differentiation. In transgenic mice where GFP is driven by the OC1.7 promoter, the GFP-expressive cells are found in a very restricted zone in the periosteum of growing mice, indicating a very restricted lineage specificity of the OC1.7 promoter. In contrast, in transgenic mice with GFP driven by the 2.3-kb fragment of rat type I collagen (Col2.3) promoter, GFP was observed in most differentiated osteoblastic cells (Kalajzic *et al.*, 2002). Others have demonstrated the use of the Col2.3 promoter in a retroviral vector system and demonstrated expression in bone marrow stromal cells induced down the osteoblastic lineage (Stover *et al.*, 2001). Based on this evidence, along with the timing of the stages (marker gene expression during osteogenic differentiation of MSCs, described above), we selected the Col2.3 promoter as the marker promoter to drive the reporter gene for imaging early stage osteogenic differentiation of hMSCs.

Similarly, there are several markers associated with different stages of chondrogenic differentiation. Upon differentiation induction, there is early expression of fibromodulin and cartilage oligomeric matrix protein. An increase in aggrecan and versican core protein synthesis defines an intermediate stage of chondrogenesis, which is followed by the expression of type II collagen (col2) and chondroadherin. Last, type X collagen, which is associated with hypertrophic cartilage, is expressed in some cells. Based on the sequential events in this pathway from the undifferentiated stem cell leading toward a mature chondrocyte, the col2 promoter was chosen as the marker promoter to drive the reporter gene for imaging chondrogenic differentiation of hMSCs in our studies.

Besides differentiation-specific marker genes, position/location specific marker genes can be used to broaden the utilization of this methodology. For example, the cardiac α-myosin heavy chain (α-MHC) can be used as a location specific marker gene for cardiac studies. Other cardiac-related (not necessarily position-specific) marker genes include cardiac troponin I, Ncx1, Anf, and Connexin 40 (Yue *et al.*, 2010). In addition, for differentiation events beyond the musculoskeletal system, genes such as Flk1, PECAM, Tie1, and VEC can be used as endothelial marker genes (Kim and von Recum, 2008) for imaging neovascular development in the context of stem cell differentiation.

3. Lenti-Viral Constructs for Event-Specific Reporter Gene Expression

3.1. Dual promoter reporter gene vector construction for imaging osteogenic differentiation

The osteogenic-induced reporter gene vector used here consists of dual fusion reporter gene, firefly luciferase (luc), and HSV thymidine kinase (tk), including HSV-tk polyadenylation sequence flanked by the 2.3-kb rat collagen type 1 alpha 1 (Col2.3) promoter. The Col2.3 promoter was a gift from Dr. David Rowe (University of Connecticut Health Center) and a

Figure 13.1 Schematic diagram of dual promoter multireporter gene construct showing relative positions of promoters including the osteoblast-specific collagen type I α (col1α1 or the 2.3-kb rat promoter Col2.3 used here) promoter and reporters. (For color version of this figure, the reader is referred to the Web version of this chapter.)

2.3-kb promoter sequence was excised by BamHI and NotI. This differentiation-inducible reporter construct was inserted into the backbone of a lenti-viral vector containing pHR8-mnd-GFP. The final dual promoter reporter gene vector, pLV-Col2.3-tk/luc-mnd-GFP, is a second-generation, self-inactivating lenti-viral vector with collagen type 1 alpha 1 promoter (Col2.3 promoter) driving tk/luc expression and the constitutive MND promoter driving GFP expression (Fig. 13.1).

1. The 2.3-kb promoter sequence is subcloned into pBluescriptII KS (pKS-Col2.3).
2. The polymerase chain reaction (PCR) product of luc and HSV-tk fusion gene including HSV-tk polyadenylation sequence from VVN-tk/luc with 5_EcoRI and 3_XhoI restriction sites was inserted into XhoI- and EcoRI-digested pKS-Col2.3 and termed "pKS-Col2.3-tk/luc". To confirm the function of this transgene plasmid, the osteoblast-like rat osteosarcoma cell line ROS 17/2.8 was used as a positive control, wherein collagen type 1 is constitutively expressed.
   a. ROS 17/2.8 cells were cultured in DMEM-HG (high glucose) media containing 10% FBS and 1% antibiotic–antimycotic (Invitrogen). The culture medium was changed every 3 days.
   b. For ROS 17/2.8 transduction, 1×10^4 ROS 17/2.8 cells per well in a 12-well plate were incubated with pCol2.3-tk/luc and Lipofectamine 2000 for 4 h, which was then replaced fresh medium.
   c. To confirm luc expression in ROS 17/2.8, the transduced ROS 17/2.8 cells were detached with cell lysis buffer 48 h later and the bioluminescence measured with a luminometer.
3. The PCR product of pCol2.3-tk/luc (from pKS-Col2.3-tk/luc) was cut with ClaI and inserted into the ClaI-digested pHR8-mnd-GFP, resulting in the final lenti-viral vector, pLV-Col2.3-tk/luc-mnd-GFP.
4. The virus particles of the dual promoter reporter gene system shown in Fig. 13.1 are produced by co-transfection of 293T/17 cells with pLV-Col2.3-tk/luc-mnd-GFP, pCMVΔR8.91 (packaging vector), and pMD.G (vesicular stomatitis virus protein G pseudotyping vector).
   a. The plasmids are transfected into 55–60% confluent 293T/17 cells using Lipofectamine 2000 according to the manufacturer's instructions.
   b. After 48 h of incubation, the supernatant containing virus particles is collected and concentrated with an Amicon Ultra-15 device.

c. Titers are determined by transduction of NIH3T3 cells. Briefly, 1×10^4 NIH3T3 cells are seeded to a 24-well plate and 0, 10, 100, or 1000 μl of supernatant including virus particles is added to the each well with 8 mg/ml polybrene at same time. After 5 days, the percentage of transduced NIH3T3 cells is determined by FACS analysis of GFP signal.
d. To confirm the function of the new vector, positive control ROS 17/2.8 cells are incubated with LV-Col2.3-tk/luc-mnd-GFP virus at MOI of 1 with 8 mg/ml polybrene for 6 h and replaced with fresh medium.
e. Forty eight hours later, ROS 17/2.8 cells are detached with trypsin–EDTA and resuspended in PBS. Luminometer assay of luc expression can be performed with D-luciferin, the substrate of luc, according to the recommendation of the vendor to plot a relationship between signal intensity and number of the cells with wild-type ROS cells as the control.

3.2. Dual promoter reporter gene vector construction for imaging chondrogenic differentiation

The chondrogenic-induced reporter gene vector consists of dual reporter gene, luc linked with HSV1-sr39tk, including BGH polyadenylation sequence flanked by 1 kb fragment of mouse collagen type 2 alpha 1 promoter (col2α1 in Fig. 13.2). The plasmid containing the mouse

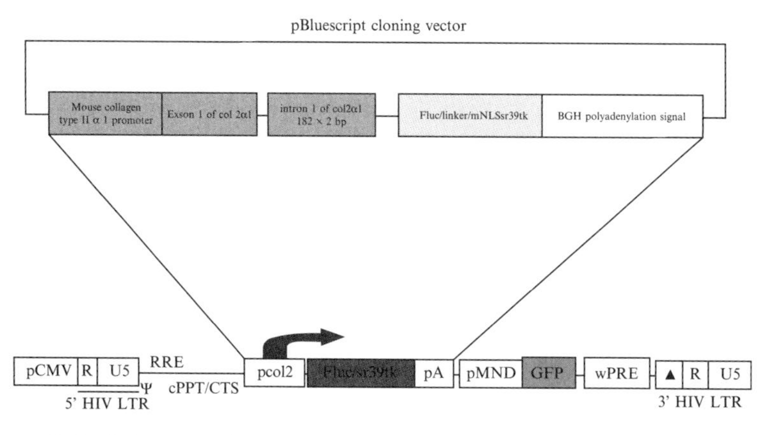

Figure 13.2 Schematic diagram of dual promoter multireporter gene construct showing relative positions of promoters including the chondrogenic-specific collagen type II α (col2) promoter and reporters. (For color version of this figure, the reader is referred to the Web version of this chapter.)

col2α1 promoter was a gift from Dr. Guang Zhou (Case Western Reserve University, Cleveland, OH). Construction of this inducible reporter into the final dual promoter lenti-viral vector with MND as the second promoter to drive GFP expression is very similar to that in figure 3.1 for osteogenic differentiation. To confirm the inducible portion or the final vector, a positive control, such as a chondrosarcoma cell line or RCS, can be used, in which col2 is constitutively active.

4. Applications of the Reporter System in Tissue Repair and Regeneration

Several labs have looked at the use of MSCs in tissue engineering for the repair of bone defects and for the attenuation of skeletal diseases, such as osteoporosis and osteogenesis imperfecta, through MSC transplantation (Gao *et al.*, 2001; Horwitz *et al.*, 1999; Mauney *et al.*, 2005). Similarly, MSCs have been increasingly used for cell transplantation as an alternative to autologous chondrocytes for cartilage repair (Kolf *et al.*, 2007; Wakitani *et al.*, 1994). Through the use of *in vivo* BLI imaging, it becomes possible to repeatedly image to monitor the onset of hMSC differentiation in an *in vitro* or *in vivo* setting. The intensity of a BLI signal generated during MSC's differentiation depends on the specific promoter activity (that drives luc expression) in response to differentiation inducers in the local microenvironment.

4.1. *In vitro* osteogenic induction and col1α1 promoter activity of transduced hMSCs

4.1.1. Transduction of human MSCs

Human hMSCs (hMSCs) were isolated from bone marrow obtained from patients undergoing hip or knee arthroplasty after patient consent and approval by the Institutional Review Board of the University Hospitals Case Medical Center following a routine procedure (Love *et al.*, 2007). 3×10^5 hMSCs in T-175 flasks were incubated with the LV-Col2.3-tk/luc-mnd-GFP virus at an MOI (multiplicity of infection) of 1 in medium containing 8mg/ml polybrene. This process was repeated three times. In each round, hMSCs were incubated with viral particles for 6h and returned to fresh hMSC growth medium for 48h. Three days after the third incubation, the cells were viewed on an Eclipse TE200 inverted microscope. At near confluence, transduced hMSCs were sorted by Becton and Dickinson (BD) FACSAria.

4.1.2. Combined differentiation factors on Col2.3 promoter expression

1.5×10^4 transduced hMSCs were seeded onto 24 well tissue culture plates and allowed to attach overnight. The following day, eight different combinations of three factors in the osteogenic medium: $10^{-7}M$ dexamethasone (Dexa), 50 μ*M* ascorbic acid-2 phosphate (AA), and 2 m*M* β-glycerophosphate (BGP) were added alone or in combination to reporter-transduced hMSCs cultured with DMEM-LG, 10% of hMSC-tested FBS, 1% of antibiotic–antimycotic solution. Medium was changed twice a week. At days 1, 4, 7, 11, 15, 20, 25, 28, and 35, hMSCs were harvested and assessed for luc activity driven by the colα1a promoter to evaluate the effect(s) of normal media, Dexa, AA, BGP or AA/BGP, Dexa/AA, Dexa/BGP, or Dexa/AA/BGP treatment on the promoter. The luc assay was performed after each harvest. The culture medium was completely removed, the hMSCs were rinsed twice with PBS and then incubated in 150 μl of lysis buffer and freeze-thawed twice. The luc activity was measured with a luminometer for 10 s (Fig. 13.3). As shown in Fig. 13.3, Dexa and AA, alone and in combination, were osteoinductive and produced higher levels of imagable (luc) signals.

4.1.3. Confirmation of expression col1α1 gene expression by RT-PCR

Total mRNA was prepared from osteogenically induced hMSCs using Trizol reagent following the manufacturer's protocol. cDNA was produced using Moloney murine leukemia virus (M-MLV) reverse transcriptase.

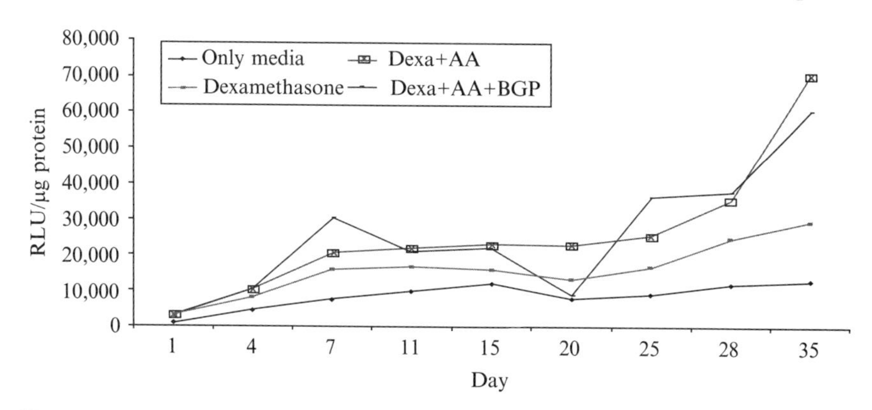

Figure 13.3 Reporter expression in transduced hMSCs during the activation of col1α1 promoter when incubated in the medium with combinations of osteogenic inducing medium components, dexamethasone (Dexa), ascorbic acid-2 phosphate (AA), and β-glycerophosphate (BGP). The lower curves with media BGP or AA/BGP were not displayed for clarity and simplicity of the display. (For color version of this figure, the reader is referred to the Web version of this chapter.)

PCR was then performed using the col1α1 primers. β-actin was amplified as a control. The amplified products obtained were analyzed by ethidium bromide stained agarose gel electrophoresis. The expression value of the *col1α1* gene was calculated by dividing the intensity of bands for col1α1 by that of β-actin for normalization.

4.1.4. Confirmation of extracellular matrix mineralization hMSCs after osteogenic induction

Alzarin red S staining is performed. Briefly, at 20 and 30days incubation in osteogenic medium, the culture medium is removed, rinsed twice with DPBS, the cells incubated a solution of 2% of Alizarin red S, pH 4.2 for 15 min at room temperature, and the cells rinsed four times with distilled water.

4.1.5. Modulation of the differentiation of hMSCs

Besides ingredients in the differentiation induction medium, the effects of other biofactors, such as parathyroid hormone (PTH), on MSCs during the bone repair process are worth studying. While PTH is known to enhance bone resorption by osteoclasts, osteoblasts also have PTH receptors and PTH has been shown to increase bone formation, especially when administered transiently (de Paula and Rosen, 2010). The dynamics of multiple PTH dosings is largely unknown (Nakazawa *et al.*, 2005). It will be interesting to study the sensitivity of the col1 promoter in response to differentiation inducers/enhancers since that will further determine the usefulness or limitation of the imaging approach. Besides PTH, bone morphogenic proteins (BMPs) that stimulate bone cell differentiation (Reddi and Reddi, 2009) can be studied in similar experiments.

For chondrogenic induction, the effects of differentiation inducers and enhancers, individually or in combination, on the activity of col2 promoters, can be studied in a similar fashion. The ingredients in the chondrogenic induction medium include 100μ*M* ascorbate 2-phosphate, 10^{-7} *M* Dexa, and 10ng/ml recombinant human TGF-β1. In addition, chondrogenic factors from the BMPs (Steinert *et al.*, 2009), IGF1 (Capito and Spector, 2007), transcription factors from the SOX family (Babister *et al.*, 2008), ligands (FGF-9 and FGF-18) to receptors such as FGFR3, and transducing molecules from the Smad family (Saraf and Mikos, 2006) can be tested in a similar way to evaluate their effect on the activity of the col2 promoter used for imaging. Finally, the effects of mechanical stimuli in a bioreactor (a brief description of bioreactor used for cartilage regeneration is given in Section 4.3 below) on col2 can be analyzed.

4.2. *In vivo* osteogenesis assays

Transduced hMSCs were flow-sorted for GFP+ cells, expanded in culture, passaged, and harvested 1 week later for implantation. The hMSCs can be implanted directly or loaded into a scaffold for implantation into a defect site. As a control, equal amounts of wild-type MSCs from the same origin were used for implantation into a matching defect site in the animal model of bone fracture. The animal was then imaged by BLI for the onset of differentiation-specific promoter frequently during the first 2 weeks and less frequently after that, when the signal intensity decreased significantly. Other imaging modalities can complement BLI to follow the late steps in the repair process. In the end, the animals were sacrificed, and the tissue of interest was harvested and processed for histology. In the following section, experimental steps are described to demonstrate the application of the new imaging technique to image osteogenic differentiation of hMSCs for bone repair. This workflow can be adapted for use in MSCs chondrogenic differentiation for cartilage repair.

4.2.1. Segmental bone defect model

A procedure described by Steven Goldstein (Orthopedic Research Lab at University of Michigan, Ann Arbor, MI) was used to create a mouse segmental defect model (Goldstein and Bonadio, 1998). The procedure was approved for practice and experimentation by the IACUC at Case Western Reserve University. Briefly, a custom-designed small stainless titanium alloy plate (10×2×1.5mm) was fixed to the femur of an adult NOD–SCID mouse using four 0.838mm diameter stainless steel screws and a small ~1.5mm defect was created in the middle using a hand drill.

4.2.2. Encapsulation of hMSCs for placement in the bone defect site

An alginate hydrogel scaffold from Eben Alsberg's Lab (Alsberg *et al.*, 2003) at Case Western Reserve University was loaded with cells and implanted into the newly created defect site. Briefly, hMSCs were suspended in 10 μl of the alginate solution at 40×10^6 cells/ml. The 10 μl sample was then pipetted into a solution of 0.1 M $CaCl_2$ for less than 5 s to cross-link the alginate. The hydrogel with encapsulated cells was then rinsed in DMEM, transferred to a fresh tube with 1 ml of DMEM, and kept on ice until implantation in the bone defect. The surgical wound was then sutured after cell implantation, and the animals were allowed to recover before imaging.

4.2.3. Imaging

BLI was performed every other day for the first 2 weeks using a cooled CCD camera, the IVIS-200 system. Each time, the animals were injected with D-luciferin according to the recommendation from the vendor, and incubated for 6 min before a 3-min BLI exposure. After that, the animals were imaged once a week and then every 2 weeks for up to 2 months. Planar X-ray scans were taken at various time points, especially toward the end to detect new bone formation at the defect site. Figure 13.4 shows the results of live animal imaging.

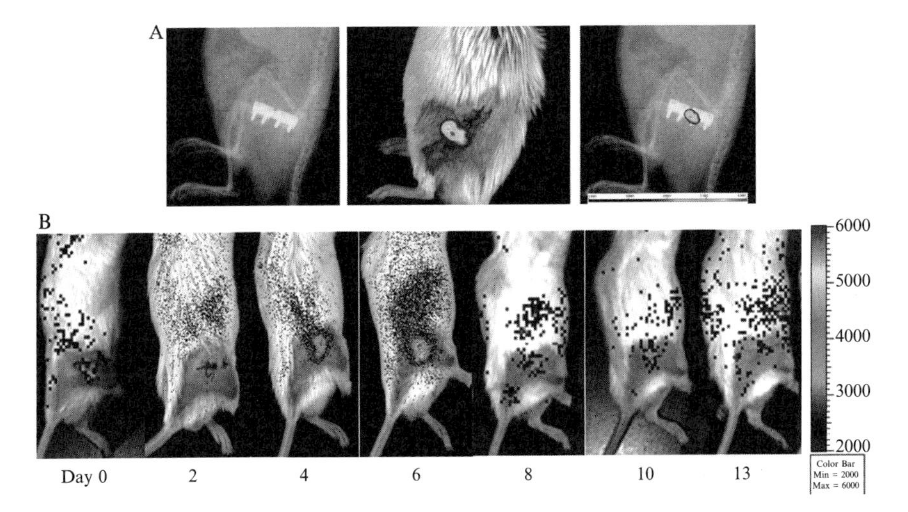

Figure 13.4 BLI of hMSC-Col2.3-tk/luc-implanted mouse. (A) X-ray image of a femur of a mouse with a freshly made bone defect (left) with all four screws visible, BLI of luc-hMSCs into femoral segmental defects (middle), overlay of X-ray and BLI; (B) Individual mice were imaged multiple times over 2 weeks. The peak time of luciferase activity was around Day 6. (For color version of this figure, the reader is referred to the Web version of this chapter.)

4.2.4. Histology

After final imaging, the removed mouse femurs were fixed and washed overnight in water, and the small alloy plate and screws were carefully removed from the femur prior to decalcification in RDO decal. The samples were processed and embedded in either paraffin or plastic. In the case of paraffin embedding, 5 μm thick sections were cut and stained with Mallory–Heidenhain or for specific markers found in the defect space. For some of the removed mouse femurs that were undecalcified, the sections were embedded in plastic, and were cut at 200 μm thickness and then ground down and stained with toluidine blue. All slides were then cover-slipped and examined by microscopy. Figure 13.5 shows one from each of the two processes.

4.3. *In vitro* chondrogenesis assays

Due to a lack of blood supply and the difficulty for progenitor cells to migrate to cartilage tissue, MSCs were predifferentiated into chondrocytes and to provide structure guidance for tissue engineering. Bioreactors are designed to create these large biomimetic osteochondral constructs in 3D. A bioreactor is a tissue engineering tool not only to provide optimal nutrient

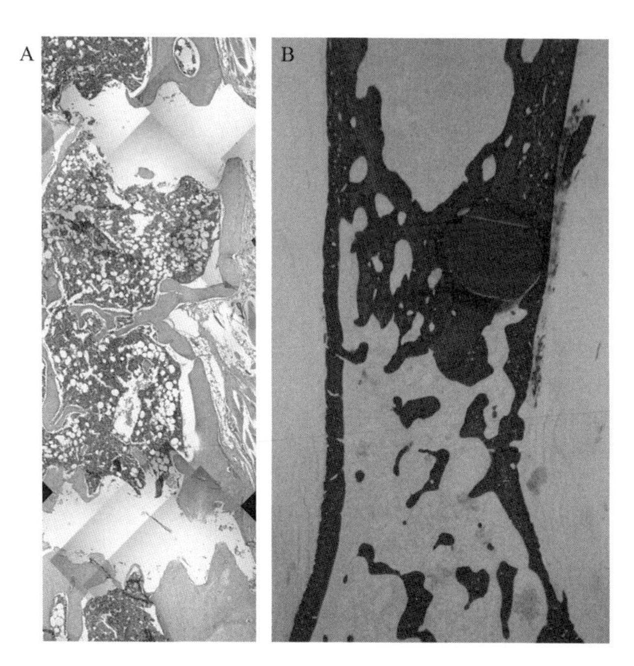

Figure 13.5 Histological evaluation on tissue samples harvested at Day 58. (A) H&E staining of a decalcified section cut from the paraffin block. The space occupied by the screws is empty; (B) Toluidine blue staining of an undecalcified section cut from the plastic embedding. One of the four metal screws is visible. (For interpretation of the references to color in this figure legend, the reader is referred to the Web version of this chapter.)

supply and waste removal but also to control environmental conditions such as pH, temperature, and biomechanical environment. There have been recent developments to closely monitor these conditions and to optimize the rate and pattern of medium flowing through the bioreactor. However, there has previously been no way of tracking of the differentiation process of the cells within the bioreactor. An imaging system was fabricated to accommodate the mini-incubator that houses the bioreactors for continuous imaging. BLI images were taken, up to once every 30min, to monitor the dynamics of col2 promoter activity in transduced hMSCs (shown in Fig. 13.2); medium (containing chondrogenic inducers along with imaging substrate, D-luciferin) flowed through the bioreactor in a steady-state fashion. MSCs aggregated into pellets were placed into the holes of the plastic locator, which can hold up to nine pellets, and then placed inside the bioreactor and imaged in the same field of view (FOV). To factor in the issue of cell retention and viability in the bioreactor over the time course, which can be weeks, a positive control was placed in the same bioreactor and imaged along with the targets. The signal from the control cells was used to normalize that from the target cells for quantitative estimation of col2 activity (via reporter gene) in the targets during chondrogenic differentiation. The positive control consists of cells from the same donor but transduced with a reporter gene driven by a constitutively active promoter. Figure 13.6 shows an image signal of the pellet made of control cells placed in the bioreactor.

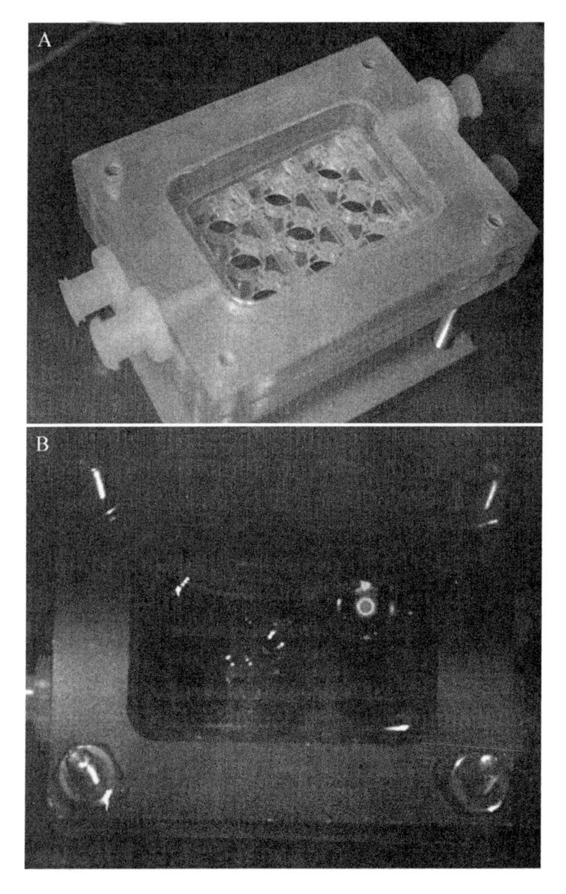

Figure 13.6 BLI of MSCs aggregated into pellets that are placed into the holes of the plastic locator, which can hold up to nine pellets (A), and placed inside the bioreactor for continues imaging (B). Images can be taken up to once every 30 min when nutritional medium (containing differentiation inducers along with imaging substrate, D-luciferin) flowed through the bioreactor in a steady-state fashion. (For color version of this figure, the reader is referred to the Web version of this chapter.)

5. Concluding Remarks

A platform imaging technology was developed to track MSCs *in vivo*. This reporter gene-based technology was tested in a mouse model of a critical-sized bone defect to monitor the initial course of osteogenic differentiation of implanted hMSCs. It was also applied to a bioreactor setting to track chondrogenic differentiation of pelleted MSCs. This novel differentiation-specific molecular imaging technology is a useful tool for studying stem cell-based repair and regeneration.

ACKNOWLEDGMENTS

We would like to thank all people listed in alphabetical order for their contributions to various aspects of this work: Eben Alsberg, Amad Awadallah, David Corn, Stan Gerson, Steve Goldstein, Yunhui Kim, Melissa D. Krebs, Donald Lennon, Seunghwan Lim, Yuan Lin, Joseph Molter, Troy Mounts, Teresa Pizzuto, Luis Solchaga, Lewis Yuan. This work is supported, in part, by a grant from the NIH (P01 AR053622).

REFERENCES

Alberius, P., and Gordh, M. (1998). Osteopontin and bone sialoprotein distribution at the bone graft recipient site. *Arch. Otolaryngol. Head Neck Surg.* **124,** 1382–1386.

Alsberg, E., *et al.* (2003). Regulating bone formation via controlled scaffold degradation. *J. Dent. Res.* **82,** 903–908.

Aubin, J. E. (1998). Bone stem cells. *J. Cell. Biochem. Suppl.* **30–31,** 73–82.

Babister, J. C., *et al.* (2008). Genetic manipulation of human mesenchymal progenitors to promote chondrogenesis using "bead-in-bead" polysaccharide capsules. *Biomaterials* **29,** 58–65.

Capito, R. M., and Spector, M. (2007). Collagen scaffolds for nonviral IGF-1 gene delivery in articular cartilage tissue engineering. *Gene Ther.* **14,** 721–732.

Caplan, A. I. (1991). Mesenchymal stem cells. *J. Orthop. Res.* **9,** 641–650.

Caplan, A. I., and Dennis, J. E. (2006). Mesenchymal stem cells as trophic mediators. *J. Cell. Biochem.* **98,** 1076–1084.

Chen, J., *et al.* (2003). Intravenous bone marrow stromal cell therapy reduces apoptosis and promotes endogenous cell proliferation after stroke in female rat. *J. Neurosci. Res.* **73,** 778–786.

Chen, F. H., *et al.* (2006). Technology insight: Adult stem cells in cartilage regeneration and tissue engineering. *Nat. Clin. Pract. Rheumatol.* **2,** 373–382.

de Paula, F. J., and Rosen, C. J. (2010). Back to the future: Revisiting parathyroid hormone and calcitonin control of bone remodeling. *Horm. Metab. Res.* **42,** 299–306.

Gao, J., *et al.* (2001). The dynamic *in vivo* distribution of bone marrow-derived mesenchymal stem cells after infusion. *Cells Tissues Organs* **169,** 12–20.

Gleave, J. A., *et al.* (2011). 99mTc-based imaging of transplanted neural stem cells and progenitor cells. *J. Nucl. Med. Technol.* **39,** 114–120.

Goldstein, S. A., and Bonadio, J. (1998). Potential role for direct gene transfer in the enhancement of fracture healing. *Clin. Orthop. Relat. Res.* S154–S162.

Gronthos, S., *et al.* (1994). The STRO-1+ fraction of adult human bone marrow contains the osteogenic precursors. *Blood* **84,** 4164–4173.

Honigman, A., *et al.* (2001). Imaging transgene expression in live animals. *Mol. Ther.* **4,** 239–249.

Honma, T., *et al.* (2006). Intravenous infusion of immortalized human mesenchymal stem cells protects against injury in a cerebral ischemia model in adult rat. *Exp. Neurol.* **199,** 56–66.

Horwitz, E. M., *et al.* (1999). Transplantability and therapeutic effects of bone marrow-derived mesenchymal cells in children with osteogenesis imperfecta. *Nat. Med.* **5,** 309–313.

Iris, B., *et al.* (2003). Molecular imaging of the skeleton: Quantitative real-time bioluminescence monitoring gene expression in bone repair and development. *J. Bone Miner. Res.* **18,** 570–578.

Jiang, Y., *et al.* (2002). Pluripotency of mesenchymal stem cells derived from adult marrow. *Nature* **418,** 41–49.

Kalajzic, Z., *et al.* (2002). Directing the expression of a green fluorescent protein transgene in differentiated osteoblasts: Comparison between rat type I collagen and rat osteocalcin promoters. *Bone* **31,** 654–660.

Kalajzic, I., *et al.* (2004). Dentin matrix protein 1 expression during osteoblastic differentiation, generation of an osteocyte GFP-transgene. *Bone* **35,** 74–82.

Kim, S., and von Recum, H. (2008). Endothelial stem cells and precursors for tissue engineering: Cell source, differentiation, selection, and application. *Tissue Eng. Part B Rev.* **14,** 133–147.

Kim, Y. H., *et al.* (2005). Reversing the silencing of reporter sodium/iodide symporter transgene for stem cell tracking. *J. Nucl. Med.* **46,** 305–311.

Koc, O. N., *et al.* (1999). Bone marrow-derived mesenchymal stem cells remain host-derived despite successful hematopoietic engraftment after allogeneic transplantation in patients with lysosomal and peroxisomal storage diseases. *Exp. Hematol.* **27,** 1675–1681.

Kolf, C. M., *et al.* (2007). Mesenchymal stromal cells. Biology of adult mesenchymal stem cells: Regulation of niche, self-renewal and differentiation. *Arthritis Res. Ther.* **9,** 204.

Lee, Z., *et al.* (2008). Imaging stem cell implant for cellular-based therapies. *Exp. Biol. Med. (Maywood)* **233,** 930–940.

Love, Z., *et al.* (2007). Imaging of mesenchymal stem cell transplant by bioluminescence and PET. *J. Nucl. Med.* **48,** 2011–2020.

Maitra, B., *et al.* (2004). Human mesenchymal stem cells support unrelated donor hematopoietic stem cells and suppress T-cell activation. *Bone Marrow Transplant.* **33,** 597–604.

Marijanovic, I., *et al.* (2003). Dual reporter transgene driven by 2.3Col1a1 promoter is active in differentiated osteoblasts. *Croat. Med. J.* **44,** 412–417.

Mauney, J. R., *et al.* (2005). Role of adult mesenchymal stem cells in bone tissue engineering applications: Current status and future prospects. *Tissue Eng.* **11,** 787–802.

Nakazawa, T., *et al.* (2005). Effects of low-dose, intermittent treatment with recombinant human parathyroid hormone (1–34) on chondrogenesis in a model of experimental fracture healing. *Bone* **37,** 711–719.

Owen, T. A., *et al.* (1990). Progressive development of the rat osteoblast phenotype *in vitro*: Reciprocal relationships in expression of genes associated with osteoblast proliferation and differentiation during formation of the bone extracellular matrix. *J. Cell. Physiol.* **143,** 420–430.

Pereira, R. F., *et al.* (1998). Marrow stromal cells as a source of progenitor cells for nonhematopoietic tissues in transgenic mice with a phenotype of osteogenesis imperfecta. *Proc. Natl. Acad. Sci. USA* **95,** 1142–1147.

Reddi, A. H., and Reddi, A. (2009). Bone morphogenetic proteins (BMPs): From morphogens to metabologens. *Cytokine Growth Factor Rev.* **20,** 341–342.

Saraf, A., and Mikos, A. G. (2006). Gene delivery strategies for cartilage tissue engineering. *Adv. Drug Deliv. Rev.* **58,** 592–603.

Sheyn, D., *et al.* (2010). Genetically modified cells in regenerative medicine and tissue engineering. *Adv. Drug Deliv. Rev.* **62,** 683–698.

Shichinohe, H., *et al.* (2004). *In vivo* tracking of bone marrow stromal cells transplanted into mice cerebral infarct by fluorescence optical imaging. *Brain Res. Brain Res. Protoc.* **13,** 166–175.

Solchaga, L. A., *et al.* (2001). Cartilage regeneration using principles of tissue engineering. *Clin. Orthop. Relat. Res.* S161–S170.

Steinert, A. F., *et al.* (2009). Enhanced *in vitro* chondrogenesis of primary mesenchymal stem cells by combined gene transfer. *Tissue Eng. Part A* **15,** 1127–1139.

Stewart, K., *et al.* (2003). STRO-1, HOP-26 (CD63), CD49a and SB-10 (CD166) as markers of primitive human marrow stromal cells and their more differentiated progeny: A comparative investigation *in vitro*. *Cell Tissue Res.* **313,** 281–290.

Stover, M. L., *et al.* (2001). Bone-directed expression of Col1a1 promoter-driven self-inactivating retroviral vector in bone marrow cells and transgenic mice. *Mol. Ther.* **3,** 543–550.

Tang, Y. L., *et al.* (2004). Autologous mesenchymal stem cell transplantation induce VEGF and neovascularization in ischemic myocardium. *Regul. Pept.* **117,** 3–10.

Toma, C., *et al.* (2002). Human mesenchymal stem cells differentiate to a cardiomyocyte phenotype in the adult murine heart. *Circulation* **105,** 93–98.

Visnjic, D., *et al.* (2001). Conditional ablation of the osteoblast lineage in Col2.3deltatk transgenic mice. *J. Bone Miner. Res.* **16,** 2222–2231.

Wakitani, S., *et al.* (1994). Mesenchymal cell-based repair of large, full-thickness defects of articular cartilage. *J. Bone Joint Surg. Am.* **76,** 579–592.

Yue, F., *et al.* (2010). Bone marrow stromal cells as an inducer for cardiomyocyte differentiation from mouse embryonic stem cells. *Ann. Anat.* **192,** 314–321.

CHAPTER FOURTEEN

Understanding the Initiation of B Cell Signaling Through Live Cell Imaging

Angel M. Davey,[1] Wanli Liu,[1] Hae Won Sohn,[1]
Joseph Brzostowski,[1] *and* Susan K. Pierce*

Contents

1. Introduction 266
2. Experiment Preparation 268
   2.1. Cell preparation 268
   2.2. Preparation of antigen-presenting planar lipid bilayers 270
   2.3. TIRF microscope design 274
3. Image Acquisition and Analysis 281
   3.1. Imaging B cell receptor oligomerization by single molecule tracking 281
   3.2. Imaging B cell receptor cluster dynamics 282
   3.3. Imaging B cell receptor signaling 286
Acknowledgments 288
References 288

Abstract

Antibody responses are initiated by the binding of antigens to clonally distributed cell surface B cell receptors (BCRs) that trigger signaling cascades resulting in B cell activation. Using conventional biochemical approaches, the components of the downstream BCR signaling pathways have been described in considerable detail. However, far less is known about the early molecular events by which the binding of antigens to the BCRs initiates BCR signaling. With the recent advent of high resolution, high speed, live cell, and single molecule imaging technologies, these events are just beginning to be elucidated. Understanding the molecular mechanisms underlying the initiation of BCR signaling may provide new targets for therapeutics to block dysregulated BCR signaling in systemic autoimmune diseases and in B cell tumors and to aid in the design of protein subunit vaccines. In this chapter, we describe the general procedures

Laboratory of Immunogenetics, National Institute of Allergy and Infectious Diseases, NIH, Rockville, Maryland, USA

*Corresponding author.
[1]These authors contributed equally.

Methods in Enzymology, Volume 506
ISSN 0076-6879, DOI: 10.1016/B978-0-12-391856-7.00038-X

for using these new imaging techniques to investigate the early events in the initiation of BCR signaling.

1. Introduction

B cells play an essential role in the adaptive immune response to many infections by producing highly specific antibodies (Abs) to antigens (Ags) expressed by invading pathogens. Ab responses are initiated by the binding of Ags to cell surface B cell receptors (BCRs) that trigger signaling cascades, leading to the proliferation and differentiation of B cells into antibody secreting cells (Rajewsky, 1996). The BCR is a transmembrane protein complex composed of a membrane form of Ab and a noncovalently associated disulphide-linked Igα and Igβ heterodimer. The BCR has no inherent kinase activity, but rather upon Ag binding, the first kinase in the signaling cascade, Lyn, is recruited to the cytoplasmic domains of Igα and Igβ which contain immunotyrosine activation motifs (ITAMs) that are phosphorylated (Reth and Wienands, 1997). The question is: by what molecular mechanism is the information that Ag has bound to the ectodomain of the BCR translated across the membrane to trigger the recruitment of Lyn? Over the last several decades, biochemical studies have resulted in the description of the components of the downstream BCR signaling pathways in great detail (Dal Porto *et al.*, 2004; DeFranco, 1997). The contribution of biochemical approaches to the understanding of the nature of BCR signaling pathways has been significant. However, as the old parable instructs, if you want to know the number of teeth in a horse's mouth you have to look. In the context of B cell biology, if you want to understand how BCRs trigger signaling you need to look at the BCR as it triggers signaling. The problem is that conventional biochemical approaches cannot provide the spatial and temporal resolution of the events that initiate BCR signaling and that are predicted to occur within seconds of Ag binding and to be highly dynamic. Recently, with the advent of high resolution, high speed, live cell, and single molecule imaging techniques, it is now possible for the first time to view these central events in B cell activation (Pierce and Liu, 2010). The technological improvements in the fluorescence microscope over the last decade, especially in the cameras and photomultipliers that detect fluorescent signals, coupled with the relative ease of creating fluorescently tagged proteins, have brought to reality the ability to capture the movement and behavior of proteins down to the level of single molecules in live cells at video or faster frame rates (Bajenoff and Germain, 2007; Balagopalan *et al.*, 2010). Just as spectacular is the fact that much of this technology can be quickly taught to the novice on instruments that can be maintained in most laboratories. The barriers to obtain spatial and temporal information for the molecules in the BCR signaling pathways are continually being chipped away, making it possible to correlate biochemical phenomenology with real-time, live cell biology.

One imaging technique that is especially applicable to questions aimed at understanding the early events near or on the plasma membrane in the initiation of B cell activation is total internal reflection fluorescence microscopy (TIRFM) (Groves *et al.*, 2008). TIRFM is a spatially limited technique in which the fluorescent signal of a specimen is confined to only ~100 nm from the coverslip. TIRFM takes advantage of a long understood, photophysical phenomenon in which light incident from a media of higher refractive index (immersion oil/glass) is totally internally reflected when it meets a media of lower refractive index (cell/sample media of specimen chamber). TIRFM uses a laser source to illuminate fluorophores. At the interface where the beam is reflected, an electromagnetic evanescent wave is produced that has the same properties as that of the incident excitation beam. The strength of the beam decays exponentially, penetrating to a depth of ~100 nm into the sample media; thus, fluorophores and autofluorescent molecules beyond the penetration depth are not excited, creating images with extremely high contrast (Groves *et al.*, 2008).

TIRFM is applicable to cells that flatten as they interact with surfaces, which recent evidence suggests is highly relevant to the B cell's encounter with Ag. Indeed, current evidence indicates that B cells likely encounter Ag, not in solution, but on the surface of Ag-presenting cells (APCs) *in vivo* (Carrasco and Batista, 2007; Junt *et al.*, 2007; Phan *et al.*, 2007; Qi *et al.*, 2006). B cells were first shown to avidly respond to Ag expressed on the surfaces of APCs *in vitro* and to form a highly organized contact area called the immunological synapse. Subsequently, B cells were shown to form immune synapses following activation with Ags anchored to planar lipid bilayers (PLBs) (Fleire *et al.*, 2006). Confocal microscopy provides the ability to image the BCR at the interface between a B cell and an APC in 3D; however, the temporal and spatial resolution is significantly limited as compared to TIRFM (Groves *et al.*, 2008). In imaging with TIRFM, the APC is replaced by coating a coverslip either directly with Ag or with a fluid PLB to which Ag is anchored. For our studies, we have chosen a fluid PLB for Ag presentation to understand the spatiotemporal dynamics of the BCR as the B cell binds to and transduces signals from Ag in the PLB (Sohn *et al.*, 2010). This model system can easily be modified to include the study of BCR coreceptors following interactions with their ligands in PLBs.

Here, we describe methods to measure the early events in the initiation of BCR signaling following the binding of Ag incorporated into PLBs. As detailed in a recent review (Pierce and Liu, 2010), studies from our lab and others indicate that a large percentage of BCRs (80%) are freely diffusing and highly mobile in the resting B cell membrane. Upon Ag binding, the BCRs first form submicroscopic oligomers through a mechanism that is dependent on the extracellular membrane proximal constant domain (Cμ4 or Cγ3) of the membrane Ig of the BCR, resulting in a sharp decrease in the long-range diffusion coefficient of the BCRs. This slowing of diffusion fits the theoretical predictions of oligomerization-induced trapping by the current picket fence model of the plasma membrane as proposed by Kusumi *et al.* (2005, 2010). Next, the immobile BCR oligomers are trapped into microscopic BCR microclusters that grow in area and in the number of BCRs as more and more oligomers accumulate. Oligomerized BCRs perturb the local membrane lipid microenvironment, leading to a transient association of the BCRs with raft lipids and then a more stable protein–protein interaction between the BCR and Lyn kinase. BCR signaling is then initiated from the microclusters via phosphorylation of the ITAMs of Igα and Igβ by Lyn kinase, and the subsequent recruitment of Syk kinase.

In this chapter, we describe how to prepare B cells and PLBs for imaging by TIRFM, provide a detailed description of the components necessary to build an in-house TIRFM system, and detail several live cell imaging protocols that we have developed in our lab to capture and analyze the early events in the initiation of BCR signaling. Since the BCR is a member of the multichain immune recognition receptor (MIRR) family that also includes the T cell receptor (TCR) and the high-affinity receptor for IgE (FcεR1), the methodologies described in this chapter for the BCR may also be applied to other MIRR family members with appropriate modifications.

2. Experiment Preparation

2.1. Cell preparation

In this section, we provide the procedures for preparing B cells for live cell imaging experiments. We highlight both human and mouse B cell preparations since both have been extensively used to study B cell signaling. We also address the use of primary cells versus cell lines, as each offer certain advantages depending on the biological question at hand.

2.1.1. Human or mouse primary B cells

Human PBMCs (peripheral blood mononuclear cells) are isolated by Ficoll density gradient centrifugation from lymphopack samples obtained from healthy donors under the appropriate institutional IRB. B cells are purified from PBMCs via negative selection magnetic cell separation using a human B cell isolation kit II from Miltenyi (Germany). Primary mouse B cells are similarly isolated by negative selection magnetic cell separation from single cell suspensions of spleens of mice. Ideally, for the study of Ag-specific responses, spleens should be obtained from transgenic (Tg) mice expressing Ag-specific BCRs, such as IgHB1-8/B1-8 Igκ$^{-/-}$ transgenic mice in which the B cells all express BCRs specific for the small hapten 4-hydroxy-3-iodo-5-nitrophenyl (NIP) (Tolar *et al.*, 2009). We have imaged mouse primary B cells either directly from spleens or after overnight culture with CpG and LPS (Calbiochem, Germany) and gene transfection (Liu *et al.*, 2010b,c).

2.1.2. B cell lines

Most of the commonly used B cell lines are amenable for live cell optical imaging studies. Linked with standard transfection procedures, B cell lines have proven to be useful tools in studies of B cell biology. We have established and characterized a variety of B cell lines by live cell imaging. To address the role of BCR affinity and isotype in BCR oligomerization and microcluster growth, we generated J558L B cell lines stably expressing Igα-YFP, Igβ, light chain Igλ1 and different versions of B1-8-IgH-CFP, an IgM and IgG version (Liu *et al.*, 2010b), a high and low affinity version (Liu *et al.*, 2010a), and mutant versions targeting the constant region (Tolar *et al.*, 2009), the transmembrane region, or the cytoplasmic domains (Liu *et al.*, 2010b). To study the perturbations in local lipid environments following BCR oligomerization, we generated CH27 B cell lines stably expressing Igα-YFP and the lipid raft probe Lyn16-CFP, or the non-lipid raft probe CFP-Ger or the Lyn kinase full length protein LynFL-CFP (Sohn *et al.*, 2006, 2008). To study the function of the FcγRIIB molecule, the human B cell line, ST486, and the mouse B cell line, A20II1.6, both negative for endogenous FcγRIIB receptor expression, were stably transfected with wild-type FcγRIIB-YFP (Liu *et al.*, 2010c).

2.1.3. Labeling B cell surface receptors

Prior to imaging experiments, it is essential to label the BCRs and other membrane receptors of interest using specific Abs conjugated to a suitable fluorophore. Due to the variation of Ab affinity and Ab fluorophore conjugation stoichiometries, it is necessary to determine the appropriate concentration of each Ab necessary to label the BCR by titration. Generally, cell surface proteins of interest are labeled by incubating approximately 1×10^6 cells with specific, fluorescently labeled Abs at 100–200 n*M* in 200 μL PBS for 10 min at 4 °C, followed by washing twice in PBS. It is important to use Ab Fabs for labeling cell surface proteins to avoid receptor crosslinking and internalization and inadvertent engagement of Fc receptors. The fluorophore-conjugated Fabs of specific Abs can be purchased from a number of vendors. If not commercially available, Fabs can be prepared from fluorescently labeled, intact Ab using a Fab micro preparation kit (Pierce, Rockford, IL), following the manufacturer's protocol. Prior to Ab fragmentation, Abs are conjugated with the desired fluorophore using AlexaFluor mAb labeling kits (Molecular Probes), following manufacturer's protocols. It is critical to select the proper Fab micro preparation kit for the subclass of the Ab to be fragmented. Ficin is commonly used to cleave IgG_1 Abs, whereas papain is used for Abs of other IgG subclasses. In addition, Protein A, G, or A/G-mediated purification of Ab Fcs should be selected to provide the best binding to the Fcs based on the Ab species and subclass. The Fab micro preparation kit protocols may yield Fab fragments in concentrations that are impractically low. In this case, the Fab fragment solution can be concentrated using a 10 kDa MWCO ultracentrifugation filtration device (Millipore, Billerica, MA), for example. All Ab fragmentations should be confirmed by SDS-PAGE in which Fab migrates to 45–50 and 25 kDa under nonreducing and reducing conditions, respectively. Appropriate SDS-PAGE controls are discussed in the Fab micro preparation kit protocol. Relatively low amounts of Fab are generated from the starting material, so silver staining offers a highly sensitive way to develop SDS-PAGE gels, detecting as little as 1 ng/band, such that it is reasonable to load only 100–200 ng of sample per lane for analysis.

2.2. Preparation of antigen-presenting planar lipid bilayers

Supported membrane bilayers have been commonly used in various functional biological studies. In B cell activation studies, Ag-containing PLBs have been used to mimic APCs. The Ags are tethered to the lipids in the PLBs and when prepared appropriately, the Ag is completely mobile and freely diffusing in the PLBs. To make fluid PLBs, an ultraclean support system is essential. For live cell imaging, glass chamber slides are generally used and this section describes the procedures for cleaning the glass supports. We also describe making the small unilamellar vesicle (SUV) stock solution used to prepare the PLBs and preparing Ag-containing PLBs.

2.2.1. Cleaning of glass items

The required materials include:

(1) *KOH/EtOH cleaning solution*: 2.1 *M* KOH, 85% EtOH;
(2) 5 mL clear glass vials with a V-shaped bottom (NextGen^{TM} V vial, Wheaton Science Products, Millville, NJ);
(3) Coplin jars with lids;
(4) *Rinsing solution*: 95% EtOH (Ethyl Alcohol 190 Proof-USP Grade);
(5) EZ-spread plating glass beads (MP Biomedicals, Solon, OH);
(6) Deionized ultrapure water, which has a resistivity of 18.2 MΩ-cm and total organic contents less than 5 parts per billion (ppb);
(7) 100% EtOH (Ethyl Alcohol 200 Proof-USP Grade).

The glass items are cleaned as follows:

1. Immerse all glass items in freshly prepared KOH/EtOH cleaning solution in a 2 L beaker for 10 min. Make sure that the vials are completely filled with the solution without air bubbles.
2. Transfer and similarly immerse the items in the rinsing solution (95% EtOH) in a 2 L beaker and moderately swirl several times.
3. Rinse the items thoroughly under a flow of ultrapure deionized water with vigorous shaking.

4. Put all the glass items in a detergent-cleaned beaker and bake at 160 °C for 1 h.
5. For vial cap washing, immerse all caps in 100% EtOH, vigorously shake for a few minutes, and then rinse three times with a copious amount of ultrapure deionized water.
6. Put the caps in a detergent-cleaned beaker and completely dry at 60 °C.

2.2.2. Preparing SUVs

The required materials include:

(1) 25 m*M* of 1,2-dioleoyl-*sn*-glycero-3-phosphocholine (DOPC) (Avanti Polar Lipids, Alabaster, AL), 10 m*M* of 1,2-dioleoyl-*sn*-glycero-3-phosphoethanolamine-cap-biotin (DOPE-cap-biotin) for the use of biotinylated Ags, or 10 m*M* of 1,2-dioleoyl-*sn*-glycero-3-[*N*(5-amino-1-carboxypentyl) imino-diacetic acid]-succinyl (Nickel Salt) (DOGS-Ni-NTA; Avanti Polar Lipids) for the use of His-tagged Ags;
(2) Hamilton syringes (Hamilton, Reno, NV);
(3) High purity chloroform (Burdick & Jackson, Muskegon, MI);
(4) Compressed argon gas;
(5) Nanostrip (OM Group Ultra Pure Chemicals Ltd, Derbyshire, UK);
(6) Water bath-type sonicator (Model: G112SP1G, Laboratory Supplies, Hicksville, NY);
(7) Ultra-Clear 13 × 51 mm ultracentrifuge tubes (Beckman Instrument Inc., Palo Alto, CA) and Beckman rotor, SW55Ti (Beckman Instruments Inc.).

SUVs are prepared as follows:

1. During SUV preparation, the glass items cleaned as described above must be used. Clean the Hamilton syringe using chloroform and 100% EtOH. In a fume hood, for each lipid to be measured fill three vials with EtOH and three vials with chloroform. In an EtOH vial, move the syringe piston up and down several times and repeat the process sequentially in the next two EtOH vials. Repeat the process in the chloroform vials. Air-dry the syringes on Kimwipes.
2. To make SUVs of DOPC and DOPE-cap-biotin to which biotinylated Ags or other molecules can be tethered, mix in a vial 1 mL of 25 m*M* DOPC and 25 μL of 10 m*M* DOPE-cap-biotin, resulting in a 100:1 M ratio of DOPC:DOPE-cap-biotin in a 5 m*M* final concentration of DOPC solution. For tethering His-tagged Ags or other molecules to SUVs, make SUVs of DOPC and DOGS-NiNTA by mixing 1 mL of 25 mM DOPC with 250 μL of DOGS-Ni-NTA, giving a 10:1 *M* ratio of DOPC:DOGS-Ni-NTA. Tighten the cap and mix by briefly shaking. Immediately fill the vials containing the stock lipid solutions with argon gas to avoid exposure to air and store the lipid stocks at −20 °C until proceeding with the next step.

3. Dry the lipid mixture with a gentle stream of argon gas (~5 psi) while slowly rotating the vial until the solution is dry and a thin film of lipid forms on the walls of the vial. Be careful not to splash the mixture up the walls of the vial. For the drying process, a Pasteur pipette can be connected to the argon tank regulator via plastic tubing. After completely evaporating the residual chloroform, dry the vial in a vacuum at room temperature (RT) overnight.
4. Hydrate the lipid film by adding 5 mL of degassed PBS freshly prepared at RT. Degas the PBS in a Nanostrip-cleaned and oven-dried side arm flask. Immediately after hydrating, fill the vial with argon gas, cap tightly, and vortex for about 30 s until the lipid film is completely dissociated from the wall and makes an opaque solution. From this step forward, keep this 5 m*M* SUV stock solution on ice.
5. Sonicate the lipid mixture using a water bath-type sonicator, maintaining a bath temperature of <4 °C. Perform the sonication in 10 min rounds until the lipid mixture becomes clear, indicating the formation of SUVs. In between rounds of sonication, pause for 10 min to avoid overheating the lipid solution.
6. To clarify the SUV solution, ultracentrifuge it using a Beckman SW55Ti rotor. First spin at 4 °C for 1 h at 46,800 × *g* and then transfer the supernatant containing the SUVs into a new tube using a sterile pipette and ultracentrifuge at 4 °C for 8 h at 54,700 × *g*.

   Note: Save aliquots of the SUV solution before each stage of ultracentrifugation to determine the concentration of the final SUV stock solution as described in Step 7.
7. Filter the SUV mixture is filtered through a 0.2 μm-syringe-type Whatman filter for further purification; the resulting concentration of lipid solution is about half of the original concentration as determined by measuring the OD at 234 nm using a UV–Vis spectrophotometer. Carefully transfer the supernatant into a syringe tube and filter through to a new 15 mL polypropylene conical tube, fill the tube with argon gas, cap, and seal with parafilm. The SUVs stock solution is ready for use.

   Note: Be sure to save an aliquot of the SUV mixture before filtration so that the OD can be compared with the OD of the mixture after filtration. The SUVs are stable for several months at 4 °C under argon gas. To minimize repeated exposure of the SUV stock to the air and to prolong stability, make aliquots in 15 mL tubes under argon and seal with parafilm.

2.2.3. Making PLBs and tethering antigens to PLBs

Here, we describe tethering the model Ag, NIP, or a surrogate Ag, anti-Ig to PLBs. The method can be generalized to attach any biotinylated or His-tagged molecule to biotin- or nickel-containing PLBs, respectively.

The required materials include:

(1) 24 × 50 mm #1.5 cover glass;
(2) 8-well Lab-Tek chamber #1.0 Borosilicate coverglass (Cat. No. 155411, Nalge Nunc International, Rochester, NY);
(3) Sylgard 164 Silicone Elastomer adhesive (Dow Corning Corp., Midland, MI);
(4) Biotinylated goat anti-mouse (or human) IgM (or IgG) $F(ab')_2$ (Jackson Immunoresearch, West Grove, PA), hapten NIP-conjugated peptide NIP1-anti-ASTGKTASACTSGASSTGS-His12 (NIP1-His12), or NIP1-His12 peptide coupled to Hylight647 through its cysteine residue (NIP1-His12-Hylight 647) (Anaspec, San Francisco, CA). The hapten–peptide conjugates are HPLC purified and verified by mass spectroscopy with >95% purity;
(5) Imaging buffer: 1× Hank's balanced salt solution (HBSS) containing 0.1% FBS (Gibco, Carlsbad, CA).

Ag is tethered to PLBs as follows:

1. Clean three Coplin jars with Nanostrip. Fill one with Nanostrip, one with deionized ultrapure water, and one with 100% EtOH.
2. Place 24 × 50 mm #1.5 coverslips into the jar containing Nanostrip for at least 1 h and then transfer the coverslips into the jar containing deionized ultrapure water. Shake the jar several times and discard, fill with water, and repeat the process 10 times with fresh ultrapure water. Then transfer the coverslips to the jar containing 100% EtOH for several minutes. Take out a coverslip and blow dry completely with argon gas. Place the coverslip on top of a monolayer of KOH/EtOH-cleaned glass beads (see Section 2.2.1).
3. Tear off the bottom coverslip of an 8-well Lab-Tek chamber and replace with a Nanostrip-cleaned coverslip using Sylgard 164. Fill the channels at the base of each well with Sylgard using an 18-G needle and syringe, place the Nanostrip-cleaned coverslip over the bottom of the chamber, and apply gentle pressure to the channels, allowing 30 min for adhesion and drying. Prepare chambers on the day of use as dust will reaccumulate and prevent the formation of high quality PLBs.
4. To make PLBs on the chamber coverglass, make a 0.1 m*M* final concentration SUV working solution in PBS, dispense 200 μL into each of the chamber wells, and wait for 10 min. Rinse the PLBs with about 20 mL of PBS per well, keeping the bilayers under the solution at all times. Fill the chambers to the top with PBS after the last rinse (each chamber holds approximately 700 μL).

   Note: It is important to maintain PLBs under hydrated conditions to avoid oxidation-induced modifications to the PLBs as reported (Plochberger *et al.*, 2010).

5. For tethering the biotinylated Ag to the PLBs, remove 400 μL of PBS from the filled wells, add 250 μL of streptavidin in PBS at 0.1 μ*M*, mix twice by gently pipetting up and down with a micropipette, leave for 10 min, and repeat the washing described in Step 4. Add biotinylated Ag at a final concentration of 100 n*M*, incubate for 20 min, and wash unbound Ag by repeating Step 4.

   Note: To remove aggregated Ags, spin the Ag solution at maximum speed for at least 10 min prior to adding it to the chambers.
6. For tethering His-tagged Ag to the PLBs, skip the step of adding streptavidin and instead directly add His-tagged Ag as described in Step 5.
7. The mobility of Ag-tethered PLBs should be assessed before imaging experiments. We check the quality of PLBs by examining the mobility of PLB-tethered, fluorophore-conjugated Ags, such as NIP1-His12-Hylight647. The mobility of fluorophore-conjugated Ag molecules on PLBs is quantified using single molecule tracking-based analysis tools as detailed in Section 3.1 or in our previous publications (Liu *et al.*, 2010a; Tolar *et al.*, 2009).
8. Immediately before imaging, exchange PBS with imaging buffer by washing the chamber with 10 mL of imaging buffer.

   Note: PLBs are less stable in the imaging buffer that contains serum, so use within 1–2 h if possible.

2.3. TIRF microscope design

The popularity of TIRFM over recent years has encouraged major microscope manufacturers to introduce sophisticated "turn-key" TIRFM systems. Such systems are a tremendous boon to core microscopy facilities but offer little room for in-house modifications in laboratories with unique requirements. Implementation of a through-lens TIRFM system to an existing inverted microscope base is relatively straightforward; it requires access to the rear excitation port (typically where the mercury arc lamp is attached), knowledge of the focal length of the microscope base's tube lens, which, in turn, allows the user to match an external lens to focus the laser beam at the back focal plane of the objective lens.

Here, we discuss some general considerations and list the components for our two modified, inverted Olympus IX81 microscope systems that were used to perform the live cell experiments described in this chapter. The components are discussed in order of excitation to image detection. The principles of TIRFM and detailed instructions on the assembly of a TIRF microscope are beyond the scope of this chapter; however, useful information can be found within the following references: Axelrod (1981, 1989, 2001, 2003, 2008); Mattheyses *et al.* (2010).

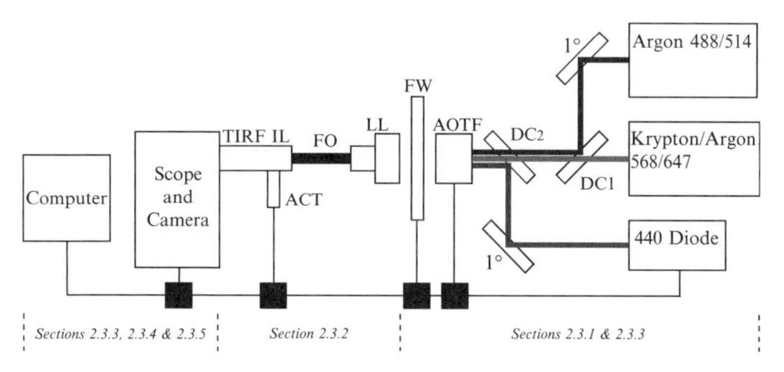

Figure 14.1 System design. Shown is a generalized diagram of one of the TIRFM rigs used in the experiments described. The pertinent sections that discuss the specific components are noted beneath the figure. All optical components to the right of the fiber optic cable (FO) are mounted on a metal breadboard with ¼″ screws on standard post mounts. Three lasers (boxes on right) are used to provide five usable excitation lines (wavelengths in nm indicated). After reflection on a primary mirror (1°), the argon laser lines are directed to the AOTF via a dichroic mirror (DC1). The krypton–argon lines pass through both DC1 and DC2. After reflection on a primary mirror (1°), the 440-nm diode laser line is directed to the AOTF via DC2. Power at the 440-nm laser head is computer controlled. Control boxes are required to interface with the computer software and are depicted as black boxes. Thin black lines at the bottom of the figure indicate computer connections. The AOTF and excitation filter wheel (FW) are linked to their respective control boxes, which in turn are coupled to the PC workstation and are controlled by MetaMorph acquisition software. The user selects the laser line via the AOTF and blocks unneeded lines with the filter wheel. The selected line is directed to the laser launch (LL) lens to allow entry into the fiber optic cable. The fiber optic cable is coupled to the TIRF illuminator (TIRF IL). The TIRF angle is controlled through the software via a motorized actuator (ACT). (For color version of this figure, the reader is referred to the Web version of this chapter.)

2.3.1. Excitation light source

Continuous wave lasers provide the illumination source for our TIRFM systems. Typical laser types include gas, diode-pumped, and straight diode; all can be used in combination. The lasers are mounted on an optical breadboard with their polarity and vertical beam heights matched so that they can be combined into a single-mode fiber optic cable, which carries the light to the TIRFM illuminator on the microscope as schematically illustrated in Fig. 14.1.

Mixed gas lasers provide an economic means to obtain multiple laser lines from a single device. Wavelength selection is accomplished using a software-controlled, acousto-optical tunable filter (AOTF). Because of the inadequate blocking power of the AOTF and the fact that gas lasers produce multiple usable (and unusable) wavelengths, extraneous excitation light must be blocked from reaching the extremely sensitive cameras used in TIRFM with "cleanup" excitation filters. On our systems, the cleanup filters are placed in a software-controlled filter wheel (FW) on the breadboard after the AOTF to provide versatility and to avoid reflection artifacts that occur if placed in the traditional location within the dichroic beam splitter housing (Fig. 14.1) (see Section 2.3.3).

Straight diode and diode-pumped solid-state lasers can be up to 10 times smaller in size and are a noise- and heat-free alternative relative to their gas-driven equivalents. Straight diode lasers have a longer life expectancy relative to gas lasers whose tubes must be replaced (usually at a third of the full cost) every 2–3 years. Straight diodes can be modulated by software control, bypassing the need for an AOTF and thus can be directly linked to a fiber optic cable either at the laser head or, as in the case of our system, mirrored directly into a fiber optic coupler. It is important to note that the square beam shape of a typical diode laser usually results in an unavoidable and significant loss of power throughput at the point of fiber optic coupling. Diode-pumped lasers use a different technology to produce their monochromatic lines and their output must be modulated through an AOTF.

2.3.1.1. Component list (also shown in Fig. 14.1)

(1) Optical table components and design (Solamere Technology Group, Salt Lake City, UT);
(2) 440 nm, 50 mW, straight diode laser (Blue Sky Research, Milpitas, CA);
(3) 488 and 514 nm, 300 mW argon gas laser, ~70 mW each line (Dynamic Laser, Salt Lake City, UT);
(4) 568 and 647 nm, 80 mW krypton/argon gas laser ~20 mW each line (Dynamic Laser);
(5) Filter wheel and control box (Applied Scientific Instrumentation, Eugene, OR);
(6) AOTF and control box (NEOS Technologies, Melbourne, FL);
(7) Fiber optic launch (OZ Optics, Ottawa, Ontario).

2.3.2. TIRFM illuminator port

The heart of the TIRF microscope is the illuminator port. While illuminator port designs can be complex, all serve the same fundamental purpose: to direct and focus the diverging beam that exits the fiber optic at the back focal plane of the objective lens. A simple two-lens configuration is used in the Olympus illuminator port and is also found in many systems designed in-house (Fig. 14.1). An internal 200 mm focal length tube lens is positioned just before the dichroic beam splitter turret and an external 100 mm lens is present in the TIRFM illuminator. The lenses are held at a fixed distance to one another. For a simplified in-house system, the external illuminator lens can be placed on a simple post mount in line with the rear microscope port. At the entrance of the illuminator, the fiber optic cable is coupled to an adjustable mount that centers the beam in the light path. The mount is manually translated horizontally along a slider to focus the beam at the back focal plane of the objective lens. The beam is collimated as it exits the objective and, if projected on the ceiling, will produce a circular spot when of good quality (see Section 2.3.3). To achieve TIR, the mount's height is lowered vertically with a manual micrometer to translate the excitation beam from the center to the side of the objective lens to angle the beam as it exits the lens. In the simplified system, a fiber optic coupling mount can be assembled on a manual axis positioner mounted to the microscope table to move the assembly in the x, y, or z plane.

The disadvantage of a single port illuminator system is that beam focus and optimal TIR angle can be set for only one excitation wavelength at a time. More sophisticated TIRFM illuminators now provide independent coupling mounts for multiple fibers that are motorized. To increase the versatility of our system, we replaced the manual TIRF angle micrometer with a motorized micrometer and replaced the internal tube and external illuminator lens with achromatic lenses of the same focal length. The motor enables rapid, software controlled angle switches that are necessary for experiments that require a specific penetration depth for all excitation wavelengths and the achromatic lenses abrogate the need to compromise focus between different excitation wavelengths. However, it is important to recognize that while small motor movements are relatively fast in multi-channel imaging experiments, extremely rapid, stream acquisitions can only be achieved by TIRFM illuminators that have multiple fiber coupling mounts if the TIR angle must be optimized between wavelengths.

2.3.2.1. Component list

(1) Fiber optic cable, type FC/APC (OZ Optics);
(2) Achromatic lens FL100 mm, JML 78903; FL200 mm, JML 32701A (JML Optical, Rochester, NY);
(3) Motorized actuator (ACT) to control TIRF angle (Thor Labs, Newton, NJ);
(4) Actuator controller (Applied Scientific Instrumentation).

2.3.3. Filters, mirrors, and lenses

Quality dichroic beam splitters and excitation, emission, and notch filters are essential for successful TIRFM and are available from a variety of manufacturers. Of note, newer, so-called sputter designs are brighter and more durable than traditional coated glass surfaces and when available, should be the glass of choice when assembling a new system. Reflection artifacts unique to TIRFM can be avoided at the outset by directly communicating with the filter manufacturer because specialty matched sets and less common designs are not always available for self-choice on websites or in catalogs.

2.3.3.1. Component list

(1) Excitation filters, dichroic beam splitters, and emission filters (Chroma Technology, Bellows Falls, VT);

(2) Notch filters (Semrock, Rochester, NY);
(3) Objective lenses: 60×/1.45 NA PlanApoN, 100×/1.45 NA PlanApo, 150×/1.45 NA UApo (Olympus, Center Valley, PA).

2.3.3.2. Excitation/cleanup filters As noted in Section 2.3.1, the excitation/cleanup filters are located on a software-controlled, motorized wheel located on the laser breadboard. Because the AOTF only blocks light to an OD of 10^4, these filters are necessary to completely block all unwanted excitation wavelengths from gas lasers and side bands from some diodes. Note that using cleanup filters comes at a cost of losing up to 5% of the excitation power before the laser is launched into the fiber optic cable.

2.3.3.3. Dichroic beam splitters For single particle imaging experiments where photons on both the excitation and emission side are at a premium, dichroic beam splitters that reflect at a single wavelength are optimal. However, for multicolor imaging, it is best to use beam splitters that are designed to reflect all excitation wavelengths required for the experiment. Switching between beam splitters to reflect the necessary wavelength to the specimen is costly in time, will likely result in a pixel shift in the image overlay, and will most likely change the TIR angle as explained below.

Reflection artifacts are a challenge in TIRFM and it is necessary to ensure that the beam splitter rests at a 45° angle in the housing cube. Standard Olympus cubes are equipped with a clip under which the rectangular beam splitter is placed. The beam splitter is fixed in place with setscrews. If the setscrews are overtorqued (even very slightly), bending of the beam splitter's surface will occur, which, in turn, disrupts the reflection of the excitation beam; this is readily observed by viewing the projected collimated beam on the ceiling. Bowed beam splitters distort the circular spot. Certain filter manufacturers who are aware of these issues now glue the beam splitter into its housing and use thicker glass substrates to compensate for stress.

2.3.3.4. Emission filters, notch filters, and emission splitters Emission filters serve two important functions: to provide the desired band or long-pass emission spectra and to block the excitation light to the camera. Under conditions where signal from the specimen is extremely low and the excitation power is high, we found it necessary to use notch filters to provide additional blocking against the excitation source to raise the signal to background ratio. Notch filters are designed to block with an extremely narrow stopband (~10 nm in width) yet allow >95% of the desired wavelengths to the camera. In our system, the notch filters are placed in the emission filter port of the beam splitter housing. The filters can be purchased without metal collars to fit multiple filters in one port.

To accommodate rapid channel switching, emission filters are best placed in a software-controlled filter wheel. If extremely fast stream imaging is required for multiple channels, the emission spectra from the specimen can be split using a dichroic beam splitter in the emission light path to simultaneously project each channel to the respective right and left half of the camera's CCD. Emission splitting devices available on the market can be threaded between a camera and filter wheel via a C-mount. The splitters provide a cassette that houses a beam splitter and ports for emission filters. Optical alignment of separated channels on the CCD takes great care and is never perfect throughout an entire field of view due to spherical aberration. Post-acquisition alignment is frequently necessary and, therefore, a good practice is to capture reference images of fixed fluorescent beads to calibrate the image analysis software. If a third-party objective lens is used on the microscope system, it is also important to check for a focus shift due to chromatic aberration by acquiring a Z-series of beads that are fluorescent in both channels. If the beads do not peak in intensity in the same Z-slice, the chromatic shift is calculated by distance between peaks and is corrected with a lens placed in either one of the emission filter ports of the splitter. If the microscope comes equipped with a focus motor, an alternative would be to offset the focus for one channel to adjust for the chromatic shift during image acquisition.

2.3.3.5. Objective lenses Depending on the manufacturer, objective lens magnification ranges from $60\times$ to $150\times$ for TIRFM. All TIRFM lenses have an extremely high numerical aperture (NA >1.4), that is necessary to achieve the critical angle required for TIR. The larger the NA, the greater the working range for critical angle adjustment.

2.3.4. Ancillary equipment

2.3.4.1. Component list

(1) ZDC infrared focus control laser (Olympus);
(2) LiveCell environmental chamber (Pathology Devices, Westminster, MD);
(3) Isolation table (Technical Manufacturing Corporation, Peabody, MA).

2.3.4.2. Focus control Because the focal plane in TIRFM is extremely narrow ($\sim$100 nm), focus control becomes a key issue for time-lapse acquisitions. Now most microscope manufacturers provide a peripheral infrared laser system to provide continuous focus using the coverslip as a point of reference. The latest generation of these autofocus laser systems can even be used during a continuous stream acquisition.

2.3.4.3. Temperature control Additional challenges are faced if the live specimen must be maintained above ambient temperature during imaging. The specimen chamber can be heated directly in a closed-chamber system, but changes (stimulation, alteration in media, and addition of cells) to the system must be made by perfusion. Open chambers with coverslip bottoms offer more versatility because changes can be made with a pipette. Open chambers are generally heated in two ways: by enclosing the microscope in a heating chamber or by using a stage-top heating chamber in conjunction with an objective lens heater. Enclosed microscopes are better at maintaining consistent temperature but are slow to heat and cool. Stage-top devices heat rapidly but a lid must be removed to gain access to the specimen, making an autofocus system very useful for time-lapse experiments. In addition, stage-top devices require the lens to be heated separately using a collar that transfers heat to the lens barrel. It is critical that the temperature of the objective lens be equilibrated to the specimen chamber; otherwise the specimen under observation will quickly drop to the temperature of the lens (an oil objective is assumed here).

2.3.4.4. Vibration An isolation table is a necessity. Building and equipment vibration can cause significant blur in TIRFM applications, especially when imaging particles in motion at fast frame rates that are below the optical resolution of the microscope. In addition, care must be taken not to translate vibration from cooling fans present in the microscope's electronic control equipment to the imaging platform by hanging wire harnesses. In some cases, satellite cooling may be required to replace an on-board camera fan.

2.3.5. Image acquisition hardware and software

2.3.5.1. Component list

(1) EMCCD Cameras: Cascade IIB 512 and Cascade II 1024 (Princeton Instruments, Trenton, NJ);
(2) Acquisition software: MetaMorph (Molecular Devices, Downingtown, PA); Analysis software: MetaMorph; Matlab (Mathworks, Natick, MA); ImagePro (MediaCybernetics, Bethesda, MD); ImageJ (rsbweb.nih.gov/ij/);
(3) Acquisition: PC workstations assembled by Molecular Devices and Solarmere Technologies; Analysis workstations (Optiplex, Dell, USA).

2.3.5.2. Camera The advent of high sensitivity/low noise electron-multiplying (EM) CCD technology has revolutionized wide-field fluorescence imaging to allow for video rate image acquisition and detection of single fluorophores in live biological specimens. The drawbacks of an EMCCD are the chip's large pixel size (13–16 μm), reducing its resolving power for certain biological structures, and its high cost. Worthy of attention is the rapidly improving, significantly less expensive sCMOS technology for scientific imaging, with chip pixel sizes ~2.5 times smaller than a typical EMCCD.

2.3.5.3. Software There are several software companies that sell image acquisition packages that can operate just about any microscope and its peripheral components and some of the microscope manufacturers offer their own software solutions. In addition, a well-supported, open-source option to consider is Micro-Manager (http://valelab.ucsf.edu/~MM/MMwiki/index.php/Micro-Manager).

2.3.5.4. Computer and data storage Because of Moore's law (http://en.wikipedia.org/wiki/Moore's_law), it is futile to make recommendations for computer specifications. However, investing in a workstation that maximizes useful RAM, processor speed, and video card capability is paramount. While the cost of hard drive space has dropped significantly, the temptation to store data on the acquisition workstation should be avoided. Acquisition programs are just beginning to take advantage of the multithreading capability of modern 64-bit systems. If the software is designed for a 32-bit system, it is unnecessary to go beyond 4 GB of RAM.

3. Image Acquisition and Analysis

In this section, we describe the methods used to acquire and analyze TIRF imaging data pertaining to the very early events in B cell signaling initiation, including BCR oligomerization, growth of BCR microclusters, and the subsequent activation of various BCR signaling-associated molecules.

3.1. Imaging B cell receptor oligomerization by single molecule tracking

To investigate the changes in the BCR upon Ag binding that result in oligomerization and microcluster formation, we tracked single BCRs in TIRFM. Our published results from these analyses indicated that on resting cells, the majority of BCRs exist as highly mobile, freely diffusing monomers that are not receptive to oligomerization (Tolar *et al.*, 2009). With Ag binding, a force is exerted that brings the BCRs into an oligomerization-receptive form such that random bumping of Ag-bound BCRs results in oligomerization through the exposed membrane-proximal constant domain (Cμ4 or Cγ3) of membrane Ig. Upon oligomerization, the diffusion of the BCR drops dramatically such that the BCRs essentially become immobilized as they are trapped within areas of confinement within the plasma membrane (Tolar *et al.*, 2009). Thus, BCR oligomerization can be measured as a change in the diffusion behavior of the BCRs, as measured by single molecule TIRF imaging.

1. B cells are labeled for single molecule TIRF imaging by incubating 1×10^6 cells with fluorescently labeled BCR-specific Fab fragments at subnanomolar concentration (25–250 p*M* or 1.25–12.5 ng/mL range) in 200 μL PBS for 10 min at 4 °C. The concentration of labeling Ab should be low enough that single BCR molecules can be visualized without the

need for photobleaching (Tolar *et al.*, 2009). Single molecules show a single-step fluorescence bleaching profile, providing a diagnostic test for the single molecule images.

2. Cells are washed three times in PBS after labeling, resuspended in 200 μL PBS, incubated with Ag-presenting lipid bilayers for 5–10 min (50 μL cells per bilayer chamber), and imaged with 5 mW of laser (at the objective lens in epifluorescence mode).
3. To acquire images of single BCR molecules, a 100 × 100 pixel Region of Interest is imaged for an exposure time of 35 ms using a streamline acquisition mode for 300 frames over approximately 10 s. A typical movie showing the mobility of single BCR molecules is provided as Supplementary Movie 1 (http://www.elsevierdirect.com/companions/9780123918567).
4. The processing of single molecule TIRF videos was detailed previously (Tolar *et al.*, 2009), with tracking and analysis performed using Matlab (Mathworks) code based on available tracking algorithms (Douglass and Vale, 2005, 2008, available at http://physics.georgetown.edu/).
5. Trajectories are visually inspected and occasional mistracking is corrected manually. The trajectories of the single BCR molecules imaged in Supplementary Movie 1 are shown in Fig. 14.2A. A 2D-Gaussian fit is used to refine the positions of the diffraction-limited spots in the trajectories. Mean square displacement (MSD) and instant diffusion coefficients for each BCR trajectory are calculated from positional coordinates as described by Douglass and Vale (2005, 2008).
6. Short-range diffusion coefficients are calculated from linear fits to MSD data of individual molecules for the time intervals 35–140 ms and plotted as cumulative probability distribution graphs, as shown in Fig. 14.2B for BCR molecules on human peripheral blood B cells placed on bilayers with or without Ag. BCR molecules with a diffusion coefficient $\leq 0.01\ \mu m^2/s$ are considered immobile.

3.2. Imaging B cell receptor cluster dynamics

Following Ag binding-induced BCR oligomerization, a microscopic structure termed the BCR microcluster is formed. BCR microclusters are believed to be the fundamental platform for initiating BCR signaling. To have a better understanding of B cell signaling activation, it is critical to image the dynamics of BCR microclusters and their coordination with downstream BCR signaling proteins. Here, we describe the general protocols for imaging the dynamic growth of BCR microclusters by two-color TIRFM using the example of NIP-specific J558L B cells expressing B1-8-High BCR, which is composed of Igα-YFP, Igβ, and B1-8-High-IgM-CFP (IgM-High J558L B cells). We summarize the fluorophores commonly used in combination for multiple-color TIRF imaging of B cell activation (Table 14.1).

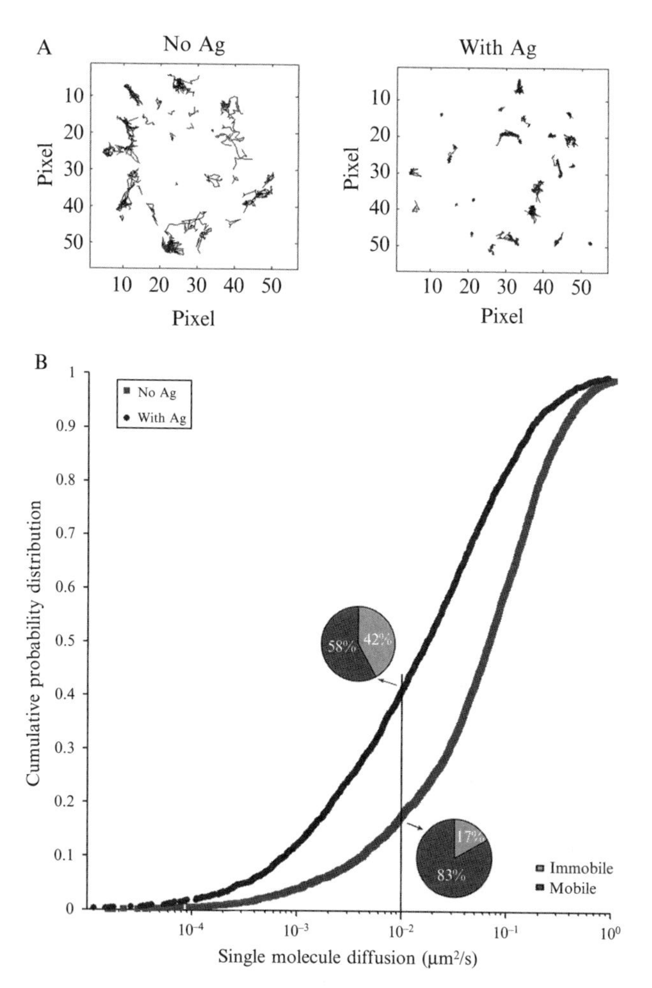

Figure 14.2 Imaging BCR oligomerization by single molecule TIRFM. (A) Trajectories of individual BCR molecules accumulated over the entire time course of Supplementary Movie 1. (B) Cumulative probability plots of the diffusion coefficients of individual BCR molecules were obtained from time-lapse TIRF movies (such as those shown in Supplementary Movie 1, http://www.elsevierdirect.com/companions/9780123918574) of human peripheral blood B cells labeled with DyLight 649-Fab anti-IgM and placed on bilayers with (blue curve) or without (red curve) goat anti-human IgM $F(ab')_2$ Ag. The trajectories used to construct each probability curve were collected from two independent experiments ($n = 1871$ with Ag; $n = 3622$ without Ag). Also given are the percent of mobile and immobile BCRs for cells with and without Ag. (See Color Insert.)

Table 14.1 Fluorophores used in multiple-color TIRFM for imaging B cell activation

Probes	Excitation (nm)	Emission (nm)
CFP	433	475
eGFP	488	507
YFP	514	527
AlexaFluor 488	499	519
AlexaFluor 568	579	603
AlexaFluor 647	652	668

This is a table of some commonly used fluorophores for TIRFM imaging. Peak excitation and emission wavelengths of each fluorophore are given. A combination of two or three colors can be used in TIRFM imaging depending on the availability of an appropriate dichroic beam splitter and excitation and emission filters.

Note: There are some analogous fluorophores that exist on the market. For example, AlexaFluor 647 is an analogous fluorophore to Dylight 649 and Hylight 647. These analogous fluorophores have comparable excitation and emission spectra and thus can be substituted in imaging experiments depending on the availability.

1. To prepare enough B cells for four imaging chambers, 1×10^6 IgM-High J558L B cells are labeled with 200 n*M* AlexaFluor 568-Fab anti-IgM in a total volume of 200 μL for 10 min at 4 °C. Cells are washed twice in PBS and resuspended in 200 μL PBS for imaging.
2. 50 microliters of IgM-High J558L B cells from Step 1 (about 0.25×10^6 B cells) are loaded onto NIP1-His12 Ag-containing PLBs, the preparation of which is detailed in Section 2.2.
3. Beginning with initial cell contact with the PLBs, the cells are examined every 2 s by collecting TIRF images of Igα-YFP and AlexaFluor 568-Fab anti-IgM over a total of 2 min (as shown in Fig. 14.3).

   Note: The acquisition is controlled by Metamorph software using its multiple dimensional acquisition mode (Molecular Devices), with a 100 ms exposure time for both images. It is also important to consider minimizing bleaching effects by titrating the output laser power.
4. The acquired TIRF image movies are analyzed and processed using a combination of Image Pro Plus (Media Cybernetics) and Matlab (Mathworks) software as described below. Similar to the analyses of single BCR molecules, as introduced in Section 3.1, Igα-YFP or AlexaFluor 568-Fab anti-IgM TIRF image movies are processed by a Matlab code based on available 2D-Gaussian positional fitting and tracking algorithms (http://physics.georgetown.edu/matlab/) as shown in Fig. 14.3. Briefly, individual BCR microclusters in each TIRF image are fit to a typical 2D Gaussian function by means of least squares. In case the profiles of some microclusters are not perfectly circular, the Gaussian function was allowed to adopt an elliptical shape (Holtzer *et al.*, 2007).

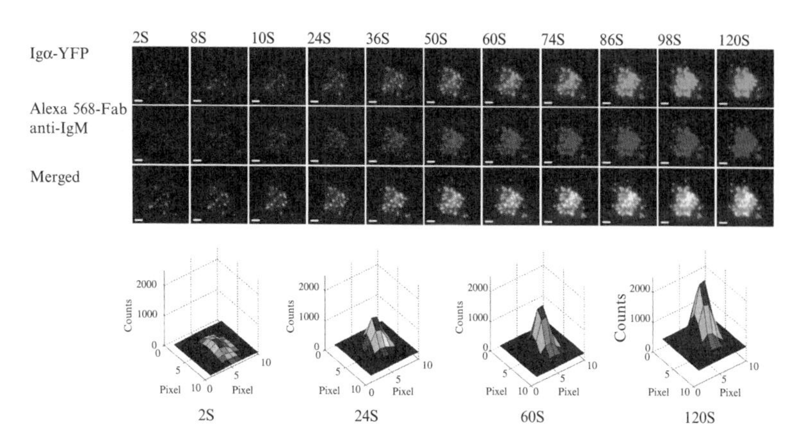

Figure 14.3 Imaging BCR cluster dynamics. Two-color time-lapse TIRF images capture the response of B cells upon initial contact with Ag-containing PLBs. Upper panels: IgM-High J558L B cells labeled with AlexaFluor 568-Fab anti-IgM were placed on PLBs containing NIP1-His12 Ag and were examined by TIRFM over 120 s by imaging Igα-YFP (green) and AlexaFluor 568-Fab anti-IgM (red). Bars, 1.5 μm. Bottom panel: pseudo-colored 2.5D Gaussian images of one representative AlexaFluor 568-Fab anti-IgM microcluster are shown for the indicated times. (For interpretation of the references to color in this figure legend, the reader is referred to the Web version of this chapter.)

$$f(x, y) = z_0 + I\frac{4\ln 2}{\pi\sigma_r^2} \times \exp\left\{-\left[4\ln 2\left(\frac{(x - x_c)^2}{\sigma_r^2/\varepsilon^2} + \frac{(y - y_c)^2}{\sigma_r^2\varepsilon^2}\right)\right]\right\}.$$

For each microcluster, the fit will yield the parameters including local background fluorescence intensity (FI) (z_0), position (x_c, y_c), integrated FI (I) indicating the brightness of the cluster, and generalized full-width at half-maximum peak height (σ_r) of the intensity distribution indicating the size of the cluster. The tracking function of the Matlab code is able to link both the FI and size information to each individual BCR microcluster trajectory.

Note: In our case, only the first 120 s of each track of the BCR microclusters from IgM-High J558L B cells or only the first 40 s of each track from B1-8 primary B cell microclusters are selected for full analyses. This is necessary to avoid microcluster tracking and 2D Gaussian fitting errors, both of which arise from BCR microclusters merging and/or overlapping at later time points of the observed processes. Tracking of BCR microclusters that are in close proximity is not feasible and 2D Gaussian fitting is reliable only for well-separated BCR microclusters. For the FI and size values of each BCR microcluster trajectory, values belonging to the same track are normalized to the first position of each track. Subsequently, the arithmetic means and standard errors of the values are calculated and plotted over the imaging time. The statistical test used to compare the kinetics of microcluster growth is performed as previously described (Baldwin *et al.*, 2007; Elso *et al.*, 2003) or is available through the online server at http://bioinf.wehi.edu.au/software/compareCurves/index.html.

Note: A second live cell imaging technique used to observe BCR microcluster dynamics is FRET-based TIRF imaging, as we have reported previously (Sohn *et al.*, 2008; Tolar *et al.*, 2005). Briefly, FRET donor CFP and acceptor YFP are coupled to the cytoplasmic domain of Igα or mIgH within the BCR complex. By examining the FRET changes upon BCR recognition of Ag using live cell time-lapse imaging, we are able to demonstrate the dynamic umbrella opening-like conformational changes within the cytoplasmic domains of BCRs comprising microclusters. We refer the reader of interest to our early publications for additional details (Sohn *et al.*, 2008; Tolar *et al.*, 2005).

3.3. Imaging B cell receptor signaling

BCR signaling is initiated from BCR microclusters. In this section, we describe how to quantify the recruitment of intracellular kinases, adaptors, and phosphatases in the BCR signaling pathway into BCR microclusters.

1. B cells are prepared as described in Section 2.1, including the labeling of any cell surface proteins of interest with specific, fluorescently tagged antibodies.
2. 2.5×10^6 cells in 50 μL of HBSS with Ca^{2+} and Mg^{2+} are added to each chamber containing prepared bilayers in 200 μL imaging buffer (see Section 2.2) and incubated for the desired length of time at 37 °C. Low buffer volumes for cells and bilayers facilitate faster cell settling, which is essential for achieving B cell interaction with the PLB over short incubation times (e.g., 2 min, which is the lower limit for stimulation by this method).
3. The chambers are washed with 2–3 mL HBSS using cut, 1 mL pipette tips for adding and removing fluid and the cells are fixed in 4% paraformaldehyde for 10 min at 37 °C, as described by Depoil *et al.* (2008).
4. Cells are washed, permeabilized with 0.1% Triton X-100 for 5 min at 20 °C, washed again and blocked with 1% BSA, 1% FCS, 1% goat serum, and 0.05% Tween-20 in PBS (blocking solution) for 30 min at RT.
5. Cells are stained in blocking solution using Abs specific to the signaling molecule of interest for 1 h at RT. Washing and blocking are repeated and the primary Ab is detected by adding an AlexaFluor-conjugated secondary Ab for 30 min at RT and then washing once more before imaging. Figure 14.4A shows representative two-color

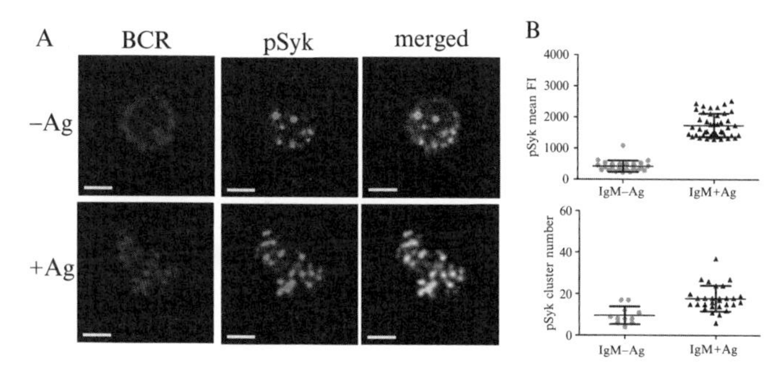

Figure 14.4 Imaging BCR signaling. (A) Two-color TIRF images show the distribution of the BCR and accumulation of pSyk on the membrane of human peripheral blood B cells that were placed on bilayers without (top panels) or with (bottom panels) goat anti-human IgM $F(ab')_2$ Ag for 10 min, fixed and labeled as described in Section 3.3. Specifically, BCRs and pSyk were visualized by imaging DyLight 649-Fab anti-IgM (red) and AlexaFluor 488-labeled pSyk (green), respectively, with TIRFM. Bars, 1.5 μm. (B) pSyk mean FI and pSyk cluster number quantified from several TIRF images of IgM-expressing B cells placed on bilayers without (gray circles) or with Ag (black triangles) as shown in (A). Each data point represents one cell analyzed in one of three independent experiments and the bars indicate the mean ± S.D. (For interpretation of the references to color in this figure legend, the reader is referred to the Web version of this chapter.)

TIRF imaging of the BCR with the signaling molecule pSyk on human peripheral blood B cells that were placed on bilayers without or with Ag for 10 min.

6. As shown for pSyk in Fig. 14.4B, the parameters used to describe the levels of BCR signaling-associated molecules include the mean fluorescence intensity (mean FI) of the signaling molecule within the contact area that the B cell makes with the Ag-presenting membrane (i.e., per unit area) and the number of signaling molecule microclusters per unit area.
7. The mean FI of the signaling molecule is measured using ImageJ software (National Institutes of Health, available at http://rsbweb.nih.gov/ij/). The image threshold should be adjusted to cover only the precise cell area (under the Image menu select Adjust Threshold). If there is more than one cell per image, select one at a time using a Region of Interest. To measure cell contact area and mean FI data, from the Analyze menu select Measure to open the Results window, choose Edit and then Set Measurements and check the boxes next to Area, Mean Gray Value, and Limit to Threshold. The Area is output as pixel area, so it must be converted to μm^2 or nm^2 using the dimensions per pixel for your specific microscope objective. Also, acquire the Mean Gray Value from a non-cell region of each image to subtract as background from the cell values.

8. To count microclusters in ImageJ, from the Analyze menu select Analyze Particles. The size should range from 0 to ∞ and circularity from 0 to 1, and check the box next to Display Results.
9. Unpaired two-tailed *t*-tests are performed for statistical comparisons (95% confidence interval). Linear regression analyses are conducted to assess the relationship between signaling molecule cluster number or mean FI and B cell contact area or BCR mean FI from fixed cell images using Prism software (GraphPad, LaJolla, CA).
10. Colocalization between the BCR and signaling molecule of interest is quantified from background-subtracted images via intensity correlation analysis as described by Li *et al.* (2004), using the WCIF plugin of ImageJ to obtain the Pearson's correlation index (Liu *et al.*, 2010c). After installing the plugin, under the Plugins menu select Colocalization Analysis and then Intensity Correlation Analysis. If the image file is a stack of images, be sure to check the box next to Current Slice Only. If there is more than one cell per image, select one at a time using a Region of Interest. The output value 'Rr' is the Pearson's correlation coefficient that should range from −1 to 1. Unpaired two-tailed *t*-tests are again performed for statistical comparisons.

Note: It is also feasible to image the dynamics of downstream BCR signaling molecules through high speed live cell imaging. For example, we are able to visualize the immobilization and recruitment of Syk molecules to the plasma membrane proximal and to BCR microclusters upon BCR recognition of Ag by imaging B cells transfected with GFP-Syk (Tolar *et al.*, 2009).

ACKNOWLEDGMENTS

We thank Dr. Rajat Varma at the National Institute of Allergy and Infectious Diseases (NIAID), National Institutes of Health (NIH) for expert advice on TIRF optics. This work has been supported by the Intramural Research Program of the NIH, NIAID.

REFERENCES

Axelrod, D. (1981). Cell-substrate contacts illuminated by total internal reflection fluorescence. *J. Cell Biol.* **89,** 141–145.

Axelrod, D. (1989). Total internal reflection fluorescence microscopy. *Methods Cell Biol.* **30,** 245–270.

Axelrod, D. (2001). Total internal reflection fluorescence microscopy in cell biology. *Traffic* **2,** 764–774.

Axelrod, D. (2003). Total internal reflection fluorescence microscopy in cell biology. *Methods Enzymol.* **361,** 1–33.

Axelrod, D. (2008). Chapter 7: Total internal reflection fluorescence microscopy. *Methods Cell Biol.* **89,** 169–221.

Bajenoff, M., and Germain, R. N. (2007). Seeing is believing: A focus on the contribution of microscopic imaging to our understanding of immune system function. *Eur. J. Immunol.* **37**(Suppl. 1), S18–S33.

Balagopalan, L., Sherman, E., Barr, V. A., and Samelson, L. E. (2010). Imaging techniques for assaying lymphocyte activation in action. *Nat. Rev. Immunol.* **11,** 21–33.

Baldwin, T., Sakthianandeswaren, A., Curtis, J. M., Kumar, B., Smyth, G. K., Foote, S. J., and Handman, E. (2007). Wound healing response is a major contributor to the severity of cutaneous leishmaniasis in the ear model of infection. *Parasite Immunol.* **29,** 501–513.

Carrasco, Y. R., and Batista, F. D. (2007). B cells acquire particulate antigen in a macrophage-rich area at the boundary between the follicle and the subcapsular sinus of the lymph node. *Immunity* **27,** 160–171.

Dal Porto, J. M., Gauld, S. B., Merrell, K. T., Mills, D., Pugh-Bernard, A. E., and Cambier, J. (2004). B cell antigen receptor signaling 101. *Mol. Immunol.* **41,** 599–613.

DeFranco, A. L. (1997). The complexity of signaling pathways activated by the BCR. *Curr. Opin. Immunol.* **9,** 296–308.

Depoil, D., Fleire, S., Treanor, B. L., Weber, M., Harwood, N. E., Marchbank, K. L., Tybulewicz, V. L., and Batista, F. D. (2008). CD19 is essential for B cell activation by promoting B cell receptor-antigen microcluster formation in response to membrane-bound ligand. *Nat. Immunol.* **9,** 63–72.

Douglass, A. D., and Vale, R. D. (2005). Single-molecule microscopy reveals plasma membrane microdomains created by protein–protein networks that exclude or trap signaling molecules in T cells. *Cell* **121,** 937–950.

Douglass, A. D., and Vale, R. D. (2008). Single-molecule imaging of fluorescent proteins. *Methods Cell Biol.* **85,** 113–125.

Elso, C. M., Roberts, L. J., Smyth, G. K., Thomson, R. J., Baldwin, T. M., Foote, S. J., and Handman, E. (2003). Leishmaniasis host response loci (lmr1-3) modify disease severity through a Th1//Th2-independent pathway. *Genes Immun.* **5,** 93–100.

Fleire, S. J., Goldman, J. P., Carrasco, Y. R., Weber, M., Bray, D., and Batista, F. D. (2006). B cell ligand discrimination through a spreading and contraction response. *Science* **312,** 738–741.

Groves, J. T., Parthasarathy, R., and Forstner, M. B. (2008). Fluorescence imaging of membrane dynamics. *Annu. Rev. Biomed. Eng.* **10,** 311–338.

Holtzer, L., Meckel, T., and Schmidt, T. (2007). Nanometric three-dimensional tracking of individual quantum dots in cells. *Appl. Phys. Lett.* **90**.

Junt, T., Moseman, E. A., Iannacone, M., Massberg, S., Lang, P. A., Boes, M., Fink, K., Henrickson, S. E., Shayakhmetov, D. M., Di Paolo, N. C., van Rooijen, N., Mempel, T. R., *et al.* (2007). Subcapsular sinus macrophages in lymph nodes clear lymph-borne viruses and present them to antiviral B cells. *Nature* **450,** 110–114.

Kusumi, A., Nakada, C., Ritchie, K., Murase, K., Suzuki, K., Murakoshi, H., Kasai, R. S., Kondo, J., and Fujiwara, T. (2005). Paradigm shift of the plasma membrane concept from the two-dimensional continuum fluid to the partitioned fluid: High-speed single-molecule tracking of membrane molecules. *Annu. Rev. Biophys. Biomol. Struct.* **34,** 351–378.

Kusumi, A., Shirai, Y. M., Koyama-Honda, I., Suzuki, K. G., and Fujiwara, T. K. (2010). Hierarchical organization of the plasma membrane: Investigations by single-molecule tracking vs. fluorescence correlation spectroscopy. *FEBS Lett.* **584,** 1814–1823.

Li, Q., Lau, A., Morris, T. J., Guo, L., Fordyce, C. B., and Stanley, E. F. (2004). A syntaxin 1, G alpha (o), and N-type calcium channel complex at a presynaptic nerve terminal: Analysis by quantitative immunocolocalization. *J. Neurosci.* **24,** 4070–4081.

Liu, W., Meckel, T., Tolar, P., Sohn, H. W., and Pierce, S. K. (2010a). Antigen affinity discrimination is an intrinsic function of the B cell receptor. *J. Exp. Med.* **207,** 1095–1111.

Liu, W., Meckel, T., Tolar, P., Sohn, H. W., and Pierce, S. K. (2010b). Intrinsic properties of immunoglobulin IgG1 isotype-switched B cell receptors promote microclustering and the initiation of signaling. *Immunity* **32,** 778–789.

Liu, W., Won Sohn, H., Tolar, P., Meckel, T., and Pierce, S. K. (2010c). Antigen-induced oligomerization of the B cell receptor is an early target of FcγRIIB inhibition. *J. Immunol.* **184,** 1977–1989.

Mattheyses, A. L., Simon, S. M., and Rappoport, J. Z. (2010). Imaging with total internal reflection fluorescence microscopy for the cell biologist. *J. Cell Sci.* **123,** 3621–3628.

Phan, T. G., Grigorova, I., Okada, T., and Cyster, J. G. (2007). Subcapsular encounter and complement-dependent transport of immune complexes by lymph node B cells. *Nat. Immunol.* **8,** 992–1000.

Pierce, S., and Liu, W. (2010). The tipping points in the initiation of B cell signalling: How small changes make big differences. *Nat. Rev. Immunol.* **10,** 767–777.

Plochberger, B., Stockner, T., Chiantia, S., Brameshuber, M., Weghuber, J., Hermetter, A., Schwille, P., and Schutz, G. J. (2010). Cholesterol slows down the lateral mobility of an oxidized phospholipid in a supported lipid bilayer. *Langmuir* **26,** 17322–17329.

Qi, H., Egen, J. G., Huang, A. Y., and Germain, R. N. (2006). Extrafollicular activation of lymph node B cells by antigen-bearing dendritic cells. *Science* **312,** 1672–1676.

Rajewsky, K. (1996). Clonal selection and learning in the antibody system. *Nature* **381,** 751–758.

Reth, M., and Wienands, J. (1997). Initiation and processing of signals from the B cell antigen receptor. *Annu. Rev. Immunol.* **15,** 453–479.

Sohn, H. W., Tolar, P., Jin, T., and Pierce, S. K. (2006). Fluorescence resonance energy transfer in living cells reveals dynamic membrane changes in the initiation of B cell signaling. *Proc. Natl. Acad. Sci. USA* **103,** 8143–8148.

Sohn, H. W., Tolar, P., and Pierce, S. K. (2008). Membrane heterogeneities in the formation of B cell receptor-Lyn kinase microclusters and the immune synapse. *J. Cell Biol.* **182,** 367–379.

Sohn, H. W., Tolar, P., Brzostowski, J., and Pierce, S. K. (2010). A method for analyzing protein-protein interactions in the plasma membrane of live B cells by fluorescence resonance energy transfer imaging as acquired by total internal reflection fluorescence microscopy. *Methods Mol. Biol.* **591,** 159–183.

Tolar, P., Sohn, H. W., and Pierce, S. K. (2005). The initiation of antigen-induced B cell antigen receptor signaling viewed in living cells by fluorescence resonance energy transfer. *Nat. Immunol.* **6,** 1168–1176.

Tolar, P., Hanna, J., Krueger, P. D., and Pierce, S. K. (2009). The constant region of the membrane immunoglobulin mediates B cell-receptor clustering and signaling in response to membrane antigens. *Immunity* **30,** 44–55.

CHAPTER FIFTEEN

A Quantitative Method for Measuring Phototoxicity of a Live Cell Imaging Microscope

Jean-Yves Tinevez,* Joe Dragavon,* Lamya Baba-Aissa,* Pascal Roux,* Emmanuelle Perret,* Astrid Canivet,* Vincent Galy,† *and* Spencer Shorte*

Contents

1. Introduction 292
2. Measuring Phototoxicity 293
    2.1. The need for a quantitative, generic, and convenient measure of phototoxicity 293
    2.2. A protocol to quantify the phototoxic effect of a microscopy system 295
    2.3. Longer image exposure times are less phototoxic 302
    2.4. Using phototoxicity curves to compare microscope-based imaging systems 302
3. Discussion 304
    3.1. A live specimen-based metrology 304
    3.2. The delivery rate of light dose matters for phototoxicity 306
    3.3. Comparing illumination modalities 307
4. Conclusion 308
Acknowledgments 308
References 308

Abstract

Fluorescence-based imaging regimes require exposure of living samples under study to high intensities of focused incident illumination. An often underestimated, overlooked, or simply ignored fact in the design of any experimental imaging protocol is that exposure of the specimen to these excitation light sources must itself always be considered a potential source of phototoxicity. This can be problematic, not just in terms of cell viability, but much more worrisome in its more subtle manifestation where phototoxicity causes

* Institut Pasteur, Imagopole, Plateforme d'imagerie dynamique, Paris, France
† UMR7622, CNRS-UPMC, 9 quai St Bernard, Paris, France

Methods in Enzymology, Volume 506
ISSN 0076-6879, DOI: 10.1016/B978-0-12-391856-7.00039-1

anomalous behaviors that risk to be interpreted as significant, whereas they are mere artifacts. This is especially true in the case of microbial pathogenesis, where host–pathogen interactions can prove especially fragile to light exposure in a manner that can obscure the very processes we are trying to observe. For these reasons, it is important to be able to bring the parameter of phototoxicity into the equation that brings us to choose one fluorescent imaging modality, or setup, over another. Further, we need to be able to assess the risk that phototoxicity may occur during any specific imaging experiment. To achieve this, we describe here a methodological approach that allows meaningful measurement, and therefore relative comparison of phototoxicity, in most any variety of different imaging microscopes. In short, we propose a quantitative approach that uses microorganisms themselves to reveal the range over which any given fluorescent imaging microscope will yield valid results, providing a metrology of phototoxic damage, distinct from photobleaching, where a clear threshold for phototoxicity is identified. Our method is widely applicable and we show that it can be adapted to other paradigms, including mammalian cell models.

1. Introduction

In our experience, running the imaging facilities at the Institut Pasteur in Paris (www.imagopole.org) managing phototoxicity is critical to long-term live cell imaging studies on infectious processes. Host cell–pathogen interactions are especially challenging when it comes to their reconstitution within the context of meaningful experimental imaging paradigms, and phototoxic effects are an abundant source of problems. For example, approaches using multidimensional live cell imaging for studies on infection have become near routine as recourse to analyze subcellular dynamics (Frischknecht and shorte, 2009; Shorte and Frischknecht, 2008). The difficulty of such approaches comes from the need to maintain spatial and temporal resolution using protocols that assure over-sampling x, y, z, and t. To achieve this, automated, high-speed acquisition aims to sample x, y "stacks" as rapidly as possible to satisfy the requirement that 3D volumes are acquired in a snapshot. Further, the 3D stack must be repeatedly sampled over time at a frequency determined by the temporal dynamics of the process being recorded. Add to this the need for multiple wavelength channels acquired at any given moment, allowing to distinguish distinct targets, which must then be colocalized, it is not uncommon to require 50–150 images to be acquired at each time point. This can amount to the need for an elevated *light budget*. Overcoming the limitations imposed by this light budget can often be the key to successful imaging. In the context of a real study, for example, following fluorescently labeled HIV virus (Arhel *et al.*, 2006), bacteria (Enninga *et al.*, 2005), parasites (Amino *et al.*, 2006, 2007; Thiberge *et al.*, 2007), or prion protein (Gousset *et al.*, 2009), the need for extensive light exposure can have substantial impact on the quality of the data, due to photobleaching that comprises the signal-to-noise ratio of the detectable signal, and phototoxicity that risks to perturb the processes under study.

A common misconception is to equate photobleaching with phototoxicity. Photobleaching is specific to fluorescence microscopy and arises due to the loss of fluorescent signal that occurs when fluorophores are excited into a state leading to an irreversible loss of signal. Phototoxicity, on the other hand, is a related phenomenon inasmuch as it may be precipitated by photobleaching of fluorophores, but not necessarily. It may also occur in the absence of fluorophore. Phototoxicity is a generalized term used as a catch-all to describe how exogenous light energy may interact with the tissue/cell metabolism (for detailed and extensive review, see Diaspro *et al.*, 2006). The term certainly refers to all those diverse processes resulting in light-induced free-radical generation, for example, from fluorescent labels, and/or light-sensitive metabolites. However, it also describes indirect effects such as localized thermal flux generation (undesired light-induced heating effects); light-induced ionizing, polarizing, and/or trapping effects; and of course unintended light-induced activation of membrane conductances. In turn, phototoxicity may result in extreme phenotypes such as cataclysmic cell death by, for example, free radicals rupturing cell membranes, and collapsing chemical and ionic compartmentalization. Such behaviors are easy to detect and reject before further analysis is performed. On the other hand, and much more problematic, phototoxicity may cause subtle effects, which are difficult to detect, or even distinguish because they do not kill the tissue, but rather subvert its functions. In the case of quantitative light microscopy, there is no ground truth inasmuch as it is the experimental device itself, the microscope, which induces these effects meaning that even careful control experiments may not be sufficient. Thus, our only remaining recourse to managing phototoxicity is to measure it and minimize risks by experimental design. Unfortunately, this is not a trivial task.

2. Measuring Phototoxicity

2.1. The need for a quantitative, generic, and convenient measure of phototoxicity

While it is rather well known, and somewhat implicit, that the impact of phototoxicity will vary with the amount of light delivered to the sample, much less clear are its underlying mechanisms. For example, while every biologist using live cell microscopy techniques will have an opinion on the subject, it is hard to know how a single-point-scanning confocal microscopy may be better or worse than, say, a multi-point-array-scanning confocal or a wide-field microscope. This uncertainty is due mainly to the difficulty associated to quantifying phototoxicity; that is to associate a number relating the impact of phototoxicity of a specific imaging configuration, which can be used to compare different imaging configurations. Indeed, there are currently no systematic methods described in the literature enabling to directly measure and thereby quantifying phototoxicity.

In efforts to measure phototoxicity, many studies have side-stepped the difficult problem by exploiting the relative ease by which photobleaching can be measured as the time-dependent decay of fluorescent intensity coming from a known quantity of fluorophore. Such studies extrapolate their conclusions to be relevant to understanding phototoxicity in live biological targets reasoning implicitly the relationship between the two phenomena (De Vos *et al.*, 2009; Hoebe *et al.*, 2008). However, this is far from satisfactory because while phototoxicity and photobleaching processes may, in many instances, be tightly coupled, they are nonetheless distinctly separate phenomena. The literature does offer some examples of efforts to directly assess phototoxicity. For example, the group of Eric Manders monitored cell rounding (a secondary indicator of apoptosis; Bortner and Cidlowski, 2002) in the illuminated field of view providing evidence supporting the claim that their method of controlled light exposure microscopy diminished phototoxicity, by reducing total exposure to incident illumination (De Vos *et al.*, 2009; Hoebe *et al.*, 2008). In a more quantitative approach, similar methods counted the proportion of dead cells appearing during the few days following illumination by a mercury lamp in an epifluorescence microscope (Zdolsek *et al.*, 1990) or the number of cells able to form a colony after irradiation in epifluorescence (Wagner *et al.*, 2010) or multi-photon imaging (König *et al.*, 1999). Cell rounding, blebbing, and/or nuclear fragmentation were taken as indicative of cell death, but no data on incident light dose were reported. Studying tobacco cells, Dixit and Cyr (2003) went further toward a metrology of phototoxicity, monitoring cell death a few days after illumination, as a function of the measured incident light dose. While all these examples use phenotypic manifestation of cell death as a measure of phototoxicity, it is rather the worst case scenario that phototoxicity may cause much more subtle effects. For instance, Nishigaki *et al.* (2006) noticed that at high incident light, the beating of the flagella of human sperm is detectably slowed, although they did not quantify the effect. During intravital imaging in living Syrian golden hamsters, Saetzler *et al.*, 1997 report that a high incident light dose can induce uncontrolled leukocyte activation and adhesion in venules and arterioles, a crucial point when studying inflammatory response. The number of activated leukocytes was measured for a given period of time and analyzed as a function of the incident light dose. In neuron calcium imaging, it was observed that the dynamic range of Ca^{2+} pulses was impaired by a strong illumination power (Ji *et al.*, 2008; Koester *et al.*, 1999).

From these previous studies, it is clear that the challenge in measuring phototoxicity is its multifarious and non-continuous nature, which is intractably linked to the particularity and specificity of the biological system under study. Indeed, phototoxicity may only ever be rather arbitrarily measured by following specific parameters in a given biological model. In this case, such parameters must be recorded in a defined experimental paradigm and should be sensitive enough to detect early effects, far upstream from cell death, but far enough downstream from the multiplicity of complex *de novo* light-induced processes so as to be robust, and reproducible over a relevant range of light dose. Toward this goal, we describe herein the rationale and method, proposing a protocol that yields a well-defined number that can be used as a quantitative measure of the tendency of an imaging system to evoke phototoxicity. We validate our approach by showing that this number is a characteristic of the illumination modality of the setup and can be used to compare different settings on the same setup, and even different setups.

2.2. A protocol to quantify the phototoxic effect of a microscopy system

In this study, we aimed to establish a method allowing phototoxicity to be compared between different imaging systems. The criteria for optimization of this method were as follows:

- *Quantitative.* The method should provide a simple readout for phototoxicity and a means to measure the relative power of light illuminating the living specimen therein. Quantification of these parameters must be reproducible and have a limited error range.
- *Generic.* Measurements must be easy to achieve and transpose readily between different imaging configurations and different imaging systems, ideally allowing to compare different illumination modalities based upon their phototoxic impact.
- *Convenient.* The method must be easily disseminated and implemented; it must be highly reproducible, lending itself ideally to become a standard amenable to be shared among different laboratories, and stable over time.

2.2.1. Experimental rationale

Caenorhabditis elegans is a free-living, transparent nematode (roundworm), about 1 mm in length, which lives in temperate soil environments. The molecular and developmental biology of *C. elegans* is well documented since over 30 years (Sulston and Horvitz, 1977). As a multicellular eukaryote, *C. elegans* is a valuable model organism offering distinct advantages. On a practical level, it is simple enough to be studied in great detail. Strains are cheap to breed and are fully viable after reanimation following long-term frozen storage. Further, from embryo to mature worm, *C. elegans* is transparent to visible light, facilitating the study of cellular differentiation and other developmental processes. Imaging techniques can be applied in the intact organism, and by consequence, the developmental fate of every single somatic cell (959 in the adult hermaphrodite; 1031 in the adult male) has been mapped out (Kimble and Hirsh, 1979; Sulston and Horvitz, 1977).

These patterns of cell lineage are largely invariant between individuals, and recent live cell studies have revealed the dynamics of cell lineage in living embryos during the first 3 h of embryogenesis (Bao *et al.*, 2008). Bao *et al.* measured the cell cycle length from the one cell stage up to seven cycles of division for 20 embryos. Remarkably, the "developmental clock" variability showed the bifurcation of the lineage tree (the timing of nuclear division events) occurred with a remarkably small standard deviation ($<4.5\%$). Presumably, this variation is due to the small variation of experimental conditions such as temperature; and the r^2 ranging 0.997–0.999 asserts that, on average, 95% of the cell divisions in the *C. elegans* embryo deviate less than 2% from the general clock. So, during embryogenesis, the *C. elegans* cell fate is tightly controlled and highly reproducible from one embryo to another.

While such studies validate the use of epifluorescence microscopy to monitor apparently healthy embryogenesis, at least during some period of time, *C. elegans* embryogenesis is actually exquisitely sensitive to epifluorescent illumination. In our own studies, we have found that increased incident illumination can result in disruption of cell division over time, specifically causing a retardation of the developmental clock. Further, this effect is easily detected by a reduction in the number of cell nuclei expected to be present at any given time during the lineage. Reasoning that the light sensitivity of *C. elegans* embryogenesis might be considered a phototoxicity paradigm, we tested whether this parameter (number of cell nuclei) was sensitive to light exposure in a dose-dependent, predictable, and reproducible manner.

2.2.2. Protocol content

2.2.2.1. Establishing imaging parameters We aimed to establish an experimental protocol to test the hypothesis that phototoxicity effects on *C. elegans* embryogenesis might manifest as a quantifiable relationship between the number of cell divisions and incident light exposure. Thus, the ideal protocol would reveal changes in the number of *C. elegans* cell divisions during early embryogenesis, attributable solely to variations in the incident light dose. Inasmuch as we hoped to use the method to compare the performance of a broad range of different imaging microscope systems, and configurations, relevant to real multidimensional imaging applications used in common experimental conditions, we designed a protocol exploiting the inherent flexibility and generic nature of a multidimensional time-lapse (Fig. 15.1). The light exposure cycle was repeated during the first 2 h of embryogenesis, starting from the onset of the first anaphase. The advantage of this approach was that it is relevant to real experimental conditions common to most live imaging experiments, and it ensured that the whole embryo

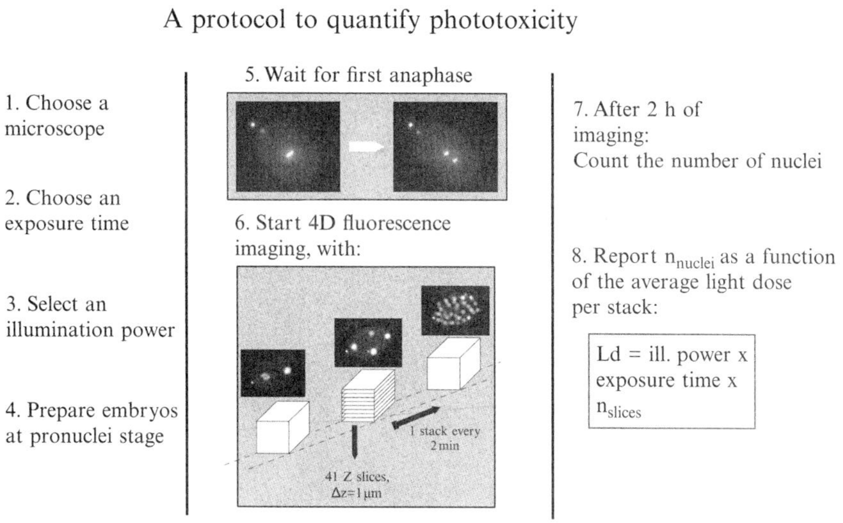

A protocol to quantify phototoxicity

1. Choose a microscope

2. Choose an exposure time

3. Select an illumination power

4. Prepare embryos at pronuclei stage

5. Wait for first anaphase

6. Start 4D fluorescence imaging, with:

1 stack every 2 min

41 Z slices, $\Delta z = 1\ \mu m$

7. After 2 h of imaging: Count the number of nuclei

8. Report n_{nuclei} as a function of the average light dose per stack:

Ld = ill. power x exposure time x n_{slices}

Figure 15.1 Summary of the protocol for the quantification of phototoxicity impact.

volume was subjected to comparable levels of light, rather than say, for example, a finite, focused, high-intensity light exposure like that used in FRAP experiments. Further, the protocol is relevant to those used for experimental imaging of *C. elegans* embryo elsewhere (Bao *et al.*, 2008, 2006).

2.2.2.2. Incident light dose measurements To compare *C. elegans*-based phototoxicity measurements on different imaging systems, or using different system configurations, it was necessary to find a means to quantify incident light delivered during the optimized acquisition protocol by measuring the light dose at the sample level. While it sounds simple, this is actually a non-trivial task. In developing the protocol presented here, we chose to report this quantity as the energy density dose deposited on the sample, when acquiring one stack, calculated as follows:

For epifluorescence wide-field microscopes and spinning-disks

- Measure the time-averaged power density at the sample level using the power set for imaging;
- Derive the light dose by multiplying the measured power by the exposure time per image, times the number of image acquisition in one stack (41 in our case).

This assumes that the whole embryo received 41 times the light energy used to image one place. Indeed, for these two systems, the whole fluorescent volume is excited, even if in the spinning-disk case, pinholes allow to collect light emitted from a single plane only.

2.2.2.3. Preparation of a selected* C. elegans *strain for imaging We chose the *C. elegans* strain AZ212 (Praitis *et al.*, 2001), which expresses the nuclear localized histone protein H2B, conjugated with eGFP (heterogeneous expression). Worms plated on Petri dishes containing agar-based growth support, supplemented for nourishment by inoculation with *Escherichia coli* bacteria, were maintained in a temperature-controlled bench incubator at 20 °C. Worm culture dishes were replated, depending on the experimental needs, every 1–4 days. H2B-GFP *C. elegans* embryos were prepared as described elsewhere (Bao *et al.*, 2008, 2006). Briefly, two to three hermaphrodite adult worms were plated on a coverslip in a droplet of M9 medium. Using a binocular stereomicroscope, the adult worms were dissected, and the eggs are dispersed. The egg-bearing coverslip was pressed onto a glass slide coated with agar 2% and sealed using valap. The sample slide set on a microscope system allowed eggs at the zygote stage to be selected for imaging.

2.2.2.4.* C. elegans *imaging and phototoxicity impact quantitation *C. elegans* embryogenesis was followed during 2 h from the first metaphase after pronuclei fusion, a non-ambiguous event in embryo development. From this precise moment, the developing embryo was subjected to an optimized light exposure protocol using an epifluorescent microscope system (ambient temperature range 20–22 °C). Conveniently, for the H2B-GFP strain used here, the first anaphase manifests as the first instant where two dense chromatin fluorescent spots can be seen and clearly separated. Two minutes before, when the fusion of the two pronuclei is complete, the cell is in metaphase and only one bright spot can be seen. The transition from metaphase to anaphase (Fig. 15.2) defines clearly an instantaneous temporal start point in the *C. elegans* embryo development from which subsequent cell division times may be measured. The occurrence of this event can be monitored with moderate fluorescent light, and it is also possible for the trained eye to see it with transmitted light.

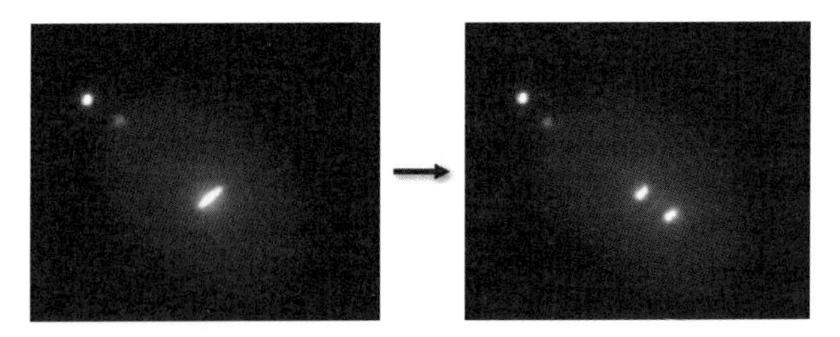

Figure 15.2 Left: *C. elegans* H2B-GFP embryo in metaphase. Chromatin appears as a single, compact, dense spot. Right: The same embryo in anaphase (2 min after Fig. 15.1). Two dense spots can now be seen. This event is used to determine precisely the onset of imaging. Images were acquired on a Zeiss wide-field epifluorescence inverted microscope, using a 63×-oil immersion objective.

Using the optimized protocol, individual embryos were exposed to repeated cycles of excitation light exposure on the imaging system in a multidimensional (x, y, z, t) acquisition protocol that was strictly fixed for all subsequent embryos. This way, a regular frequency of light exposure starting from the onset of the first anaphase was maintained until 2 h after this event (equivalent to seven cell division cycles in the non-phototoxic case). Thus, the whole embryo volume was exposed to excitation light by virtue of a z-stack acquisition, 41 z-axis "slices" (spaced by 1 μm) collected as rapidly as possible, once every 2 min during the 2 h. Depending on the imaging system, the frame exposure time was chosen according to an expected optimal range, and then fixed throughout, and for all subsequent embryos. Inasmuch as the image-data *per se* were not used in any subsequent analysis, it did not matter what image quality resulted from under-, or over-exposure of the detector. On the other hand, while the rhythm and frequency of light delivery was fixed for all embryos in any given analysis, the total light exposure was graded by varying the incident power at the light source. At the end of each run, the total number (N) of cells distributed throughout the embryo volume was counted (Fig. 15.3).

2.2.3. Impact of phototoxicity on *C. elegans* embryogenesis

Under conditions where no phototoxicity occurred, and at 20–22 °C, the total number of nuclei after 2 h is expected to be 46–54. Many embryos were imaged using exactly the same experimental conditions, and acquisition protocol, but gradually increased incident illumination resulted in data that could be represented as the number of nuclei versus light dose. Thus, without changes in the rate of light exposure, a clear effect of increased

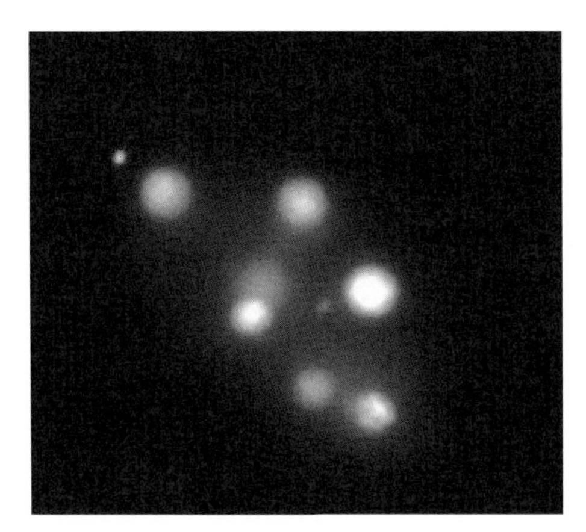

Figure 15.3 The same embryo that of Fig. 15.2, 2 h after the onset of anaphase. For the very high incident light dose used for imaging, the measured number of cells at this stage was only seven.

excitation light intensity was to diminish the number of cells, consistent with the expected adverse effect of excitation light exposure on cell division (Fig. 15.4). Indeed, the pattern of data from many embryos revealed that low levels of incident excitation illumination caused little phototoxicity, but at a threshold caused a sharp decline in the number of nuclei, above which this diminution increased in a manner proportional to increased light power, until embryogenesis was completely and cataclysmically arrested (Fig. 15.4). This sigmoidal relationship between the number of nuclei and incident light power was reproducible, and predictable, entirely consistent with our expectations, that the decrease is attributable to photodamage, quantifiable with respect to the incident light dose. On this curve, a single point represents the imaging of a *C. elegans* embryo for at least 2 h. Finally, to study the effect of increasing light dose, it was necessary to increase incident light dose over a broad range of relevant intensities using multiple embryos for each experimental run. The data pooled from at least 10 embryos allowed a conclusion on the threshold of light dose wherein the given microscopy system retarded embryogenesis, the point of inflexion being deemed as the "phototoxicity threshold." More precisely, the data are fit by a sigmoid, the simplest curve reflecting the observed distribution of points in the phototoxicity curve. From this fit, one can then derive the location of the inflexion point, with a 95% confidence interval, which we define as the phototoxicity threshold of a system under some specific imaging conditions (Fig. 15.4).

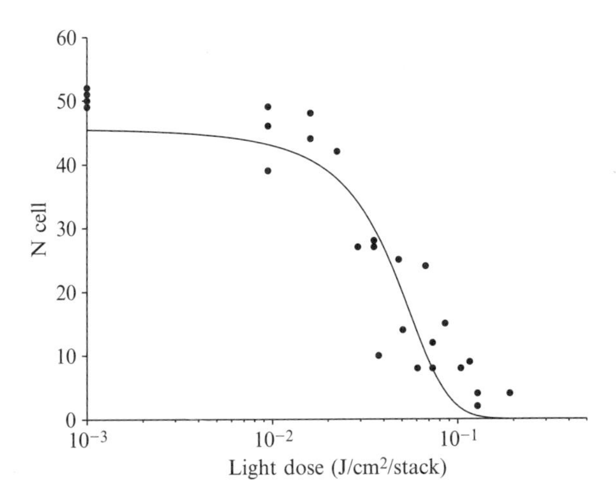

Figure 15.4 An example of a phototoxicity curve measured on a wide-field microscope. Dots are individual measurements; straight line is a fit by a sigmoid. The fit yields a phototoxicity threshold of 4.27×10^{-2} J/cm^2/stack (95% confidence interval is 3.65×10^{-2} to 4.88×10^{-2} J/cm^2/stack).

Parenthetically, data point variance may have arisen due to the difficulty to avoid small ambient temperature variations, as suggested in another study (Bao *et al.*, 2008). In the current studies, our environmental control was sufficient to limit a temperature range within a few degrees (21 ± 1 °C). On this point, it is important to stipulate the importance of environmental temperature. In separate studies, we found that at temperatures below 18 °C, *C. elegans* embryogenesis is significantly retarded. For example, at 17–18 °C and without illuminating the sample, we found that $N = 21–25$, compared to a range of 46–54 expected with a temperature of 20–22 °C. On the other hand, embryogenesis was accelerated such that $N = 60–72$ at temperatures in the range 23–25 °C, whereas above this range, embryogenesis lasted only a few cell divisions.

2.2.4. Understanding the meaning of phototoxicity curves

The shape of the phototoxicity threshold curves was predictable inasmuch as it could be understood as the accumulation of three distinct phases according to different light dose ranges (see Fig. 15.4):

2.2.4.1. Neutral phototoxicity at low-energy light dose The number of cells measured 2 h after imaging does not depend on the incident light leading us to the conclusion that in this dynamic context, no phototoxicity occurs. That is to say that if there is any phototoxicity occurring we do not detect it, either because our method is not sufficiently sensitive or the intrinsic dynamic mechanisms of the embryo (e.g., cytoplasmic mechanisms for free-radical management) are sufficient to minimize phototoxicity. It is noteworthy that the worm is a terrestrial animal that lives in soil and feeds on the bacteria growing on vegetative detritus, as such it may be equipped with some capacity to protect itself from occasional exposure to potentially hazardous ultraviolet sunlight, but presumably with only limited capacity able to deal with acute, and not chronic, exposure. In any case, at low light doses, the measured number of cells is the same whether we illuminate the sample or not. Consequently, this part of the curve is a plateau at $N = 46–54$ cells. The existence of an energy range where there is no visible phototoxic effect was observed in other studies (De Vos *et al.*, 2009; Dixit and Cyr, 2003).

2.2.4.2. Cumulative phototoxicity at intermediate range Above a certain light dose level, the timing of embryogenesis is impaired, observed as a temporal retardation in embryogenesis. Consequently, the number of cells measured 2 h after the onset of anaphase decays with increasing light dose. The fact that moderate incident light dose above a certain threshold can generate a dose-dependent phototoxic damage is consistent with previous studies showing single cell induced-phototoxic effects (Ji *et al.*, 2008; Koester *et al.*, 1999; Nishigaki *et al.*, 2006) and other studies in cell populations (Dixit and Cyr, 2003; König *et al.*, 1999; Saetzler *et al.*, 1997; Wagner *et al.*, 2010;

Zdolsek *et al.*, 1990). We also report that in this range of energy, the more incident light, the more damage is accrued in an apparently linear manner as evidence by the characteristic decay of the phototoxicity curve.

2.2.4.3. *Cataclysmic phototoxicity at high-energy light dose*

For very high incident energy, severe damage is predicted, resulting in a total and complete arrest of embryo development, as soon as after the first or second cell division. Consequently, the phototoxicity curve goes through a point of inflexion, leading to a nadir at $N = 2$–4 cells. In this range of energy, our protocol cannot distinguish between different amounts of damage, for it reports only a constant number of cells. However, given that all relevant, physiological dynamic function is lost, the value of further analysis for the purpose of achieving optimum conditions for dynamic cellular imaging is obviated. At this stage, we are restricted to study the processes of photon-induced pathophysiology.

2.3. Longer image exposure times are less phototoxic

To further validate the developed method and to test its sensitivity, we next used it to measure phototoxicity under conditions where changes in the image acquisition regime should give predictable effects. We compared exactly the same image acquisition protocol except the camera exposure time used to acquire each image in each stack was increased from 100 to 500 ms. The effect upon phototoxicity curves, given the same total light dose, demonstrates that the latter would be less phototoxic due to the lower rate of light delivery (Fig. 15.5). Indeed, the shift of the phototoxicity curve to the right indicates that the phototoxicity threshold is higher under the image acquisition regime using longer exposure times. That is to say empirically that our method revealed that delivery of the same total amount of light energy, except more slowly, is less phototoxic than delivery of the same total light energy in less time. This observation is consistent with the expectation that the *rate* and *intensity* of light delivery are important factors in increasing risk of phototoxic damage to live cells (Amino *et al.*, 2007).

2.4. Using phototoxicity curves to compare microscope-based imaging systems

Given some constraints, an exciting perspective of our method for measuring phototoxicity is that it offers an easily distributed, common standard (H2B-GFP *C. elegans*) and protocol allowing direct comparison of different microscope systems anywhere. Indeed, precisely reproducing the protocol generates standard phototoxicity curves based on definite variables, which are independent from intensity-based image analysis or image quality. The curves using these well-defined, robust variables (total light dose energy

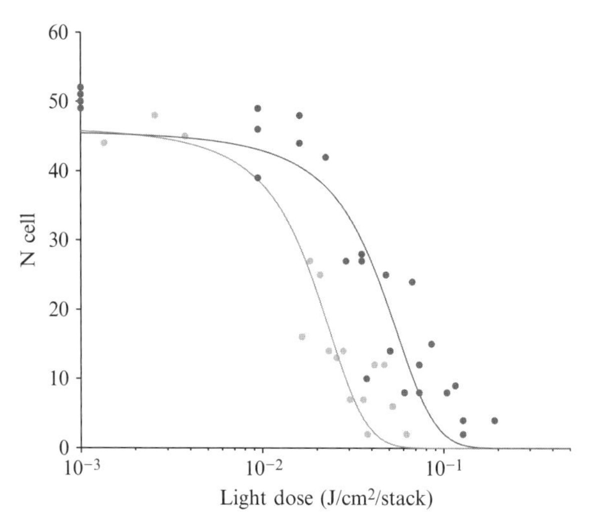

Figure 15.5 Two phototoxicity curves, acquired on the same wide-field system, differing only in the exposure time per slice. Orange: exposure time is 100 ms. Green: 500 ms. The phototoxicity thresholds are, respectively, 1.85×10^{-2} [1.57×10^{-2} to 2.13×10^{-2}] and 4.27×10^{-2} [3.65×10^{-2} to 4.88×10^{-2} J/cm^2/stack (value and (95% confidence interval)]. (See Color Insert.)

versus the number of nuclei) can therefore be compared, no matter their original source or time, and may be used to compare microscopy systems, by simple curve overlay.

Using this approach, Fig. 15.6 compares phototoxicity on two setups so far:

- A spinning-disk (an Andor Revolution system mounted on an inverted Zeiss AxioVert 200M and based on a CSU10 head) and
- A wide-field epifluorescence inverted microscope.

Total light energy doses reported use of measurements made from behind the objective and are not corrected for transmittance. Further, because we used objectives from the same series, comparative results hold valid. Comparatively, the two systems gave phototoxicity curves consistent with the expected characteristic properties. Namely, while some point dispersion inherent to biological noise was observed both phototoxicity curves resembled the characteristic shape predicted. In particular, a range of light dose with no measurable phototoxic effect was observed, and as the light dose was increased, this was accompanied by a dose-dependent phototoxicity effect culminating with developmental arrest. Our results demonstrate *a priori* that the spinning-disk illumination has a stronger phototoxic impact than plain wide-field illumination, yielding a phototoxicity threshold

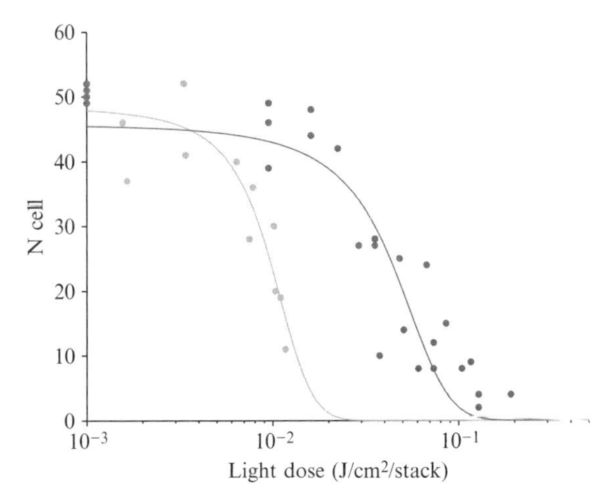

Figure 15.6 Two phototoxicity curves for two different systems using the same objective. Blue: a spinning-disk microscope using a Yokagawa CSU10 head. Green: a wide-field microscope. The phototoxicity thresholds are, respectively, 9.58×10^{-3} [8.58×10^{-3} to 1.06×10^{-2}] and 4.27×10^{-2} [3.65×10^{-2} to 4.88×10^{-2}] J/cm^2/stack. (See Color Insert.)

of approximately 1×10^{-2} J/cm^2/stack for the spinning-disk, versus 4×10^{-2} J/cm^2/stack for the wide-field microscope.

3. Discussion

3.1. A live specimen-based metrology

In Life Sciences, microscopes are mainly used to gather information on live specimen. It is not surprising then that assessing the performance of these microscopes involve using a live sample, particularly when it comes to quantifying the impact of imaging on live specimen. Measuring other performance criteria may not require stepping in the complexity of dealing with live samples. The rate of photobleaching, for instance, focuses on a fluorophore and its environment. To measure it, one can very conveniently purchase commercially available probes coupled to latex beads that are easy to mount and can be stored reliably over an extended period of time without sample deterioration. Phototoxicity, on the other hand, is an effect observed on a living sample, and as such can only be quantified by taking a measurement on some dynamic property of this sample. Two key points are linked to this simple observation.

First, a proper live sample must be defined as a standard for measurements. Contrary to common imaging situations, these metrology experiments are performed to get insight on the performance of microscopy systems, not on the live specimen, which can therefore be selected on the basis of convenience. Second, a measurable dynamic feature able to relate the impact of phototoxicity must be sought. Its baseline (its value when there are no phototoxic effects on the sample) must be known or measurable. Ideally, its numerical value is a robust, deterministic function of the incident light dose for a given system, safe from typical biological variations, so that the same measure taken several times with the same input gives a distribution with a very small deviation.

Various authors have proposed different models to quantify phototoxicity. Several studies are conducted based on monitoring the survival of cultured cells after a given light dose exposure (Dixit and Cyr, 2003; König *et al.*, 1999; Wagner *et al.*, 2010; Zdolsek *et al.*, 1990). Individual cell viability is assessed by a live-dead viability kit or direct observation of the cell's morphology. Saetzler *et al.* (1997) monitor leukocyte rolling and adhesion in arterioles and venules as a function of incident light dose during intravital imaging of golden hamsters. Two groups follow the dynamics of Ca^{2+} pulses, followed by fluorescence, elicited in rodent brain neurons as a measure of phototoxicity (Ji *et al.*, 2008; Koester *et al.*, 1999). All these methods give adequate and meaningful results, but are not trivial to disseminate. Maintaining any culture cell line, not mentioning living animals, requires equipment, people training, and administrative authorizations that can be a strain for a non-dedicated lab. Yet, a living sample-based metrology tool affordable by laboratories specialized in the development of imaging devices or industrial manufacturers would be beneficial to the imaging domain.

In this view, we choose to build our phototoxicity protocol using *C. elegans* embryo. The specific strain used here can be ordered from The Caenorhabditis elegans Genetics Center. The manipulation and mounting for imaging require only a basic stereomicroscope. Finally, their care and maintenance require very little expertise and equipment. Prior to this study, our lab had no expertise in or dedicated equipment for *C. elegans*. We sent some protocol kits to two remote labs, with no prior experience as well, and had them carry successful measurements on their systems. This demonstrates the portability of the protocol.

We chose to report the number of cells counted after 2 h of imaging for several reasons. We want to make a statement on the imaging effects that occur for long-term time-lapse 3D experiments, for which subtle artifacts can occur over a long period of time. We then must follow the embryogenesis over a time long enough to be meaningful for these experiments. Counting the number of cells is trivial in our case; in the strain we chose the nuclei are labeled with eGFP, we can simply use the last time point in the acquisition to make the measurement. Also, the number of cells 2 h post first anaphase is always tractable (50 for the baseline), yet it is large enough to allow for detectable drops when phototoxicity kicks in.

Another dynamic feature could have been chosen; for instance, measuring the time it takes for the embryo to reach the 28-cell stage. A strong phototoxicity effect would then manifest itself from having a much longer time to reach this stage. We chose, however, to count the number of cells after a given time, because it enables the extended use of this protocol:

- First, it can be used on non-imaging devices or on imaging devices for which a proper *C. elegans* embryo image cannot be formed (e.g., at low magnification or without detector). Since the delivery of the light (input) and the counting of the cells (measure) are decoupled, the embryo can be placed onto the device to be tested for the 2 h, then brought onto another suitable imaging device to count the number of cells.
- Second, it is not limited to measure damage caused by light. An embryo can be placed in a hostile environment (chemicals, radiations, etc.) for 2 h, then again brought to an imaging device to count cells.

3.2. The delivery rate of light dose matters for phototoxicity

We present a standard (H2B-GFP *C. elegans*) and method to measure directly the relationship between light intensity and phototoxicity on any imaging microscope system, configured to acquire data in any given optimized image acquisition regime. This allows us to estimate the "photon-light budget" available to any given experimental regime. Demonstrating its broad utility, we show the method applied to two relatively simple experimental paradigms: (i) changing image exposure times and (ii) changing the imaging system. In the first example, we clearly distinguished the phototoxicity cost that accompanies the need to use shorter exposure times with increased excitation intensities. This exigency is often stipulated by the need, for example, to image fast cellular events. Interestingly, our results suggest that the *rate* at which photons are delivered is critical to their eventual phototoxic impact. Along these same lines, a second series of experiments revealed spinning-disk microscopy to be more phototoxic than conventional wide-field epifluorescence illumination.

In their study on tobacco cells, Dixit and Cyr (2003) observed that for the same incident dose, the photodamage was more pronounced for a higher rate of illumination. That is, illuminating the sample with shorter exposures but higher power generates more photodamage than the converse. With the spinning-disk confocal illumination, we are in a situation where a point at the sample level receives light energy by short intense bursts. The pixel dwell-times for spinning-disk confocal are on the order of microseconds. This is to be compared with the 500 ms continuous illumination in the epifluorescence case. This rate-dependence might be due to the presence of "phototoxicity buffers" in the cell. The predominant process for phototoxicity generated by fluorescent imaging is thought to be a type II reaction (Zdolsek, 1993) where reactive oxygen species (ROS) are generated by the energy transfer of an excited fluorophore in a triplet state to a molecule of oxygen, exciting it to a singlet state. The singlet oxygen can interact with other molecules within the cell and generate phototoxicity effects (Saetzler *et al.*, 1997). Natural ROS scavengers, such as glutathione, ascorbate, and tocopherol, allow the cell to cope with a certain amount of photodamage with no impact (De Vos *et al.*, 2009; Dixit and Cyr, 2003). If we suppose now that these scavengers have a regeneration rate with a timescale similar that of the photon illumination rate, then superior rates of photon delivery may overcome the protective metabolism leading to irreversible, cumulative photodamage. For a given incident light dose, a slower photon delivery rate, for example, imaging at longer exposure times with lower illumination power will be protective, allowing free-radical scavenger mechanisms to regenerate and cope with ROS. We assume in either regime the same amount of ROS is generated, but the lower illumination rate is permissive to metabolic management by ROS scavengers. We speculate that this is the reason why spinning-disk confocal compares unfavorably in this study. Our conclusion that increased illumination rate explains why spinning-disk confocal reported a decreased phototoxicity threshold (i.e., was more phototoxic) compared to wide-field is entirely consistent with our wide-field microscope data showing shorter (100 ms) exposure times are more phototoxic than long (500 ms) exposure times.

3.3. Comparing illumination modalities

It is often stated that spinning-disks are the optimal system when it comes to photobleaching (Vermot *et al.*, 2008), ergo photodamage, a conclusion challenged by the results presented here. However, it must be noted that this optimality statement is based on the fact that these systems are equipped with very sensitive cameras, and offer optical sectioning, that enable the use of a much lower light dose to reach an acceptable SNR (Inoue and Inoue, 2002). These two options are seldom used on conventional wide-field microscopes, and an experimenter trying to optimize the image quality instead of the incident light dose will likely find a spinning-disk to be optimal.

Because it involves solely the light dose measured at the sample level, our protocol characterizes the impact of the illumination path and delivery modality only. It does not take into account the image quality that depends on the detection path. This is a desirable feature for a metrology tool: it allows the investigation and optimization of one part of the system independently of the repercussion of other parts. However, practical cases of imaging strive at obtaining the best image quality under the constraint of minimal photodamage. The method proposed here does not take image quality into account, and we aim to reconcile these two aspects in a future work.

4. Conclusion

We present a method and standard providing a non-ambiguous and singular readout reporting the phototoxicity of any imaging microscope system. The method yields a phototoxicity threshold value, which can be used to compare configuration settings in the same modality or even other. The protocol was engineered to be portable and convenient, so that it can be used in any lab. Using this approach, we demonstrate spinning-disk confocal imaging to be characterized by a lower phototoxicity threshold (i.e., is more phototoxic) than a wide-field epifluorescence microscope, a result we interpret as due to the critical dependence of phototoxicity upon the rate of photon illumination. Consistent with this view, we show that in wide-field epifluorescence microscopy the same total light dose delivered by shorter exposure times is associated with higher phototoxicity (i.e., a lower phototoxicity threshold) than the same light dose delivered by longer exposure times.

ACKNOWLEDGMENTS

We thank Johan Henriksson for a stimulating discussion. This work was funded by the European Commission FP7 Health (project "LEISHDRUG", www.leishdrug.org, SLS) and ICT (project "MEMI", www.memi-fp7.org, SLS), the Conny-Maeva Foundation (USA), and the Institut Pasteur Paris. J.D. received a fellowship from the Pasteur-Foundation (New York).

REFERENCES

Amino, R., *et al.* (2006). Quantitative imaging of Plasmodium transmission from mosquito to mammal. *Nat. Med.* **12,** 220–224.

Amino, R., *et al.* (2007). Imaging malaria sporozoites in the dermis of the mammalian host. *Nat.Protoc.* **2,** 1705–1712.

Arhel, N., *et al.* (2006). Quantitative four-dimensional tracking of cytoplasmic and nuclear HIV-1 complexes. *Nat. Methods* **3,** 817–824.

Bao, Z., *et al.* (2006). Automated cell lineage tracing in Caenorhabditis elegans. *Proc. Natl. Acad. Sci. USA* **103**(8), 2707–2712.

Bao, Z., *et al.* (2008). Control of cell cycle timing during C. elegans embryogenesis. *Dev. Biol.* **318**(1), 65–72.

Bortner, C. D., and Cidlowski, J. A. (2002). Apoptotic volume decrease and the incredible shrinking cell. *Cell Death Differ.* **9**(12), 1307–1310.

De Vos, W. H., *et al.* (2009). Controlled light exposure microscopy reveals dynamic telomere microterritories throughout the cell cycle. *Cytometry A* **75**(5), 428–439.

Diaspro, A., *et al.* (2006). Photobleaching. *In* "Handbook of Biological Confocal Microscopy," (J. B. Pawley, ed.) 3rd edn, pp. 690–702.

Dixit, R., and Cyr, R. (2003). Cell damage and reactive oxygen species production induced by fluorescence microscopy: Effect on mitosis and guidelines for non-invasive fluorescence microscopy. *Plant J.* **36**(2), 280–290.

Enninga, J., *et al.* (2005). Secretion of type III effectors into host cells in real time. *Nat. Methods* **2**(12), 959–965.

Frischknecht, F., and Shorte, S. (eds.) (2009). Imaging host-pathogen interactions. *Biotechnol. J. (Special Edition)* **4**(6), 773–948, Wiley-Blackwell.

Gousset, K., *et al.* (2009). Prions hijack tunnelling nanotubes for intercellular spread. *Nat. Cell Biol.* **3,** 328–336.

Hoebe, R. A., *et al.* (2008). Quantitative determination of the reduction of phototoxicity and photobleaching by controlled light exposure microscopy. *J. Microsc.* **231**(1), 9–20.

Inoué, S., and Inoué, T. (2002). Direct-view high-speed confocal scanner: The CSU-10. *Methods Cell Biol.* **70,** 87–127.

Ji, N., *et al.* (2008). High-speed, low-photodamage nonlinear imaging using passive pulse splitters. *Nat. Methods* **5**(2), 197–202.

Kimble, N., and Hirsh, D. (1979). The postembryonic cell lineages of the hermaphrodite and male gonads in *Caenorhabditis elegans*. *Dev. Biol.* **70**(2), 396–417.

Koester, H. J., *et al.* (1999). Ca2+ fluorescence imaging with pico- and femtosecond two-photon excitation: Signal and photodamage. *Biophys. J.* **77**(4), 2226–2236.

König, K., *et al.* (1999). Pulse-length dependence of cellular response to intense near-infrared laser pulses in multiphoton microscopes. *Opt. Lett.* **24**(2), 113–115.

Nishigaki, T., *et al.* (2006). Stroboscopic illumination using light-emitting diodes reduces phototoxicity in fluorescence cell imaging. *Bio Techniques* **41**(2), 191–197.

Praitis, V., *et al.* (2001). Creation of low-copy integrated transgenic lines in *Caenorhabditis elegans*. *Genetics* **157**(3), 1217–1226.

Saetzler, R. K., *et al.* (1997). Intravital fluorescence microscopy: Impact of light-induced phototoxicity on adhesion of fluorescently labeled leukocytes. *J. Histochem. Cytochem.* **45** (4), 505–513.

Shorte, S., and Frischknecht, F (eds.) (2008). Imaging Cellular and Molecular Biological Functions, Springer-Verlag, Heidelberg, Germany.

Sulston, J. E., and Horvitz, H. R. (1977). Post-embryonic cell lineages of the nematode, Caenorhabditis elegans. *Dev. Biol.* **56**(1), 110–156.

The Caenorhabditis elegans Genetics Center page on AZ212 strain. https://cgcdb.msi.umn.edu/strain.php?id=5544.

Thiberge, S., *et al.* (2007). In vivo imaging of malaria parasites in the murine liver. *Nat. Protoc.* **2,** 1811–1818.

Vermot, J., *et al.* (2008). Fast fluorescence microscopy for imaging the dynamics of embryonic development. *HFSP J.* **2**(3), 143–155.

Wagner, M., *et al.* (2010). Light dose is a limiting factor to maintain cell viability in fluorescence microscopy and single molecule detection. *Int. J. Mol. Sci.* **11**(3), 956–966.

Zdolsek, J. M. (1993). Acridine orange-mediated photodamage to cultured cells. *APMIS* **101**(2), 127–132.

Zdolsek, J. M., *et al.* (1990). Photooxidative damage to lysosomes of cultured macrophages by acridine orange. *Photochem. Photobiol.* **51**(1), 67–76.

CHAPTER SIXTEEN

High Content Screening of Defined Chemical Libraries Using Normal and Glioma-Derived Neural Stem Cell Lines

Davide Danovi, Amos A. Folarin, Bart Baranowski, *and* Steven M. Pollard

Contents

1. Introduction 312
   1.1. Neural stem cells *in vivo* and *in vitro* 312
   1.2. Glioblastoma-derived neural stem cells 313
   1.3. High content chemical screening 314
2. Protocol: Chemical Screening Using Human NS and GNS Cells 315
   2.1. Notes and troubleshooting 319
3. Protocol: Automated Quantitation of Cellular Responses 320
   3.1. Notes and troubleshooting 322
   3.2. Future scope and directions 324
   3.3. Equipment and software 326
   3.4. Culture media 326
Acknowledgments 327
References 327

Abstract

Small molecules with potent biological effects on the fate of normal and cancer-derived stem cells represent both useful research tools and new drug leads for regenerative medicine and oncology. Long-term expansion of mouse and human neural stem cells is possible using adherent monolayer culture. These cultures represent a useful cellular resource to carry out image-based high content screening of small chemical libraries. Improvements in automated microscopy, desktop computational power, and freely available image processing tools, now means that such chemical screens are realistic to undertake in individual academic laboratories. Here we outline a cost effective and versatile

Samantha Dickson Brain Cancer Unit, Department of Cancer Biology, UCL Cancer Institute, University College London, London, United Kingdom

Methods in Enzymology, Volume 506
ISSN 0076-6879, DOI: 10.1016/B978-0-12-391856-7.00040-8

time lapse imaging strategy suitable for chemical screening. Protocols are described for the handling and screening of human fetal Neural Stem (NS) cell lines and their malignant counterparts, Glioblastoma-derived neural stem cells (GNS). We focus on identification of cytostatic and cytotoxic "hits" and discuss future possibilities and challenges for extending this approach to assay lineage commitment and differentiation.

1. Introduction

1.1. Neural stem cells *in vivo* and *in vitro*

The term neural stem cell is used to define cells capable of both long-term self-renewal and neuronal/glial differentiation. A variety of phenotypically distinct cell types that fulfill these criteria have been identified in both the developing and adult mammalian central nervous system. In the developing embryo, neuroepithelial cells and radial glia appear sequentially and are the ancestors of a diversity of mature neurons, astrocytes, and oligodendrocytes (Merkle and Alvarez-Buylla, 2006). Neural Stem cells also exist within the walls of the forebrain ventricles and the hippocampus and can generate specific neuronal subtypes throughout life (reviewed in Kriegstein and Alvarez-Buylla, 2009). There is also heterogeneity in the positional identity displayed by NS cells in both embryo and adult, reflected in expression of specific subsets of sequence specific transcription factors (Alvarez-Buylla *et al.*, 2008). Thus, although markers, such as the intermediate filament Nestin, and the transcription factor Sox2 are commonly used to identify neural stem and progenitor cells (Ellis *et al.*, 2004; Lendahl *et al.*, 1990), it should be remembered that "Neural Stem cell" is a broad term used to describe a range of distinct cell states.

It is possible to expand cells with Neural Stem cell properties in cell culture, where continuous self-renewal and multipotent differentiation capacity is clearly evident. *In vitro* expansion removes the dynamic and diverse environmental cues that cells are exposed to *in vivo*. While artificial in nature, cell culture does provide a tractable model system which has proved useful for elucidation of molecular and cellular mechanisms that control Neural Stem cell behavior (Johe *et al.*, 1996; Qian *et al.*, 1998). One of the most popular culture regimes for expanding Neural Stem cells is a suspension culture technique. Primary neural progenitor cultures are exposed to a serum-free neurobasal media supplemented with the mitogen EGF, and after several days "neurospheres" emerge capable of neuronal, astrocyte, and oligodendrocyte differentiation (Reynolds and Weiss, 1992; Reynolds *et al.*, 1992). While this methodology has proved useful for studies of Neural Stem cells, there are many technical and conceptual caveats when relying on the neurosphere assay to infer and quantitate endogenous stem cells (reviewed in Pastrana *et al.*, 2011; Reynolds and Rietze, 2005).

We have previously built on these and other earlier studies and combined the advantages of adherent culture with the long-term clonal expansion that is possible using EGF plus FGF-2 as mitogens (Conti *et al.*, 2005; Pollard *et al.*, 2006b; Sun *et al.*, 2008). Tissue-restricted "NS" cell lines can be established from fetal or adult germinal tissues or following differentiation of embryonic stem (ES) cells and have radial glia-like characteristics (Conti *et al.*, 2005; Pollard *et al.*, 2006a).

Adherent culture provides several important experimental advantages over suspension culture which have been discussed elsewhere (Pollard and Conti, 2007). Most notably, the spontaneous differentiation and cell death that occurs in suspension culture is suppressed by the more uniform culture environment, and stem cells can be readily monitored at single cell resolution—a prerequisite for imaging based studies. Thus, adherent NS cell lines and their related glioblastoma neural stem cell (GNS) counter parts (described below) are especially suited to genetic and chemical screening strategies.

NS cell lines express concomitantly the following repertoire of transcription factors: Sox2, Olig2, Nkx2.2, Pax6, Nfia, and Mash1 (Ascl1); they also display many hallmarks of radial glia, such as RC2, BLBP, and GLAST (Pollard and Conti, 2007). This marker profile suggests NS cell cultures resemble a late forebrain ventral radial glia-like progenitor with tripotent differentiation capacity. This combination of transcription factors is acquired rapidly *in vitro*, partly in response to the activity of exogenous FGF (Gabay *et al.*, 2003; Pollard *et al.*, 2008) and consequently likely restricts the neuronal differentiation potency to GABAergic interneurons—regardless of the stage or position of the parental primary culture. Thus, whether the *in vitro* "NS cell state" precisely corresponds to a specific *in vivo* progenitor remains unclear (Conti and Cattaneo, 2010).

1.2. Glioblastoma-derived neural stem cells

Astrocytomas are the major class of primary brain cancer in humans (Furnari *et al.*, 2007). The hypothesis that these tumors arise from and are sustained by cells with neural stem cell character is currently an active area of research (reviewed in Stiles and Rowitch, 2008). The most aggressive form of astrocytoma (WHO classification, Grade IV) also known as glioblastoma multiforme (GBM) is often associated with hyperactivity of the EGFR signaling pathway. These tumors display cellular heterogeneity, including both immature stem and progenitor cell markers, such as Nestin, Olig2, and Sox2 (Annovazzi *et al.*, 2011; Dahlstrand *et al.*, 1992; Ligon *et al.*, 2004). Thus, GBMs may be viewed as a neoplasia driven by the unconstrained self-renewal of neural stem-like cells.

GBM biopsies can give rise to neurospheres under appropriate culture conditions (Galli *et al.*, 2004; Hemmati *et al.*, 2003; Singh *et al.*, 2003), importantly similar "tumor initiating cells" can also be isolated acutely and therefore seem not to result from exposure to the culture environment (Singh *et al.*, 2004). These cancer stem cell cultures can be maintained long-term *in vitro* while still retaining tumor initiation capacity and differentiation profiles mirroring the original human disease (Galli *et al.*, 2004; Lee *et al.*, 2006; Pollard *et al.*, 2009). This contrasts with cells expanded using "classic" serum containing media, which seems to select for a highly artificial cell type with significant genetic and epigenetic changes unrelated to the human disease (Lee *et al.*, 2006).

Many hallmarks of GBM are properties shared with human NS cell lines, such as dependence on EGFR signaling, expression of *OLIG2*, efficient glial differentiation, and their motility/infiltration *in vitro* and *in vivo*. Importantly, one key difference is that fetal NS cells do not initiate tumors upon transplantation—as they lack the genetic and epigenetic disruptions of their tumorigenic counterparts. Thus, comparisons between normal fetal NS cell lines and their malignant counterparts may help identify clinically relevant pathways. Chemical screening for small molecules that specifically disrupt GNS cell self-renewal may have value in neurooncology.

1.3. High content chemical screening

Identification of small molecules that affect the behavior of stem cells or cancer-derived stem cells is possible through screening of libraries of compounds and then assessing cellular responses. A balance has to be struck between the size of the chemical library and the practicalities of capturing phenotypic information. The larger the library, the more time-consuming and costly will be the collection and analysis of phenotypic data. High content cell screening (HCS) is the term used to describe cell based assays that extract rich phenotypic information from either fixed or live cells following exposure to libraries of small molecules or in genetic screens. This contrasts with high throughput screens, which usually assess hundreds of thousands of compounds using a single biochemical assay. Defined libraries of pharmacologically active cell permeable small molecule libraries are rapidly becoming available from a range of commercial suppliers or academic laboratories. For example, we have previously screened the NIH clinical collection (Pollard *et al.*, 2009) and kinase inhibitor libraries (Danovi *et al.*, 2010), while others have used a library of pharmacologically active compounds (Diamandis *et al.*, 2007). As the specific molecular target is usually known, this simplifies identification of the specifically affected biological pathway, a considerable advantage compared to screening of novel chemical libraries.

HCS is particularly suitable as a discovery tool in stem cell biology where the goal is to uncover specific effects on cell fate (e.g., differentiation, death, proliferation, senescence) and where cellular heterogeneity within the responding population is commonplace. Cell based phenotypic screens make use of automated microscopy, or related image capture technologies, to extract phenotypic information. For example, commercially available imaging systems can quantitate marker expression following fixation and immunocytochemistry of cells at the experimental endpoint. These have been used recently to search for agents that modulate ES cell differentiation, such as induction of neuronal or cardiomyocyte differentiation (reviewed in Xu *et al.*, 2008).

A complementary approach is to make use of live cell imaging to monitor the kinetics of cellular responses and the fates of individual cells. This is the approach that our laboratory has taken recently (Danovi *et al.*, 2010; Pollard *et al.*, 2009), and for which detailed methods are described in this chapter. The advantages of this strategy are the wealth of phenotypic information and kinetic data that can be extracted. This is vital in stem cell biology where differentiation is often nonuniform and asynchronous. By tracking cells over time, it is possible to monitor the behavior of transient intermediate progenitor cells—information which is lost using solely endpoint assays. However, endpoint immunocytochemistry for specific markers can still be carried out and combined with the live cell experiment to provide complementary phenotypic data.

Live cell imaging generates large datasets of many thousands of phase-contrast and/or fluorescent images from which quantitative information and time-lapse movies need to be extracted. Improved image analysis software and computational power now make processing of many thousands of TIFF files achievable at minimal cost using desktop workstations. A free software package that is particularly user-friendly, but flexible and powerful, is the software named CellProfiler and CellProfiler Analyst, developed by scientists at the Broad Institute, USA, which is generously made freely available to the scientific community (Carpenter *et al.*, 2006; Jones *et al.*, 2008; Logan and Carpenter, 2010). Here, we describe a strategy and detailed methods for using human NS cells and GNS cells with a live cell imaging system (Incucyte) and downstream processing with CellProfiler (Fig. 16.1). The costs and labor requirements for such a screen should be within reach of individual academic laboratories and could readily be extended to other types of stem cell or used in genetic screens.

2. Protocol: Chemical Screening Using Human NS and GNS Cells

We describe below protocols for live cell imaging-based chemical screening with both NS cells and GNS cells in 96-well plate format. This approach enables screening of approximately 1000 compounds in single replicate per week. We use the Incucyte imaging platform (Essen Bioscience, USA). This

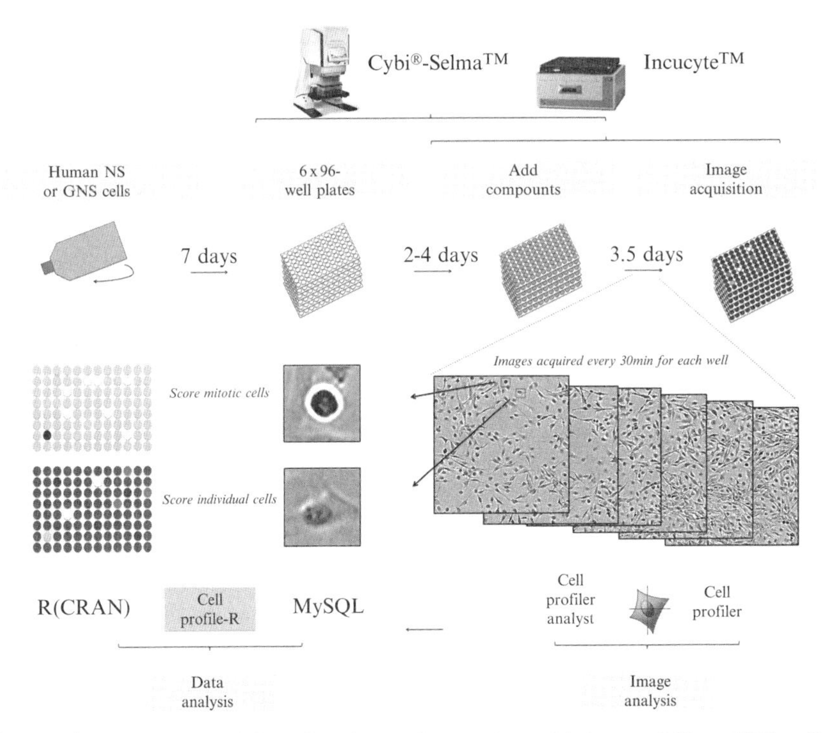

Figure 16.1 Typical workflow for chemical screening with human NS or GNS cell lines. For handling of cells and compounds we use a benchtop 96-well format liquid handling device CyBi®-Selma™ (CyBio). 6 × 96-well plates are screened for 3.5 days using an Incucyte™ (Essen Bioscience) to capture phase contrast images at 60 min intervals. Images are subsequently processed, segmented, and analyzed (CellProfiler). Specific morphologies (e.g., mitotic cells, see inset) can be identified through use of machine learning based training algorithms, with the CellProfiler Analyst software (Jones, 2008).

instrument is housed within a standard tissue culture incubator and contains a movable camera linked to a 10 × objective that can capture images of the same field for each well in a 96-well plate (or 384-well plate) at user defined time intervals. Downstream image processing and quantitative analysis are then carried out to determine total cell number and can be extended to score changes in cell morphology to quantify numbers of mitosis or differentiated cells using CellProfiler and Cell Profiler Analyst software.

The capacity of the Incucyte is six 96-well plates. We typically run parallel triplicate screens of two 96-well plates in each run. NS cells and GNS cells have a doubling time of 3–5 days, and therefore we maintain parallel T75 cultures so that cells are continuously available for new screens. Immunocytochemistry end-point assays can also be incorporated at the end of each screening run and can be readily processed using the downstream image analysis, providing a wealth of phenotypic information. The Incucyte FLR system can also acquire GFP fluorescence images of live cells, and we discuss this application in the final section.

(1) *Cell culture of human NS and GNS cells*
   a. *Cell lines.* New human NS cell and GNS cell lines can be derived using our previously described protocols (Conti *et al.*, 2005; Pollard *et al.*, 2009; Sun *et al.*, 2008). Or alternatively, previously, characterized cell lines are available on request.
   b. *Thaw cells.* Cryovials containing 0.5 ml of $\sim$0.5–1.0 $\times$ 10^6 human NS and GNS cells can be stored at -80 °C for several months. A single vial is rapidly thawed in a water bath at 37 °C; cells are pelleted and replated into a T25 flask containing 10 ml fresh prewarmed complete media (CM) supplemented with Laminin at 1 μg/ml and EGF/FGF-2 (10 ng/ml).
   c. *Detach cells.* Cells will grow to confluence within 2–3 days and are then transferred into a T75 flask as follows: wash gently with 5 ml PBS; immediately aspirate PBS and replace with 1 ml of 1 $\times$ Accutase solution (Sigma); leave cells for 3–5 min in the incubator, checking detachment on microscope every 2 min until around half the cells have detached; tap the flask by knocking on bench to dislodge the remaining cells.
   d. *Harvest and replate cells.* Immediately flood the flask with 10 ml of prewarmed wash media (WM), and resuspend cells smoothly by drawing up and down in a 10 ml pipette two or three times; transfer cells to a 20 ml universal tube and spin for 5 min at 1200 rpm. Resuspend the cell pellet and plate cells (typically 1–2 million cells) in 10 ml of CM plus laminin and EGF/FGF (as in "b") into a T75 flask and place in the incubator. Cells will reach confluence within 3–5 days (depending on cell line) and can then be used for screening or cryopreserved.
   e. *Cryopreservation.* If cells are no longer required for screening they are cryopreserved as follows: harvest a cell pellet (as described above), resuspend the pellet in 1.5 ml of "freeze media" (CM plus DMSO freshly prepared to a final concentration of 10%). Transfer 0.5 ml to each of three cryovials, label, and store at -80 °C for up to several months (or transfer to liquid nitrogen for long-term storage).

(2) *Preparation of 6 $\times$ 96-well plates*
   a. *Prepare cells and media.* Prepare in a 50 ml falcon tube, 35 ml of CM for 6 $\times$ 96-well plates (this is the capacity of the Incucyte system). We typically dispense 3000 cells/well (60,000 cells/ml). Thus, 2 $\times$ 10^6 cells are required for a full run (a near confluent T75 will contain around 5 $\times$ 10^6 cells). Add fresh growth factors and laminin as described above.

b. *Deposit cells*. For handling of cells and compounds, we use a bench-top 96-well format liquid handling device (CyBi Selma, CyBio). Program the liquid handling device to dispense cells (50 μl/well) using sterile tips. Gently shake the trough containing cells to ensure consistent plating density. Disperse cells to ensure even plating by gently shaking each plate and place in the incubator for cells to attach and expand for 2 days.

c. *Commence imaging run to collect "untreated" images*. Go to the "schedule upcoming scans" menu on the Incucyte device. Take three images before addition of compounds (these serve as the untreated "control" images). During the image collection, ensure to include as many experimental details as possible (cell line, plate numbers are included in the plate labels). This is valuable metadata that can be extracted from the filename during the image analysis.

d. *Add compounds and commence screening*. For 6 × 96-well plates, the minimum capture time is 30 min intervals.

e. Check that images and cells are of good quality and consistent density (typically ~100 cells/field). Take the plates to the tissue culture hood for addition of compounds.

(3) *Addition of compounds and library screening*

a. *Prepare library*. Handling of the primary chemical library should be reduced to a minimum. We recommend plating an aliquot of an appropriate dilution into "daughter" plates.

b. Thaw the daughter library plate(s) for a few hours or overnight at 4 °C. Using a 96-well liquid handling device, transfer typically 1–2 μl into a dilution of PBS. For example, for the kinase inhibitor library we carried out a screen at 100 n*M*.

c. Aspirate compounds (typically 1–2 μl) from daughter plates and add to the plates.

d. Insert the plates back into the Incucyte device. Allow 15–20 min equilibration of the plates to ensure condensation does not affect the autofocus and resume image acquisition.

e. Acquire images for 3–4 days. Longer periods would require readdition of the compounds, whereas shorter periods do not generally provide sufficient time for changes in proliferation to be scored accurately.

f. Stop acquisition and retrieve plates for immediate fixation and processing for immunocytochemistry (if required).

g. Archive the single experiment and transfer images to an image processing workstation.

(4) *Immunocytochemistry end-point assays (optional)*

a. To maximize phenotypic information gained from the screen, the plates can be fixed at the experimental endpoint (3.5 days) and processed for immunomarker analysis.

b. Remove media from plates by shaking off into a sink.

c. Immediately add 50 μl of a freshly prepared solution of 4% paraformaldehyde to each well (process one plate at a time to ensure they do not dry out).
d. Leave 10 min to fix cells.
e. Discard the paraformaldehyde (PFA) by shaking off and replace with 50 μl of 1 × PBS supplemented with 0.1% Triton X-100 (PBS-T). Repeat this step to remove any residual PFA. Plates can be stored up to several weeks at 4 °C.
f. Block for 30 min in blocking solution (PBS-T, plus 1% BSA, and 3% goat serum).
g. Incubate with primary antibody in block solution overnight at 4 °C. A range of primary antibodies can be used to assess NS cells and differentiation status and have been described elsewhere (Conti *et al.*, 2005; Pollard *et al.*, 2006b, 2009).
h. Wash twice with PBS-T.
i. Incubate with secondary antibody and DAPI for approximately 1 h.
j. Wash twice with PBS-T.
k. Collect images on an automated immunofluorescence microscope or using plate imaging scanner.

2.1. Notes and troubleshooting

The wash medium (WM) and complete medium (CM) are modified from a previously described "N2B27" formulation (Pollard *et al.*, 2006a), and the full composition is described in the materials section at the end of this chapter. EGF and FGF-2 are stored in 50 μl aliquots at − 20 °C (at 100 and 50 μg/ml stock solutions in PBS, respectively) and are added freshly to the flasks. We recently found that addition of laminin as a culture supplement (1 μg/ml) can effectively promote monolayer growth of NS and GNS cell lines. This reduces the labor-intensive "precoating" of plates with laminin and also reduces costs. This also improves substantially the consistency of the cultures. However, some cell lines cells are prone to grow as clumps or detach as spheres. Increasing the concentration of laminin in the culture media to 2 μg/ml can resolve this.

Human NS and GNS cell cultures vary to some extent in their doubling time, so these guidelines serve as a starting point. In our experience, fetal NS cells expand more consistently if kept at high densities; therefore, we do not split cells at greater than 1:3. This usually means splitting cells once every week. Fresh growth factors are added every 3–4 days.

The protocols described require availability of a liquid handling device. We have used a 96-well CyBi®-Selma™ (CyBio). If a dedicated liquid handling device is not available, then manually handling the cells and depositing the compounds can be achieved using a standard 12-well multichannel pipette although we have found this to be less consistent than an automated system.

The Incucyte system is compatible with 384-well format. However, screens are based on images acquired typical every 30–60 min with 96-well plates. 384-well requires four times greater time intervals to enable each well to be imaged, which in our experience is too long to monitor the kinetics of response to compounds or for tracking of individual cell fates. Access to a dedicated liquid handling device is essential with 384-well plates.

3. Protocol: Automated Quantitation of Cellular Responses

The protocols described above generate a wealth of phase-contrast images (in total ~50 K TIFF image files; 6 plates × 96-wells × 24 h × 3.5 days) per screening run. Quantitative data automatically generated from the Incucyte system software provide an estimate of cell confluence. However, changes in cell morphology such as flattening of cells or line-to-line variations can significantly affect confluence measures. Instead, to exploit the full wealth of information generated by phase-contrast images, we have made use of additional freely available image processing and analysis tools. The quantitation of total cell number can be easily achieved using the procedures described below and can be "trained" to take account of line specific morphologies. Morphological changes such as rounded up cells in mitosis or emergence of differentiated features can be scored.

Our analysis is carried out using CellProfiler (www.cellprofiler.org), although other similar packages are available including Cell Cognition (Held *et al.*, 2010) and the R high-content packages EBImage (Pau *et al.*, 2010). Images are collected with the Incucyte, processed to identify and measure cell morphometric parameters, and then data analyzed using a dedicated R script and CellProfiler Analyst (Fig. 16.1).

The analysis is implemented by using a pipeline of modules which are parameterized to carry out particular functions such as "image loading" and "metadata acquisition," "preprocessing," "object segmentation," "object measurement," and "population of the database" (CSV, SQLite, MySQL). After each image is loaded, it is first segmented and "objects" (typically cells) are identified in order to be indexed, counted, and analyzed. The morphological characteristics of each identified cell are stored in a database table. Pipeline measurement modules such as "MeasureObjectSizeShape" and "MeasureObjectIntensity" enable a wide range of morphological parameters to be measured (e.g., Area, Eccentricity, MajorAxisLength, MinorAxisLength, Orientation, Perimeter, Solidity, Intensity) in an unbiased fashion.

Two tables are initially stored in the database. Each image analyzed yields a row in the *per_image* table, and each object segmented yields a row in the *per_object* table, associated columns in each table store the relevant parameters. Endpoint immunofluorescent images can also be processed through the same pipelines.

CellProfiler Analyst is a software tool that complements CellProfiler, enabling a user to train and classify individual cells using morphological data generated from CellProfiler (Jones *et al.*, 2008). CellProfiler Analyst provides a user-friendly interface for supervised machine learning to quickly train classifiers, computational tools that discriminate between training sets of objects by defining "rules" (i.e., complex algorithms) that distinguish between attributes of each set. A simple graphic user interface is provided for binning objects into user-defined classes. In our studies, this method has proved useful for scoring each cell within the field to quantitate number of mitoses (Davide Danovi and Steve M. Pollard, unpublished).

(1) *Extract images*
 a. Open Incucyte Graphic User Interface.
 b. Double click on the plate icon (this opens a separate window).
 c. Select "Export movie or image set..." from the Utilities pane.
 d. Choose time-points. Name prefix and Target folder and "Metamorph ND" format.
 e. Ensure the filename contains the appropriate metadata.
 f. Repeat steps a–c for each of the six folders (6-well plates).
 g. Images are now in TIFF format and ready for processing/analysis.

(2) *Execute the following CellProfiler modules to identify cells and measure morphological parameters*
 a. LoadImages (handle different channels and capture metadata from file/folder names) IdentifyPrimaryObjects (one per each type of object you wish to count)
 b. MeasureImageAreaOccupied (image level measurement)
 c. MeasureObjectSizeShape (object level measurements)
 d. ExportToDatabase (cache results). These morphological parameters can be scored for each object segmented in the pipeline and are cached in the *per_object* table, image-wise parameter averages of the objects in each image are also cached in the *per_image* table. Additionally included in the *per_image* table is any metadata for example Cell Type, Plate, Well, Time, etc.

(3) *Train cell profiler analyst*
 a. Open Cell Profiler Analyst
 b. Load set of data, using the "properties" file generated by CellProfiler
 c. Select the appropriate number of bins
 d. Retrieve objects.

e. Allocate objects to bins to form a "training set"
f. Train rules.
g. Repeat steps d-f until scoring accuracy converges.

(4) *Example of R script for data analysis*

a. *Loading database per_image table into R*

```
drv < - dbDriver("MySQL")
con < - dbConnect(drv,host = "localhost",dbname =
        databaseName, user = databaseUser,
        pass = databasePassword)
per.image < - dbGetQuery(con,statement = "select *
                    fromper_image")
```

b. *Data remodeling (to show count of metaphase cells over time)*

```
cellcounts.tab < - read.csv(file = CPA.outputFile.csv*)
image.counts < - merge(per.image, cellcounts.tab,
  by.x = "ImageNumber", by.y = "ImageNumber")
plate.map < - read.csv("list-of-well-names")
dat.exp < - split.data.frame(image.counts, image.
  counts$"plate.map"**)
md < - melt(dat.exp[[i]], id = c("Image_Metadata_
  Hour", "Image_Metadata_Well"), measure.vars =
  c("Metaphase.Cell.Count"))
cd < - cast(md,Image_Metadata_Hour ~ Image_Meta-
  data_Well)
```

*CellProfiler Analyst counts summarized in delimited file from "Score All" operation.

**Grouping variable for screen compound, alternatively well index.

c. *Visualize morphological changes across a screen*

Unsupervised clustering of image-wise parameter averages by well (matrix X) can be carried out using a heatmap (gplots package) for hierarchical clustering:
heatmap.2(X)
Or principle component analysis:

```
m < - prcomp(X)
biplot(m)
```

3.1. Notes and troubleshooting

Incucyte: An internal algorithm for the Incucyte device can measure the area of a field covered by cells (confluence). This provides a simple means to quickly assess the success of the screening run and to identify any outliers. All the control wells should show similar curves and the cells should increase in confluence, reaching at least 50% more than the initial plating confluence. Low initial plating confluences are advantageous as variations in initial confluence from well to well are minimized and the range of signal detection is favorable. As in other screening methods, the "power" of the screen is assessed by the z' factor (Zhang *et al.*, 1999).

CellProfiler: The computational analysis of many thousands of TIFFs is potentially a bottleneck. However, modern multi-core processors now widely available can process a full Incucyte screening experiment overnight. We have used the latest Xeon processors that contain six cores (*x*2/per workstation) and CellProfiler jobs can be batched to each core providing an order of magnitude increase in throughput. Archived experiments store compressed images that are not immediately available to the users. Images can be derived from archives when convenient and in different formats such as TIFF or JPEG. Single experiments (e.g., one round of a 6 plate screen) need to be archived separately. We have used the "MetaMorph ND File Sequence" which gives more stability in the derivation of big archives. To avoid problems associated with retrieval of images from large archive (i.e., multiple screening runs archived together), we instead recommend archiving of single experiments.

CellProfiler Analyst: Using CellProfiler Analyst, we are able to estimate the number of mitotic cells on each image. First, several morphological parameters are measured on each of the objects in a CellProfiler pipeline and then CellProfiler Analyst is used to generate a classifier to identify cells undergoing metaphase. Mitotic cells are readily identified by their characteristic rounded morphology, a phase bright "halo," and typically a central dark band from the condensed chromosomes (Fig. 16.1). We also take advantage of the classification to denote an additional bin of objects which we deem to be erroneously segmented parts of the image, such as cell debris. We found that cell protusions and filopodia were sometimes segmented as an individual object/cell. We therefore scored cells using three bins, classifying either as "mitotic cell," "interphase cell," and "non-cell"; numbers of each class in each image are exported to a spreadsheet for data analysis.

Training is an iterative process, where initial training sets are binned from a random sample, then an interim classifier is constructed. This is then used to return more objects of each class, and a new classifier is trained and the process is repeated till accurate scoring is achieved. It is more effective to begin training with a small number of highly stereotypical mitotic objects, then train a classifier using five rules; using this classifier to select further metaphase objects into the training set, and place the remaining objects into other bins, this process is repeated incrementing to 20, 50, 100 rules. As each object in the (*per_object*) table is stored with information describing it, this information enables the software to determine objects of particular classes. Object counting can then proceed. The pipeline should therefore include measurements on the cells that provide functional information to carry out the classification (e.g., intensity, area, shape, etc.). The more relevant information is captured, the better the classification will perform.

Once the sets are trained, images can be selected and scored either singularly or as a whole. With the "score image" operation, the user can chose an image and check whether objects are appropriately segmented and classified. The image-wise totals of each class of objects from a "score all" operation can also be stored in the database or as a delimited file; additional options exist to capture the individual object classifications by setting the *class_table* in the CellProfiler properties file. Alternatively, the counts are typically summarized as number of objects belonging to each class per image.

The database can be drawn on for summarization, statistical analysis and graphical presentation. The data generated in these types of analyzes are too large to be easily handled using spreadsheet applications. Instead, we have found it extremely useful to carry out the necessary data analysis within R, a freely available statistical analysis application. Using this software, it is relatively easy to interface with the SQLite or MySQL database tables, and reorganize this data in a meaningful way for graphical presentation or statistical analysis. A number of package extensions to R are of particular use. Importing the databases into R can be handled using the database interface functions defined in RMySQL, RSQLite extension packages. An additional set of utility functions for working with CellProfiler/CellProfiler Analyst data in R have been included by us as an R package (http://code.google.com/p/cellprofile-r/).

Collating and restructuring the data is an important step. Typically, it is necessary to restructure the data and the utility functions (*melt* and *cast*) in the reshape package can be used for example to express counts/image over time. Using functions in R one can assess changes induced by compounds across a whole screening dataset. Data from the object or image tables can be analyzed and visualized by performing unsupervised clustering on the matrix of cells/images × morphological parameters (e.g., hierarchical clustering or principle component analysis) to show cells or images with common morphology or extreme outliers relative to the untreated control wells/cells. Looking for morphological outliers in this way is a useful method for identification of "hits".

3.2. Future scope and directions

The Incucyte FLR is equipped with fluorescence image capture capability and therefore an obvious extension of protocols described here is to combine phase image capture with live cell imaging of GFP reporter genes that are activated by gene-specific promoters. Although human NS and GNS cells can be genetically modified. Mouse NS cells can be generated readily from any ES cell line or fetal/adult mice. Further, mouse NS cells proliferate and differentiate more rapidly than their human counterparts, providing greater scope for effects on self-renewal and differentiation to be identified within a 3–4-day experiment. Thus, we have initially piloted such studies using mouse reporter cell lines.

Our preliminary results from a fluorescence based screen using a mouse NS cell with eGFP reporter controlled by an oligodendrocyte lineage marker have raised several important issues. The setup for image analysis of fluorescently labeled images is similar to the phase image analysis described above. As the proportion of fluorescent cells may often be of greater interest than total counts, it is possible to acquire concurrent phase and fluorescent images using the IncucyteFLR and, using these pairs of images, evaluate the percentage of fluorescently labeled cells.

We have found that the autofluorescence of the media can compromise fluorescence image capture. This is due to the riboflavin supplements within the DMEM:F12 basal media. We find that if EMEM is used instead there is a dramatic reduction in background autofluorescence. Also the longer time intervals required for fluorescence image capture can result in phototoxicity. This limits the scope of these studies as individual cells cannot therefore be tracked. Providing counts of fluorescently labeled cells (e.g., GFP) requires some consideration to how these are to be normalized across datasets as absolute levels of fluorescence and background levels can vary between runs. Each run should be therefore internally controlled. We have also experienced an image-wide "halo" where background intensity increases radially, this effect seems particularly prominent at early timepoints and does decrease over time as background fluorescence is bleached. It is possible to mitigate the effects of the halo by using CellProfiler's "CorrectIlluminationCalculate" and "CorrectIlluminationApply" modules. While it is in principle possible to get clean signal from the labeled cells, in practice we have observed a number of possible sources of artifact, these include fragments, debris, and autofluorescent compounds from the screen, careful use of the thresholding options can be used to eliminate these or reduce them greatly.

Further applications of these approaches are the tracking of cells to gain insights into motility. Such single cell tracking requires shorter time intervals, typically 5–10 min. An extension of the tracking is the incorporation of both tracking and differentiation markers to create individual lineage trees for multiple cells within each well. Such studies have proved possible using small sample fields and should be scalable (Qian *et al.*, 1998; Ravin *et al.*, 2008).

Over the next few years there should be continued improvements in image acquisition technologies, analysis tools, and computational power. Moreover, our ability to engineer multilineage transgenic reporters makes it foreseeable that multilineage stem cell and differentiation reporters, alongside lineage tracking, motility, cell death, and mitosis may all be accurately quantified in living cells. Such assays will be extendable to both chemical and genetic screening of stem cells and should identify a plethora of new laboratory reagents, pathways, and potential drugs for use in biomedical applications.

3.3. Equipment and software

Cell counter Cell Viability Analyzer, VI-Cell XR (Beckman-Coulter).

Multichannel pipette (e.g., 20–200 μl for the cells and 1–20 μl for the compounds)

Cell dispenser and compound dispenser	CyBi®-SELMA™, CyBio
Live image based device	Incucyte HD™ and Incucyte FLR™, Essen Bioscience
Image processing workstation	Dell T5500, 2 × Intel Xeon 2.66 GHz (12 cores), 24 GB DDR3 RAM, 1 GB ATI FirePro graphics card, 2 × 2TB internal hard drive
Imaging software	CellProfiler and CellProfiler Analyst, Broad Institute (website: www.cellprofiler.org)
Data analysis	R (CRAN), Additional R Packages: RMySQL, RSQLite, reshape, gplots

3.4. Culture media

Complete media comprises:

DMEM/Ham's F-12 liquid with L-glutamine	(PAA; E15-813)
Non Essential Amino Acids (NEAA)	(PAA; M11-003 100 ml)
HEPES buffer	(PAA; S11-001, 100 ml)
Penicillin/Streptomycin	(PAA; P11-010, 100 ml)
N2 Supplement	(PAA; F005-004, 5 ml)
B27 Supplement	(Invitrogen 17504-044)
Glucose (2 *M*)	(Sigma, G8644-100ML)
2-Mercaptoethanol	(Invitrogen 31350-010 20 ml)
BSA (aliquot into 10 ml stocks)	(Invitrogen 15260-037, 100 ml)

To make up 10 × 500 ml bottles.

- Thaw all components
- Use fresh stocks of everything; must be prepared sterile in TC hood.
- To each 500 ml bottle of DMEM/F-12
- Add 7.25 ml of sterile glucose
- Add 5 ml of NEAA
- Add 5 ml of Pen/Strep
- Add 2.25 ml of HEPES
- Add 800 μl of BSA solution (75 mg/ml)

- Add 0.5 ml 2-mercaptoethanol
- Add 5 ml of B27
- Add 2.5 ml of N2
- Take 10 ml aliquot for sterility check with Tryptose solution 1:1
- Freeze 10 bottles at -20 °C

Wash media comprises: all components of the CM except the N2 and B27 supplements.

Note: for live fluorescence image capture DMEM:F12 should be replaced with Eagles modified essential media EMEM (Sigma, M2645-10L).

Other reagents:

EGF	(PeproTech EC Ltd, 315-09)
FGF-2	(PeproTech EC Ltd, 100-18B)
Laminin	(Sigma, L2020)
Accutase	(Sigma, A6964)

ACKNOWLEDGMENTS

We thank Stefan Stricker and Christine Ender for helpful comments on the chapter. This work was supported by grants from Cancer Research UK and the Samantha Dickson Brain Tumor Trust.

REFERENCES

Alvarez-Buylla, A., Kohwi, M., Nguyen, T. M., and Merkle, F. T. (2008). The heterogeneity of adult neural stem cells and the emerging complexity of their niche. *Cold Spring Harb. Symp. Quant. Biol.* **73,** 357–365.

Annovazzi, L., Mellai, M., Caldera, V., Valente, G., and Schiffer, D. (2011). SOX2 expression and amplification in gliomas and glioma cell lines. *Cancer Genomics Proteomics* **8,** 139–147.

Carpenter, A. E., Jones, T. R., Lamprecht, M. R., Clarke, C., Kang, I. H., Friman, O., Guertin, D. A., Chang, J. H., Lindquist, R. A., Moffat, J., Golland, P., and Sabatini, D. M. (2006). Cell Profiler: Image analysis software for identifying and quantifying cell phenotypes. *Genome Biol.* **7,** R100.

Conti, L., and Cattaneo, E. (2010). Neural stem cell systems: Physiological players or *in vitro* entities? *Nat. Rev. Neurosci.* **11,** 176–187.

Conti, L., Pollard, S. M., Gorba, T., Reitano, E., Toselli, M., Biella, G., Sun, Y., Sanzone, S., Ying, Q. L., Cattaneo, E., and Smith, A. (2005). Niche-independent symmetrical self-renewal of a mammalian tissue stem cell. *PLoS Biol.* **3,** e283.

Dahlstrand, J., Collins, V. P., and Lendahl, U. (1992). Expression of the class VI intermediate filament nestin in human central nervous system tumors. *Cancer Res.* **52,** 5334–5341.

Danovi, D., Falk, A., Humphreys, P., Vickers, R., Tinsley, J., Smith, A. G., and Pollard, S. M. (2010). Imaging-based chemical screens using normal and glioma-derived neural stem cells. *Biochem. Soc. Trans.* **38,** 1067–1071.

Diamandis, P., Wildenhain, J., Clarke, I. D., Sacher, A. G., Graham, J., Bellows, D. S., Ling, E. K., Ward, R. J., Jamieson, L. G., Tyers, M., and Dirks, P. B. (2007). Chemical genetics reveals a complex functional ground state of neural stem cells. *Nat. Chem. Biol.* **3,** 268–273.

Ellis, P., Fagan, B. M., Magness, S. T., Hutton, S., Taranova, O., Hayashi, S., McMahon, A., Rao, M., and Pevny, L. (2004). SOX2, a persistent marker for multipotential neural stem cells derived from embryonic stem cells, the embryo or the adult. *Dev. Neurosci.* **26,** 148–165.

Furnari, F. B., Fenton, T., Bachoo, R. M., Mukasa, A., Stommel, J. M., Stegh, A., Hahn, W. C., Ligon, K. L., Louis, D. N., Brennan, C., Chin, L., DePinho, R. A., and Cavenee, W. K. (2007). Malignant astrocytic glioma: Genetics, biology, and paths to treatment. *Genes Dev.* **21,** 2683–2710.

Gabay, L., Lowell, S., Rubin, L. L., and Anderson, D. J. (2003). Deregulation of dorsoventral patterning by FGF confers trilineage differentiation capacity on CNS stem cells *in vitro*. *Neuron* **40,** 485–499.

Galli, R., Binda, E., Orfanelli, U., Cipelletti, B., Gritti, A., De Vitis, S., Fiocco, R., Foroni, C., Dimeco, F., and Vescovi, A. (2004). Isolation and characterization of tumorigenic, stem-like neural precursors from human glioblastoma. *Cancer Res.* **64,** 7011–7021.

Held, M., Schmitz, M. H., Fischer, B., Walter, T., Neumann, B., Olma, M. H., Peter, M., Ellenberg, J., and Gerlich, D. W. (2010). Cell Cognition: Time-resolved phenotype annotation in high-throughput live cell imaging. *Nat. Methods* **7,** 747–754.

Hemmati, H. D., Nakano, I., Lazareff, J. A., Masterman-Smith, M., Geschwind, D. H., Bronner-Fraser, M., and Kornblum, H. I. (2003). Cancerous stem cells can arise from pediatric brain tumors. *Proc. Natl. Acad. Sci. USA* **100,** 15178–15183.

Johe, K. K., Hazel, T. G., Muller, T., Dugich-Djordjevic, M. M., and McKay, R. D. (1996). Single factors direct the differentiation of stem cells from the fetal and adult central nervous system. *Genes Dev.* **10,** 3129–3140.

Jones, T. R., Kang, I. H., Wheeler, D. B., Lindquist, R. A., Papallo, A., Sabatini, D. M., Golland, P., and Carpenter, A. E. (2008). Cell profiler analyst: Data exploration and analysis software for complex image-based screens. *BMC Bioinformatics* **9,** 482.

Kriegstein, A., and Alvarez-Buylla, A. (2009). The glial nature of embryonic and adult neural stem cells. *Annu. Rev. Neurosci.* **32,** 149–184.

Lee, J., Kotliarova, S., Kotliarov, Y., Li, A., Su, Q., Donin, N. M., Pastorino, S., Purow, B. W., Christopher, N., Zhang, W., Park, J. K., and Fine, H. A. (2006). Tumor stem cells derived from glioblastomas cultured in bFGF and EGF more closely mirror the phenotype and genotype of primary tumors than do serum-cultured cell lines. *Cancer Cell* **9,** 391–403.

Lendahl, U., Zimmerman, L. B., and McKay, R. D. (1990). CNS stem cells express a new class of intermediate filament protein. *Cell* **60,** 585–595.

Ligon, K. L., Alberta, J. A., Kho, A. T., Weiss, J., Kwaan, M. R., Nutt, C. L., Louis, D. N., Stiles, C. D., and Rowitch, D. H. (2004). The oligodendroglial lineage marker OLIG2 is universally expressed in diffuse gliomas. *J. Neuropathol. Exp. Neurol.* **63,** 499–509.

Logan, D. J., and Carpenter, A. E. (2010). Screening cellular feature measurements for image-based assay development. *J. Biomol. Screen.* **15,** 840–846.

Merkle, F. T., and Alvarez-Buylla, A. (2006). Neural stem cells in mammalian development. *Curr. Opin. Cell Biol.* **18,** 704–709.

Pastrana, E., Silva-Vargas, V., and Doetsch, F. (2011). Eyes wide open: A critical review of sphere-formation as an assay for stem cells. *Cell Stem Cell* **8,** 486–498.

Pau, G., Fuchs, F., Sklyar, O., Boutros, M., and Huber, W. (2010). EBImage—An R package for image processing with applications to cellular phenotypes. *Bioinformatics* **26,** 979–981.

Pollard, S. M., and Conti, L. (2007). Investigating radial glia *in vitro*. *Prog. Neurobiol.* **83,** 53–67.

Pollard, S. M., Benchoua, A., and Lowell, S. (2006a). Neural stem cells, neurons, and glia. *Methods Enzymol.* **418,** 151–169.

Pollard, S. M., Conti, L., Sun, Y., Goffredo, D., and Smith, A. (2006b). Adherent neural stem (NS) cells from fetal and adult forebrain. *Cereb. Cortex* **16**(Suppl 1), i112–i120.

Pollard, S. M., Wallbank, R., Tomlinson, S., Grotewold, L., and Smith, A. (2008). Fibroblast growth factor induces a neural stem cell phenotype in fetal forebrain progenitors and during embryonic stem cell differentiation. *Mol. Cell. Neurosci.* **38,** 393–403.

Pollard, S. M., Yoshikawa, K., Clarke, I. D., Danovi, D., Stricker, S., Russell, R., Bayani, J., Head, R., Lee, M., Bernstein, M., Squire, J. A., Smith, A., and Dirks, P. (2009). Glioma stem cell lines expanded in adherent culture have tumor-specific phenotypes and are suitable for chemical and genetic screens. *Cell Stem Cell* **4,** 568–580.

Qian, X., Goderie, S. K., Shen, Q., Stern, J. H., and Temple, S. (1998). Intrinsic programs of patterned cell lineages in isolated vertebrate CNS ventricular zone cells. *Development* **125,** 3143–3152.

Ravin, R., Hoeppner, D. J., Munno, D. M., Carmel, L., Sullivan, J., Levitt, D. L., Miller, J. L., Athaide, C., Panchision, D. M., and McKay, R. D. (2008). Potency and fate specification in CNS stem cell populations *in vitro*. *Cell Stem Cell* **3,** 670–680.

Reynolds, B. A., and Rietze, R. L. (2005). Neural stem cells and neurospheres—Re-evaluating the relationship. *Nat. Methods* **2,** 333–336.

Reynolds, B. A., and Weiss, S. (1992). Generation of neurons and astrocytes from isolated cells of the adult mammalian central nervous system. *Science* **255,** 1707–1710.

Reynolds, B. A., Tetzlaff, W., and Weiss, S. (1992). A multipotent EGF-responsive striatal embryonic progenitor cell produces neurons and astrocytes. *J. Neurosci.* **12,** 4565–4574.

Singh, S. K., Clarke, I. D., Terasaki, M., Bonn, V. E., Hawkins, C., Squire, J., and Dirks, P. B. (2003). Identification of a cancer stem cell in human brain tumors. *Cancer Res.* **63,** 5821–5828.

Singh, S. K., Hawkins, C., Clarke, I. D., Squire, J. A., Bayani, J., Hide, T., Henkelman, R. M., Cusimano, M. D., and Dirks, P. B. (2004). Identification of human brain tumor initiating cells. *Nature* **432,** 396–401.

Stiles, C. D., and Rowitch, D. H. (2008). Glioma stem cells: A midterm exam. *Neuron* **58,** 832–846.

Sun, Y., Pollard, S., Conti, L., Toselli, M., Biella, G., Parkin, G., Willatt, L., Falk, A., Cattaneo, E., and Smith, A. (2008). Long-term tripotent differentiation capacity of human neural stem (NS) cells in adherent culture. *Mol. Cell. Neurosci.* **38,** 245–258.

Xu, Y., Shi, Y., and Ding, S. (2008). A chemical approach to stem-cell biology and regenerative medicine. *Nature* **453,** 338–344.

Zhang, J. H., Chung, T. D., and Oldenburg, K. R. (1999). A simple statistical parameter for use in evaluation and validation of high throughput screening assays. *J. Biomol. Screen.* **4,** 67–73.

CHAPTER SEVENTEEN

High-Throughput Screening in Primary Neurons

Punita Sharma,* D. Michael Ando,*,† Aaron Daub,*,‡,§
Julia A. Kaye,*,§ *and* Steven Finkbeiner*,§,¶

Contents

1. Introduction 332
2. Sample Preparation 333
   2.1. Outbred, inbred, or hybrid mouse 333
   2.2. Dissection of mouse cortices and primary neuron cultures 335
   2.3. Cellular and subcellular labeling 337
   2.4. Transfection of primary neurons 340
3. Image Acquisition 341
   3.1. Plate management 341
   3.2. Our HTS microscopy platform 343
   3.3. Data storage and retrieval systems 344
4. Data Analysis 346
   4.1. Image analysis pipeline 346
   4.2. Image analysis software packages used in our laboratory 352
5. Statistical Approaches to HCS and Multivariate Data 352
6. Future Directions 356
7. Concluding Remarks 357
Acknowledgments 358
References 358

Abstract

Despite years of incremental progress in our understanding of diseases such as Alzheimer's disease (AD), Parkinson's disease (PD), Huntington's disease (HD), and amyotrophic lateral sclerosis (ALS), there are still no disease-modifying

* Gladstone Institute of Neurological Disease, San Francisco, California, USA
† Biomedical Sciences Graduate Program, University of California, San Francisco, California, USA
‡ Medical Scientist Training Program and Program in Bioengineering, University of California, San Francisco, California, USA
§ Taube-Koret Center for Huntington's Disease Research, The Hellman Family Foundation Program in Alzheimer's Disease Research and the Consortium for Frontotemporal Dementia Research San Francisco, San Francisco, California, USA
¶ Departments of Neurology and Physiology, Graduate Programs in Neuroscience, Biomedical Sciences, and Cell Biology, University of California, San Francisco, California, USA

Methods in Enzymology, Volume 506
ISSN 0076-6879, DOI: 10.1016/B978-0-12-391856-7.00041-X

therapeutics. The discrepancy between the number of lead compounds and approved drugs may partially be a result of the methods used to generate the leads and highlights the need for new technology to obtain more detailed and physiologically relevant information on cellular processes in normal and diseased states. Our high-throughput screening (HTS) system in a primary neuron model can help address this unmet need. HTS allows scientists to assay thousands of conditions in a short period of time which can reveal completely new aspects of biology and identify potential therapeutics in the span of a few months when conventional methods could take years or fail all together. HTS in primary neurons combines the advantages of HTS with the biological relevance of intact, fully differentiated neurons which can capture the critical cellular events or homeostatic states that make neurons uniquely susceptible to disease-associated proteins. We detail methodologies of our primary neuron HTS assay workflow from sample preparation to data reporting. We also discuss the adaptation of our HTS system into high-content screening (HCS), a type of HTS that uses multichannel fluorescence images to capture biological events *in situ*, and is uniquely suited to study dynamical processes in living cells.

1. Introduction

High-throughput screening (HTS) is the iterative testing of different substances in a common assay. A screen is generally considered high throughput if it can assay >10,000 assays (wells) per day. HTS allows a researcher to quickly conduct millions of tests and to rapidly identify relevant modifier genes, proteins, or compounds involved in a specific biological pathway. The results of these screens typically identify new drug targets or drug activities at a single target and can unveil structure–function relationships in small-molecule "hits" and functional clustering within biological pathways.

The first HTS methods were developed by the pharmaceutical industry (Hertzberg and Pope, 2000) and were *in vitro* assays measuring molecular interactions by florescence, luminescence, or absorbance readouts (Inglese *et al.*, 2007; Macarron, 2006; Macarron and Hertzberg, 2009). A smooth transition from hits generated *in vitro* to efficacious compounds in more complex disease models—in cells, tissues, and most critically in animal models—has often been hard to accomplish (Houston and Galetin, 2008; Zhang *et al.*, 2000). Cell- and animal-based assays have the advantage of identifying compounds within the complex environments of cells and tissues but are more costly, difficult to miniaturize, and tend to have lower throughput due to their complexity. Diseases that affect the brain add another layer of complications. Mature neurons differentiated *in vivo* must be derived from primary sources and are difficult to transfect. Thus, most primary large-scale screens use neuroblastoma cell lines. Better culturing and transfection protocols, however, now make HTS with primary neurons more feasible, and the increased biological and clinical relevance (Daub *et al.*, 2009; Nolan, 2007) is worth the extra effort and expense. We have developed a HTS method for primary neurons that is applicable for large-scale testing, ranging from compound libraries to whole-genome RNA interference (RNAi).

Our laboratory established primary neuron models of a variety of neurodegenerative diseases, including Huntington's disease (HD), Parkinson's disease, Alzheimer's disease, amyotrophic lateral sclerosis, and frontotemporal dementia, with the aim of using these models to understand disease mechanisms and identify therapeutics.

This chapter will focus on our work with HD. HD results from a mutation that causes an expanded polyglutamine (polyQ) tract in the N-terminal region of the huntingtin (htt) protein (Brouillet *et al.*, 1999). Htt polyQ expansions >40 glutamines result in progressive neurodegeneration and death. The disease is characterized by microscopic inclusion bodies (IBs) of aggregated htt and neuronal death. Importantly, from a drug-discovery perspective, we developed both a primary neuron model of HD (Arrasate *et al.*, 2004) that recapitulates many key features of the human disease (Miller *et al.*, 2010) and an automated microscope system (Arrasate and Finkbeiner, 2005) that returns to precisely the same neuron after arbitrary intervals, even after cells have been removed from the microscope stage. Using the primary neuron model and the automated microscope, we observed that IBs reduce both htt levels and the risk of neuronal death (Arrasate *et al.*, 2004; Finkbeiner *et al.*, 2006). To resolve this, we also employed a statistical approach called survival analysis that provides a numerical readout of the risk of different parameters (e.g., levels of htt) on the eventual death of the neuron.

Here we describe how we adapted the primary neuron model, automated microscopy, and survival analysis into a HTS workflow (Fig. 17.1) to create a platform that can be used for drug screening, drug testing, and small-scale protein or genetic screens and large-scale genome-wide RNAi screens. We detail our methods for sample preparation, primary neuron transfection, assay plate management, along with our custom-built image acquisition, data handling, data mining, and data reporting platform. These approaches can also be applied to primary neuron models of other diseases.

2. Sample Preparation

2.1. Outbred, inbred, or hybrid mouse

The optimal genetic background for a HTS depends heavily on the purpose of the screen. There is no *a priori* reason to designate any particular background as "standard" since each background has its own pros and cons. For our assay development, we used the inbred C57BL/6 strain. This inbred

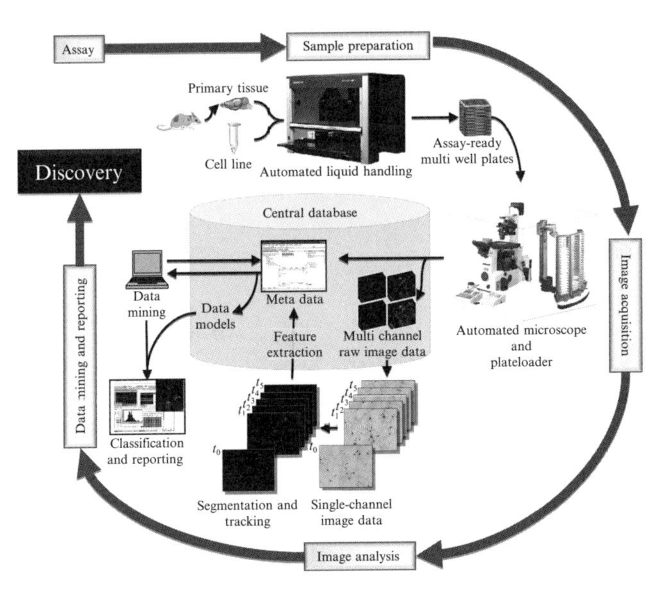

Figure 17.1 Schematic of our HTS System. Workflow of our high-throughput screening system for live-cell imaging showing the entire process from tissue preparation to data mining and reporting (From Daub *et al.*, 2009). (For color version of this figure, the reader is referred to the Web version of this chapter.)

strain is well characterized genetically and behaviorally, making it convenient for examining mutations developed or backcrossed onto the strain. Inbred strains also provide a greater level of consistency due to the lack of genetic drift.

However, recessive traits that affect behavioral, anatomical, and/or physiological phenotypes in inbred strains can limit the generalizability of screening results (Staats, 1985). Each strain has its own set of alleles and polymorphisms, whose relevance to those in humans might be unknown and might confound therapy development. Finally, it is highly likely that the inbred strain may not only be a key determinant of the possible hits that can be detected in the screen but also the efficacy of the hits for further therapeutics. While developing our assays, we found that primary neurons from inbred mice are modestly but significantly less viable (perhaps due to a lack of outbred vigor), which might affect certain assays.

Two alternatives of inbred strains are hybrid (F1 progeny from a cross between two inbred strains) or outbred strains. Hybrid backgrounds have less penetrance of recessive traits, but may still have confounding genetic abnormalities (Spyropoulos *et al.*, 2003). Outbred strains, in which large gene pools supposedly more accurately mimic the range of genetic polymorphisms and phenotypic variations that one would find in humans, may make them a preferred choice when developing assays for human diseases.

The disadvantage with outbred strains, though, is that variability in results can occur over time due to genetic drifts, and depending on the strength of the phenotype, such variability can sometimes entirely obscure an effect. Thus, the choice of the primary neuron source is critical to the biology tested by the assay.

2.2. Dissection of mouse cortices and primary neuron cultures

Required equipment and reagents

- 5% CO_2, 37 °C temperature, 95% relative humidity incubator (for coating plates and growing cells)
- Dissection microscope
- Dissection tools (two forceps, scissors, chemical spatula; Fine Science Tools Cat. No. 11295-10, 14060-09, 10099-15; autoclave sterilized)
- *Neurobasal medium*: 500-mL Neurobasal (Invitrogen, 21103-049) with 10-mL B27 (Invitrogen, 17504), 200 m*M* Glutamax (Invitrogen, 35050), and penicillin/streptomycin mix (100 U mL^{-1}/100 μg mL^{-1}; GIBCO, 15140-122). Mix, sterile filter, and store at 4 °C. Use within 1 week
- *OptiMEM/glucose*: OptiMEM I reduced serum medium (Invitrogen, 31985) with 20 m*M* glucose (Sigma, G8270). Mix, sterile filter, and store at 4 °C. Use within 1 week
- *Dissociation medium/kyuneric acid (DM/KY)*: For DM, add 81.8 m*M* Na_2SO_4, 30 m*M* K_2SO_4, 5.8 m*M* $MgCl_2$, 0.25 m*M* $CaCl_2$, 1 m*M* HEPES, 20 m*M* glucose, 0.001% phenol red, 0.16 m*M* NaOH. Adjust the pH to 7.5–7.6 and make up to 1 L with distilled water. For 10 × KY solution, gradually add 10 m*M* KY to 1 L distilled water containing 0.0025% phenol red, 5 m*M* HEPES, and 100 m*M* $MgCl_2$. Use the color of the phenol red to titrate the pH of the solution back up to about 7.4 as the acid dissolves. If necessary, use 1 N NaOH to titrate pH. Mix, sterile filter, and store at 4 °C. Use within 2 months
- *Papain solution*: Make fresh for each dissection. Add 0.2 mg/mL L-cysteine (Sigma, W326305) in 10 mL of DM/KY. Adjust pH to 7.5–7.6 with 1 N NaOH. Mix, sterile filter, and keep in a 37 °C water bath. At 10 min before the dissection is finished, add 100 U of papain (Worthington Biochemical 3.4.22.2)
- *Trypsin inhibitor solution*: Make fresh for each dissection. Add 15 mg/mL trypsin inhibitor (Sigma, T2011) in 10 mL of DM/KY. Adjust the pH to 7.5–7.6 with 1 N NaOH. Mix, sterile filter, and keep in a 37 °C water bath

Method

1. *One day before dissection*: Add 0.05 mg/mL poly D-lysine (Sigma, P4707) to sterile water. Mix thoroughly. To each 96-well plate (Techno Plastic

Products, TPP92696), add the solution (100 μL/well). Leave the plates in the incubator at least overnight. However, we have noticed better cell adhesion if plates are left in the incubator for several nights.

2. *On the day of dissection*: wash the plates twice with sterile water (200 μL/well for each wash); remove the final wash; and return them liquid-free back in the incubator. Liquid-free plates can be sealed in a plastic film and stored at 4 °C for up to a month.
3. Euthanize a time-pregnant female mouse (with embryos at E18–20). Clean the belly of the mouse with 70% sterile alcohol and, using cesarean section, remove the uterus into a 10-cm culture dish containing ice-cold DM/KY and keep on ice.
4. This step uses ice-cold DM/KY and all steps are carried out by placing the dishes on ice. Remove embryos and decapitate into another 10-cm dish containing DM/KY. For each head, remove the skin, open the skull, and scoop out the entire brain into another 10-cm dish containing DM/KY. Intermittently, swirl the dish with the dissected brains to prevent local buildup of lactic acid.
5. The rest of the dissection is also carried out on ice. Place a 20-cm dish containing ice under a dissection microscope and then place the dish containing the brains on top of the ice. For each brain, remove the meninges and superficially separate the cortical hemispheres to locate the capillary network of the choroid plexus of the lateral ventricles. Split the cortices exactly along the plane of the choroid plexus to give the lateral half of the cortical hemisphere. Dissect away the striatum and hippocampus to give just the cortex. Cut the cortex into three to four smaller pieces and transfer into a 15-mL falcon tube containing ice-cold DM/KY and keep on ice.
6. Before finishing the dissections of the last one to two brains, add papain to the papain solution and place both papain solution and trypsin inhibitor into a 37 °C water bath.
7. Sterile filter the papain and trypsin inhibitor solutions. Carefully aspirate out the excess DM/KY from the falcon tube with a pipette, leaving behind the cortices in minimal DM/KY. Add 10 mL of papain solution to the dissected tissue and incubate in a 37 °C water bath for 15 min, gently mixing every 5 min.
8. Aspirate out papain solution, add 10 mL of trypsin inhibitor solution, and incubate at 37 °C for another 15 min, again gently mixing every 5 min. Aspirate the trypsin inhibitor solution and wash twice with 10 mL of Optimem/glucose solution (kept at 37 °C). Aspirate the Optimem/glucose solution.
9. Add 5 mL of Optimem/glucose. Triturate gently a few times with a 5-mL sterile pipette until the solution turns slightly cloudy. Allow the cortices to settle to the bottom of the falcon tube and then take the cloudy supernatant and transfer it to an empty 50-mL falcon tube. Add 5 mL of new Optimem/glucose to the cortices, repeat the trituration, and transfer until

most of the cortices are dissociated. Allow everything to settle in the 50-mL conical tube, and using a P1000 pipette, suck up and discard the debris at the bottom of the tube. Gently mix the cell suspension.

10. To obtain the cell count, pipette out 10 μL of the cell suspension into an eppendorf tube containing 10 μL of DM/KY and 10 μL of Trypan blue (Sigma, T8154) and count using a cytometer. Dilute the cells with Optimem/glucose solution to a final count of 0.6 million cells/mL.
11. Plate 120 μL (~70,000 cells) per well into a 96-well plate. Swirl the plate gently to make sure the neurons are evenly distributed and leave the plate in the sterile hood at room temperature for 1 h for the neurons to adhere to the plate. While plating intermittently, swirl the 50-mL falcon tube to keep the cells evenly distributed.
12. Check cell adherence and health under a microscope and transfer the plate to the incubator for another 1 h. Replace the Optimem/glucose with 200 μL Neurobasal Medium (prewarmed at 37 °C) and return the plate to the incubator until the day of transfection.

2.3. Cellular and subcellular labeling

Successful HTS in a live-cell assay is critically dependent upon the ability to fluorescently label the relevant cell type, subcellular region, or physiological state being investigated. Hence, the choice of the most appropriate fluorescence method for an assay depends primarily on the biology being investigated. Two recent reviews by Giepmans *et al.* (2006) and Lavis (2011) discuss the wide variety of methods to fluorescently label cells for different applications. Though we do briefly introduce the two major fluorescence labeling methods, fluorescent molecules and genetically encoded fluorescent proteins (FPs), we focus this section on discussing the primary neuron fluorescence labeling methods routinely used for HTS in our laboratory.

Labeling cellular features was first achieved by using chemical features of certain fluorescent molecules (dyes) which enabled them to have an intrinsic affinity for an organelle, subcellular region, or cell type. Live-cell labeling with MitoTracker, LysoTracker, the DNA-binding molecule DRAQ5 and lipophilic dyes like the membrane stain DiI have been favorite tools for cell biologists for decades. The development of genetically encoded FPs has further expanded the cell biologist's toolbox by providing a way to label practically any protein in the proteome. Unlike fluorescent dyes, FPs can be expressed in specific cell types with spatial and temporal control via genetic promoters or fusion to a tag. Targeted mutagenesis of FPs has further led to a series of derivatives with enhanced functionalities such as colors that range from blue to far-red, increased fluorescence brightness, faster maturation time, better tolerance of N-terminal fusions, and increased photostability. A brief comparison of the pros and cons of fluorescent dyes and FPs is outlined in Table 17.1.

Table 17.1 Advantages and disadvantages of fluorescent dyes and proteins

Advantages	Disadvantages
Fluorescent dyes	
No need to genetically modify the sample	Specificity determined by chemical environment
Narrow Ex and Em spectra for greater chromatic resolution with conventional filter-based imaging	Less persistent reducing utility for long-term experiments
Easier to design modifications in fluorescence	Dyes can accumulate in organelles such as lysosomes
Photostability and quantum yield can be outstanding	Degradation products and effects often unknown
Fluorescent proteins	
Easy to specifically target	Absorption and emission spectra tend to be broad
Continual production of new label allows labeling for indefinite periods	Genetic manipulation of sample
Fully biodegradable	Photostability and quantum yield varies with the protein
Amenable to engineering to report complex biology	

Ex, Excitation; Em, Emission.

Our laboratory extensively uses many common FPs to track neuron lifetimes (Arrasate *et al.*, 2004; Tsvetkov *et al.*, 2010), follow the aggregation dynamics of mutant proteins (Arrasate *et al.*, 2004; Barmada *et al.*, 2010; Miller *et al.*, 2010, Tsvetkov *et al.*, 2010), and measure protein turnover (Mitra *et al.*, 2009). In our HTS assays, we prefer FPs to fluorescent dyes for two reasons. First, since our assay measures lifetimes of individual neurons, the ability to transfect neurons once and then track expression throughout the entire assay (days) increases both throughput and decreases error and variability caused by daily application of dyes. Second, low-transfection efficiency (typical of primary neurons and largely considered a deterrent for using primary neurons in HTS) is in fact an advantage in our system because it leads to a few dispersed transfected neurons surrounded by many untransfected cells. This facilitates detection and tracking of individual neurons during automated image analysis. Fluorescent dyes that label all neurons in the well confounds individual cell detection and tracking. Other standard considerations that we typically follow when choosing the appropriate FP for our HTS assays are outlined in Table 17.2.

It is important when using multiple fluorophores to label different cellular components or proteins within the same cell that the excitation

Table 17.2 Considerations for the development of neuronal HCS assays

1. Brightness
   - Quantum yield of fluorescent proteins and dyes
   - Maturation of fluorescent proteins
2. Photostability
3. Spectral overlap for multicolor applications
4. Efficiency of introduction into primary neurons
5. Fidelity of cellular and subcellular labeling
   - Localization
   - Tendency for artifacts such as aggregation
   - Local environment effects such as pH on fluorescence

and emission spectra of each fluorophore are carefully analyzed to avoid excitation cross talk or emission bleed-through from spectral overlap. When using full-spectrum light sources, the excitation and emission filters must also be matched with the excitation and emission spectra of the fluorophores to enable optimal transmittance of the signal and reflection of undesired light. Finally, the quantum yield of the fluorophore (the probability that the excited fluorophore will emit a photon) and the resistance to photobleaching are important characteristics for maximizing reliable and consistent signal detection throughout the assay. In Table 17.3, we have listed the most routinely used FPs for three-channel HTS assays in our lab based on the above considerations.

A common morphology marker used in our laboratory is a cytosolic FP that labels the entire neuron. Bright fluorophores such as enhanced GFP (EGFP), mApple or Venus require low exposure times thus increasing the assay speed and reducing phototoxicity and photobleaching. These FPs show fluorescence throughout the cell body and neurites, enabling detection and tracking of individual neurons with ease and reliability. They can also be used as reporters of transfection efficiency by measuring percentage of expressing cells and reporters of cell health by measuring neurite extension, retraction, or blebbing.

Table 17.3 Fluorescent proteins routinely used for multiplexing in our lab

Fluorescent channel	Fluorescent protein
Cyan	Cerulean
Yellow	mCitrine or Venus
Red	mApple

Two routinely used strategies in our laboratory for labeling subcellular compartments within a neuron are genetic fusions of FPs with minimal signaling sequences that localize to specific subcellular compartments (e.g. ER or mitochondria) or fusions with functional proteins (e.g. PSD-95, MAP2, Tau) to localize to synapses, dendrites, or axons. We find that there are two main concerns with using fusions to signaling peptides or proteins for subcellular labeling. First, the fidelity of the fused tag can be an issue due to poor intrinsic localization of the signal peptide or altered trafficking of the fused protein. Immunocytochemistry has helped us verify that the localization of the fluorescently tagged protein is equivalent to the localization of its endogenous counterpart. Second, toxicity introduced into the experiment from a fused tag can significantly affect assays where cell lifetimes under different conditions are being compared. We have empirically determined toxicity from our fusion proteins and take special care to titrate their expression to a level that yields good signal with minimal toxicity.

2.4. Transfection of primary neurons

Consistent, reliable, and automated transfection of primary neurons is critical for the success of a HTS assay. Many methods of transfecting primary neurons exist. Transfection by calcium phosphate gives efficiencies of 0.5–5% of transfected cells (Dudek *et al.*, 2001). While we routinely use calcium phosphate transfection for primary neurons, we find that it lacks the reproducibility required for a HTS, and it requires the researcher to manually track the development of the calcium phosphate precipitate under a microscope, making the method difficult for complete automation. However, Lipofectamine 2000 (a lipid-based transfection reagent; Invitrogen) gives us similar transfection efficiencies and is amenable to complete automation (see Dalby *et al.*, 2004 for optimization of Lipofectamine 2000 transfections). Others have reported even higher transfection efficiencies of primary neurons with Lipofectamine 2000 (Ohki *et al.*, 2001). Our 5% transfection efficiency with Lipofectamine 2000 allows us to track ~50 neurons in a 3 × 3 grid of images from a single well which is sufficient to give us statistical significance in our assays while minimizing toxicity.

Our HTS involves cotransfection of multiple plasmids, each carrying a different fluorescent tag and reporting a different assay variable. For transfecting multiple plasmids, we use plasmids of the same backbone that show a high rate of cotransfection in our assays (Arrasate *et al.*, 2004). We regularly transfect four to six day *in vitro* primary neurons, though we have also obtained close to optimal transfection efficiencies with older neurons. We incubate neurons with the lipid/DNA complex for 15-30 min (much less time than Invitrogen's suggested incubation time). This optimization of incubation time leads to comparable transfection efficiency with significantly decreased toxicity. Below, we outline the Lipofectamine 2000 primary neuron transfection method used in our laboratory.

1. Place OptiMEM and Neurobasal/KY (450 mL Neurobasal + 50 mL of 10 × KY) in a 37 °C water bath 1 h before transfection.
2. Aspirate the Neurobasal media from cells and replace with 100 μL Neurobasal/KY per well. Replace the plate into the incubator and place the Neurobasal media (conditioned Neurobasal media) aspirated from the cells into 37 °C water bath.
3. Prepare DNA/OptiMEM solution by adding the appropriate plasmid DNA (50 ng/well for bright fluorophores and 0.1–0.2 μg/well for most other DNA) to make up 25 μL of OptiMEM + DNA solution for each well.
4. Prepare Lipofectamine 2000/OptiMEM solution by adding 0.5 μL Lipofectamine 2000 to 24.5 μL OptiMEM for each well. Incubate at room temperature for 5 min.
5. Add DNA/OptiMEM solution to the Lipofectamine 2000/OptiMEM solution dropwise and vortex a single 1–2 s pulse gently. Incubate at room temperature for 20 min.
6. Add 50 μL of the DNA/Lipofectamine 2000/OptiMem solution dropwise to each well. Gently swirl the plate and incubate at room temperature for 15 min. Aspirate the solution and wash with 200 μL Neurobasal medium. Aspirate the previously added Neurobasal medium and wash a second time with 200 μL fresh Neurobasal medium.
7. Combine 1:1 conditioned Neurobasal medium and fresh Neurobasal medium and filter sterilize. Replace with 200 μL of the 1:1 conditioned Neurobasal medium. Return the plate to the incubator until the start of image acquisition.

3. Image Acquisition

3.1. Plate management

Plate management during HTS varies significantly with the assay and typically includes at least three key components: multiwell plate-transporting robots that move plates through the workflow, liquid-handling workstations to dispense appropriate liquids with precision, and bar-coding devices that label and track individual plates throughout the screening workflow. A wide variety of commercially available instruments for each of these tasks come with varying ranges of precision, ease-of-integration, and required user interaction to operate. A caveat of incorporating multiple instruments into a screening workflow is that a failure of one instrument can be catastrophic, as it can hold up the entire pipeline.

3.1.1. Plate-transporting robots

For HTS, robotic arms along with multiwell plate stackers automate the loading of plates precisely and continuously into our microscope. To enable round-the-clock imaging, we integrated a KiNEDx robotic arm (Peak Robotics, Colorado Springs, CO., KX-300-435-TGP), which has customized grippers that load and unload plates of transfected neurons from stackers onto the microscope stage fitted with a customized plate holder. Having a robot arm with a built-in absolute encoder is important because if an emergency stop is triggered during the screening, then the robot will be able to know its last position and resume the screening from there instead of having to restart the run from the beginning.

3.1.2. Liquid-handling workstations

Liquid-handling workstations replace manual liquid pipetting. They are timesaving, use parallel sample preparation, and provide the precision required for HTS assays. Commercially available liquid-handling systems vary in ease and extent of integration with other equipment. Some systems provide additional screen-related functionalities, such as library reformatting, cherry picking, or pin-transfer. For all our HTS-related liquid-handling tasks, we use the MicroLab Starlet workstation (Hamilton, Reno, NV.). We chose the MicroLab Starlet due to its ease of integration with a robotic arm, the large number of plate locations on the deck, and the flexible scripting language. We use an eight channel head with compressed O-ring expansion for loading and seating tips and capacitance liquid level detection for reliable pipetting. The MicroLab Starlet is also equipped with tilting capabilities to ensure complete aspiration of liquids. We developed protocols in the workstation for many screening-related tasks, from the mundane (e.g., coating and washing plates) to the arduous (e.g., lipid transfections of primary neurons).

3.1.3. Bar-coding devices

Bar coding enables the researcher to manage and track multiwell plates in HTS. Some devices offer wider functionality and flexibility by being compatible with many bar-code symbologies and consist of a reader, printer, and applicator which can read or label the plate. The plate bar code and well number enable forward and backward tracking of data from individual plates, wells, and cells. It provides exceptional security in data tracking, minimizes errors, and allows real-time data exchange. Most commercial multiwell plates for HTS come preprinted with a bar code (standard or user defined). A bar-code reader is typically incorporated at various points in a HTS system depending upon the complexity of the workflow.

3.2. Our HTS microscopy platform

Two key challenges slowed the automation of live-cell fluorescence microscopy-based assays. First, formidable technical issues arose with the integration of the appropriate hardware and software controls to pass full control of image acquisition to a computer. Second, reducing the information rich image datasets from the automated runs into meaningful readouts is still an on-going challenge.

In this section, we describe our solution to the first challenge of creating a software and hardware system to fully automate image acquisition. We based our system on two principles: flexibility in imaging fluorophores within an assay and the ability to adapt the system to assays with different timescales. For example, maximizing our selection of filter sets and filter cubes has been a strong component of maintaining flexibility, and integrating the microscope with a robotic arm and controlled environment chamber ensured that it can handle assays lasting hours or days.

Our system is based on a Nikon Eclipse TE2000E-PFS microscope (Nikon, Melville, NY) with an epifluorescence illuminator and a transmitted light adapter. Motorized components control focus and position of the objectives, positions of filter cube turret, and operation of condenser turret. The microscope is equipped with the Perfect Focus System (Nikon, Melville, NY), which uses reflected light from a near-IR LED to maintain focus at a preset distance below the surface of a multiwell plate. To balance our need for acquiring a larger field of view and high-resolution images, we use a Nikon 20X Plan Fluor ELWD objective, with a 0.45 numerical aperture (NA) and a 6.9–8.2 mm working distance. This higher NA objective collects more light, enabling detection of dim signals, but the depth of field is narrower, so portions of a single cell can be out of focus because they are located above or below the focal plane. An intense and spatially even illumination from Lambda LS illuminator (Sutter Instrument Company, Novato, CA), which houses a 10-position 25-mm excitation wheel and a 300W CERMAX xenon arc lamp with an output range of 340–700 nm (PerkinElmer, Waltham, MA), is carried with a 3-mm liquid-light guide (Sutter Instruments) to the microscope. A separate 10-position 25-mm emission wheel (Sutter Instruments) is situated in front of the CCD camera. Hard-coated DAPI/FITC/Texas Red, C/Y/R, and DAPI/FITC/TRITC/CY5 filter sets from Chroma provide precise matching of the filter combination with the fluorophore across a wide range of the spectrum. A Lambda 10-3 controller (Sutter Instruments) rapidly switches excitation and emission filter wheels when assaying multiple FPs in the same image field and also controls a *Smart*Shutter (Sutter Instruments). Images are collected with a CoolSnap HQ cooled CCD camera (Photometrics, Tucson, AZ) attached to the microscope through the basement port, to maximize the optical efficiency of the light path. Our camera has an LVDS interface that provides 12-bit images with 1392 × 1040 pixels. The CCD has a pixel size of 6.45 × 6.45 μm and is cooled to −30 °C to reduce thermal noise. Automated image acquisition during the HTS assay is controlled by custom-made scripts in Image-Pro Plus software (Media Cybernetics, Bethesda, MD). A multiwell plate is loaded from a plate stacker by a KiNEDx KX-300-435-TGP plate-transporting robot (Peak Robotics, Colorado Springs, CO) with a custom-made gripper onto a plate holder that fits into a MS-2000 XY (Applied Scientific Instrumentation, Eugene, OR) automated closed loop stage. The plate holder contains a computer-controlled actuator that positions the plate in the top left corner once it is loaded by the robot. This stage provides a 120 × 110 mm range of travel, and with the selection of a 25.40-mm lead screw pitch gives an *XY* axis resolution of 88 nm and speed of 7 mm/s. A linear encoder provides extra resolution and accuracy in the stage movements.

The microscope, robotic arm, and plate stackers are enclosed within a custom-built controlled environment chamber (Technical Instruments, Burlingame, CA) that maintains 37 °C and 5% CO_2. The temperature is regulated by a model 300353, and the CO_2 is regulated by a model AC100 (World Precision Instruments, Sarasota, FL) detector.

Once data acquisition is completed, the data are exported to a data storage and retrieval system. From then on, the image data moves through the analysis, statistics, mining, and reporting pipelines, which are described in the subsequent sections.

3.3. Data storage and retrieval systems

Storing high-resolution images from high-content screen (HCS) data sets poses significant challenges because the files are large (Megapixel CCDs with 12–16 bit depth can result in megabytes worth of data for each image), and thousands of images can be acquired from a single 96-well plate. Multichannel, longitudinal acquisitions therefore result in tens of gigabytes of raw image data for each plate and terabytes for full screens. The time spent transferring data from acquisition hardware to servers can be lengthy. If errors occur, data can be lost. Before starting an HCS experiment, proper infrastructure should be in place to adequately manage the resulting data.

Two general approaches can be used to organize images. For relatively simple screens, a hierarchical folder structure can be saved on a server with a root folder for each experiment and subfolders for further levels of organization. Our root folders use a name with the date, an experiment descriptor, and subfolders for each fluorescence channel. Each image file name has descriptors so individual images are unambiguously placed in the appropriate folder, and analysis programs can parse the image filenames and logically group them. In longitudinal experiments, we have chosen to embed the date, a unique plate identifier, time point, well, montage index, and fluorescence channel into the filename separated by underscores. Ideally, a more detailed description of each image is contained in the image metadata to allow other viewers to understand exactly what is contained in the image and how it was acquired. Although admirable efforts are being made to standardize image formats and metadata (Linkert *et al.*, 2010), adoption of these standards by HCS imaging platforms will likely take time. The second approach involves a relational database. Databases are optimal for storing data from larger screens and are the preferred method for labs with the resources to create and manage them. The up-front cost of time and money is greater, but they offer advantages for querying information, grouping data across multiple experiments, and carrying out retrospective analyses for secondary endpoints.

Multiple open source database management programs are available. The most popular are based on the MySQL or PostgreSQL specifications. OMERO (Open Microscope Environment Remote Objects) is a combined client–server platform based on PostgreSQL that can be used for image visualization, management, and analysis. It is easy to implement data management solution for labs that want the advantages of a database but do not want to develop one on their own. MySQL or PostgreSQL directly offer more flexibility and customization but require significantly more technical expertise in database development. Almost all of the high-content imaging platforms have their own proprietary database software that interfaces directly with the imaging hardware. Labs that purchase a bundled HCS system can take advantage of these data management solutions, but it is important to know exactly how flexible and extensible they are and whether they export data into portable file formats, such as delimited text files. These common file formats can be used to transfer data to new imaging systems upon technology upgrades and can be critical for sharing data among labs.

Finally, raw image files can be stored in a hierarchical folder structure, and analysis programs can extract information from these images and save the resulting data into a database. In this approach, the images are not directly contained in the database, but fields within the database can point to the locations of the images on the server. Proprietary data flow programs, such as Pipeline Pilot, facilitate easy reading and writing of images from folders and can interface with databases for storage of analysis data.

Whatever method of storage and retrieval is chosen, there should be a clear vision of how the data relate to each other and what the most intuitive overarching structure is for containing the data. In longitudinal experiments where we track individual neurons over time, we use the data structure represented in Fig. 17.2. Once a database is created that is adequately flexible to store the results from of a large number of experiments, data mining programs can compare data across experiments, generate hypotheses, and

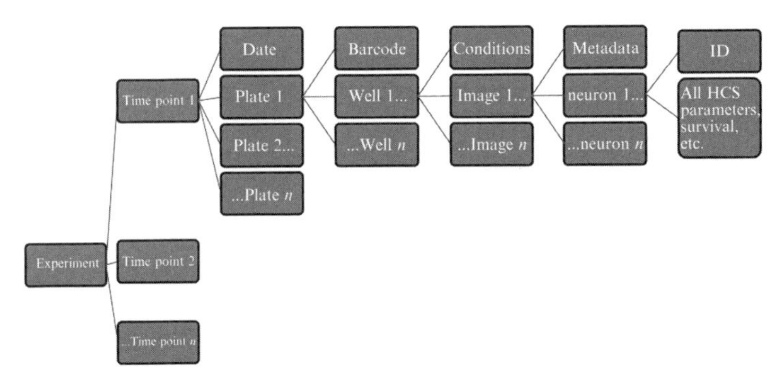

Figure 17.2 Schematic showing the data structure for our data storage and retrieval system. The data are organized based on the experiment and are broken down until each neuron has a unique cell ID and its own morphological and intensity data.

even retrospectively test these hypotheses with existing data. Data sources can be queried through structured query language (SQL) scripts, and the resulting tables can be imported into analysis programs. R Statistics (www.r-project.org) can interface directly with databases through the RODBC package. Researches can then use the full suite of R statistical capabilities to analyze the query results. Alternatively, highly powerful proprietary software packages, such as Spotfire (TIBCO, Palo Alto, CA) and Pipeline Pilot, provide a more accessible way to visualize data and generate reports.

4. Data Analysis

4.1. Image analysis pipeline

Automated image acquisition systems integrated with laboratory automation produce image datasets that are too large for manual processing and require computer analysis programs that are either fully automated or require minimal user intervention. The purpose of image analysis in biology is to identify biologically relevant objects and extract quantitative measurements from these objects. Computer-based image analysis provides an objective means of quantification that is independent of potentially more biased manual interpretation. Automated analysis can also be more sensitive, consistent, and accurate due to the large sample sizes and the extraction of multiple measurements.

The challenge has been to automate the sophisticated image recognition needed for neuron recognition.

The design of our image analysis system is based on two general principles: modularity of image analysis and multiple levels of analysis.

4.1.1. Modularity of image analyses

Automated imaging is used for a variety of imaging assays in our laboratory. Thus, multiple algorithms across various image analysis software systems are implemented, depending on the type of biology in question and quality of images. Modularity in each of the image analysis steps provides the flexibility necessary to adapt to each analysis without requiring duplication of accomplished work. Thus, individual tunable algorithms are assembled into a variety of "pipelines" to adapt to a specific assay or experiment. These pipelines can be adjusted and expanded by adding new algorithms and adjusting parameters of existing algorithms to increase sensitivity and specificity of the analysis.

Here, we list the typical steps we take during the analysis of data from our HTS assay.

The input for our automated analysis work flow starts with one or more fluorescence-channel images for each image field (Fig. 17.3).

1. *Image preprocessing*: The first stage of an analysis pipeline involves image preparation for subsequent analyses. It consists of assembling the acquired set of images into a larger mosaic image that is then processed for flattening uneven field illumination. These functions are performed by built-in filters available with most image analysis software. Some processes tend to alter signal information. We try to acquire high-quality images and work with raw data when doing intensity or similar measurements which would be altered by image processing (Fig. 17.3).
2. *Image registration*: The goal of an image registration algorithm, given two or more images from a time-lapse series, is to estimate a mapping that will bring the images into alignment. Each image in our time lapse is acquired ~12–24 h apart and typically contains 5–15 neurons. Difficulties in registration result from cells moving, dying, or changing in intensity between times, and additional shifts can be introduced by microscope stage hysteresis. This reduces the intensity and spatial overlap

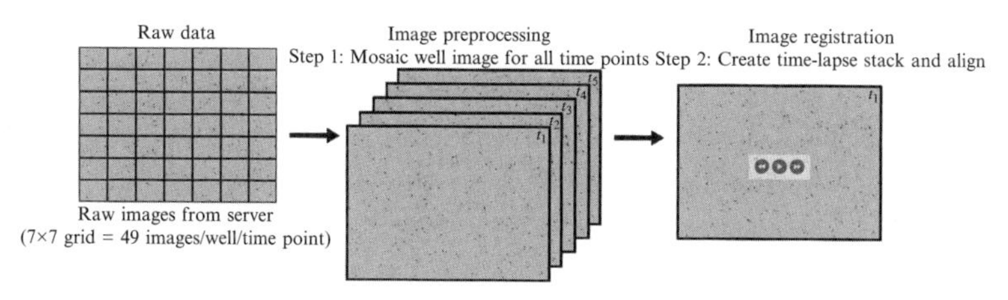

Figure 17.3 Workflow of image preprocessing and registration. Raw data (7 × 7 grid = 49 images) from each well are mosaiced to give a single image for each well at each time point. These images are then stacked together to make a time lapse. Each image in the time lapse is then registered to remove any hysteresis introduced from microscope stage movements.

of the positive objects (cells) in adjacent frames. Small misalignments or hysteresis in the stage is easier for an image registration algorithm to rectify if changes in composition between each image in a sequence are modest and incremental: image registration methods are either intensity-based, feature-based, or some blend of both (see Hill *et al.*, 2001, for a comprehensive overview). We register images with algorithms from a number of different image analysis programs. MultiStackReg, an ImageJ plugin that uses an iterative cross-correlation alignment method (Thevenaz *et al.*, 1998), has proved most efficacious for us in a wide variety of image stacks. However, images that contain no, few, or many out-of-focus cells tend to fail registration.

3. *Image segmentation*: Once the image is preprocessed and registered, cells are recognized using various segmentation algorithms. Segmentation is partitioning of an image into sets of pixels (segments) that correspond to distinct objects. This part of the image analysis process is the most difficult and probably the most crucial step for the success of the analysis.

   We follow each neuron throughout the assay period and use disappearance of the cell body as a surrogate for neuronal cell death in our longitudinal studies (Arrasate *et al.*, 2004). To aid segmentation, we seed the cells at a low-medium density (~70,000 cells/well). Because primary neurons are difficult to transfect, the transfected cells are well dispersed, reducing segmentation errors. The first time point (~24 h posttransfection) shows dim cells because most FPs take 24–48 h to accumulate enough to achieve steady state levels that can be detected with our illumination setting and camera. By the second to third time points, intensities of most cells stabilize except for a few dim cells that show low or delayed expression and survival differences begin to emerge between different assay conditions. From the 3–4th frames onward, we look for further separation of survival trends to achieve statistical significance.

   Such dynamic time-lapse images pose a significant challenge for an accurate segmentation, as it needs to be (1) flexible to handle changing cell intensities, shapes, and sizes; (2) specific to consistently select live cells; and (3) accurate to report the correct lifetime of each cell. In our experience, none of the generic segmentation algorithms from standard image analysis software were adequate. They either reported too many false positives, corrupting the analysis with the presence of nonexistent cells, or detected fewer cells than were present, causing subsequent analyses to be biased toward more clearly distinguishable cells. Thus, we developed a transiently transfected primary neuron segmentation algorithm using Image-Pro Plus.

   We describe here one of our most commonly used HTS segmentation pipelines, which detects and tracks only cell bodies in time-lapse images (Fig. 17.4). We tend to use survival for primary screening. Our other segmentation approaches give more detailed morphological read-outs during secondary and tertiary screening.

The pipeline uses a series of filters and operations. Since cell bodies and particulate debris show up as spots and neurites as elongated "thread-like" structures, we first use a spot locate filter to exclude the neurites. This filter consists of a mean-filter that "smoothes" the image by averaging the pixel intensities in a 9-pixel square (3 × 3), followed by a watershed

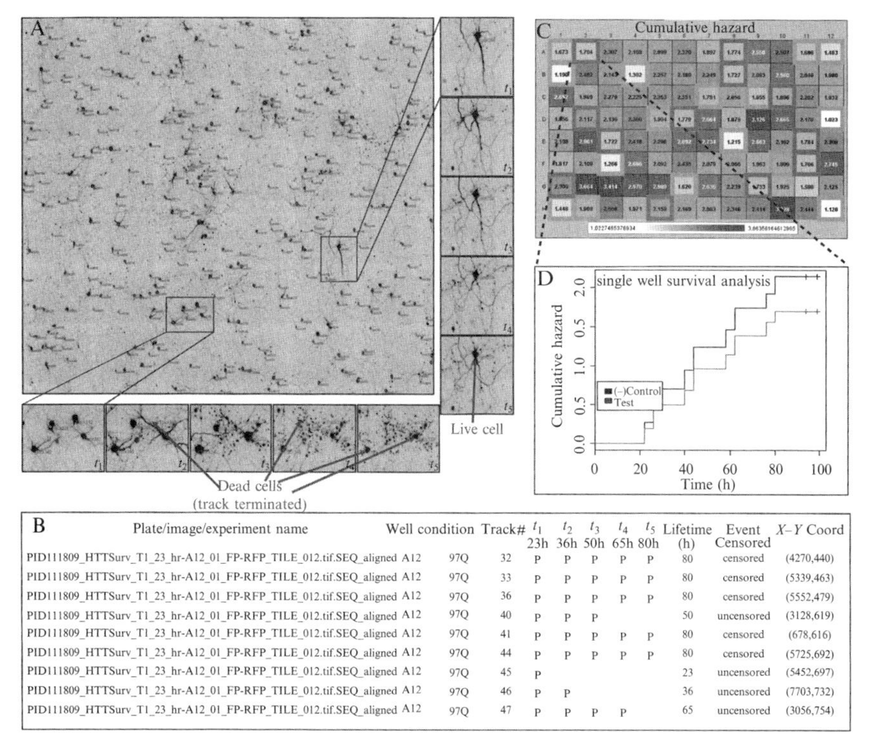

Plate/image/experiment name	Well	condition	Track#	t_1 23h	t_2 36h	t_3 50h	t_4 65h	t_5 80h	Lifetime (h)	Event Censored	X–Y Coord
PID111809_HTTSurv_T1_23_hr-A12_01_FP-RFP_TILE_012.tif.SEQ_aligned	A12	97Q	32	P	P	P	P	P	80	censored	(4270,440)
PID111809_HTTSurv_T1_23_hr-A12_01_FP-RFP_TILE_012.tif.SEQ_aligned	A12	97Q	33	P	P	P	P	P	80	censored	(5339,463)
PID111809_HTTSurv_T1_23_hr-A12_01_FP-RFP_TILE_012.tif.SEQ_aligned	A12	97Q	36	P	P	P	P	P	80	censored	(5552,479)
PID111809_HTTSurv_T1_23_hr-A12_01_FP-RFP_TILE_012.tif.SEQ_aligned	A12	97Q	40	P	P	P			50	uncensored	(3128,619)
PID111809_HTTSurv_T1_23_hr-A12_01_FP-RFP_TILE_012.tif.SEQ_aligned	A12	97Q	41	P	P	P	P	P	80	censored	(678,616)
PID111809_HTTSurv_T1_23_hr-A12_01_FP-RFP_TILE_012.tif.SEQ_aligned	A12	97Q	44	P	P	P	P	P	80	censored	(5725,692)
PID111809_HTTSurv_T1_23_hr-A12_01_FP-RFP_TILE_012.tif.SEQ_aligned	A12	97Q	45	P					23	uncensored	(5452,697)
PID111809_HTTSurv_T1_23_hr-A12_01_FP-RFP_TILE_012.tif.SEQ_aligned	A12	97Q	46	P	P				36	uncensored	(7703,732)
PID111809_HTTSurv_T1_23_hr-A12_01_FP-RFP_TILE_012.tif.SEQ_aligned	A12	97Q	47	P	P	P	P		65	uncensored	(3056,754)

Figure 17.4 Image segmentation, cell tracking, feature extraction, and data reporting. (A) Shows a typical montaged image showing the first time point after image segmentation. Unique cell track numbers are shown in red next to each cell. Two regions within this image have been zoomed in to show cell lifetime detection from the entire time lapse. (B) A typical delimited text file showing the arrangement of data output and feature extraction by the analysis program. (C) Shows a representative heat map of a control plate where a known modifier (blue outlined wells) was spotted throughout the plate in the midst of positive control wells (green outlined wells). White shaded wells indicate longer survival and dark red shaded wells indicate reduced survival. The numbers represent the cumulative death in each well at the end of the experiment. (D) A survival curve from a modifier well showing decreased hazard (increased survival) compared to the negative control wells. (See Color Insert.)

transformation to separate spots from the background. This detects the cell bodies as they are usually larger than the scattered debris from dead cells. A top hat filter is then used to distinguish neurons from debris. The algorithm uses three parameters (i.e., top radius, brim radius, and height) to relate to the object's appearance. The "top" of the "hat" is set at the maximum expected spot radius. The brim radius is often taken to be the shortest expected distance to the neighboring spot. The height of the top above the brim is set to the minimum intensity that a spot must rise above its immediate background. For this, the spot intensity is the average within the top, and the immediate background intensity is the average within the brim. This averaging reduces noise by decreasing the variance in the estimation of a noisy object and its background and thus improves the robustness and performance of this filter from removing cell debris. Some very dim objects in the images are hard to decipher as being cells or debris without seeing the next time point in the time-lapse stack. We decided to use a simple thresholding operation to eliminate these very dim signals. Intensity statistics are calculated for the entire image, and then dim cells and background are eliminated by selecting pixels with values in the top fifth percentile. This approach provides flexibility to researchers in designing their assays (like choosing from a variety of fluorophores and having flexible image acquisition of each time point within the time lapse) and improves the efficacy of the segmentation method across a wide range of assay conditions. Finally, to locate and count cells, we binarize the image, converting the cell bodies so they have a pixel intensity value equal to 1 and the background has a value equal to 0. The resulting black and white image is further processed so that each cell is identified as an independent object with its own numerical identifier (Fig. 17.4).

4. *Cell tracking*: Our longitudinal assays require us to follow individual neurons over time. Cell-tracking algorithms are designed to track each individual positive object (cell) by linking the corresponding object in successive frames of a time-lapse sequence and designating it as the same cell from the previous frame or a new cell. Optimal image registration and segmentation of the time-lapse stack are prerequisites for reliable cell tracking. Poor segmentation will cause the cell-tracking algorithm to yield nonsensical tracks in which tracks from correctly segmented cells in one frame are connected to different cells or falsely segmented cells (i.e., debris) in the next frame.

   Cell-tracking algorithms can be categorized into three main types. The centroid cell-tracking algorithm calculates the center-of-mass (centroid) of the object of interest. It performs best when all objects move in exactly the same way between consecutive frames, relative to each other and irrespective of changes in their shapes and intensities. The Gaussian cell-tracking algorithm directly fits Gaussian curves to the intensity profile and performs best when the intensities of the objects are same between consecutive frames, even if their movement is random relative to each other. In our assays, each time-lapse sequence contains many neurons, and between successive images, cells normally move randomly (at very modest speeds), their intensities change, cells disappear (die), new cells show up (delayed expression), and cells move in and out of the image due to the limited field of view of the microscope (Fig. 17.4A). For us, the third type of cell-tracking algorithm, a cross-correlation algorithm, performed best. It compares an image to a matrix of pixels (user defined) of a successive image. The matrix is shifted relative to the image in 1-pixel increments. For each increment, a correlation value is calculated that describes how well the values in the matrix match those of the image, and the program determines the shift that yields the maximum correlation value. This algorithm is computationally more intensive than the other two and significantly slows the analysis, depending on the size of the matrix selected.

5. *Feature extraction*: Feature extraction computes numerical descriptors of objects identified by segmentation. Features are classified into either low-level features (e.g., area, location, perimeter, and intensity) that are extracted directly from the image or high-level features (e.g., shape, texture, and projections) that are computed from the low-level features. The end result of extraction is typically an extensive set of features, commonly called a feature vector, on a per-cell or whole-image basis. We use objects with unique identifiers from the segmented binary image as regions of interest to allow extraction of numerical descriptors for low-level features like cell location (*x*-, *y*-coordinates), maximum diameter (major axis), minimum diameter (minor axis), ratio of major to minor axis, size (area), surface area (perimeter), average intensity, and intensity variance for each cell. Based on these features, we compute descriptors for the high-level feature, such as cell viability (scored as live or dead), cell censored (moved out of the image field), cell lifetime, cell migration between time points, inclusion formation, and number of inclusions on a per-cell basis (Fig. 17.4B). These cell descriptors are exported as comma-separated values that can then be either viewed (with programs like Microsoft Excel) or advanced further into the data analysis pipeline.

4.1.2. Multiple levels of analyses

Depending on whether the data is from a handful of plates (small-scale biological assay) or from a HTS assay, we analyze the data at three different population levels: well-level, multiwell and plate-level, and multiplate level. We use Pipeline Pilot (Accelrys, San Diego, CA) to automate our workflow and visualize results for each type of analysis. At the root of the data, a well is considered an individual experiment and is composed of a population of neurons with a known condition. The software automatically builds well information from the designated user-defined plate format (test wells, positive control wells, and negative control wells; Fig. 17.4C and D). For multiwell and plate-level analysis, the software compares data across multiple wells on a plate. In this way, we can easily identify potential biases, for instance, edge effects, outlier wells, and general issues with cell health. Heat maps are very informative when assessing these types of plate-level effects and can easily identify wells of interest or control wells that have not passed quality control measures. For multiplate and experiment level analysis, the software collects data for replicate wells on different plates and is useful for increasing the sample size when the effect size is small. Furthermore, the full power of HCS datasets comes from mining a central database where data from hundreds to thousands of plates are stored. Existing datasets can be reanalyzed to answer new questions and meta-analyses can be carried out by pooling large amounts of data that have accumulated over many years.

4.2. Image analysis software packages used in our laboratory

HCS imaging platforms come equipped with software for image acquisition control, image analysis, statistical tools, data visualization tools, and a database for storing and retrieving data. These algorithms are adaptable to screening in different cell lines with different markers. They are meant to be used with minimal training and do not require the user to develop his own tools or to be able to program. Drawbacks of such ready-to-go systems are the source code is protected and the user does not know the methodology of the different steps, cannot modify them, or fully understand their functionality. Also, proprietary software licenses might not be affordable for all users. The existing platforms can be divided into two categories: open source and proprietary. Open source softwares used in our laboratory are ImageJ and CellProfiler, and the proprietary software includes MATLAB, Metamorph, Pipeline Pilot, and Image-Pro (Table 17.4).

5. Statistical Approaches to HCS and Multivariate Data

The power of HCS datasets lies in the ability to parameterize cellular morphologies, intensities, and textures with multiple variables, allowing the identification of interesting phenotypes with minimal bias and greater sensitivity. Biases are minimized because rather than *a priori* deciding on a single feature that is thought to differ between two cell populations, many possible features and combinations of features are compared in parallel, and the most distinguishing features are statistically identified through variable reduction methods. Collinet *et al.* (2010) used multiparametric techniques

Table 17.4 Image analysis softwares used in our laboratory

Open source	
CellProfiler	MATLAB-based open source image analysis package (compatible with Mac OS X, Windows, and Unix) for cellular HCS (Carpenter *et al.*, 2006). Specifically designed to bridge the gap between developer tools, such as MATLAB and the proprietary software for HCS, and offers ~50 modules for typical image analysis steps with user-friendly GUIs. MATLAB is not required to run the application, but MATLAB users can access the source code, expand the capacities of the program with new code, and modify existing code to adapt to specific problems. It accepts many conventional image formats, and new ones can be encoded. A pipeline is created using the various modules to automate the analysis. CellProfiler is relatively new, but a user community is growing; so new modules may appear at an increasing rate. As CellProfiler was designed for HCS, distributed computing is feasible, so that clusters can be used.
ImageJ	Java-based open source image analysis package (compatible with Mac OS X, Windows, Unix, and Linux) has a large user community and over 300 plugins (modules). Plugins accept most image formats, and new ones can be coded in. Macros are created using the various plugins to automate the analysis. Our laboratory uses FIJI, an ImageJ distribution along with a plugin updater with a graphical user interface and many of the necessary plugins preinstalled and organized in a coherent structure. Using custom macros, we use the image registration capabilities of ImageJ and have also semiautomated image analysis for our autophagy, RNA granule tracking, and neuronal branching assays.
Commercial/proprietary	
MATLAB	Commercially available development environment (compatible with Mac OS X, Windows, Unix, and Linux) is widely used in engineering for a wide range of applications and can also be used for image analysis (with appropriate toolboxes). Although Matlab is proprietary, applications with Matlab languages are freely available (Cellprofiler (Carpenter *et al.*, 2006), CellC (Selinummi *et al.*, 2005)). It easily interfaces with other softwares and accepts most conventional image formats. As a developer tool, it is highly flexible but also requires time to optimize the image analysis process. It can handle distributed computing to enable clusters for parallel and faster analysis.

(*Continued*)

Table 17.4 (*Continued*)

Image-Pro Plus	Commercially available image acquisition and analysis package (compatible with Windows only). It has multichannel and multiwell plate image acquisition formats. To give us more control and flexibility of our acquisition settings, we have written macros for automating the entire image acquisition process. We routinely use it for analyzing our HTS data. It accepts some common image formats and easily interfaces with other software. Some of our analysis pipeline is controlled by a simple Perl script that calls Image Pro plus at appropriate times and feeds in and out the appropriate images. It has options to link to databases to retrieve and store image data and has a moderate-sized user community.
Metamorph	Commercially available image acquisition and analysis package (compatible with Windows only). The modules are designed for biology-specific applications. It has multichannel and multiwell plate image acquisition formats. We routinely use it for image acquisition and to give us more control and flexibility of our acquisition settings, we have written journals for automating the entire image acquisition process. Also has a moderate-sized user community.
Pipeline pilot	Commercially available program for dataflow management, image analysis, and reporting that provides an easy way to build pipelines for customized data analysis (compatible with Windows only). It can read all standard and some proprietary file formats and has the advantage of being able to run ImageJ, Matlab, Python, and R scripts. Dataflow pipelines are created through sequential addition of drag-and-drop components acting at different points in the analysis. Data can be written in a variety of formats, including storage within databases or saved in delimited text files. Users can connect remotely to a Pipeline Pilot server through a web browser and process images with any saved pipeline.

to identify differences in endocytic uptake of two types of cargo. In a similar manner, Loo *et al.* (2007) determined differential response of cancer cells to drug treatment and could predict drug class and on- and off-target effects from the multiparametric signature. Support vector machine (SVM) algorithms were used to determine hyperplanes of maximal separation between cell populations in multivariate space, and SVM recursive feature elimination was used to reduce the number of variables. All algorithms were implemented in Matlab. These examples are some of the first to use multichannel, fluorescence datasets to quantify cellular phenotypes.

As described above, our lab extracts baseline characteristics in the form of a few variables from multichannel fluorescence images of neurons to determine their effect on future outcomes (e.g., htt IB formation or cell death). We track individual cells over time to capture dynamic changes and compare cell populations by Cox proportional hazards (CPH), a statistical model for determining the effects of multiple variables on time-to-event outcomes (Klein and Moeschberger, 2003). Analysis programs analyze the images and produce a delimited text file containing the variables and outcome measures (Fig. 17.4B). R (www.r-project.org), a comprehensive and widely adopted open source statistics program, then imports the file and a Cox model is fit to the data with the R survival package. The CPH model is created as follows:

Required R packages: splines and survival

```
# This code fits a CPH model to data from two constructs
# to determine how the constructs
# effect neuron survival
# Load splines package
library(splines)
# Load survival package
library(survival)
# Create a color vector for plotting
Colrs <- palette()
# Import csv file into dataframe
exp.data <- read.csv("NeuronSurvivalData.csv");
# Create dataframe with only columns needed for build-
# ing CPH model
# Assumes that "Lifetime", "EventCensored", and
# "Condition" are column headings in the .csv file.
cox.data<-data.frame(time=exp.data$Lifetime,
event=exp.data$EventCensored, group=as.factor(exp.
data$Condition))
# Fit a Kaplan Meier model to the data and plot survival
# curves.
km.model < -survfit(Surv(time,as.logical(event)) ~
group,data = cox.data)
plot(km.model,main="Kaplan Meier Curve for Survival",
xlab="Time [hrs]",col=colrs)
# Fit a CPH model based on Condition and Lifetime
cph.model <- coxph(Surv(time,as.logical(event)) ~
group,data = cox.data)
cph.pval <- summary(cph.model)$coef[,5]
```

```
cph.HR <- summary(cph.model)$coefficients[,2]
# Print out a summary of the CPH model
summary(cph.model)
```

The standard test for determining if a particular low-throughput assay is amenable to HTS is calculation of the Z'-factor (Zhang *et al.*, 1999). This number is calculated as a signal detection window between positive and negative controls scaled by the dynamic range. A Z'-factor >0.5 has historically indicated that there is a sufficient detection window for carrying out a screen. However, this factor was derived for biochemical assays, and because of the inherent higher variability of cell-based assays, laboratories now carry out screens where the Z'-factor is >0. We found that summarizing our time-dependent analysis with descriptive statistics (as used in the Z'-factor) did not capture the increase in sensitivity that is gained by tracking cells over their entire lifetime. We therefore chose to use statistical tests that are calculated from information contained in the entire survival curve. To estimate false-positive and false-negative hit rates, we use control plates with one column each of positive and negative controls and then spot known positive controls randomly throughout the test wells of the plate. We next compose CPH models for all the test wells and compute a hazard ratio between each individual well and the negative control column. *p*-Values are calculated from the hazard ratios, and if the well is significantly less toxic than the negative control column, we label the well as a hit. The decision to screen is based on acceptable positive and negative likelihood ratios (LR^+ and LR^-, respectively) that are computed from up to four replicate plates. The stringency of the test (i.e., the threshold of the *p*-value) can be used to adjust the rates of false positives and negatives. Typically, in primary screens, false negatives are worse than false positives because false positives will eventually fail in secondary screens, whereas false negatives are lost completely.

6. Future Directions

Embryonic stem (ES) cells and induced pluripotent stem (iPS) cells are other model systems with great potential for HCA–HTS neurological disease assays. Because human neurological diseases are the result of tens to hundreds of risk factors that jointly and in different combinations lead to a clinical phenotype, ES cells and iPS cells may more realistically model such multifactorial diseases than primary neurons from rodent tissues. Further, murine models contain thousands of small base-pair changes that differ between human cells. Thus, studying disease states in rodent tissues may miss key components of neuropathology.

Because iPS cells are derived from human tissues, they carry all of the associated endogenous genetic components, which will likely more accurately reflect the biology of human disease. They also combine the physiological relevance of primary cells with the ease of culturing of immortalized cell lines.

The discovery that human fibroblasts taken from patients' with neurological disorders can be genetically reprogrammed into iPS cells (Takahashi and Yamanaka, 2006; Takahashi *et al.*, 2007; Yu *et al.*, 2007) and differentiated into neurons that can retain biochemical and pathological deficits observed in the actual disease (Ebert *et al.*, 2009; Lee *et al.*, 2009; Marchetto *et al.*, 2010; Seibler *et al.*, 2011) will revolutionize our ability to model neurodegenerative diseases in a culture dish. In addition, these findings have opened up possibility of screening for drug therapeutics in a disease-specific, "human-centric" platform. Thus, HCA–HTS assays could use patient-derived cell lines to model multifactorial neurological diseases and rapidly generate potential hits with very high therapeutic values.

Our lab is differentiating iPS cells into specific cell types affected in HD, such as MSNs (Aubry *et al.*, 2009), motor neurons that are affected in amytrophic lateral sclerosis and spinal muscular atrophy (Ebert *et al.*, 2009; Dimos *et al.*, 2008) and dopaminergic neurons that are lost in Parkinson's disease (Seibler *et al.*, 2011; Wernig *et al.*, 2008). We optimized these differentiations to a 96-well format and are subjecting these specific cell-types to our automated imaging time-lapse microcopy to track cell death, neuron dynamics, and cellular morphologies such as neurite and process length over time. We are investigating phenotypes of iPS lines generated from patients with HD that would be suitable for screening purposes. Finally, we are generating new iPS cell lines from individuals with HD, Amyotrophic lateral sclerosis, and Parkinson's disease that will be used in our HCA–HTS. Our goals are to subject neurological disease-specific iPS cells to HCA–HTS to create more well-suited platforms for drug screening to identify much needed biotherapeutics for neurodegenerative disorders.

7. Concluding Remarks

Although other systems achieve higher throughput than the transiently transfected live primary neuron screening system we have described in this chapter, the assay has greater sensitivity than commercially available systems due to the longitudinal approach and the ability to follow individual neurons over long periods of time. This significantly reduces the need for a large number of cells to detect statistical significance in our assay. As the hardware and software technology for HTS progresses, we are striving to combine the consistent high-quality images with speed increases in automated image acquisition and analysis to potentially allow on-the-fly analysis of images that then informs the acquisition process.

Our current HTS assay is focused on measuring cell lifetime. However, multiple levels of information can be extracted, which requires the integration of information from a number of channels (such as bright field and fluorescence) and hundreds of cell attributes, such as size, shape, intensity, neurites, lifetime, or inclusion formation. This multiplexing improves the assay's relevance because the dataset allows much more sophistication in data and image analysis and interpretation of the cellular information. The successful development of such HCA–HTS primary neuron assays for neurodegenerative disorders promises more efficient identification of hits from a primary screen with high disease relevance and therapeutic–predictive value, thereby reducing extensive testing in a range of secondary assays.

ACKNOWLEDGMENTS

We thank members of the Finkbeiner laboratory for helpful discussions. K. Nelson provided administrative assistance, and G. Howard edited the manuscript. This work was supported by National Institutes of Health Grants 2R01 NS039746 and 2R01 NS045191 from the National Institute of Neurological Disorders and Stroke and by Grant 2P01 AG022074 from the National Institute on Aging, the J. David Gladstone Institutes, and the Taube–Koret Center for Huntington Disease Research (to S. F.); a California Institute for Regenerative Medicine (CIRM) fellowship Grant T2-00003 (to P. S); a CIRM fellowship Grant TG2-01160 (to J.K.); and the National Institutes of Health–National Institute of General Medical Sciences University of California, San Francisco Medical Scientist Training Program, (to A. D). The animal care facility was partly supported by a National Institutes of Health Extramural Research Facilities Improvement Project (C06 RR018928).

REFERENCES

Arrasate, M., and Finkbeiner, S. (2005). Automated microscope system for determining factors that predict neuronal fate. *Proc. Natl. Acad. Sci. USA* **102,** 3840–3845.

Arrasate, M., Mitra, S., Schweitzer, E. S., Segal, M. R., and Finkbeiner, S. (2004). Inclusion body formation reduces levels of mutant huntingtin and the risk of neuronal death. *Nature* **431,** 805–810.

Aubry, L., Peschanski, M., and Perrier, A. L. (2009). Human embryonic-stem-cell-derived striatal graft for Huntington's disease cell therapy. *Med. Sci. (Paris)* **25,** 333–335.

Barmada, S. J., Skibinski, G., Korb, E., Rao, E. J., Wu, J. Y., and Finkbeiner, S. (2010). Cytoplasmic mislocalization of TDP-43 is toxic to neurons and enhanced by a mutation associated with familial amyotrophic lateral sclerosis. *J. Neurosci.* **30,** 639–649.

Brouillet, E., Conde, F., Beal, M. F., and Hantraye, P. (1999). Replicating Huntington's disease phenotype in experimental animals. *Prog. Neurobiol.* **59,** 427–468.

Capenter, A. E., Jones, T. R., Lamprecht, M. R., Clarke, C., Kang, I. H., Friman, O., Guertin, D. A., Chang, J. H., Lindquist, R. A., Moffat, J., Golland, P., and Sabatini, D. M. (2006). Cell Profiler: Image analysis software for identifying and quantifying cell phenotypes. *Gen. Biol.* **7,** R100.

Collinet, C., Stoter, M., Bradshaw, C. R., Samusik, N., Rink, J. C., Kenski, D., Habermann, B., Buchholz, F., Henschel, R., Mueller, M. S., Nagel, W. E., Fava, E., *et al.* (2010). Systems survey of endocytosis by multiparametric image analysis. *Nature* **464,** 243–249.

Dalby, B., Cates, S., Harris, A., Ohki, E. C., Tilkins, M. L., Price, P. J., and Ciccarone, V. C. (2004). Advanced transfection with Lipofectamine 2000 reagent: Primary neurons, siRNA, and high-throughput applications. *Methods* **33,** 95–103.

Daub, A., Sharma, P., and Finkbeiner, S. (2009). High-content screening of primary neurons: Ready for prime time. *Curr. Opin. Neurobiol.* **19,** 537–543.

Dimos, J. T., Rodolfa, K. T., Niakan, K. K., Weisenthal, L. M., Mitsumoto, H., Chung, W., Croft, G. F., Saphier, G., Leibel, R., Goland, R., *et al.* (2008). 'Induced pluripotent stem cells generated from patients with ALS can be differentiated into motor neurons'. *Science* **321**(5893), 1218–1221.

Dudek, H., Ghosh, A., and Greenberg, M. E. (2001). Calcium phosphate transfection of DNA into neurons in primary culture. *Curr. Protoc. Neurosci.* 3.11.1–3.11.6.

Ebert, A. D., Yu, J., Rose, F. F., Jr., Mattis, V. B., Lorson, C. L., Thomson, J. A., and Svendsen, C. N. (2009). Induced pluripotent stem cells from a spinal muscular atrophy patient. *Nature* **457,** 277–280.

Finkbeiner, S., Cuervo, A. M., Morimoto, R. I., and Muchowski, P. J. (2006). Disease-modifying pathways in neurodegeneration. *J. Neurosci.* **26,** 10349–10357.

Giepmans, B. N., Adams, S. R., Ellisman, M. H., and Tsien, R. Y. (2006). The fluorescent toolbox for assessing protein location and function. *Science* **312,** 217–224.

Hertzberg, R. P., and Pope, A. J. (2000). High-throughput screening: New technology for the 21st century. *Curr. Opin. Chem. Biol.* **4,** 445–451.

Hill, D. L., Batchelor, P. G., Holden, M., and Hawkes, D. J. (2001). Medical image registration. *Phys. Med. Biol.* **46,** R1–R45.

Houston, J. B., and Galetin, A. (2008). Methods for predicting in vivo pharmacokinetics using data from in vitro assays. *Curr. Drug Metab.* **9,** 940–951.

Inglese, J., Johnson, R. L., Simeonov, A., Xia, M. H., Zheng, W., Austin, C. P., and Auld, D. S. (2007). High-throughput screening assays for the identification of chemical probes. *Nat. Chem. Biol.* **3,** 466–479.

Klein, J. P., and Moeschberger, M. L. (2003). Survival Analysis: Techniques for Censored and Truncated Data. 2nd edn. Springer, New York.

Lavis, L. D. (2011). Histochemistry: Live and in color. *J. Histochem. Cytochem.* **59,** 139–145.

Lee, H., Park, J., Forget, B. G., and Gaines, P. (2009). Induced pluripotent stem cells in regenerative medicine: An argument for continued research on human embryonic stem cells. *Regen. Med.* **4,** 759–769.

Linkert, M., Rueden, C. T., Allan, C., Burel, J. M., Moore, W., Patterson, A., Loranger, B., Moore, J., Neves, C., Macdonald, D., Tarkowska, A., Sticco, C., *et al.* (2010). Metadata matters: Access to image data in the real world. *J. Cell Biol.* **189,** 777–782.

Loo, L. H., Wu, L. F., and Altschuler, S. J. (2007). Image-based multivariate profiling of drug responses from single cells. *Nat. Methods* **4,** 445–453.

Macarron, R. (2006). Critical review of the role of HTS in drug discovery. *Drug Discov. Today* **11,** 277–279.

Macarron, R., and Hertzberg, R. P. (2009). Design and implementation of high-throughput screening assays. *Methods Mol. Biol.* **565,** 1–32.

Marchetto, M. C., Carromeu, C., Acab, A., Yu, D., Yeo, G. W., Mu, Y., Chen, G., Gage, F. H., and Muotri, A. R. (2010). A model for neural development and treatment of Rett syndrome using human induced pluripotent stem cells. *Cell* **143,** 527–539.

Miller, J., Arrasate, M., Shaby, B. A., Mitra, S., Masliah, E., and Finkbeiner, S. (2010). Quantitative relationships between huntingtin levels, polyglutamine length, inclusion body formation, and neuronal death provide novel insight into Huntington's disease molecular pathogenesis. *J. Neurosci.* **30,** 10541–10550.

Mitra, S., Tsvetkov, A. S., and Finkbeiner, S. (2009). Protein turnover and inclusion body formation. *Autophagy* **5,** 1037–1038.

Nolan, G. P. (2007). What's wrong with drug screening today. *Nat. Chem. Biol.* **3,** 187–191.

Ohki, E. C., Tilkins, M. L., Ciccarone, V. C., and Price, P. J. (2001). Improving the transfection efficiency of post-mitotic neurons. *J. Neurosci. Methods* **112,** 95–99.

Seibler, P., Graziotto, J., Jeong, H., Simunovic, F., Klein, C., and Krainc, D. (2011). Mitochondrial Parkin recruitment is impaired in neurons derived from mutant PINK1 induced pluripotent stem cells. *J. Neurosci.* **31,** 5970–5976.

Selinummi, J., Seppala, J., Yli-Harja, O., and Puhakka, J. A. (2005). Software for quantification of labeled bacteria from digital microscope images by automated image analysis. *Biotechniques* **39**(6), 859–863.

Spyropoulos, D. D., Bartel, F. O., Higuchi, T., Deguchi, T., Ogawa, M., and Watson, D. K. (2003). Marker-assisted study of genetic background and gene-targeted locus modifiers in lymphopoietic phenotypes. *Anticancer Res.* **23,** 2015–2026.

Staats, J. (1985). Standardized nomenclature for inbred strains of mice: Eighth listing. *Cancer Res.* **45,** 945–977.

Takahashi, K., and Yamanaka, S. (2006). Induction of pluripotent stem cells from mouse embryonic and adult fibroblast cultures by defined factors. *Cell* **126,** 663–676.

Takahashi, K., Tanabe, K., Ohnuki, M., Narita, M., Ichisaka, T., Tomoda, K., and Yamanaka, S. (2007). Induction of pluripotent stem cells from adult human fibroblasts by defined factors. *Cell* **131,** 861–872.

Thevenaz, P., Ruttimann, U. E., and Unser, M. (1998). A pyramid approach to subpixel registration based on intensity. *IEEE Trans. Image Process.* **7,** 27–41.

Tsvetkov, A. S., Miller, J., Arrasate, M., Wong, J. S., Pleiss, M. A., and Finkbeiner, S. (2010). A small-molecule scaffold induces autophagy in primary neurons and protects against toxicity in a Huntington disease model. *Proc. Natl. Acad. Sci. USA* **107,** 16982–16987.

Wernig, M., Zhao, J. P., Pruszak, J., Hedlund, E., Fu, D., Soldner, F., Broccoli, V., Constantine-Paton, M., Isacson, O., and Jaenisch, R. (2008). Neurons derived from reprogrammed fibroblasts functionally integrate into the fetal brain and improve symptoms of rats with Parkinson's disease. *Proc. Natl. Acad. Sci. USA* **105,** 5856–5861.

Yu, J., Vodyanik, M. A., Smuga-Otto, K., Antosiewicz-Bourget, J., Frane, J. L., Tian, S., Nie, J., Jonsdottir, G. A., Ruotti, V., Stewart, R., Slukvin, I. I., and Thomson, J. A. (2007). Induced pluripotent stem cell lines derived from human somatic cells. *Science* **318,** 1917–1920.

Zhang, J. H., Chung, T. D., and Oldenburg, K. R. (1999). A simple statistical parameter for use in evaluation and validation of high throughput screening assays. *J. Biomol. Screen.* **4,** 67–73.

Zhang, J. H., Chung, T. D., and Oldenburg, K. R. (2000). Confirmation of primary active substances from high throughput screening of chemical and biological populations: A statistical approach and practical considerations. *J. Comb. Chem.* **2,** 258–265.

CHAPTER EIGHTEEN

Live Imaging Fluorescent Proteins in Early Mouse Embryos

Panagiotis Xenopoulos, Sonja Nowotschin, *and* Anna-Katerina Hadjantonakis

Contents

1. Introduction 362
2. Genetically Encoded FPs for Live Imaging Morphogenetic Events in the Early Mouse Embryo 363
   2.1. Multispectral FPs 363
   2.2. Fusion tags to FPs for labeling and tracking 366
   2.3. Photomodulatable FPs 369
3. Tools for Live Cell Imaging 372
4. Methodology 378
   4.1. Microscope setup 378
   4.2. Collecting and culturing early mouse embryos 379
   4.3. Imaging early mouse embryos 382
5. Conclusions 384
Acknowledgments 385
References 385

Abstract

Mouse embryonic development comprises highly dynamic and coordinated events that drive key cell lineage specification and morphogenetic events. These processes involve cellular behaviors including proliferation, migration, apoptosis, and differentiation, each of which is regulated both spatially and temporally. Live imaging of developing embryos provides an essential tool to investigate these coordinated processes in three-dimensional space over time. For this purpose, the development and application of genetically encoded fluorescent protein (FP) reporters has accelerated over the past decade allowing for the high-resolution visualization of developmental progression. Ongoing efforts are aimed at generating improved reporters, where spectrally distinct as well as novel FPs whose optical properties can be photomodulated, are exploited for live imaging of mouse embryos. Moreover, subcellular tags in combination with using FPs allow for the visualization of multiple subcellular

Developmental Biology Program, Sloan-Kettering Institute, New York, New York, USA

Methods in Enzymology, Volume 506
ISSN 0076-6879, DOI: 10.1016/B978-0-12-391856-7.00042-1

characteristics, such as cell position and cell morphology, in living embryos. Here, we review recent advances in the application of FPs for live imaging in the early mouse embryo, as well as some of the methods used for *ex utero* embryo development that facilitate on-stage time-lapse specimen visualization.

1. Introduction

During their gestation, embryos undergo coordinated complex morphogenetic changes as they develop from single fertilized eggs into neonates having all major organ systems in place for supporting adult life (Fig. 18.1). Embryonic development comprises of a dynamic three-dimensional orchestration of cellular interactions, including proliferation, apoptosis, movement, and differentiation, which occur in a temporally and spatially regulated manner (Arnold and Robertson, 2009; Nowotschin and Hadjantonakis, 2010; Rossant and Tam, 2009). Defining these processes requires time-lapse visualization of the embryo as development progresses. Thus, live imaging provides an essential tool for acquiring spatio-temporal information of the dynamic molecular mechanisms and cellular behaviors driving embryonic development *in vivo*.

The mouse is the premier mammalian model organism used in embryological studies because of its genetic and physiological similarities to humans, as well as the ease with which its genome can be manipulated and analyzed. Key morphogenetic events in mouse and hence mammalian development have been studied for a few decades now, however recent advances in live imaging technologies have brought these studies to another level where these events can be visualized and examined at higher resolution and in utero. Today, protocols have been established and optimized that allow live imaging of mouse embryos *ex utero*, thus providing a way to study mouse embryonic development live and at high resolution (Garcia *et al.*, 2011a,b,c,d; Nowotschin *et al.*, 2010; Udan and Dickinson, 2010).

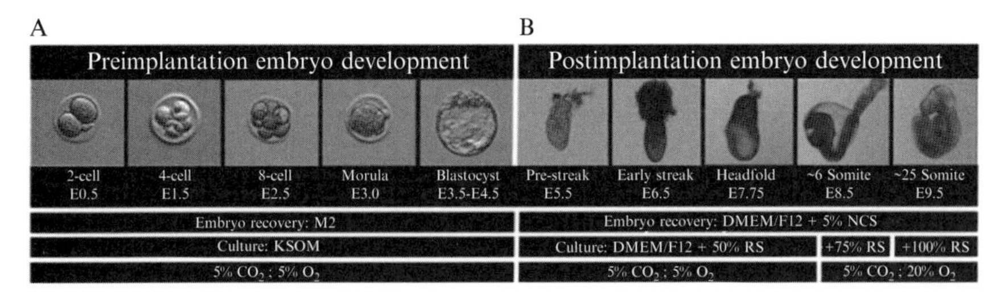

Figure 18.1 Development of (A) preimplantation and (B) postimplantation embryos. Requirements for dissection and culture media as well as gas content are shown for each embryonic stage. NCS, newborn calf serum; RS, rat serum; E, embryonic day (adapted from Nowotschin *et al.*, 2010; with permission from Academic Press/Elsevier).

Fluorescence emitted from excited fluorophores has been exploited for cellular imaging and for the observation of single or groups of cells in developing embryos. Injection of vital dyes, such as the lipophilic tracers, DiI, or DiO (Serbedzija *et al.*, 1992), inorganic semiconductor nanocrystals, known as quantum dots (Dubertret *et al.*, 2002), and genetically encoded fluorescent proteins (FPs) (Nowotschin *et al.*, 2009b) have been used as fluorophores to label cells in embryos. Of these, FPs have been most prominent for live cell imaging because of their high signal-to-noise ratio, minimal toxicity, non-reliance on the availability of antibodies for the protein of interest, and ease of use. FPs, such as the green fluorescent protein (GFP) and its variants, as well as the orange and red FPs are ideal reporters for live imaging since they can be expressed under the control of promoter/*cis*-regulatory regions driving sufficient levels of expression of the FP for visualization. Moreover, FPs can be fused to any protein of interest to investigate both protein dynamics *in vivo* as well as to provide subcellular segmentation (Nowotschin *et al.*, 2009b). The use of FPs for visualizing mouse embryonic development in live *ex utero* cultured mouse embryos (Hadjantonakis *et al.*, 2003) has led to novel insights of the intrinsic cell behaviors underlying tissue morphogenesis and cell lineage specification in mouse embryos (Kwon *et al.*, 2008; Plusa *et al.*, 2008).

In this chapter, we provide an overview of some of the most commonly used FPs and their application for live imaging in mouse embryos. We also discuss the necessary tools for live imaging, such as standard confocal microscopy techniques available at most institutions. Finally, we provide a brief description of the methods and conditions that are routinely used in our and other laboratories for *ex utero* culturing and time-lapse imaging of mouse embryos.

2. Genetically Encoded FPs for Live Imaging Morphogenetic Events in the Early Mouse Embryo

2.1. Multispectral FPs

Currently, many FPs have been isolated and are readily available to use for live imaging applications (Table 18.1). Among them, the GFP and its variants are the most commonly used FPs to study the complex cell behaviors and organization of the embryo and the adult animal (Nowotschin *et al.*, 2009b) due to their nontoxicity in eukaryotic cells. GFP and its variants have been widely applied in live cell imaging regimes either expressed in their native form throughout the cellular cytoplasm or, as will be discussed in the next section, fused in frame to a protein of interest so as to function as tags allowing for the visualization of specific protein

Table 18.1 Characteristics of commonly used fluorescent proteins

		Characteristics			Used for live imaging in mouse embryos		
	FPs	Absorbance (nm)	Emission (nm)	Conformation status	Cytoplasmic	Tagged	References[a]
GFP variant	ECFP	433	475	Monomer	–	–	
	Cerulean	433	475	Monomer	–	✓	Stewart *et al.* (2009)
	EGFP	488	507	Monomer	✓	✓	Hadjantonakis and Papaioannou (2004), Kwon *et al.* (2006)
	EYFP	512	528	Monomer	–	✓	Fraser *et al.* (2005)
	mVenus	512	528	Monomer	–	✓	Rhee *et al.* (2006)
Orange, red, and far-red	tdTomato	554	581	Dimer	–	✓	Trichas *et al.* (2008)
	Katushka	574	602	Dimer	✓	–	Dieguez-Hurtado *et al.* (2011)
	mRFP1	584	607	Monomer	✓	✓	Long *et al.* (2005), Hayashi-Takanaka *et al.* (2009)
	mCherry	587	610	Monomer	✓	✓	Viotti *et al.* (2011), Nowotschin *et al.* (2009a)

[a] Examples of characteristic references describing the usage of corresponding FPs in live or fixed mouse embryos.

localization, cell division, cell tracking, cell death, and cell morphology (Hadjantonakis and Papaioannou, 2004; Rhee *et al.*, 2006). Moreover, cell-type-specific resolution within the tissues of the embryo can be achieved when these FPs are placed under the control of *cis*-regulatory elements (Ferrer-Vaquer *et al.*, 2010; Kwon and Hadjantonakis, 2007; Kwon *et al.*, 2006, 2008; Monteiro *et al.*, 2008). Additional temporal control can be achieved by expressing the FP in a genetically inducible regime, as in genetically inducible fate mapping approaches (Joyner and Zervas, 2006). Further, the use of multiple spectrally distinct FPs enables the simultaneous study of several cell characteristics (Livet *et al.*, 2007).

GFP emits green light when excited with ultraviolet (UV) or blue light. The engineering of GFP resulted in several spectral variants, the blue and cyan FPs (e.g., BFP, CFP, and Cerulean), the yellow FPs (e.g., YFP and Venus) which emit blue and yellow light, respectively, as well as in improved versions of the originally isolated wild-type GFP protein with brighter fluorescence and greater photostability (Cubitt *et al.*, 1995; Heim *et al.*, 1995). The most popular version for use in mammalian systems is the enhanced GFP (EGFP). Apart from GFP and its variants, spectrally distinct FPs that emit orange, red, and far-red light are sought after for imaging in tissue specimens (Table 18.1). FPs with emission in the long-wavelength spectrum are advantageous in respect of their reduced cell toxicity, less background autofluorescence, deeper tissue penetration, which is important for larger specimens such as mouse embryos or whole animal imaging, as well as for their ease of covisualization with FPs possessing short-wavelength emission spectra. Though the list of RFPs is fairly short, some RFPs have already been used successfully in transgenic reporter mice (Kwon and Hadjantonakis, 2009; Viotti *et al.*, 2011) and live imaging of mouse embryos. When expressed under the control of *cis*-regulatory elements, RFPs like mRFP and mCherry ('m' stands for monomeric) have been shown to provide cell-type-specific resolution during development of the early mouse embryo (Kwon and Hadjantonakis, 2009; Poche *et al.*, 2009; Viotti *et al.*, 2011). Transgenic mice ubiquitously expressing mRFP1 and mCherry in the cytoplasm (Fink *et al.*, 2010; Long *et al.*, 2005) have shown that the cytoplasmic expression of these RFPs is developmentally neutral. However, though mRFP1 is a monomeric variant of the tetramer DsRed, ubiquitous expression of histone (H2B) or myristoylated (myr) fusions of mRFP1 or mCherry are in some instances not developmentally neutral and thus incompatible with normal development in mice (Nowotschin *et al.*, 2009a). Nevertheless, live imaging short-term lineage tracing experiments by injecting mRNA for H2B-RFP and myr-RFP in 8-cell stage embryos have been performed (Yamanaka *et al.*, 2010). Another variant of mRFP1, the tandem dimer (td) (td)Tomato, characterized by its high brightness and photostability as well as its amenability to fusion tags has been used for live imaging in mice (Muzumdar *et al.*, 2007; Trichas *et al.*, 2008). Fusions to mCherry, such as histone H2B as a nuclear tag, have been shown to display a satisfactorily bright signal (Abe *et al.*,

2011; Egli *et al.*, 2007). Though, similar to mRFP1, constitutive widespread expression in mice of H2B-mCherry has proven to be toxic, its expression under specific *cis* regulatory elements has been shown to be non-teratogenic (Nowotschin *et al.*, 2009a).

As already discussed, FPs with excitation and emission spectra in the far-red region of the spectrum can be advantageous when combined with GFP and its variants as spectral separation as well as deeper tissue penetration is crucial (Shcherbo *et al.*, 2007). The dimer Katushka is currently the brightest FP among those with emission maxima above 620 nm to have been used in the mouse embryo. In fact, the first transgenic mouse line conditionally expressing Katushka was recently reported (Dieguez-Hurtado *et al.*, 2011). Tissue-specific expression of Katushka was ubiquitous and strong without displaying any toxic effects in this transgenic line; nevertheless, usage of Katushka fused to protein tags has not been reported yet. New dimeric variants of Katushka with further red-shift emission as well as reduced cytotoxicity and brighter signal were recently reported. They were tested in tissue culture cells and *Xenopus* embryos, opening exciting possibilities for their use in mouse embryos (Shcherbo *et al.*, 2010). However, the dimeric nature of Katushka and its variants has limited its potential for incorporation in protein fusions for subcellular localization.

Monomeric versions of far-red FPs have been developed, such as mKate, lacking the brightness of Katushka. However, the recently developed successor of mKate, mKate2, and a pseudo-monomeric Katushka have both been reported to work well with fusion tags while exhibiting bright fluorescence when imaged live in cells and transgenic *Xenopus* embryos (Shcherbo *et al.*, 2009). Moreover, the development of nuclear-localized variants of mKate were recently reported that allowed observing tumor cells in living mice (Piatkevich *et al.*, 2010). The development of novel monomeric red and far-red FPs are eagerly anticipated and should allow for their fusion with subcellular tags (Piatkevich and Verkhusha, 2010).

2.2. Fusion tags to FPs for labeling and tracking

The use of FP fusion proteins has provided a powerful tool to investigate mouse embryonic development in three dimensions through live imaging (Table 18.2) (Kwon and Hadjantonakis, 2007; Hadjantonakis *et al.*, 2003). For example, fusions of FPs to the human histone H2B are bound to active chromatin and localize to the nucleus thus enabling tissue segmentation, and the visualization of cellular characteristics, such as division, movement, or apoptosis, as well as the tracking of daughter cells during developmental progression (Hadjantonakis and Papaioannou, 2004; Kanda *et al.*, 1998). Histone fusions do not only facilitate tracking of cells, but also the subsequent segmentation of the spatially separated nuclei in cell lines and within the complex 3D structure of the pre- or postimplantation embryo during image

Table 18.2 Fusion tags to fluorescent proteins that have been used for live imaging in mouse embryos

Localization	Tags	Characteristics	References[a]
Nucleus	H2B	Human histone	Hadjantonakis and Papaioannou (2004)
Plasma membrane	GPI	Glycosylphosphatidylinositol anchor	Rhee *et al.* (2006)
	myr	Myristoyl anchor	Rhee *et al.* (2006), Abe *et al.* (2011)
	Lyn	Tyrosine kinase	Abe *et al.* (2011), Nowotschin and Hadjantonakis (2010)
	RAS	GTPase	Abe *et al.* (2011)
	GAP43	Axonal membrane protein	Abe *et al.* (2011)
	pDisplay	Membrane targeting vector (Invitrogen)	Abe *et al.* (2011)
Mitochondria	Mito	Cytochrome C oxidase subunit VIII A	Abe *et al.* (2011)
Golgi	Golgi	β-1,4-Galactosyltransferase 1	Abe *et al.* (2011)
Microtubules	Tuba	α-Tubulin 2	Abe *et al.* (2011)
	hEMTB	Human microtubule associate protein 7	Abe *et al.* (2011)
	EB1	APC binding protein	Abe *et al.* (2011)
Cytoskeletal proteins	Actin	β-Actin	Abe *et al.* (2011)
	Moesin	ERM family protein (linker between plasma membrane and actin cytoskeleton)	Abe *et al.* (2011)
Focal adhesions	Paxillin	Focal adhesion-associated adaptor protein	Abe *et al.* (2011)

[a] Examples of characteristic references describing the usage of corresponding FPs in live or fixed mouse embryos.

data analysis (Fig. 18.2). Cell-type-specific combined with nuclear-specific resolution has been achieved by placing H2B-FP reporters under defined promoter and other *cis*-regulatory elements. For example, H2B-GFP driven by regulatory elements of the *Pdgfra* locus has been useful for lineage tracking in the preimplantation embryo (Artus *et al.*, 2010; Plusa *et al.*, 2008). In another study, H2B-GFP driven by TCF/Lef bound *cis*-regulatory elements have provided a nuclear-localized, single-cell reporter that can be used to live image the behavior and fate of cells responsive to Wnt signaling (Ferrer-Vaquer *et al.*, 2010). Moreover, a recent report using H2B-GFP fusion reporter under the regulation of the epsilon-globin *cis*-regulatory elements allowed visualization of the first committed hematopoietic progenitors within the developing mouse embryo (Isern *et al.*, 2011).

Fusions of FPs to plasma membrane proteins such as glycosylphosphatidylinositol (GPI) or myristoylation (myr) and farnesylation (CAAX) tagging allows the visualization of cell morphology *in vivo* (Table 18.2) (Muzumdar *et al.*, 2007; Rhee *et al.*, 2006). Additional tags targeting FPs to the plasma membrane were recently reported (Table 18.2) (Abe *et al.*, 2011). Importantly, these plasma membrane localized fusions can be used in combination with H2B fusions enabling the visualization of multiple cellular characteristics, such as nuclear position as well as cell morphology (Fig. 18.3); these combinations have been used in embryonic stem cell cultures and in mouse

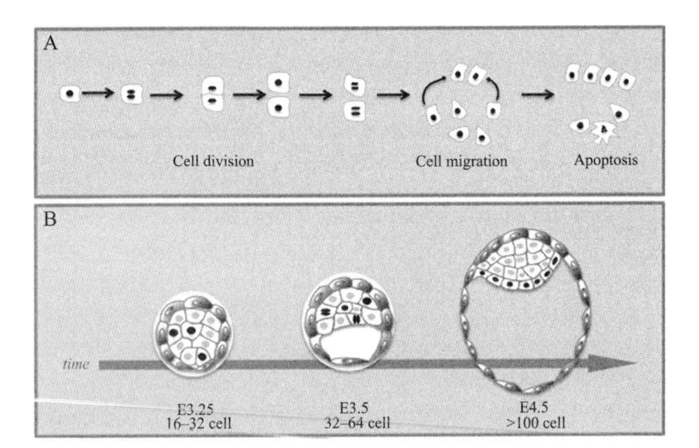

Figure 18.2 Histone fusions to FPs provide nuclear labeling. (A) Schematic representation of cells expressing a histone (H2B) fusion to fluorescent protein. The H2B-FP remains bound to chromatin, thus providing nuclear labeling. Cells expressing this fusion reporter can be tracked as they divide, migrate, or undergo apoptosis. (B) Cartoon for time-lapse imaging of preimplantation mouse embryo expressing H2B-FP under the control of the *Pdgfra-α* locus. This reporter labels the progenitors of the primitive endoderm lineage that lies adjacent to the blastocoel cavity in the E4.5 stage preimplantation embryo. Dynamic cell processes, such as division, migration, and apoptosis can be tracked by using this reporter (Plusa *et al.*, 2008).

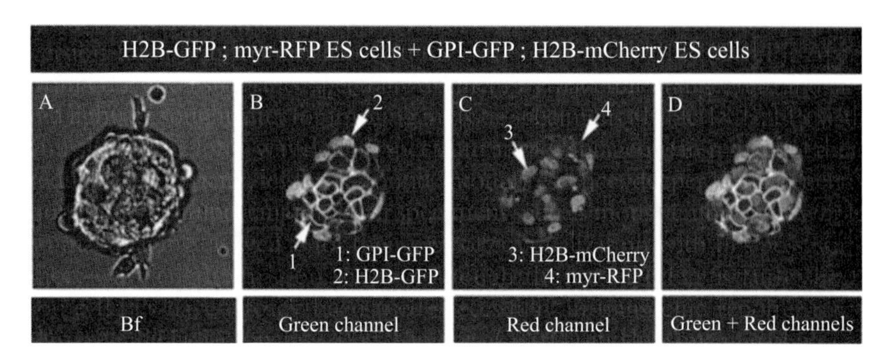

Figure 18.3 Live imaging of multi-tagged mouse ES cell colonies. 2D images of an embryonic stem (ES) cell colony comprising two distinct transgenic cell populations expressing H2B-GFP ; myr-RFP or H2B-mCherry ; GPI-GFP. 1: GPI-GFP; 2: H2B-GFP; 3: H2B-mCherry; 4: myr-RFP. Bright field (Bf) image (A), green channel (B), red channel (C), and merge of green and red channel (D).

embryos (Abe *et al.*, 2011; Nowotschin *et al.*, 2009a; Stewart *et al.*, 2009; Trichas *et al.*, 2008). Recently, a mouse reporter line was developed that allows conditional expression of nuclear localized mCherry expression combined with plasma membrane GFP expression; this dual reporter was shown to be nontoxic for live imaging embryonic development and provides a powerful tool to mark and track distinct populations of cells *in vivo* (Shioi *et al.*, 2011).

In addition, tags have recently been used to target FPs to subcellular locations, such as mitochondria, Golgi apparatus, microtubules, and actin filaments as well as focal adhesion points (Abe *et al.*, 2011). This study also reported dual labeling of nucleus and Golgi apparatus, as well as nucleus and plasma membrane (Abe *et al.*, 2011).

2.3. Photomodulatable FPs

Traditionally, labeling and tracking cells in live mouse embryos were performed using invasive techniques such as dye injections and tissue grafts, as well as binary transgenic strategies (Joyner and Zervas, 2006; Kinder *et al.*, 1999, 2001; Nagy, 2000). Photomodulatable FPs can now be used to non-invasively label and track cells or proteins within the complex 3D structure of the mouse embryo (Table 18.3) (Nowotschin and Hadjantonakis, 2009a,b). This can be achieved at a specific embryonic region of interest and in a defined spatio-temporal resolution. There are two categories of photomodulatable FPs that change properties after excitation with UV or blue light: the photoactivatable FPs (PA-FPs) that change from a non-fluorescent to a fluorescent state and the photoconvertible FPs (PC-FPs) that change fluorescence absorbance and emission spectra and thus, convert from one color

Table 18.3 Characteristics of commonly used photomodulatable fluorescent proteins

		Characteristics		Used for live imaging in mouse embryos		
	FPs	Change of fluorescence	Conformation status	Cytoplasmic	Tagged	References[a]
Photoactivatable	PAGFP	No fluorescence → green	Monomer	✓	–	Plachta *et al.* (2011)
	KFP	No fluorescence ↔ red	Tetramer	–	–	
	Dronpa	No fluorescence ↔ green	Monomer	–	–	
Photoconvertible	PS-CFP2	Cyan → green	Monomer	–	–	
	Kaede	Green → red	Tetramer	✓	–	Tomura *et al.* (2008)
	KikGR	Green → red	Tetramer	✓	✓	Nowotschin and Hadjantonakis (2009b), Abe *et al.* (2011)

[a] Examples of characteristic references describing the usage of corresponding FPs in live or fixed mouse embryos.

to another (Table 18.3). To date, photomodulatable FPs have been used successfully in live chick, *Drosophila*, zebrafish, and *Xenopus* embryos (Hatta *et al.*, 2006; Murray and Saint, 2007; Stark and Kulesa, 2005; Wacker *et al.*, 2007). However, their usage in mouse embryos has only recently started to be applicable (discussed below).

Photoactivatable GFP was developed as a GFP variant with a single residue substitution that remains in a non-fluorescent state. However, PA-GFP yields a 100-fold increased green light fluorescence upon exposure to short-wavelength light as it irreversibly converts to a fluorescent state (Fig. 18.4). PA-GFP has been of special interest because of its monomeric nature and its potential to have no toxicity in live embryos. An example of PA-GFP application in mouse embryos has been to examine postnatal neocortex development (Gray *et al.*, 2006). Moreover, PA-GFP was recently used to investigate lineage commitment in live preimplantation mouse embryos (Plachta *et al.*, 2011). Use of other PA-FPs, such as PA-mCherry or the kindling FP (KFP) and Dronpa proteins, has not yet been reported for imaging applications in live mouse embryos (Table 18.3).

Regarding PC-FPs, transgenic mice constitutively expressing the photoconvertible Kaede, that emits green light and converts to a red fluorescent state upon exposure to UV or violet light, have been generated and used to study cell movements from lymphoid organs to other tissues (Tomura *et al.*, 2008). Moreover, an *in utero* photoconversion procedure was recently

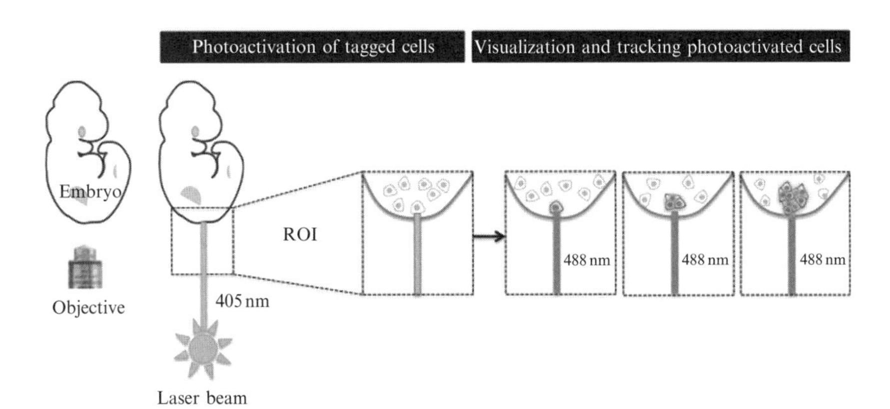

Figure 18.4 Schematic representation of photoactivation in an embryo constitutively expressing a photoactivatable fluorescent protein fused to a histone. Initially, all cells express the fluorescent protein, which remains however in a non-fluorescent state. Upon excitation of certain cells (within a region of interest, ROI) with short-wavelength high-power (UV or blue) laser, the fluorescent protein switches to a fluorescence emitting state. These photoactivated cells can then be tracked over time. Fusion of the fluorescent protein to a histone protein (e.g., H2B) provides single-cell resolution, and cell tracking of labeled nuclei. (For interpretation of the references to color in this figure legend, the reader is referred to the Web version of this chapter.)

reported by using the Kaede FP (Imai *et al.*, 2010). Another PC-FP that has been used in embryonic stem cells and mouse embryos is the Kikume Green-Red protein (KikGR). KikGR exhibits similar photoconversion properties as Kaede and can be used to label and track cells in embryonic stem cell cultures and in live embryos. KikGR was compared to PA-GFP and two other PC-FPs, Kaede, and PS-CFP2 (which converts from cyan to green fluorescence) for applicability in murine embryonic stem cell lines and live mouse embryos (Nowotschin and Hadjantonakis, 2009b). It was shown that KikGR is most suitable for cell labeling and lineage studies because it is developmentally neutral, bright, and undergoes rapid and complete photoconversion. A recent study reported the development of a transgenic mouse with broad expression of KikGR from the Ubiquitin C promoter (Griswold *et al.*, 2011). Although KikGR does not work well when fused to nuclear tags (perhaps due to its tetrameric nature), a recently reported conditionally expressed KikGR reporter with plasma membrane localization was developed, however its level of expression was weak (Abe *et al.*, 2011). The recent development of additional PC-FPs that can convert from red to green or from orange to far-red fluorescence has opened the way for more options to be used for live imaging in the mouse embryo in the near future (Kremers *et al.*, 2009; Piatkevich and Verkhusha, 2010).

3. Tools for Live Cell Imaging

"Seeing is believing" and in order to visualize FP reporters, appropriate microscopy equipment is required. Several optical imaging modalities are widely available (Table 18.4). Advances in imaging techniques have accelerated over the past decade and each type of microscope system provides certain advantages as well as limitations regarding live cell imaging (Table 18.4) (Walter *et al.*, 2010). First, conventional widefield fluorescence microscopy (Lichtman and Conchello, 2005) was considered a powerful tool to observe whole embryos since it provides fast acquisition and flexible excitation at low cost. However, physical destruction could occur as a consequence of widespread exposure and out-of-focus light could interfere during image acquisition because of the thickness of the specimen. Image processing and deconvolution techniques can partially resolve out-of-focus interference; however, these methods have been applied efficiently for smaller specimens, such as bacteria. By contrast, distortion originating from the multiple layers of cells within the thick structure of an embryo is not easy to categorize and formulate, thus precluding use of deconvolution formulae as the system of choice for visualizing details within live embryos using widefield fluorescence.

Table 18.4 Types of confocal microscope used for live imaging mouse embryos

System configuration	Advantages	Disadvantages
Point laser scanning confocal	Most commonly used, best out-of-focus exclusion, best resolution	Slow, decreased sensitivity, high risk of phototoxicity, expensive
Slit laser scanning confocal	Fastest confocal, low risk of phototoxicity, high sensitivity	Low resolution
Spinning disk confocal	Fast, most gentle confocal, low risk of phototoxicity	Low resolution, not ideal for very thick specimens, works best with only one fluorophore
Light sheet based fluorescent	Ideal for very thick specimens, low risk of phototoxicity	Need for rotating device, huge amount of data are generated
Two- and multiphoton	Low risk of phototoxicity	Parameters need to be checked before experiment, red and far-red FPs cannot be used, difficult to perform multiphoton experiments

By contrast, confocal microscopy excludes light originating from outside the focal plane and thus, provides optical sectioning. This allows observation within a complex specimen such as an embryo at subcellular and spatiotemporal resolution (Conchello and Lichtman, 2005). The different optical sections (usually referred to as a *z*-stack) can then be reconstructed into a 3D projection with appropriate software. For this reason, confocal fluorescence microscopy has become the tool of choice when visualizing live embryos. A list of several types of confocal technologies suitable for live embryo imaging is discussed next.

Point laser scanning (comparable to Zeiss LSM700 or 710 and Leica SP5 systems). This is the most popular modality. It passes a single point of excitation laser light through a narrow cylinder across a specimen. The laser light illuminates all points of the specimen within the cylinder perpendicular to a specific focal plane. After excitation, fluorophores emit light that then passes through a pinhole. The pinhole functions to exclude light gathered from above and below the focal plane and provides a physical device to out-of-focus light. Thus, point scanning is less prone to be inaccurate because the pinhole function is not based on mathematical formulae (in contrast to deconvolution). A point scanning confocal is usually the system of choice in most laboratories because it provides good resolution and out-of-focus suppression as well as the highest multispectral flexibility. Point scanning with multispectral flexibility allows imaging of multiple probes simultaneously and is the most commonly used approach for performing photoactivation and photoconversion experiments. However, the process of image acquisition is relatively slow because of the time needed for the laser beam to scan the entire sample, which raises concerns about phototoxicity and photobleaching. For imaging live mouse embryos, it is preferable to increase the scan frequency and set up bidirectional *xy* scanning, thus resulting in a lower risk of phototoxicity and photobleaching of the specimen. It is worth noting that the slower acquisition obtained with a point scanning confocal could be problematic for imaging fast biological dynamics in live specimens because there will always be a temporal difference from the pixel in the top left corner of the image to the bottom right pixel. Another point that should be taken into consideration regarding point laser scanning microscopy is the detectors used. The typical detector used is a Photomultiplier Tube (PMT). A PMT is less sensitive detector than a CCD camera based system. These devices are preferred because they acquire light quickly and allow the scanning speeds to be much faster than if they were acquired with a CCD. Therefore, commonly used point scanning confocal devices provide the best resolution, when thick specimens are visualized. The trade off is sensitivity, meaning that weaker fluorescent signals and thinner specimens do not resolve as well in point scanning modalities. Another disadvantage of point scanning confocal is its high cost due to the need for several lasers, each specific for only a subset of fluorophores.

Slit scanning (comparable to Zeiss LSM7LIVE systems). This modality allows for more rapid image acquisition because it passes a fine slit, rather than a point, of laser light across the specimen. A pinhole is again used for limiting the gathering of light originating only from the focal plane. Even though this type of system does not provide better resolution or suppression of out-of-focus light compared to point confocal, it is the system of choice when observing rapid processes (e.g. microtubule or cilia movement) and when samples are sensitive to phototoxicity or photobleach. Moreover, slit scanning (as well as spinning disk confocals, discussed next) use CCD detectors instead of PMTs for detection and do not exclude as much out-of-focus light as point scanning does, allowing for higher sensitivity, especially when weaker signals or thinner specimens are visualized.

Spinning disk-Nipkow type (comparable to Perkin-Elmer UltraView RS5 and Leica SD6000 systems). These systems are used in our laboratory for imaging pre- and early postimplantation embryos, as well as embryonic stem cells. This modality uses a pair of rotating disks each containing thousands of pinholes. Laser light is passed through these holes and is then projected to the specimen. The emitted light returns through the same holes providing a high-quality confocal image of the entire field of view. In this way the spinning disk confocal collects multiple points simultaneously rather than scanning a single point at a time, allowing for faster image acquisition. Spinning disk confocals, although much faster than point scanning confocals, are not considered faster than slit scanning but they do provide the least invasive imaging of the sample compared to all modalities discussed so far. This gentle acquisition is attributed to the fact that the whole field is excited by laser illumination and detected on a CCD chip. The illumination is not focused at high power on any one spot on the sample. Many different CCD options allow for extremely low laser power imaging for long-term time lapses, hence providing a powerful tool for reduced phototoxicity in live cell imaging. However, image resolution is decreased compared to the slower point laser scanning systems. Moreover, spinning disk confocal acquisition is well suited for imaging single fluorophores, but it does not exhibit great multispectral flexibility. Additionally, a spinning disk confocal may not be the most ideal system for performing photoactivation and photoconversion experiments. There are two additional considerations regarding the use of a spinning disk: the multiple pinholes allow light from one point in the sample to appear in multiple pinholes when imaging deeper into a sample. Therefore, the optical sectioning is only discrete to approximately 20–40 μm of penetration depth. After this depth, the images are closer to standard epifluorescence than confocal sections. It should also be noted that the disk contains multiple pinholes that are not adjustable to different apertures for different magnification objectives. That means the spinning disk (at least the Yokogawa versions) is only taking ideal optical sections with a 100 × objective. At 63 × the optical section is equivalent to 2 Airy Units (a pinhole of 1 Airy Unit gives the best signal-to-noise ratio), at 40 × to 4 Airy Units and the lower the magnification selected, the closer the performance of the spinning disk to that of a standard fluorescence. Some spinning disk systems offer multiple disks with different size apertures depending on the magnification desired.

Light sheet-based fluorescence microscopes (LSFM, SPIM, DSLM) (Hell, 2003; Reynaud *et al.*, 2008; Santi, 2011). These modalities were further developed from spinning disk confocal systems and allow optical sectioning without illumination of the entire specimen (in contrast to point scanning confocal where the entire specimen is illuminated even though only a single focal plane is observed). As a consequence, phototoxic damage and photobleaching of the specimen is considerably reduced. The principle is to illuminate an *xy* plane at varying angles to the objective imaging the sample. This is achieved by moving the sample so that the optical sheet penetrates the sample at different angles. This modality becomes especially attractive when very thick specimens need to be visualized and other fluorescence techniques cannot be applied. The challenge is that the specimen has to be mounted in a device that can be rotated through 360°. A disadvantage for this sort of imaging is that it acquires a large amount of data which a computer needs to be able to render. Nevertheless, these methodologies are very attractive for imaging live embryos; to date these modalities have been used for zebrafish and *Drosophila* embryonic development and it is only a matter of time until they are used for mice (Huisken *et al.*, 2006; Keller *et al.*, 2008, 2010).

Super resolution fluorescence microscopy. These modalities encompass recent advances in light microscopy providing increased spatial resolution and allowing the visualization of unobserved details of biological functions and processes. Two of these systems, photoactivated localization microscopy (PALM) (Betzig *et al.*, 2006) and stochastic optical reconstruction microscopy (STORM) (Rust *et al.*, 2006) are not currently suited for imaging samples such as embryos. Both techniques rely on photomanipulation of a molecule within the specimen to activate fluorescence and then switch it off again to allow the concentration or localization of the original point of fluorescence. The challenge lies in the fact that the photoactivation needs to take place at discrete x, y, and z coordinates. In a thick specimen, it would be impossible to identify the z plane from which the signal arises. However, the STED (stimulated emission depletion) technique could, in principle, be used for embryo imaging (Hell, 2003). The principle behind STED is to use a laser at one wavelength defocused slightly to compress or shrink fluorescent emission at a second wavelength thus providing higher resolution than a typical confocal image. The challenge is that two lasers of exact wavelengths must be used, and to date STED has only been demonstrated with certain fluorophores. It is also worth noting that to date only two color imaging is possible, and the cost of these systems is relatively high.

Two- and multiphoton microscopy (Helmchen and Denk, 2005; So *et al.*, 2000). Similarly to confocal systems multiphoton microscopes gather data originating from the focal plane within the specimen. Unlike confocals these microscopes only illuminate the focal plane rather than the entire specimen and they do not require pinholes. Therefore, multiphoton modalities are not considered part of confocal optical technologies. The basic principle for these systems is that the fluorophore is excited by two long-wavelength photons that have to strike it simultaneously. The fluorophore will then emit light of shorter wavelength than that of the excitation light, which is a major difference compared to conventional fluorescence microscopy. Based on this principle, fluorophores from outside the focal plane should not be excited; that makes two- and multiphoton systems powerful tools for live imaging with reduced instances of specimen phototoxicity and photobleaching these microscopes are more costly and not as easy to use and imaging parameters have to be arranged prior to the start of experiment. Also, many red and far-red FPs cannot be used with these types of systems because of their long-wavelength excitation spectra. Moreover, it is more difficult to perform multi-fluorophore experiments with multiphoton systems because of the broad and overlapping excitation spectra. Nevertheless, a recent report showed application of two-photon microscopy for live imaging of preimplantation mouse embryos, providing a framework for application of this powerful technology for imaging mouse embryos at different stages and with various reporters (McDole *et al.*, 2011).

For live cell imaging, systems should be preferably fitted on an inverted microscope base. The maintenance of an environment with a stable temperature, humidity, and gas concentration, which is absolutely essential for proper embryo culture, is difficult to achieve on an upright microscope configuration.

The conditions of stable temperature, humidity, and gas concentration are achieved with use of an environmental chamber that encloses the microscope (Fig. 18.5). These are commercially available but because of their high cost, in-house environmental chambers can also be manufactured.

Confocal-based microscopes should be equipped with appropriate objectives, usually dry 5×, 10×, and 20× (with the latter also can be used multi-immersion), multi-immersion 40× and oil 63×. The 5× objective is used to scan the field of view and locate the embryos. The 10× objective can be used for low-magnification 3D time-lapse imaging. The 20× and 40× objectives are the most suitable for high-magnification 3D time-lapse image acquisition. The 63× objective is rarely used because it is too high a magnification to be suitable for whole embryo imaging. It is also worth considering different types of objectives and which might be most appropriate for live imaging. For example, Carl Zeiss provides two types of objectives: the Neofluar (Neo) lenses and the Plan Apochromat (Plan Apo) objectives. The Neo and Apo objectives provide different degrees of optical corrections (fluorite and apochromatic aberration correction respectively); the Apo objectives enjoy the

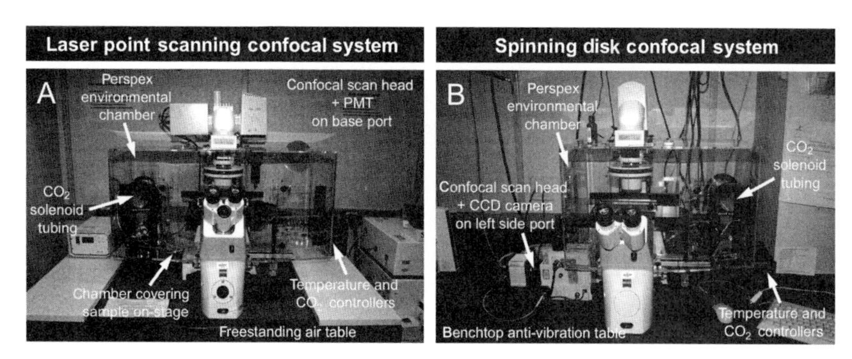

Figure 18.5 Microscope set up for static on-stage live imaging of mouse embryos. (A) Inverted point laser scanning confocal microscope and (B) inverted spinning disk confocal microscope. Both systems include an environmental chamber that encloses the stage and optics carrier and provides a heated, humidified, and gassed environment for on-stage embryo culture. Both microscopes are positioned on an anti-vibration air table (adapted from Nowotschin *et al.*, 2010; with permission from Academic Press/Elsevier).

highest correction for spherical and chromatic aberrations. Therefore, Neo objectives acquire a slightly thicker optical slice than the Apo series, meaning that when imaging through different z planes, each individual image is not as crisp but it is often brighter. However, the Apo lenses acquire a finer optical slice and each individual image is crisper. Specifically, the Apo lenses for the $63\times$ objective have been developed for live cell imaging, providing optimal focus stability for time-lapse experiments.

Finally, a computer workstation(s) running appropriate software are needed for image acquisition and analysis. Software packages used for image acquisition often come with microscope systems and are available from Zeiss, Perkin-Elmer (Volocity), Molecular Devices (Metamorph). Software packages for image analysis are available from Bitplane (Imaris), Perkin-Elmer (Volocity), and Molecular Devices (Metamorph) websites to name a few. Image analysis can also be performed using open source applications such as ImageJ (http://rsbweb.nih.gov/ij/).

4. Methodology

4.1. Microscope setup

For live imaging of *ex utero* cultured embryos, inverted microscope systems are preferable. The microscope stage must be enclosed within an appropriate environmental chamber that will provide conditions closely resembling *in utero* development (Fig. 18.5). The chamber allows for embryo culturing with constant temperature at 37 °C as well as with humidity and stable gas content. The precise gaseous mixture comprising $CO_2/O_2/N$ used for embryo culture depends on the embryonic stage (Fig. 18.1). Usually 5% CO_2 is provided for all stages. Regarding oxygen concentrations, 5% O_2 is used for preimplantation embryos and early postimplantation embryos (E5.5–E8 stages), and 20% O_2 for later postimplantation embryos (E8.5–E10.5; Fig. 18.1).

The incubator should be turned on and allowed to come to temperature at least 1 h prior to starting an imaging experiment. Embryos within cultured media should be placed on glass coverslip chambers or culture dishes containing cover glass bottoms. Efficient imaging can only be acquired when the glass thickness does not exceed 1.5 mm (MatTek dishes or Lab-Tek coverslip chambers). The culture dish should be covered with embryo-quality mineral oil to prevent evaporation during on-stage culture.

In general, UV and laser light originating from other excitation spectra could be harmful to living embryos. Therefore, parameters for image acquisition will need to be optimized to minimize toxicity to embryos. Setting up these imaging adjustments depends on the confocal system being used. For example, when using slower point scanning confocal systems, laser power, and exposure time should be reduced whereas the size of the optical sections and the scan frequency can be increased to prevent phototoxicity and photobleaching. Additionally, these adjustments greatly depend on the fluorophore that is used. Finally, imaging parameters also depend on the tissue or developmental stage being visualized.

4.2. Collecting and culturing early mouse embryos

Protocols have been developed for *ex utero* culture of pre- and postimplantation mouse embryos. We briefly overview the steps that need to be followed for collecting and preparing on-stage (static) cultures for live imaging preimplantation as well as postimplantation embryos. We also describe how to prepare roller cultures as a way to achieve the most optimal *ex utero* conditions for the development of postimplantation embryos.

Collection and on-stage culture of preimplantation embryos. After fertilization, mouse embryos float within the upper reproductive tract, making the process of sample recovery much simpler as compared to later postimplantation stages of development, which requires dissection. Culture conditions for preimplantation embryos are well established and require appropriate media, stable temperature, and gas content to resemble *in utero* conditions (Nagy *et al.*, 2003). Briefly, preimplantation embryos are recovered in M2 media and cultured in KSOM media (both are commercially available by Millipore, Specialty Media). Both these media can also be made in-house (Nowotschin *et al.*, 2010). In more detail:

(i) Before starting the dissection of embryos, M2 and KSOM media are prewarmed at 37 °C. The glass bottom of a MatTek dish is covered with 2% Bacto agar (BD Medical) supplemented with 0.9% NaCl in order to prevent embryos sticking to the glass. Around 300-500 microliters of KSOM are then placed on the agar surface and covered with mineral oil to prevent evaporation. The dish should be placed for at least 30 min in a humidified incubator with a stable temperature of 37 °C, gassed with 5% CO_2.
(ii) The dish should not remain outside the incubator for long because changes in temperature and pH and exposure of the embryo in KSOM to atmospheric conditions are deleterious for culture.
(iii) After sacrificing a pregnant female (Nagy *et al.*, 2003), the oviduct (when recovering E0.5–E2.5 embryos) or the uterus (when recovering E3.5–E4.5 embryos) is dissected and placed in prewarmed M2 media.
(iv) The dish is then placed under a stereomicroscope. The oviduct/uterus are flushed with prewarmed M2 media. A blunt 30-gauge needle is inserted into the oviduct infundibulum for flushing the oviduct. A 1 mL syringe with a 30- or 26-gauge needle is used for flushing the uterus.

(v) Embryos are collected using a mouth pipette attached to a Pasteur pipette that has been pulled over a bunsen burner (Nagy *et al.*, 2003). They are then transferred through several microdrops of KSOM (to wash off residual M2 media) and finally placed in equilibrated KSOM in the MatTek dish. It is essential that the zona pellucida of preimplantation embryos recovered after stage E2.5 has to be removed before they are placed on the MatTek dish; this can be performed by transferring the embryos through several microdrops of Acid-Tyrodes (Millipore, Specialty Media). A higher density of embryos enhances their development, thus embryos should be preferably transferred together.

(vi) Embryos placed within the equilibrated KSOM microdrop can then be visualized (Fig. 18.6). The steps followed for live imaging are discussed in the next section. Under these on-stage culture conditions, development occurs *ex utero* from zygote to blastocyst stages (Nagy *et al.*, 2003).

Collection and on-stage culture of postimplantation embryos. Around embryonic stage E4.0, embryos begin to implant into the maternal uterus and continue their development with a continuous physical connection with

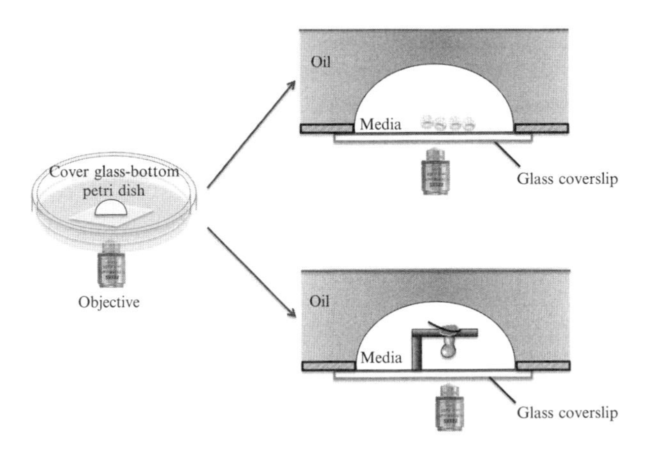

Figure 18.6 On-stage culture set-up and live imaging of pre- and postimplantation mouse embryos. After dissection, embryos are placed in culture media on glass-bottom dishes covered with mineral oil. The dishes are then positioned on the heated, humidified, and gassed microscope stage. Upper lateral section: preimplantation embryos are placed together in a drop of culture media covered with mineral oil. Bottom lateral section: postimplantation embryos are immobilized for on-stage culture using chamber gaskets. A region (for example the ectoplacental cone, see Nagy *et al.*, 2003) is pierced with an eyelash, so that the embryo is immobilized and remains suspended in culture media in the hole of the gasket. (For color version of this figure, the reader is referred to the Web version of this chapter.)

the uterine wall. Briefly, postimplantation embryos are dissected free of the uterine tissues in culture media and then cultured in either roller culture conditions (for optimum *ex utero* culture) or on-stage (static) conditions. The latter being necessary for live imaging and acquisition of time-lapse data.

(i) Dissecting media for collecting postimplantation embryos are 95% DMEM/F12 (1:1) (Invitrogen) supplemented with 5% newborn calf serum (Lonza). After sacrificing the pregnant female, the uterus is removed and placed in a dish containing prewarmed dissecting media.
(ii) The uterus is placed under a stereomicroscope where embryos are dissected. Dissection details are provided in Nagy *et al.*, 2003. Imperative for their *ex utero* development is that embryos are rapidly dissected and remain undamaged during dissection.
(iii) *Ex utero* embryo culture media are DMEM/F12 (1:1) (Invitrogen) supplemented with GLUTAMAX (Invitrogen), 1% penicillin/streptomycin (Invitrogen/Ginco), and rat serum (Harlan Bioproducts or made in-house; Nowotschin *et al.*, 2010). The percentage of rat serum varies depending on the embryonic stage (compare with Fig. 18.1). Prior to dissection, culture media are placed in a culture dish, covered with embryo-tested mineral oil and incubated for at least 1 h in a humidified incubator providing a stable temperature of 37 °C and 5% CO_2.
(iv) Embryos should be transferred with a transfer pipette into a dish with prewarmed culture media; and only the smallest amount of dissection media should be transferred along with the embryos into this dish. The dish should be incubated in an incubator providing 37 °C temperature and 5% CO_2 for 15–20 min for embryos to equilibrate in culture conditions, before they are set up for on-stage culture.
(v) Prior to setting up an on-stage culture, the chamber enclosing the microscope stage should be switched on and prewarmed at 37 °C for at least 1 h. The dish prepared in the previous step is then transferred to the microscope stage and CO_2 is immediately provided. Embryos are then ready for imaging, details of which are discussed in the next section (Fig. 18.6).

    Note: For some experiments, the embryo needs to be immobilized, so that specific regions are in close proximity to the objective and can be visualized (Fig. 18.6). To do this, embryos are suspended in culture using either suction holding pipette or in modified culture dishes (e.g., with a CoverWell chamber gasket). An eyelash can be inserted through a region not being imaged to position the embryo within the hole of the gasket (Fig. 18.6). After covering the media with mineral oil and transferring the dish to the microscope stage incubator (set at 37 °C and providing 5% CO_2), immobilized embryos are ready to be visualized.

Roller culture of postimplantation embryos. Roller culture conditions, even though not suitable for time-lapse imaging, provide the best conditions for *ex utero* development of postimplantation mouse embryos. These conditions provide stable temperature and continuous gassing while the embryos move constantly within media.

- The roller culture incubator is prewarmed at 37 °C prior to setting up the culture. Also, culture media have to be appropriately mixed based on the embryonic stage (Fig. 18.1) and transferred into the roller culture bottles within the incubator to be equilibrated with the appropriate gas composition for at least 1 h prior to starting the culture.
- After collecting postimplantation embryos in dissecting media, embryos are transferred to roller culture bottles containing culture media, with approximately 1 mL of media per one embryo. Only the smallest amount of dissecting media should be transferred into the culture along with the embryos.
- If using a close system, culture tubes have then to be regassed and sealed and be placed on the rotator that is located within the incubator (details provided in Nowotschin *et al.*, 2010).
- Culture media have to be regassed every 6 h (unless a roller culture system with a constant gassing is used) and may need to be replaced with freshly equilibrated media every 24 h.

  Note: *Live imaging of early postimplantation mouse embryos* (E5.5–E6.0). Alternative media compositions have recently been reported for improved culture of very early postimplantation stage embryos (E5.5–E6.0) (Srinivas, 2010). Briefly, embryos are collected in prewarmed M2 media and stored for up to 1–2 h at 37 °C. The culture media used consist of a 1:1 mix of heat-inactivated mouse serum and CMRL medium (Invitrogen) supplemented with L-glutamine. DMEM can be used instead of CMRL, but CMRL appears to give more consistent results (Srinivas, 2010). After dissection, E5.5–E6.0 embryos are transferred in an equilibrated drop of culture media placed on cover glass-bottomed dish and covered with mineral oil. Embryos can then be visualized in static culture (on-stage) as described previously for pre- and postimplantation embryos.

4.3. Imaging early mouse embryos

For live imaging mouse embryos, laser power and exposure time should be decreased as much as possible to reduce the risk of photodamage. To prevent phototoxicity and ensure proper development, decreasing the frequency of scans and perhaps increasing the size of optical sections and scan speed also helps retaining embryo viability. For example, when using a point laser scanning confocal, bidirectional scan in "single track" mode, where all channels are acquired at once, can be performed to achieve faster acquisition. It is preferable to test these imaging parameters when visualizing the same fluorescent reporters in wild-type embryos rather than directly experimenting on embryos of a hard to come by mutant. We usually prefer to image every 15–20 min optical sectioning of 2–4 μm, acquiring z-stacks of up to 150 μm. However, imaging adjustments are determined empirically and depend on the microscope system used (e.g., point scanning vs. spinning disk), developmental stage of the embryo (e.g., the curvature of the embryo at certain postimplantation stages can increase the optical sections needed), and the brightness of the FP reporter. Bright fluorescent reporters are preferable for live imaging because the risk of phototoxicity and photobleaching of the fluorophore is reduced. It is also preferable to know of the exact excitation and emission maxima values of the FP reporter used. The wavelength of excitation should be close to the excitation maximum and the emission filters should capture as much of the emission spectrum as possible (ensuring a high signal-to-noise ratio).

Embryonic drift or movement during growth occurs and is problematic during time-lapse imaging acquisition. For example, preimplantation embryos float freely in the drop of culture media, whereas postimplantation embryos are buoyant, which makes them susceptible to small currents in the culture media and their heavier ectoplacental cone sinks toward the bottom of the dish often reorienting the embryo. To accommodate drifting, it is a good idea to set up a larger z-stack by extending the optical sectioning by several micrometers above and beneath the sample, in order to continuously visualize the embryo even if it slightly moves or grows. Specifically for imaging preimplantation embryos, the amount of media in the drop (too much or too little) can also affect embryo drift; it is worth trying to place several embryos together as it helps to immobilize them. For postimplantation embryos, only suction-holding pipette offers significant control over the orientation of the embryo. The choice of objectives is important; imaging with low magnification (e.g., 5 ×) could be advantageous when there is embryonic drift because the image field view may not be dramatically affected. Using such a low magnification does not permit high-resolution acquisition for single-cell resolution and thus, image analysis (e.g., cell tracking) can be problematic. Imaging with a higher magnification (e.g., 20 × and 40 ×) is preferred for image analysis, but it may require monitoring the embryonic drift and adjusting the focus every few hours throughout the experiment. Alternatively, after time-lapse imaging is complete, drift can sometimes be corrected using software for image analysis.

It is imperative to check for fluctuations in gas flow and temperature during an experiment, as they may have deleterious effects on development of the embryo or time-lapse image acquisition. Also, morphologic characteristics (such as the heart beating, etc.) of the embryo should be monitored during the imaging process. Finally, when a time-lapse imaging experiment is complete the final morphological status of the embryo should be evaluated to ensure that development occurred normally, and that any defects occurred resulting from phototoxicity and/or photodamage.

5. Conclusions

Live imaging FPs *in vivo* has provided an exciting and rapidly advancing method to study the dynamic processes occurring in the early mouse embryo. However, even though the most recently developed genetically encoded and photomodulatable FPs have been tested successfully in embryos of other model organisms, such as *Xenopus*, *Drosophila*, and zebrafish, their use for live imaging in mice has been limited. This could be attributed to the time-consuming process for the generation of mouse transgenesis, as well as the higher levels of fluorescence needed for proper imaging and the increased risk of phototoxicity. Therefore, it is preferable to use the most recently developed and brightest FPs for generating reporter expressing mice. Bright FPs will allow for the usage of low laser power and faster image acquisition and thus low risk of phototoxicity. It is also imperative for a live fluorescent reporter to be developmentally neutral. Many FPs with oligomeric conformation cannot be fused to subcellular tags because they result in toxicity for the embryo. Therefore, development of monomeric variants for these FPs would be very desirable. Nevertheless, even the generation of monomeric FPs fused to tags can be problematic.

Despite the difficulties that can be encountered when generating FP reporter-expressing, strains of mice, the exciting studies that can be performed using FPs are definitely worth the risk. The field is rapidly moving forward; for example, the multicolored approach exploited in BrainBow mice for labeling and tracking individual cells within a population is attractive for live imaging and tracking individual cells or clones of cells *in vivo*, providing an alternative to fusion tags for certain applications (Livet *et al.*, 2007; Snippert *et al.*, 2010). Moreover, the generation of "fluorescent timers" that have been tested in live imaging applications in *Caenorhabditis elegans* and *Xenopus* embryos hold great potential for use in mice (Chen *et al.*, 2010; Terskikh *et al.*, 2000). These specific FPs can be used to monitor both activation and downregulation of target promoters and thus, trace time-dependent expression (Chudakov *et al.*, 2010; Piatkevich and Verkhusha, 2010). This type of approach would be powerful when investigating fluctuations in protein levels over time *in vivo*.

We have provided a general overview of the state of the field of live imaging FPs in mouse embryos. The field is rapidly moving forward, with new and improved reporters being continually generated. We thus apologize to the many investigators, whose work has not been discussed due to space constraints.

ACKNOWLEDGMENTS

We thank Paul Carman (Carl Zeiss Microimaging) for advice on microscope systems. We thank Marilena D. Papaioannou, Silvia Munoz-Descalzo and Minjung Kang for comments on the manuscript. Work in our laboratory is supported by the Human Frontiers Research Program (HFSP), National Institutes of Health (NIH), and New York State Stem Cell Science (NYSTEM). S.N. is supported by a development grant from the Muscular Dystrophy Association (MDA).

REFERENCES

Abe, T., Kiyonari, H., Shioi, G., Inoue, K. I., Nakao, K., Aizawa, S., and Fujimori, T. (2011). Establishment of conditional reporter mouse lines at ROSA26 locus for live cell imaging. *Genesis* **49,** 579–590.

Arnold, S. J., and Robertson, E. J. (2009). Making a commitment: Cell lineage allocation and axis patterning in the early mouse embryo. *Nat. Rev. Mol. Cell Biol.* **10,** 91–103.

Artus, J., Panthier, J. J., and Hadjantonakis, A. K. (2010). A role for PDGF signaling in expansion of the extra-embryonic endoderm lineage of the mouse blastocyst. *Development* **137,** 3361–3372.

Betzig, E., Patterson, G. H., Sougrat, R., Lindwasser, O. W., Olenych, S., Bonifacino, J. S., Davidson, M. W., Lippincott-Schwartz, J., and Hess, H. F. (2006). Imaging intracellular fluorescent proteins at nanometer resolution. *Science* **313,** 1642–1645.

Chen, M. R., Yang, S., Niu, W. P., Li, Z. Y., Meng, L. F., and Wu, Z. X. (2010). A novel fluorescent timer based on bicistronic expression strategy in Caenorhabditis elegans. *Biochem. Biophys. Res. Commun.* **395,** 82–86.

Chudakov, D. M., Matz, M. V., Lukyanov, S., and Lukyanov, K. A. (2010). Fluorescent proteins and their applications in imaging living cells and tissues. *Physiol. Rev.* **90,** 1103–1163.

Conchello, J. A., and Lichtman, J. W. (2005). Optical sectioning microscopy. *Nat. Methods* **2,** 920–931.

Cubitt, A. B., Heim, R., Adams, S. R., Boyd, A. E., Gross, L. A., and Tsien, R. Y. (1995). Understanding, improving and using green fluorescent proteins. *Trends Biochem. Sci.* **20,** 448–455.

Dieguez-Hurtado, R., Martin, J., Martinez-Corral, I., Martinez, M. D., Megias, D., Olmeda, D., and Ortega, S. (2011). A Cre-reporter transgenic mouse expressing the far-red fluorescent protein Katushka. *Genesis* **49,** 36–45.

Dubertret, B., Skourides, P., Norris, D. J., Noireaux, V., Brivanlou, A. H., and Libchaber, A. (2002). *In vivo* imaging of quantum dots encapsulated in phospholipid micelles. *Science* **298,** 1759–1762.

Egli, D., Rosains, J., Birkhoff, G., and Eggan, K. (2007). Developmental reprogramming after chromosome transfer into mitotic mouse zygotes. *Nature* **447,** 679–685.

Ferrer-Vaquer, A., Piliszek, A., Tian, G., Aho, R. J., Dufort, D., and Hadjantonakis, A. K. (2010). A sensitive and bright single-cell resolution live imaging reporter of Wnt/ss-catenin signaling in the mouse. *BMC Dev. Biol.* **10,** 121.

Fink, D., Wohrer, S., Pfeffer, M., Tombe, T., Ong, C. J., and Sorensen, P. H. (2010). Ubiquitous expression of the monomeric red fluorescent protein mCherry in transgenic mice. *Genesis* **48,** 723–729.

Fraser, S. T., Hadjantonakis, A. K., Sahr, K. E., Willey, S., Kelly, O. G., Jones, E. A., Dickinson, M. E., and Baron, M. H. (2005). Using a histone yellow fluorescent protein fusion for tagging and tracking endothelial cells in ES cells and mice. *Genesis* **42,** 162–171.

Garcia, M. D., Udan, R. S., Hadjantonakis, A. K., and Dickinson, M. E. (2011a). Live imaging of mouse embryos. *Cold Spring Harb. Protoc.* DOI: 10.1101/pdb.top10404.
Garcia, M. D., Udan, R. S., Hadjantonakis, A. K., and Dickinson, M. E. (2011b). Preparation of postimplantation mouse embryos for imaging. *Cold Spring Harb. Protoc.* (4), DOI: 10.1101/pdb.prot5594.
Garcia, M. D., Udan, R. S., Hadjantonakis, A. K., and Dickinson, M. E. (2011c). Preparation of rat serum for culturing mouse embryos. *Cold Spring Harb. Protoc.* DOI: 10.1101/ pdb.prot5593.
Garcia, M. D., Udan, R. S., Hadjantonakis, A. K., and Dickinson, M. E. (2011d). Time-lapse imaging of postimplantation mouse embryos. *Cold Spring Harb. Protoc.* DOI: 10.1101/pdb.prot5595.
Gray, N. W., Weimer, R. M., Bureau, I., and Svoboda, K. (2006). Rapid redistribution of synaptic PSD-95 in the neocortex *in vivo*. *PLoS Biol.* **4,** 2065–2075.
Griswold, S. L., Sajja, K. C., Jang, C. W., and Behringer, R. R. (2011). Generation and characterization of iUBC-KikGR photoconvertible transgenic mice for live time-lapse imaging during development. *Genesis* **49,** 591–598.
Hadjantonakis, A. K., Dickinson, M. E., Fraser, S. E., and Papaioannou, V. E. (2003). Technicolour transgenics: Imaging tools for functional genomics in the mouse. *Nat. Rev. Genet.* **4,** 613–625.
Hadjantonakis, A. K., and Papaioannou, V. E. (2004). Dynamic *in vivo* imaging and cell tracking using a histone fluorescent protein fusion in mice. *BMC Biotechnol.* **4,** 33.
Hatta, K., Tsujii, H., and Omura, T. (2006). Cell tracking using a photoconvertible fluorescent protein. *Nat. Protoc.* **1,** 960–967.
Hayashi-Takanaka, Y., Yamagata, K., Nozaki, N., and Kimura, H. (2009). Visualizing histone modifications in living cells: Spatiotemporal dynamics of H3 phosphorylation during interphase. *J. Cell Biol.* **187,** 781–790.
Heim, R., Cubitt, A. B., and Tsien, R. Y. (1995). Improved green fluorescence. *Nature* **373,** 663–664.
Hell, S. W. (2003). Toward fluorescence nanoscopy. *Nat. Biotechnol.* **21,** 1347–1355.
Helmchen, F., and Denk, W. (2005). Deep tissue two-photon microscopy. *Nat. Methods* **2,** 932–940.
Huisken, J., Swoger, J., Del Bene, F., Wittbrodt, J., and Stelzer, E. H. (2004). Optical sectioning deep inside live embryos by selective plane illumination microscopy. *Science* **305,** 1007–1009.
Imai, J. H., Wang, X., and Shi, S. H. (2010). Kaede-centrin1 labeling of mother and daughter centrosomes in mammalian neocortical neural progenitors. *Curr. Protoc. Stem Cell Biol.* DOI: 10.1002/9780470151808.sc05a05s15.
Isern, J., He, Z., Fraser, S. T., Nowotschin, S., Ferrer-Vaquer, A., Moore, R., Hadjantonakis, A. K., Schulz, V., Tuck, D., Gallagher, P. G., and Baron, M. H. (2011). Single-lineage transcriptome analysis reveals key regulatory pathways in primitive erythroid progenitors in the mouse embryo. *Blood* **117,** 4924–4934.
Joyner, A. L., and Zervas, M. (2006). Genetic inducible fate mapping in mouse: Establishing genetic lineages and defining genetic neuroanatomy in the nervous system. *Dev. Dyn.* **235,** 2376–2385.
Kanda, T., Sullivan, K. F., and Wahl, G. M. (1998). Histone-GFP fusion protein enables sensitive analysis of chromosome dynamics in living mammalian cells. *Curr. Biol.* **8,** 377–385.
Keller, P. J., Schmidt, A. D., Santella, A., Khairy, K., Bao, Z., Wittbrodt, J., and Stelzer, E. H. (2010). Fast, high-contrast imaging of animal development with scanned light sheet-based structured-illumination microscopy. *Nat Methods* **7,** 637–642.
Keller, P. J., Schmidt, A. D., Wittbrodt, J., and Stelzer, E. H. (2008). Reconstruction of zebrafish early embryonic development by scanned light sheet microscopy. *Science* **322,** 1065–1069.

Kinder, S. J., Tsang, T. E., Quinlan, G. A., Hadjantonakis, A. K., Nagy, A., and Tam, P. P. (1999). The orderly allocation of mesodermal cells to the extraembryonic structures and the anteroposterior axis during gastrulation of the mouse embryo. *Development* **126,** 4691–4701.

Kinder, S. J., Tsang, T. E., Wakamiya, M., Sasaki, H., Behringer, R. R., Nagy, A., and Tam, P. P. (2001). The organizer of the mouse gastrula is composed of a dynamic population of progenitor cells for the axial mesoderm. *Development* **128,** 3623–3634.

Kremers, G. J., Hazelwood, K. L., Murphy, C. S., Davidson, M. W., and Piston, D. W. (2009). Photoconversion in orange and red fluorescent proteins. *Nat. Methods* **6,** 355–358.

Kwon, G. S., Fraser, S. T., Eakin, G. S., Mangano, M., Isern, J., Sahr, K. E., Hadjantonakis, A. K., and Baron, M. H. (2006). Tg(Afp-GFP) expression marks primitive and definitive endoderm lineages during mouse development. *Dev. Dyn.* **235,** 2549–2558.

Kwon, G. S., and Hadjantonakis, A. K. (2007). Eomes::GFP—A tool for live imaging cells of the trophoblast, primitive streak, and telencephalon in the mouse embryo. *Genesis* **45,** 208–217.

Kwon, G. S., and Hadjantonakis, A. K. (2009). Transthyretin mouse transgenes direct RFP expression or Cre-mediated recombination throughout the visceral endoderm. *Genesis* **47,** 447–455.

Kwon, G. S., Viotti, M., and Hadjantonakis, A. K. (2008). The endoderm of the mouse embryo arises by dynamic widespread intercalation of embryonic and extraembryonic lineages. *Dev. Cell* **15,** 509–520.

Lichtman, J. W., and Conchello, J. A. (2005). Fluorescence microscopy. *Nat. Methods* **2,** 910–919.

Livet, J., Weissman, T. A., Kang, H., Draft, R. W., Lu, J., Bennis, R. A., Sanes, J. R., and Lichtman, J. W. (2007). Transgenic strategies for combinatorial expression of fluorescent proteins in the nervous system. *Nature* **450,** 56–62.

Long, J. Z., Lackan, C. S., and Hadjantonakis, A. K. (2005). Genetic and spectrally distinct *in vivo* imaging: Embryonic stem cells and mice with widespread expression of a monomeric red fluorescent protein. *BMC Biotechnol.* **5,** 20.

McDole, K., Xiong, Y., Iglesias, P. A., and Zheng, Y. (2011). Lineage mapping the preimplantation mouse embryo by two-photon microscopy, new insights into the segregation of cell fates. *Dev. Biol.* **355,** 239–249.

Monteiro, R. M., de Sousa Lopes, S. M., Bialecka, M., de Boer, S., Zwijsen, A., and Mummery, C. L. (2008). Real time monitoring of BMP Smads transcriptional activity during mouse development. *Genesis* **46,** 335–346.

Murray, M. J., and Saint, R. (2007). Photoactivatable GFP resolves Drosophila mesoderm migration behaviour. *Development* **134,** 3975–3983.

Muzumdar, M. D., Tasic, B., Miyamichi, K., Li, L., and Luo, L. (2007). A global double-fluorescent Cre reporter mouse. *Genesis* **45,** 593–605.

Nagy, A. (2000). Cre recombinase: The universal reagent for genome tailoring. *Genesis* **26,** 99–109.

Nagy, A., Gertsenstein, M., Vintersten, K., and Behringer, R. (2003). Manipulating the Mouse Embryo. *A Laboratory Manual.* Cold Spring Harbor Laboratory Press, Cold Spring Harbor, NY.

Nowotschin, S., Eakin, G. S., and Hadjantonakis, A. K. (2009a). Dual transgene strategy for live visualization of chromatin and plasma membrane dynamics in murine embryonic stem cells and embryonic tissues. *Genesis* **47,** 330–336.

Nowotschin, S., Eakin, G. S., and Hadjantonakis, A. K. (2009b). Live-imaging fluorescent proteins in mouse embryos: Multi-dimensional, multi-spectral perspectives. *Trends Biotechnol.* **27,** 266–276.

Nowotschin, S., Ferrer-Vaquer, A., and Hadjantonakis, A. K. (2010). Imaging mouse development with confocal time-lapse microscopy. *Methods Enzymol.* **476,** 351–377.

Nowotschin, S., and Hadjantonakis, A. K. (2009a). Photomodulatable fluorescent proteins for imaging cell dynamics and cell fate. *Organogenesis* **5,** 135–144.

Nowotschin, S., and Hadjantonakis, A. K. (2009b). Use of KikGR a photoconvertible green-to-red fluorescent protein for cell labeling and lineage analysis in ES cells and mouse embryos. *BMC Dev. Biol.* **9,** 49.

Nowotschin, S., and Hadjantonakis, A. K. (2010). Cellular dynamics in the early mouse embryo: From axis formation to gastrulation. *Curr. Opin. Genet. Dev.* **20,** 420–427.

Piatkevich, K. D., Hulit, J., Subach, O. M., Wu, B., Abdulla, A., Segall, J. E., and Verkhusha, V. V. (2010). Monomeric red fluorescent proteins with a large Stokes shift. *Proc. Natl. Acad. Sci. USA* **107,** 5369–5374.

Piatkevich, K. D., and Verkhusha, V. V. (2010). Advances in engineering of fluorescent proteins and photoactivatable proteins with red emission. *Curr. Opin. Chem. Biol.* **14,** 23–29.

Plachta, N., Bollenbach, T., Pease, S., Fraser, S. E., and Pantazis, P. (2011). Oct4 kinetics predict cell lineage patterning in the early mammalian embryo. *Nat. Cell Biol.* **13,** 117–123.

Plusa, B., Piliszek, A., Frankenberg, S., Artus, J., and Hadjantonakis, A. K. (2008). Distinct sequential cell behaviours direct primitive endoderm formation in the mouse blastocyst. *Development* **135,** 3081–3091.

Poche, R. A., Larina, I. V., Scott, M. L., Saik, J. E., West, J. L., and Dickinson, M. E. (2009). The Flk1-myr::mCherry mouse as a useful reporter to characterize multiple aspects of ocular blood vessel development and disease. *Dev. Dyn.* **238,** 2318–2326.

Reynaud, E. G., Krzic, U., Greger, K., and Stelzer, E. H. (2008). Light sheet-based fluorescence microscopy: More dimensions, more photons, and less photodamage. *Hfsp J.* **2,** 266–275.

Rhee, J. M., Pirity, M. K., Lackan, C. S., Long, J. Z., Kondoh, G., Takeda, J., and Hadjantonakis, A. K. (2006). *In vivo* imaging and differential localization of lipid-modified GFP-variant fusions in embryonic stem cells and mice. *Genesis* **44,** 202–218.

Rossant, J., and Tam, P. P. (2009). Blastocyst lineage formation, early embryonic asymmetries and axis patterning in the mouse. *Development* **136,** 701–713.

Rust, M. J., Bates, M., and Zhuang, X. (2006). Sub-diffraction-limit imaging by stochastic optical reconstruction microscopy (STORM). *Nat. Methods* **3,** 793–795.

Santi, P. A. (2011). Light sheet fluorescence microscopy: A review. *J. Histochem. Cytochem.* **59,** 129–138.

Serbedzija, G. N., Bronner-Fraser, M., and Fraser, S. E. (1992). Vital dye analysis of cranial neural crest cell migration in the mouse embryo. *Development* **116,** 297–307.

Shcherbo, D., Merzlyak, E. M., Chepurnykh, T. V., Fradkov, A. F., Ermakova, G. V., Solovieva, E. A., Lukyanov, K. A., Bogdanova, E. A., Zaraisky, A. G., Lukyanov, S., and Chudakov, D. M. (2007). Bright far-red fluorescent protein for whole-body imaging. *Nat. Methods* **4,** 741–746.

Shcherbo, D., Murphy, C. S., Ermakova, G. V., Solovieva, E. A., Chepurnykh, T. V., Shcheglov, A. S., Verkhusha, V. V., Pletnev, V. Z., Hazelwood, K. L., Roche, P. M., Lukyanov, S., Zaraisky, A. G., *et al.* (2009). Far-red fluorescent tags for protein imaging in living tissues. *Biochem. J.* **418,** 567–574.

Shcherbo, D., Shemiakina, I. I., Ryabova, A. V., Luker, K. E., Schmidt, B. T., Souslova, E. A., Gorodnicheva, T. V., Strukova, L., Shidlovskiy, K. M., Britanova, O. V., Zaraisky, A. G., Lukyanov, K. A., *et al.* (2010). Near-infrared fluorescent proteins. *Nat. Methods* **7,** 827–829.

Shioi, G., Kiyonari, H., Abe, T., Nakao, K., Fujimori, T., Jang, C. W., Huang, C. C., Akiyama, H., Behringer, R. R., and Aizawa, S. (2011). A mouse reporter line to conditionally mark nuclei and cell membranes for *in vivo* live-imaging. *Genesis* **49,** 570–578.

Snippert, H. J., van der Flier, L. G., Sato, T., van Es, J. H., van den Born, M., Kroon-Veenboer, C., Barker, N., Klein, A. M., van Rheenen, J., Simons, B. D., and Clevers, H. (2010). Intestinal crypt homeostasis results from neutral competition between symmetrically dividing Lgr5 stem cells. *Cell* **143,** 134–144.

So, P. T., Dong, C. Y., Masters, B. R., and Berland, K. M. (2000). Two-photon excitation fluorescence microscopy. *Annu. Rev. Biomed. Eng.* **2,** 399–429.

Srinivas, S. (2010). Imaging cell movements in egg-cylinder stage mouse embryos. *Cold Spring Harb. Protoc.* DOI: 10.1101/pdb.prot5539.

Stark, D. A., and Kulesa, P. M. (2005). Photoactivatable green fluorescent protein as a single-cell marker in living embryos. *Dev. Dyn.* **233,** 983–992.

Stewart, M. D., Jang, C. W., Hong, N. W., Austin, A. P., and Behringer, R. R. (2009). Dual fluorescent protein reporters for studying cell behaviors *in vivo*. *Genesis* **47,** 708–717.

Terskikh, A., Fradkov, A., Ermakova, G., Zaraisky, A., Tan, P., Kajava, A. V., Zhao, X., Lukyanov, S., Matz, M., Kim, S., Weissman, I., and Siebert, P. (2000). "Fluorescent timer": Protein that changes color with time. *Science* **290,** 1585–1588.

Tomura, M., Yoshida, N., Tanaka, J., Karasawa, S., Miwa, Y., Miyawaki, A., and Kanagawa, O. (2008). Monitoring cellular movement *in vivo* with photoconvertible fluorescence protein "Kaede" transgenic mice. *Proc. Natl. Acad. Sci. USA* **105,** 10871–10876.

Trichas, G., Begbie, J., and Srinivas, S. (2008). Use of the viral 2A peptide for bicistronic expression in transgenic mice. *BMC Biol.* **6,** 40.

Udan, R. S., and Dickinson, M. E. (2010). Imaging mouse embryonic development. *Methods Enzymol.* **476,** 329–349.

Viotti, M., Nowotschin, S., and Hadjantonakis, A. K. (2011). Afp::mCherry, a red fluorescent transgenic reporter of the mouse visceral endoderm. *Genesis* **49,** 124–133.

Wacker, S. A., Oswald, F., Wiedenmann, J., and Knochel, W. (2007). A green to red photoconvertible protein as an analyzing tool for early vertebrate development. *Dev. Dyn.* **236,** 473–480.

Walter, T., Shattuck, D. W., Baldock, R., Bastin, M. E., Carpenter, A. E., Duce, S., Ellenberg, J., Fraser, A., Hamilton, N., Pieper, S., Ragan, M. A., Schneider, J. E., *et al.* (2010). Visualization of image data from cells to organisms. *Nat. Methods* **7,** S26–S41.

Yamanaka, Y., Lanner, F., and Rossant, J. (2010). FGF signal-dependent segregation of primitive endoderm and epiblast in the mouse blastocyst. *Development* **137,** 715–724.

CHAPTER NINETEEN

Methods for Three-Dimensional Analysis of Dendritic Spine Dynamics

Enni Bertling, Anastasia Ludwig, Mikko Koskinen, *and* Pirta Hotulainen

Contents

1. Introduction 392
2. Methods 393
   2.1. Primary cultures and transient transfections 393
3. Imaging Conditions and Procedures 395
4. Illustrative Experiments 397
   4.1. Experiment 1: Latrunculin B treatment at DIV 14 397
   4.2. Experiment 2: cLTP at DIV 21 398
5. Quantitative Methods for Analyzing Dendritic Spines 398
   5.1. Spine morphology analysis 398
   5.2. Spine motility analysis 401
Acknowledgments 404
References 404

Abstract

Dendritic spines are small bulbous expansions that receive input from a single excitatory synapse. Although spines are often characterized by a mushroom-like morphology, they come in a wide range of sizes and shapes, even within the same dendrite. In a developing brain, spines exhibit a high degree of structural and functional plasticity, reflecting the formation and elimination of synapses during the maturation of neuronal circuits. The morphology of spines in developing neurons is affected by synaptic activity, hence contributing to the experience-dependent refinement of neuronal circuits, learning, and memory. Thus, understanding spine dynamics and its regulation is of central importance to studies of synaptic plasticity in the brain. The challenge has been to develop a computer-based assay that will quantitatively assess the three-dimensional change in spine movements caused by various stimuli and experimental conditions. Here, we provide detailed protocols for cell plating, transient transfections, and time-lapse imaging of dendritic spines. For the analysis of dendritic spine dynamics, we present two methods based on quantitative three-dimensional measurements.

Neuroscience Center, University of Helsinki, Helsinki, Finland

Methods in Enzymology, Volume 506
ISSN 0076-6879, DOI: 10.1016/B978-0-12-391856-7.00043-3

1. Introduction

The human brain consists of 100 billion neurons interconnected into functional neuronal circuits that underlie our behavior, thoughts, emotions, dreams, and memories. Precise control of synaptic development and connectivity is critical for an accurate neural network activity and normal brain function. Most excitatory synapses in the mammalian brain are formed at tiny dendritic protrusions named dendritic spines (Bourne and Harris, 2008).

Dendritic spines typically have a spherical head and a narrow neck (Harris and Stevens, 1989; Noguchi *et al.*, 2005), but individual spines come in a wide range of sizes and shapes (Matus, 2000). Spines can be roughly divided into three categories based on their morphology: thin, filopodia-like protrusions (thin spines); short spines without a well-defined spine neck (stubby spines); and spines with a large bulbous head (mushroom spines) (Bourne and Harris, 2008) (Fig. 19.1). Experimental evidence has shown that the heterogeneity of spine morphology accounts for functional differences at the synaptic level (Kasai *et al.*, 2003; Yuste and Bonhoeffer, 2001). The functional and structural plasticity of spines is believed to be the basis of learning and memory (Kasai *et al.*, 2010). Spine malformation is a symptom of some human neuropsychiatric disorders (Penzes *et al.*, 2011). Thus, understanding spine dynamics and its regulation is of central importance to studies of brain function and disease.

Since the first experiments demonstrated that spines are motile structures (Fischer *et al.*, 1998), a number of studies have analyzed spine motility *in vitro* and *in vivo*. Spines of developing neurons can elongate, shrink, change diameter, or change position in space relative to the spine neck or dendrite. In mature neurons, the predominant spine motility is morphing of spine heads. Certain conditions (e.g., induction of long-term potentiation

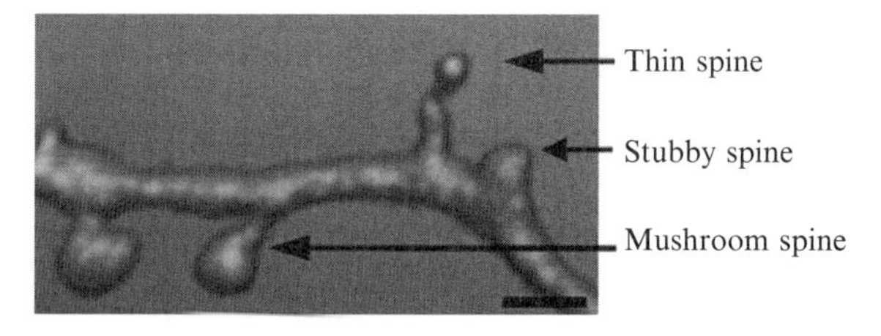

Figure 19.1 Spine types. Spines can be divided into three structural classes according to their morphology. Thin spines are filopodia-like protrusions with either small or undetectable heads. Mushroom spines have a well-defined structure with a distinct thin neck and round head. Stubby spines do not have a neck, and they usually appear as a spherical expansion on the dendrite. Scale bar, 1 μm. (For color version of this figure, the reader is referred to the Web version of this chapter.)

(LTP), one of the major cellular mechanisms underlying learning and memory) can cause a remarkable change in spine motility (Kasai *et al.*, 2003; Yuste and Bonhoeffer, 2001). Often, changes in dendritic spine dynamics can be clearly seen when visually comparing spine motility recordings before and after the experimental stimulus. However, the challenge has been to develop an assay that will quantitatively assess the change in spine movements caused by various stimuli and experimental conditions. Most of the previously used analysis methods have been based on 2D manual measurements due to the lack of available 3D image analysis software (Dunaevsky *et al.*, 1999; Lendvai *et al.*, 2000; Majewska and Sur, 2003; Nestor *et al.*, 2011; Oray *et al.*, 2004, 2006; Tashiro and Yuste, 2004, 2008). The drawbacks of manual measurements are that they are very time-consuming, thus limiting the number of analyzed objects, and they have a high potential for bias and inconsistency due to the experimentalist. Further, a significant amount of information can be lost in 2D measurements of highly three-dimensional networks of dendrites and spines. In particular, the tiny filopodia often go undetected in manual 2D measurements.

Here, in the first approach, we take advantage of the NeuronStudio software, which has been developed to detect and measure dendritic spines from 3D confocal stacks (Rodriguez *et al.*, 2008). In the second approach, we use Bitplane's Imaris software, which allows more general measurements of 3D and 4D datasets. Three-dimensional analysis requires high-quality images with good brightness/contrast, high resolution, and low background. To acquire such images, we use dissociated hippocampal neuron cultures instead of brain slices. The successful cultivation and transfection of these neuronal cultures are key elements required for advanced microscopy techniques. Here, we provide our protocols for cell plating and transient transfection. We also present two protocols for the analysis of spine dynamics as well as experiments illustrating the implementation of the protocols.

2. Methods

2.1. Primary cultures and transient transfections

The key element of a successful spine imaging experiment is healthy neuronal cultures with a sufficient number of neurons expressing an adequate level of fluorescent protein. To achieve this goal, we developed a protocol for primary neuron cultivation and transient transfection. This protocol is provided below, and it is a modification of the classical procedure published in Banker and Goslin (1998). Most of the steps are common for mouse and rat cultures; however, if there is a difference, it is marked with the MOUSE and RAT signs.

We dissect hippocampi from mice at embryonic day 16 or 17 or from rats at embryonic day 17. We cultivate neurons in 5% CO_2 at 37 °C in the incubator, in serum-free conditions, using a Neurobasal medium supplemented with B27 (Gibco) and L-glutamine (Invitrogen). Some cultures can also be supplemented with antibiotics (Penicillin–Streptomycin (Lonza)). The culturing procedure consists of preparation of coverslips and dishes, culture plating and maintaining, and transient transfection of neurons before the imaging experiment.

2.1.1. Preparation of coverslips and dishes

Glass coverslips or glass-bottom dishes (MatTek) used for culturing have to be sterile and treated with polyornithine or polylysine to improve cell adhesion. Neurons do not attach to uncoated glass surfaces.

1. Take the sterile coverslips (sterilized in an oven or in EtOH) or sterile MatTek dishes.
2. Place the coverslips into 4-well plates.
3. Wash the coverslips/dishes twice with 300 μl PBS.
4. Add 300 μl poly-D-L-ornithine MOUSE (Sigma) (use 0.5 mg/ml) or poly-L-lysine RAT (Sigma) (use 0.1 mg/ml), and leave in the incubator overnight (o/n).
5. On the following day, wash the coverslips/dishes twice with PBS and once with Neurobasal. Keep in the incubator o/n or until neuron plating.

2.1.2. Plating and maintaining mouse and rat hippocampal neuron cultures

1. Dilute the cells so that you have 100,000 cells in 500 (for coverslips) or 300 μl (for MatTek dishes with a 14 mm diameter opening in the center of a 35 mm culture dish). Mix the cells well, and add 500 or 300 μl neuron mix (=100,000 cells) to each coverslip or MatTek dish. When using MatTek dishes, it is important to add cells carefully to the center of the glass. Take care that the medium does not go over the opening borders but stays as a drop. Handle dishes carefully.
2. Grow cells in 37 °C, 5% CO_2. Change half of the medium every 3 days.

2.1.3. Transient transfection of cultured E17 hippocampal neurons

According to our experience, neurons cultured for 11–16 days *in vitro* (DIV) are optimal for transfections. The younger the neurons are, the better the transfection efficiency. However, the survival after transfection increases with neuron age. The protocol presented here results in adequate transfection efficiency for imaging (~1%).

1. Collect the growth medium from the cells, and store in the incubator.
2. If the culture medium contained antibiotics, gently wash the cells twice with warm Neurobasal.

3. Add to the cells Neurobasal + 10 mM $MgCl_2$. $MgCl_2$ increases neuronal survival.
4. Keep cells in Neurobasal + 10 mM $MgCl_2$ 1–2 h in the incubator.
5. For one (1) well of 24-well plate:
   a. Add to one tube 50 μl Neurobasal + 2 μl Lipofectamine 2000.
   b. Add to another tube 50 μl Neurobasal + 0.5–1 μg DNA (and/or 300 ng siRNA).
   c. Incubate for 5–7 min at RT.
   d. Mix the contents of the tubes together.
   e. Incubate for 20–25 min at RT.
6. Add the transfection mixture dropwise to the cells.
7. Incubate for 2–4 h at +37 °C. For older neurons, increase the incubation time to 4–6 h.
8. Wash 3× with warm Neurobasal.
9. Replace the transfection media with the original growth medium.

Tip 1: If transfection disturbs the cells, it may be preferable to refresh the medium 1 day before transfection.

Tip 2: Neurons are very sensitive to pH; thus, you can keep Neurobasal + 10 mM $MgCl_2$ in the incubator in a petri dish 2–6 h or o/n before addition to cultures. The incubation of the medium ensures that the pH does not change dramatically.

Tip 3: Before starting more advanced imaging experiments, learn how to determine if the cells are healthy under a microscope. We try to help to distinguish between healthy and nonhealthy cells by providing images of them (Fig. 19.2). Figure 19.2A presents a healthy neuron with well-spread and branched dendrites. The spines on the dendrites exhibit normal density and typical morphology. Figure 19.2B presents neuron which we do not include in our analysis. Dendrites of this neuron are short and spines are difficult to detect. This neuron expresses GFP at such high levels that it precipitates (bright dots on dendrites). Figure 19.2C and D present details which can be used to determine if cell is healthy. Figure 19.2C shows blebbing and 2D shows "holes" in diffuse labeling. Both of these marks indicate that cells are not healthy. It is strongly recommended to look first at your control cells before analyzing mutated or manipulated cells. If control cells have suffered from cultivation or transfection, discard your experiment.

3. Imaging Conditions and Procedures

For imaging, we use a Leica TCS SP5 confocal microscope equipped with a temperature-controlled chamber and CO_2 supply. To achieve stable temperature maintenance, we use a microscope cage incubator around the microscope. Both the temperature and the CO_2 levels should

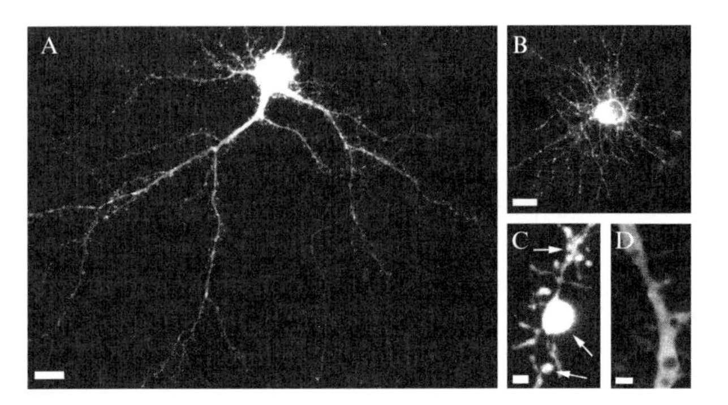

Figure 19.2 The condition of the cultured cells has a profound influence on the reliability of all live-cell imaging assays. When starting an experiment, the cell morphology should be examined. The control cells should always be checked first to confirm normal morphology, fluorescent protein expression levels, and developmental status. (A) A healthy looking E17 rat hippocampal neuron expressing GFP at DIV14. (B) An E17 rat hippocampal neuron at DIV14 showing an abnormal dendritic tree with dense, short neurites with little branching and cell polarization. This neuron expresses GFP at such high levels that it precipitates (bright dots on dendrites). The precipitation disturbs normal cellular function and normal diffusion of the fluorescent protein. (C) A dendrite of an E17 rat hippocampal neuron expressing GFP at DIV14 where blebbing of the membrane has started to occur (white arrows show blebbing areas). This can happen to cells during imaging if the imaging conditions are imperfect. (D) A dendrite of an E17 rat hippocampal neuron where "holes" can be seen in the GFP fluorescence in the dendrite. These darker areas are not normal for these cells and should therefore lead to discarding of the cell. Scale bars, (A, B) 10 μm, (C, D) 1 μm.

be measured as close as possible to the cell plate to detect the real values. Our motility recordings are performed in culture medium at 37 °C and with 5% CO_2. In some experiments (e.g., chemical LTP (cLTP)), we use perfusion to wash solutions in and out but do not use it during recordings to improve stability of the sample. It is important to use prewarmed equipment, media, drugs, pipet tips, etc., to avoid temperature fluctuations that may cause a substantial focus drift. After setting the cell plate to the microscope, we normally wait for 30 min before starting the time-lapse series. In some microscope setups, the autofocus can at least partially overcome the focus drift problem.

We typically analyze spine motility of dissociated hippocampal neurons at DIV 10–21. Neurons should be transfected with a diffuse fluorescent protein 1–2 days before the experiment. We recommend using GFP mainly due to its high fluorescence intensity. GFP also has other important properties for spine motility analysis: it localizes diffusively, it fills the spines well, and it is quite inert and nontoxic for the cells.

The images for analysis are 4D datasets consisting of 3D stacks of optical sections collected at several time points. To visualize dendritic spines, we use a 63×/0.90 NA dipping water objective. At each time point, we collect a 6–9-μm-deep *Z* stack above and below the plane of interest to visualize all dendritic structures. The voxel (3D pixel) size in our images (*x*, *y*, *z*) is between 74 × 74 × 130 and 130 × 130 × 300 nm. Spine analysis requires high-quality images with small voxel size and low background. Overexposure of images should be avoided, especially when images are taken to quantitative image analysis. The microscope settings, such as gain and offset, should be maintained throughout the entire experiment in order to maintain comparability of the images.

The time interval of recording depends on the rate of motility of the area of interest. For analyses of the spine motility of DIV14-21 hippocampal neurons, we typically scan 10–25 optical stacks with 1–2-min intervals. However, the time resolution should be adjusted for each system separately. It is important to keep laser exposure of the cells reasonable. The longer the recording time is, the longer the interval between frames should be. Laser exposure can also be decreased by keeping the laser power as low as possible and by increasing the scan speed.

4. Illustrative Experiments

4.1. Experiment 1: Latrunculin B treatment at DIV 14

The shape and dynamics of spines are determined by the cytoskeleton, and the major cytoskeletal protein in dendritic spines is filamentous (F-) actin (Furuyashiki *et al.*, 2002; Matus *et al.*, 1982; Racz and Weinberg, 2004). Actin filaments are polar structures with one end (plus or "barbed" end) growing more rapidly than the other (minus or "pointed" end). In dendritic spines, the barbed ends push the plasma membrane and induce spine shape changes (Honkura *et al.*, 2008; Hotulainen *et al.*, 2009). A rough method to study actin-induced spine dynamics is to treat cells with latrunculin A or latrunculin B. Latrunculin sequesters free actin monomers and thus blocks the polymerization of the filaments. The rate of disappearance of filamentous actin correlates with the rate of depolymerization of actin filaments. Thus, treating dendritic spines with latrunculin blocks the morphing of spine heads and leads to the thinning of spines (Korkotian and Segal, 2001, Nestor *et al.*, 2011).

One day before imaging, DIV 13 hippocampal neurons are transfected with GFP. The experiment starts with the control recording consisting of eight optical stacks taken at intervals of 1 min. Once the control recording is ready, we treat the cells with 5 μ*M* latrunculin B by mixing it with the culture medium. Ten to thirty minutes after latrunculin B application, we take another set of eight optical stacks.

4.2. Experiment 2: cLTP at DIV 21

During the experiment, the culture was subjected to the cLTP protocol. The cLTP protocol includes brief (10 min) treatment of the dissociated neurons with bicuculline (40 μ*M*), strychnine (1 μ*M*), and glycine (200 μ*M*). The treatment leads to a blockade of $GABA_A$ and glycine receptors and the strong activation of synaptic NMDA receptors (Liao *et al.*, 2001; Lu *et al.*, 2001). Functionally, cLTP results in the increase of the frequency and amplitude of miniature excitatory postsynaptic currents reflecting a synaptic insertion of AMPA receptors. Structurally, 1 h after cLTP, we and others (Fortin *et al.*, 2010; Kopec *et al.*, 2006; Park *et al.*, 2006) observe increases in spine size and reduced spine motility.

In this representative experiment, we used GFP-expressing hippocampal neurons at DIV 21. The control recording consists of 11 optical stacks taken at intervals of 1 min, which corresponds to a 10-min movie. Once the control recording is ready, the cLTP solution is washed in. Ten minutes later, it is washed out. One hour after the onset of the cLTP protocol, another time-lapse recording is made.

5. Quantitative Methods for Analyzing Dendritic Spines

5.1. Spine morphology analysis

5.1.1. Overview

The approach presented in this chapter is based on the 4D imaging of a GFP-expressing spine-rich neuron and the subsequent comparison of 3D-traced spines between images taken at different time points. We use NeuronStudio software for automatic tracing and reconstruction of dendrites and spines from confocal image stacks (Rodriguez *et al.*, 2008) (Fig. 19.3A). Using the Rayburst algorithm, the software measures the 3D width of the head and the maximum distance between the spine tip and the surface of the dendrite (Fig. 19.3B). The maximum distance refers to spine length. In the approach described here, we follow the morphological changes of same spines before and after the treatment (Fig. 19.3C). This results in two types of information: (1) We can compare the spine head width and spine length before and after the treatment (Fig. 19.3D) and (2) we can compare the spine head width and spine length between two time points and calculate the percent change from one frame to another (Fig. 19.3E). This refers to the spine dynamics.

The biggest advantage of using NeuronStudio-based tracing of dendrites and spines is that the methods based on automatic tracing are not sensitive to the drift or movement of dendrites, because each dendrite can be separately traced in every stack. This allows to compare spines from very long

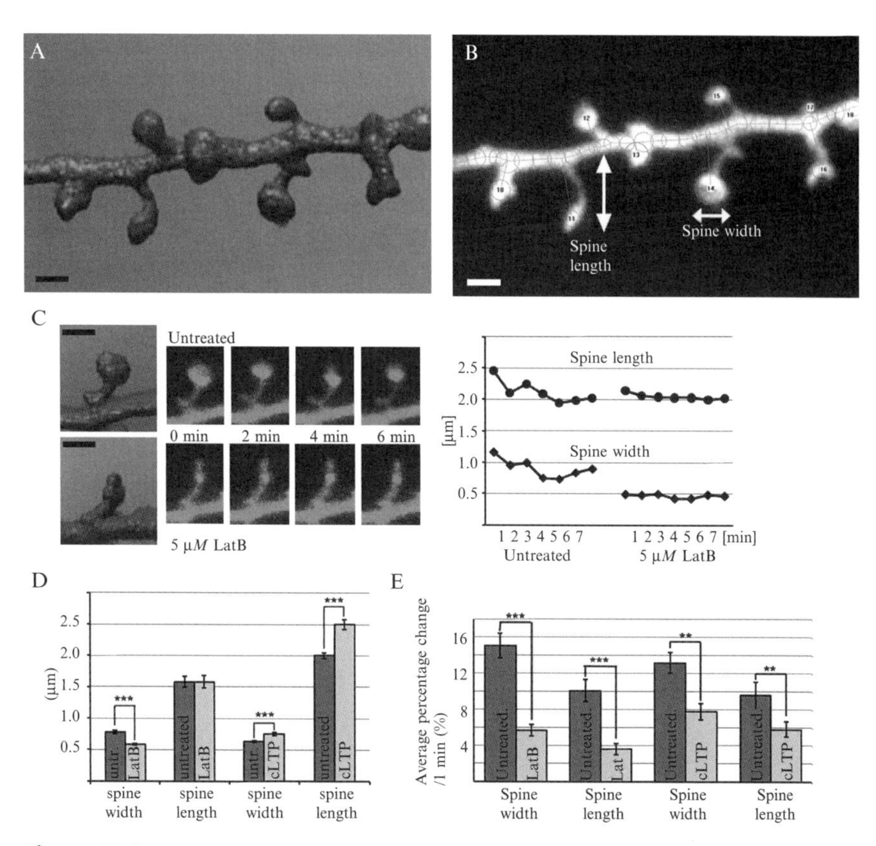

Figure 19.3 Spine morphology analysis. (A) NeuronStudio software uses a three-dimensional structure for spine head width and spine length measurements. (B) Dendrites and spines are traced automatically. Spine length is measured from the surface of the dendrite to the furthest point of the head. Spine head width (spine width) is measured from the spine head rendered as a sphere. (C) Spine parameters before and after latrunculin B treatment are measured from the confocal images acquired with 1-min intervals. (D) The average spine head width is reduced after latrunculin B treatment, but the spine length is not altered (20 spines). After cLTP, both parameters are increased (32 spines). (E) The average percentage change per 1 min can be considered a value for dynamics. Both treatments reduce the average percentage change in spine head width and spine length representing reduced spine dynamics. Scale bars, 1 μm. Bars are means ± SEM, $***p < 0.001$; $**p < 0.01$ (student's *t*-test). (For color version of this figure, the reader is referred to the Web version of this chapter.)

recordings or more separated time points. As the NeuronStudio analysis results in accurate 3D measurements of width of spine heads and spine length, this method is useful in detecting changes in spine morphology. Further, the use of computer-based measuring enables the analysis of hundreds of spines in a relatively short time.

The main weakness in using NeuronStudio is that as the automatic detection of the spine head determines the spine head as a sphere, small changes in the spine head shape (e.g., morphing) are not detected. Further, very harsh treatments causing spines to largely lose their normal appearance can make automatic spine detection impossible.

The main technical challenge in the use of NeuronStudio software is that the quality of the images must be very high. To achieve good brightness/contrast with low background, the expression level of GFP (or other fill color) must be relatively high. However, the transfection efficiency should be optimized so that the transfected cells are not too close to each other. Otherwise, neurite tracing is difficult.

The software algorithm detects the spines of relatively young neurons better (DIV 12–16), because in that case, the spine density is not too high, and spines can be detected by the program as separate structures. Although the analysis is automatic, every analyzed dataset should be revised by a person blind to the experimental conditions.

5.1.2. Image analysis

NeuronStudio software is freely available and can be found at http://research.mssm.edu/cnic/tools-ns.html. A manual for NeuronStudio can be found at http://research.mssm.edu/cnic/help/ns/index.html. After opening the dataset in NeuronStudio, the voxel size must be entered into the program. The noise in the images can be reduced by image filtering. All images used for comparison should be prepared in exactly the same way. Also the parameters dividing spines to different classes should be kept same thorough the experiment. At first, the dendrite is traced by a neurite tool. The correctness of the tracing should be checked by the experimentalist. Stubby spines should be carefully attended to, as the program sometimes adds them to the neurite diameter. After neurite tracing, the spines can be detected and edited using the spine tool. The software uses the 3D image for calculation (Fig. 19.3A), whereas on the screen, we see only the maximum projection. This must be kept in mind when revising the automatically detected spines. We recommend checking the spine tracing from the 3D image as well as from XZ and YZ projections. The spine IDs (identity numbers) need to be corrected to be the same as in the previous stacks. After spine tracing, the text file of measured spine parameters can be imported to a suitable program (e.g., Microsoft Office, Excel) and analyzed.

From the illustrative experiments, we measured the spine head width and the spine length of 20 (latrunculin B) or 32 (cLTP) spines from five control stacks and from five stacks recorded after treatment. The time frames of a single spine (the same one in both time series) before and after latrunculin B application as well as its head width and length changes over time are shown in Fig. 19.3C. After latrunculin B application, the mean value of the spine head width was reduced from 0.78 ± 0.026 to 0.58 ± 0.014 μm, whereas the spine length remained the same (1.58 ± 0.095 and 1.58 ± 0.084 μm, respectively). In contrast, cLTP treatment increased both the spine head width (from 0.64 ± 0.014 to 0.75 ± 0.026 μm) and spine length (from 2.0 ± 0.048 to 2.5 ± 0.076 μm) (Fig. 19.3D). In order to measure spine dynamics, we calculated percentual change of spine head width and spine length between images taken with 1-min intervals. We averaged values from five images, resulting in a value representing the mean change of spine size in 1 min, which refers to spine dynamics. When comparing control values to values after latrunculin B or cLTP treatments, the changes of spine dynamics can be evaluated. Both treatments significantly reduced the spine dynamics measured either as percentage change of spine head width (latrunculin B: from 15.1% to 5.7% cLTP: from 13.2% to 5.7%) or spine length (latrunculin B: from 10.1% to 3.6%, cLTP: from 9.6% to 5.8%) (Fig. 19.3E). The spine head width and spine length values as well as percentage change of spine head width and spine length between images are directly comparable between datasets. Thus, the data of several experiments can be directly pooled together without further normalization.

5.2. Spine motility analysis

5.2.1. Overview

The approach presented in this chapter is based on the time-lapse imaging of a fluorescently labeled dendritic branch and subsequent voxel to voxel comparison between images taken at two time points (Fig. 19.4A). Those voxels that remain at the same intensity represent a stable part of the dendrite, and those voxels that change intensity belong to motile structures. In mature neurons, spines have the fastest motility rate in comparison to other dendritic structures. If the interval between time points is relatively short, most of the displaced voxels correspond to spines. Practically, voxel to voxel comparison can be done in several ways depending on the available image analysis software. Software options such as image subtraction and colocalization analysis are the most useful.

The advantage of this method is that voxel to voxel comparison allows detecting small changes in spine morphology that do not cause significant change in the width or the length of the spines (compare to Section 5.1). This method is best suited to detect spine morphing, a type of motility typical for heads of mature spines. The method also requires little image handling. It does not involve selection or manual outline of spines and thus allows the analysis of large sets of data as well as minimizes the error introduced by experimenter bias.

The weakness of this approach is that it cannot directly compare the motility of spines in different imaging fields. The method is based on the quantification of the percentage of displaced voxels in the entire dataset (Fig. 19.4B and C). This percentage depends not only on spine motility but

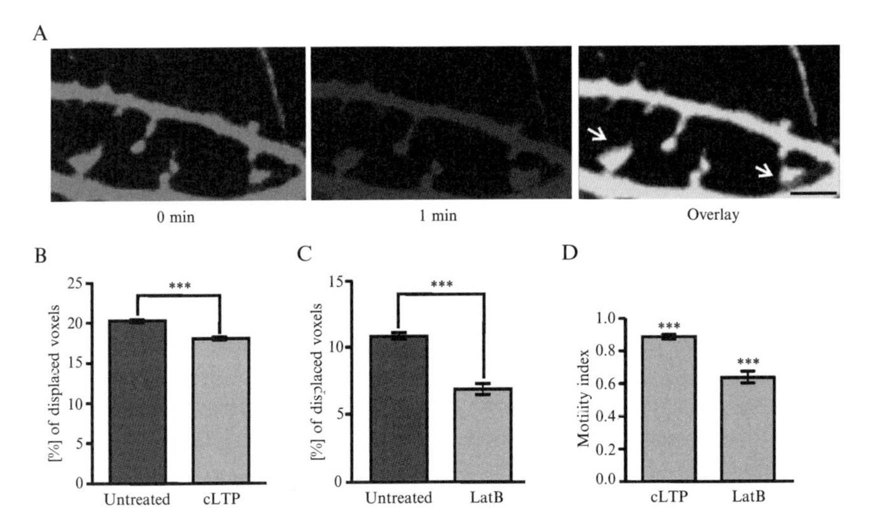

Figure 19.4 Spine motility analysis. (A) A representative image of a dendritic branch with spines scanned at two time points with 1-min intervals. An overlay panel shows stable areas in yellow and motile areas in either red or green. Note spine heads morphing marked with arrows. Scale bar, 2 μm. (B) Quantification of cLTP experiment. (C) Quantification of latrunculin B experiment. (D) Spine motility index for cLTP and latrunculin B experiments. Bars are means ± SEM; $^{***}p < 0.001$ (B and C—Mann–Whitney test, D—one sample *t*-test). (See Color Insert.)

also on the proportion between motile (spines) and stable (shafts) elements in the analyzed dendritic branch. Accordingly, this method can be used to analyze the change in spine motility, comparing the same imaging field before and after an experimental treatment. However, it cannot be used to compare different imaging fields directly. The motility values collected in several experiments and/or under several experimental conditions can be pooled together and compared only after normalization to the motility value in control conditions (Fig. 19.4D).

There are two technical problems that might strongly affect the result: the noise and the movement of the specimen during recording. High noise levels could add a substantial value to the observed voxel displacement. The noise in the images should be minimized by using brightly labeled cells, optimal imaging conditions (low gain, small pinhole, and slow scanning speed), and line averaging during image acquisition. After the experiment, the noise can be reduced by image deconvolution or Gaussian filtering. Movement of the specimen should be avoided at all costs. Firm chamber fixation, minimal or no perfusion, and steady temperature conditions are the keystones of the stable imaging experiment. Recordings with horizontal movement of the imaging field or with focus drift should be discarded.

If stability is a problem, it is recommended to make two to three recordings per experimental condition to be able to choose the most stable one during the analysis step.

It should be noted that this method is not suitable for experimental conditions that make dendrites flexible and thus more motile. In this case, the majority of the displaced pixels do not correspond to spines, and spine motility cannot be assessed using the colocalization approach.

5.2.2. Image analysis

We use Bitplane's Imaris software that allows manipulating 3D and 4D datasets. During the analysis step, the datasets are imported to Imaris. Then, all necessary image manipulations are performed, the noise is minimized by Gaussian filtering or deconvolution, and images are thresholded if needed. Importantly, all images used for comparison should be prepared in exactly the same way.

The next step is swapping time and channels. This procedure converts a dataset with 11 time points and 1 color channel into a dataset with 1 time point and 11 color channels. The channels of the latter dataset correspond to the time points of the former one. After swapping time and channels, colocalization analysis can be used to compare the voxel intensities of channels.

Colocalization analysis is an integrated function of Imaris software. It includes thresholding the source channels and building a new colocalization channel that contains only the voxels that represent the colocalization result. Typically, we manually select the threshold for the first two channels (time points 0 and 1 min) and then compare consecutive pairs of channels using the same threshold. The preview window allows for visualization of the source channel and the colocalization result while selecting the threshold value. It is helpful to change the color of the source channels to green and red so that colocalized structures look yellow (Fig. 19.4A). The threshold value should be equal for both source channels and should be selected so that the colocalization result depicted in white matches the yellow areas. For the sake of clarity, the colocalization channel is omitted from the figure. After each comparison, a new colocalization channel is built. The procedure results in 10 new channels containing voxels that were not displaced during the 1-min intervals between the consecutive time points. For each colocalization channel, Imaris provides a table with colocalization statistics. The data from this table can be exported as a text file and used for analysis in a statistical program of choice.

From the colocalization statistical data obtained in the previous step, we take the percentage of the source channel volume above the threshold that was colocalized. This value reflects how many voxels in the dendritic branch of interest remained stable as compared to the previous time point. Now, by subtracting the percentage of stable voxels from 100%, we can calculate the percentage of displaced voxels. Then, the mean percentage of voxels displaced during 1 min is found by averaging the percentage of displaced voxels calculated for each pair of consecutive time points in the dataset (Fig. 19.4B and C). As mentioned above, the percentage of displaced voxels varies significantly between datasets. Therefore, to pool the data of several experiments together or to compare different treatments, the percentage of displaced voxels should be normalized after the treatment to that before the treatment. This operation gives us a motility index. A motility index of 1 corresponds to the spine motility of untreated neurons (Fig. 19.4D).

For the analysis of the illustrative experiments, we measured and calculated the percentage of displaced voxels. The cLTP treatment decreased the percentage of displaced voxels from 20.3% to 18.0% (Fig. 19.4B). In the line with the spine morphology analysis, the latrunculin B treatment had a significantly stronger effect on motility, decreasing the percentage of displaced voxels from 11.4% to 7.35% (Fig. 19.4C). This resulted in the motility index for cLTP treatment of 0.89, while the motility index for latrunculin B treatment was 0.63.

ACKNOWLEDGMENTS

Pavel Uvarov is acknowledged for valuable discussions that led to the optimization of the motility analysis method. Maria Vartiainen, Kimmo Tanhuanpää, and Olaya Llano are acknowledged for their critical reading of and valuable comments regarding the manuscript. Outi Nikkilä and Seija Lågas are acknowledged for primary neuronal cells. Imaging was performed at the Light Microscopy Unit, Institute of Biotechnology. We would like to thank the Finish Academy (SA 1125867), the University of Helsinki, the Neuroscience Center, and the Helsinki Biomedical Graduate School for funding.

REFERENCES

Banker, G., and Goslin, K. (1998). *Culturing Nerve Cells.* 2nd edn., pp. 339–370. MIT press, Cambrige, UK.

Bourne, J. N., and Harris, K. M. (2008). Balancing structure and function at hippocampal dendritic spines. *Annu. Rev. Neurosci.* **31,** 47–67.

Dunaevsky, A., Tashiro, A., Majewska, A., Mason, C., and Yuste, R. (1999). Developmental regulation of spine motility in the mammalian central nervous system. *Proc. Natl. Acad. Sci. USA* **96,** 13438–13443.

Fischer, M., Kaech, S., Knutti, D., and Matus, A. (1998). Rapid actin-based plasticity in dendritic spines. *Neuron* **5,** 847–854.

Fortin, D. A., Davare, M. A., Srivastava, T., Brady, J. D., Nygaard, S., Derkach, V. A., and Soderling, T. R. (2010). Long-term potentiation-dependent spine enlargement requires synaptic Ca2+-permeable AMPA receptors recruited by CaM-kinase I. *J. Neurosci.* **35,** 11565–11575.

Furuyashiki, T., Arakawa, Y., Takemoto-Kimura, S., Bito, H., and Narumiya, S. (2002). Multiple spatiotemporal modes of actin reorganization by NMDA receptors and voltage-gated Ca2+ channels. *Proc. Natl. Acad. Sci. USA* **99,** 14458–14463.

Harris, K. M., and Stevens, J. K. (1989). Dendritic spines of CA 1 pyramidal cells in the rat hippocampus: Serial electron microscopy with reference to their biophysical characteristics. *J. Neurosci.* **9,** 2982–2997.

Honkura, N., Matsuzaki, M., Noguchi, J., Ellis-Davies, G. C., and Kasai, H. (2008). The subspine organization of actin fibers regulates the structure and plasticity of dendritic spines. *Neuron* **57,** 719–729.

Hotulainen, P., Llano, O., Smirnov, S., Tanhuanpää, K., Faix, J., Rivera, C., and Lappalainen, P. (2009). Defining mechanisms of actin polymerization and depolymerization during dendritic spine morphogenesis. *J. Cell Biol.* **185,** 323–339.

Kasai, H., Matsuzaki, M., Noguchi, J., Yasumatsu, N., and Nakahara, H. (2003). Structure-stability-function relationships of dendritic spines. *Trends Neurosci.* **26,** 360–368.

Kasai, H., Fukuda, M., Watanabe, S., Hayashi-Takagi, A., and Noguchi, J. (2010). Structural dynamics of dendritic spines in memory and cognition. *Trends Neurosci.* **33,** 121–129.

Kopec, C. D., Li, B., Wei, W., Boehm, J., and Malinow, R. (2006). Glutamate receptor exocytosis and spine enlargement during chemically induced long-term potentiation. *J. Neurosci.* **7,** 2000–2009.

Korkotian, E., and Segal, M. (2001). Regulation of dendritic spine motility in cultured hippocampal neurons. *J. Neurosci.* **21,** 6115–6124.

Lendvai, B., Stern, E. A., Chen, B., and Svoboda, K. (2000). Experience-dependent plasticity of dendritic spines in the developing rat cortex *in vivo*. *Nature* **404,** 876–881.

Liao, D., Scannevin, R. H., and Huganir, R. (2001). Activation of silent synapses by rapid activity-dependent synaptic recruitment of AMPA receptors. *J. Neurosci.* **16,** 6008–6017.

Lu, W. Y., Man, H. Y., Ju, W., Trimble, W. S., MacDonald, J. F., and Wang, Y. T. (2001). Activation of synaptic NMDA receptors induces membrane insertion of new AMPA receptors and LTP in cultured hippocampal neurons. *Neuron* **1,** 243–254.

Majewska, A., and Sur, M. (2003). Motility of dendritic spines in visual cortex *in vivo*: Changes during the critical period and effects of visual deprivation. *Proc. Natl. Acad. Sci. USA* **100,** 16024–16029.

Matus, A. (2000). Actin-based plasticity in dendritic spines. *Science* **290,** 754–758.

Matus, A., Ackermann, M., Pehling, G., Byers, H. R., and Fujiwara, K. (1982). High actin concentrations in brain dendritic spines and postsynaptic densities. *Proc. Natl. Acad. Sci. USA* **79,** 7590–7594.

Nestor, M. W., Cai, X., Stone, M. R., Bloch, R. J., and Thompson, S. M. (2011). The actin binding domain of βI-spectrin regulates the morphological and functional dynamics of dendritic spines. *PLoS One* **6,** e16197.

Noguchi, J., Matsuzaki, M., Ellis-Davies, G. C., and Kasai, H. (2005). Spine-neck geometry determines NMDA receptor-dependent Ca2+ signaling in dendrites. *Neuron* **46,** 609–622.

Oray, S., Majewska, A., and Sur, M. (2004). Dendritic spine dynamics are regulated by monocular deprivation and extracellular matrix degradation. *Neuron* **44,** 1021–1030.

Oray, S., Majewska, A., and Sur, M. (2006). Effects of synaptic activity on dendritic spine motility of developing cortical layer V pyramidal neurons. *Cereb. Cortex* **16,** 730–741.

Park, M., Salgado, J. M., Ostroff, L., Helton, T. D., Robinson, C. G., Harris, K. M., and Ehlers, M. D. (2006). Plasticity-induced growth of dendritic spines by exocytic trafficking from recycling endosomes. *Neuron* **5,** 817–830.

Penzes, P., Cahill, M. E., Jones, K. A., VanLeeuwen, J.-E., and Woolfrey, K. M. (2011). Dendritic spine pathology in neuropsychiatric disorders. *Nat. Neurosci.* **14,** 285–293.

Racz, B., and Weinberg, R. J. (2004). The subcellular organization of cortactin in hippocampus. *J. Neurosci.* **24,** 10310–10317.

Rodriguez, A., Ehlenberger, D. B., Dickstein, D. L., Hof, P. R., and Wearne, S. L. (2008). Automated three-dimensional detection and shape classification of dendritic spines from fluorescence microscopy images. *PLoS One* **3,** e1997.

Tashiro, A., and Yuste, R. (2004). Regulation of dendritic spine motility and stability by Rac1 and Rho kinase: Evidence for two forms of spine motility. *Mol. Cell. Neurosci.* **26,** 429–440.

Tashiro, A., and Yuste, R. (2008). Role of Rho GTPases in the morphogenesis and motility of dendritic spines. *Methods Enzymol.* **439,** 285–302.

Yuste, R., and Bonhoeffer, T. (2001). Morphological changes in dendritic spines associated with long-term synaptic plasticity. *Annu. Rev. Neurosci.* **24,** 1071–1089.

CHAPTER TWENTY

Imaging Cell Competition in *Drosophila* Imaginal Discs

Shizue Ohsawa,* Kaoru Sugimura,† Kyoko Takino,* *and* Tatsushi Igaki*,‡

Contents

Abstract

Cell competition is a process in which cells with higher fitness ("winners") survive and proliferate at the expense of less fit neighbors ("losers"). It has been suggested that cell competition is involved in a variety of biological processes such as organ size control, tissue homeostasis, cancer progression, and the maintenance of stem cell population. By advent of a genetic mosaic technique, which enables to generate fluorescently marked somatic clones in *Drosophila* imaginal discs, recent studies have presented some aspects of molecular mechanisms underlying cell competition. Now, with a live-imaging technique using *ex vivo*-cultured imaginal discs, we can dissect the spatiotemporal nature of competitive cell behaviors within multicellular communities. Here, we describe procedures and tips for live imaging of cell competition in *Drosophila* imaginal discs.

1. Introduction

Cell–cell interactions in multicellular organisms play crucial roles in coordination of cell proliferation, differentiation, and cell death during normal development and homeostasis. "Cell competition" is a form of cell–cell

* Department of Cell Biology, G-COE, Kobe University Graduate School of Medicine, Kobe, Japan
† Institute for Integrated Cell-Material Sicences (iCeMS), Kyoto University iCeMS Complex 2, Sakyo-ku, Kyoto, Japan
‡ PRESTO, Japan Science and Technology Agency (JST), Saitama, Japan

Methods in Enzymology, Volume 506
ISSN 0076-6879, DOI: 10.1016/B978-0-12-391856-7.00044-5

interaction in which cells with higher fitness ("winners") survive and proliferate at the expense of neighboring cells with lower fitness ("losers") (Morata and Ripoll, 1975). Loser cells, but otherwise viable cells, undergo apoptosis when confronted with winner cells. Winner cells then proliferate to occupy the space, resulting in the size of the population unchanged. Thus, cell competition is a process in which fitter cells are selected among otherwise homogeneous population. It has been suggested that cell competition is involved in a variety of biological processes such as organ size control, tissue homeostasis, cancer progression, and the maintenance of stem cell population (Baker, 2011; Johnston, 2009; Morata and Martin, 2007; Moreno, 2008).

Studies in *Drosophila* imaginal discs have indicated that the "cellular fitness" can be determined by several factors: (i) the amount of ribosomal proteins, (ii) the expression level of a proto-oncogene *myc*, (iii) the activity of the Hippo pathway, and (iv) the integrity of apico-basal polarity. Cells with heterozygous mutation in any one of the ribosomal protein genes (called *Minute* mutants) are eliminated from the tissue as losers when confronted with wild-type cells (Morata and Ripoll, 1975; Moreno *et al.*, 2002; Simpson, 1979). Cells expressing higher level of *Drosophila myc* (*dmyc*) become winners when confronted with cells with relatively lower *dmyc* expression (de la Cova *et al.*, 2004; Moreno and Basler, 2004). Similarly, cells with inactivated Hippo tumor-suppressor pathway components behave as "super-competitors" that eliminate neighboring wild-type cells (Tyler *et al.*, 2007). In addition, epithelial cells seem to compete the integrity of the polarized structure with one another; cells with disrupted apico-basal polarity are eliminated as losers when confronted with normally polarized wild-type cells (Bilder, 2004; Hariharan and Bilder, 2006). This elimination of polarity-deficient cells by cell competition seems to work as an "intrinsic tumor suppression" (Igaki, 2009) (see below).

Most cancers originate from epithelium. Loss of apico-basal polarity in epithelial cells is frequently associated with cancer progression (Bissell and Radisky, 2001; Fish and Molitoris, 1994). Similarly, mutant flies deficient for evolutionarily conserved apico-basal polarity genes such as *scribble* (*scrib*) or *discs large* (*dlg*) develop overgrown tumors in their imaginal epithelium (Bilder, 2004; Hariharan and Bilder, 2006). Intriguingly, when surrounded by wild-type cells, these polarity-deficient mutant cells do not overgrow but are eliminated from the tissue (Brumby and Richardson, 2003; Igaki *et al.*, 2006, 2009; Pagliarini and Xu, 2003; Woods and Bryant, 1991). This suggests that normal epithelium possesses an intrinsic tumor-suppression mechanism that eliminates oncogenic polarity-deficient cells by cell competition. The polarity-deficient cells that are confronted with wild-type cells result in c-Jun N-terminal kinase (JNK)-dependent cell death (Brumby and Richardson, 2003), which is triggered by endocytic activation of Eiger (*Drosophila* tumor necrosis factor, TNF) (Igaki *et al.*, 2009). Surrounding normal cells also activate Eiger–JNK signaling, which does not cause cell death but enhances the elimination of neighboring polarity-deficient cells by activating the PVR-ELMO–Mbc-mediated engulfment pathway (Ohsawa *et al.*, 2011). The JNK activation in polarity-deficient cells has also been shown to be triggered by Eiger expressed in hemocytes (Cordero *et al.*, 2010). Thus, Eiger–JNK signaling plays a central role in competition between polarized and nonpolarized epithelial cells. Interestingly, *Minute* clones are still eliminated by cell competition in the absence of *eiger* (Ohsawa *et al.*, 2011), suggesting that competitive interactions in different cellular contexts are regulated by different mechanisms.

About 30 years after the first discovery (Morata and Ripoll, 1975), progress has been made in understanding the molecular mechanism of cell competition with some technical advances especially for producing fluorescently labeled somatic clones in imaginal discs (Lee and Luo, 1999; Xu and Rubin, 1993). Recent studies have identified both positive (JNK, hid, Flower, Eiger, and the engulfment genes) and negative (Dpp, the Hippo pathway components, and Sparc) regulators of cell competition (Brumby and Richardson, 2003; de la Cova *et al.*, 2004; Igaki *et al.*, 2009; Li and Baker, 2007; Moreno *et al.*, 2002; Neto-Silva *et al.*, 2010; Ohsawa *et al.*, 2011; Rhiner *et al.*, 2010; Tyler *et al.*, 2007; Ziosi *et al.*, 2010). Yet, the underlying mechanism by which these regulators act, if exists, in a common cell competition pathway and the upstream mechanisms that regulate these factors still remain to be elucidated. Spatiotemporal analysis in live tissues could provide new insights into understanding cell competition. Here, we describe procedures and tips for live imaging of cell competition using *ex vivo*-cultured *Drosophila* imaginal discs.

2. Live Imaging of Cell Competition

Live imaging, in conjunction with genetic manipulations, is a powerful tool for dissecting dynamic cellular processes within multicellular communities. Here, we overview a recently established live-imaging system for visualizing cell competition in *ex vivo*-cultured *Drosophila* imaginal discs.

2.1. Preparation of chambers

The chambers for culturing imaginal discs are prepared by assembling following materials (Fig. 20.1). Glass bottom dishes for inverted microscopes are also commercially available.

For upright microscopes (chamber A)

- 60-mm petri dish
- 35-mm dish bottom
- Glass ring as a spacer (e.g., height: 4 mm: outside diameter: 20 mm)
- Nontoxic glue paste SILPOT 184 W/C (DOW CORNING TORAY)

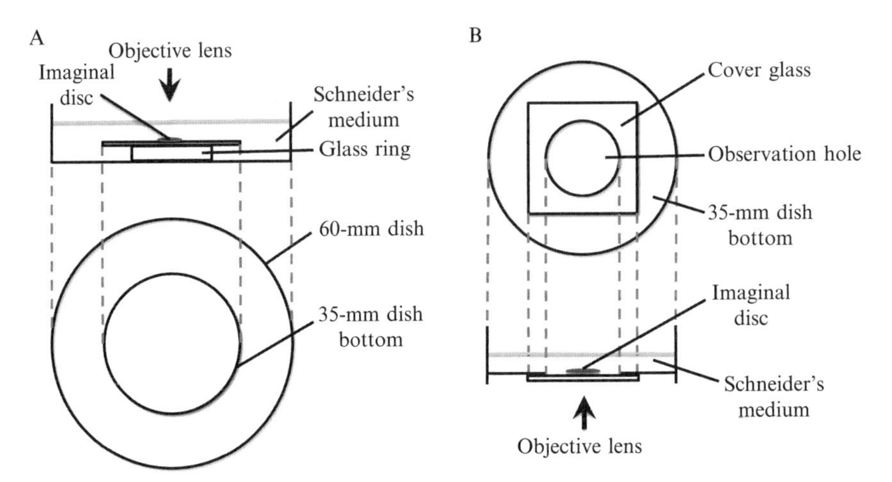

Figure 20.1 Chamber preparation for live imaging of imaginal discs using an upright (A) or an inverted (B) confocal laser scanning microscope.

For inverted microscopes (chamber B)

– 35-mm glass bottom dish (MATSUNAMI Glass No. 1S #D111300)

The commercially available glass bottom dishes should be rinsed with 2 ml of 100% ethanol for 2–3 times to remove toxicity before use.

2.2. Imaginal disc culture

As cell competition is a phenomenon observed in proliferating imaginal epithelium, the cultured discs must be kept with constant cell proliferation rate during the imaging. In addition, the imaginal discs must be tightly attached to the dish/glass to prevent out-of-focus imaging. The following procedure allows to perform live imaging of cell competition for at least 3 h in proliferating imaginal epithelium.

Imaginal discs (eye-antennal discs or wing discs) with genetically manipulated somatic clones (labeled with fluorescent proteins) are dissected out from third-instar larvae in phosphate-buffered saline (PBS) and are set on the prepared chambers as follows:

1. Dissect larvae in PBS and take imaginal discs out from the larvae using forceps.
2. Place the dissected imaginal disc, with peripodial-side up, on a chamber with a drop of PBS. PBS facilitates nonspecific adhesion of imaginal disc to dish/glass in the chamber.
3. Pour Schneider's *Drosophila* medium (GIBCO) containing heat-inactivated 10% FBS into the chamber and culture the disc at room temperature.

2.3. Live imaging of imaginal discs

The imaginal disc culture is subjected to the time-lapse imaging using an upright or inverted confocal laser scanning microscope. Here, we describe an example of live imaging of cell competition between polarized and nonpolarized cells in eye-antennal discs using both upright and inverted microscopes.

Clones of cells mutant for *scrib* were induced in eye-antennal discs using the genetic mosaic technique (Lee and Luo, 1999; Xu and Rubin, 1993). Homozygous (*scrib/scrib*; polarity deficient) and heterozygous (*scrib/+*; polarized) mutant clones were visualized by 2×Ubi-GFP and 1×Ubi-GFP (green), respectively, and Eiger-overexpressing polarized cells were visualized by actin-Gal4/UAS-myrRFP (magenta). The eye-antennal disc was cultured in chamber A, and images were acquired at 5-min interval for 3 h using an upright confocal microscope (FV1000, Olympus) equipped with an Olympus 60×/NA1.1 LUMFL water-immersion objective lens. A polarity-deficient *scrib*$^{-/-}$ cell (arrowhead) is eliminated by cell competition when confronted with polarized cells (Fig. 20.2 and Movie 20.1) (http://www.elsevierdirect.com/companions/9780123918567) (Ohsawa *et al.*, 2011). Similar images were also obtained using an inverted confocal microscope (LSM700, Zeiss) equipped with an Zeiss Plan-Apochromat 40×/NA1.3 oil-immersion objective lens (data not shown).

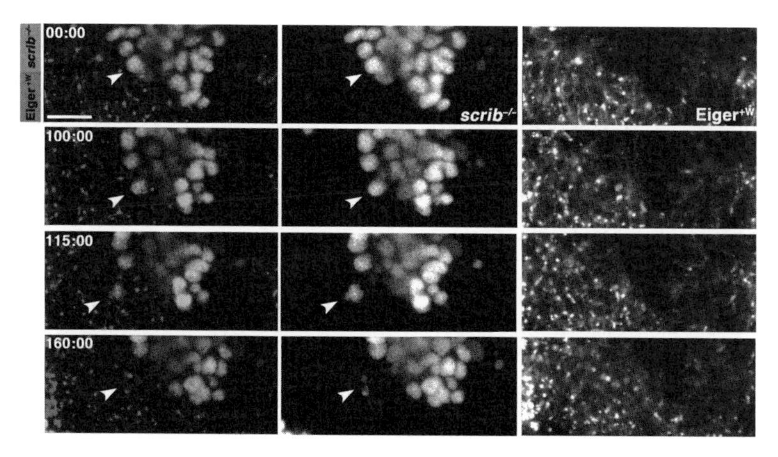

Figure 20.2 Live imaging of cell competition between polarized and nonpolarized cells in imaginal epithelium. Homozygous *scrib* mutant cells (labeled with 2×Ubi-GFP; "losers") are eliminated when confronted with normally polarized Eiger-expressing cells (labeled with actin-Gal4/UAS-myrRFP; "winners") in an *ex vivo*-cultured eye-antennal disc. A polarity-deficient *scrib*$^{-/-}$ cell is eliminated from the tissue after it is incorporated into "polarized" cell population (arrowheads). Eiger was overexpressed in normal cells to enhance cell competition. Four frames of Movie 20.1 are shown. Image processing was done with ImageJ. Scale bar, 10 μm. Genotypes is: *y,w, eyFLP1; G454, Act > y+ > Gal4, UAS-myrRFP/UAS-Eiger*$^{+W}$*; FRT82B/FRT82B, ubi-GFP, Tub-Gal80, scrib*^{1}. This figure was modified from the data previously published in Ohsawa *et al.* (2011).

REFERENCES

Baker, N. E. (2011). Cell competition. *Curr. Biol.* **21,** R11–R15.

Bilder, D. (2004). Epithelial polarity and proliferation control: Links from the Drosophila neoplastic tumor suppressors. *Genes Dev.* **18,** 1909–1925.

Bissell, M. J., and Radisky, D. (2001). Putting tumours in context. *Nat. Rev. Cancer* **1,** 46–54.

Brumby, A. M., and Richardson, H. E. (2003). Scribble mutants cooperate with oncogenic Ras or Notch to cause neoplastic overgrowth in Drosophila. *EMBO J.* **22,** 5769–5779.

Cordero, J. B., Macagno, J. P., Stefanatos, R. K., Strathdee, K. E., Cagan, R. L., and Vidal, M. (2010). Oncogenic Ras diverts a host TNF tumor suppressor activity into tumor promoter. *Dev. Cell* **18,** 999–1011.

de la Cova, C., Abril, M., Bellosta, P., Gallant, P., and Johnston, L. A. (2004). Drosophila myc regulates organ size by inducing cell competition. *Cell* **117,** 107–116.

Fish, E. M., and Molitoris, B. A. (1994). Alterations in epithelial polarity and the pathogenesis of disease states. *N. Engl. J. Med.* **330,** 1580–1588.

Hariharan, I. K., and Bilder, D. (2006). Regulation of imaginal disc growth by tumor-suppressor genes in Drosophila. *Annu. Rev. Genet.* **40,** 335–361.

Igaki, T. (2009). Correcting developmental errors by apoptosis: Lessons from Drosophila JNK signaling. *Apoptosis* **14,** 1021–1028.

Igaki, T., Pagliarini, R. A., and Xu, T. (2006). Loss of cell polarity drives tumor growth and invasion through JNK activation in Drosophila. *Curr. Biol.* **16,** 1139–1146.

Igaki, T., Pastor-Pareja, J. C., Aonuma, H., Miura, M., and Xu, T. (2009). Intrinsic tumor suppression and epithelial maintenance by endocytic activation of Eiger/TNF signaling in Drosophila. *Dev. Cell* **16,** 458–465.

Johnston, L. A. (2009). Competitive interactions between cells: Death, growth, and geography. *Science* **324,** 1679–1682.

Lee, T., and Luo, L. (1999). Mosaic analysis with a repressible cell marker for studies of gene function in neuronal morphogenesis. *Neuron* **22,** 451–461.

Li, W., and Baker, N. E. (2007). Engulfment is required for cell competition. *Cell* **129,** 1215–1225.

Morata, G., and Martin, F. A. (2007). Cell competition: The embrace of death. *Dev. Cell* **13,** 1–2.

Morata, G., and Ripoll, P. (1975). Minutes: Mutants of Drosophila autonomously affecting cell division rate. *Dev. Biol.* **42,** 211–221.

Moreno, E. (2008). Is cell competition relevant to cancer? *Nat. Rev. Cancer* **8,** 141–147.

Moreno, E., and Basler, K. (2004). dMyc transforms cells into super-competitors. *Cell* **117,** 117–129.

Moreno, E., Basler, K., and Morata, G. (2002). Cells compete for decapentaplegic survival factor to prevent apoptosis in Drosophila wing development. *Nature* **416,** 755–759.

Neto-Silva, R. M., de Beco, S., and Johnston, L. A. (2010). Evidence for a growth-stabilizing regulatory feedback mechanism between Myc and Yorkie, the Drosophila homolog of Yap. *Dev. Cell* **19,** 507–520.

Ohsawa, S., Sugimura, K., Takino, K., Xu, T., Miyawaki, A., and Igaki, T. (2011). Elimination of oncogenic neighbors by JNK-mediated engulfment in Drosophila. *Dev. Cell* **20,** 315–328.

Pagliarini, R. A., and Xu, T. (2003). A genetic screen in Drosophila for metastatic behavior. *Science* **302,** 1227–1231.

Rhiner, C., Lopez-Gay, J. M., Soldini, D., Casas-Tinto, S., Martin, F. A., Lombardia, L., and Moreno, E. (2010). Flower forms an extracellular code that reveals the fitness of a cell to its neighbors in Drosophila. *Dev. Cell* **18,** 985–998.

Simpson, P. (1979). Parameters of cell competition in the compartments of the wing disc of Drosophila. *Dev. Biol.* **69,** 182–193.

Tyler, D. M., Li, W., Zhuo, N., Pellock, B., and Baker, N. E. (2007). Genes affecting cell competition in Drosophila. *Genetics* **175,** 643–657.

Woods, D. F., and Bryant, P. J. (1991). The discs-large tumor suppressor gene of Drosophila encodes a guanylate kinase homolog localized at septate junctions. *Cell* **66,** 451–464.

Xu, T., and Rubin, G. M. (1993). Analysis of genetic mosaics in developing and adult Drosophila tissues. *Development* **117,** 1223–1237.

Ziosi, M., Baena-Lopez, L. A., Grifoni, D., Froldi, F., Pession, A., Garoia, F., Trotta, V., Bellosta, P., and Cavicchi, S. (2010). dMyc functions downstream of Yorkie to promote the supercompetitive behavior of hippo pathway mutant cells. *PLoS Genet.* **6,** e1001140.

CHAPTER TWENTY-ONE

Live Cell Imaging of the Oviduct

Sabine Kölle

Contents

Abstract

In the oviduct, the integrity of oocyte and sperm transport, fertilization, and early embryonic ontogenesis is essential for successful reproduction. Up to now, most of the knowledge on oocyte and sperm transport, gamete interaction and embryonic development has in most cases been gained exclusively by *in vitro* studies. In addition, especially the mechanisms of gameto–maternal interaction and embryo–maternal communication in the oviduct are still unknown. Recent techniques of live cell imaging and digital videomicroscopy allow for the first time to provide actual new insights in the mechanisms of sperm transport, sperm storage, oocyte transport, fertilization, gameto–maternal interaction and embryo–maternal crosstalk under near *in vivo* conditions. Detailed knowledge of these important events in the oviduct is the prerequisite to develop new therapeutic concepts for subfertility and infertility and to increase the success rates of the actual techniques of assisted reproduction (ART). Additionally the effects of drugs and hormones used in ART can be effectively studied using a functional oviductal epithelium. The guidelines for live cell imaging in the oviduct presented here should enable researches to establish a functional digital analysis system which allows to study physiological and pathological events in the oviduct under near *in vivo* conditions.

Department of Urology, Ludwig Maximilians University of Munich, Munich, Germany

Methods in Enzymology, Volume 506
ISSN 0076-6879, DOI: 10.1016/B978-0-12-391856-7.00045-7

1. Introduction

The oviduct is the place where the pick-up and transport of the oocyte, the transport of the spermatozoa, the fertilization, as well as the development and the transport of the early embryo occur (Greve and Callesen, 2001; Menezo and Guerin, 1997). Successful fertilization only takes place when the oocyte and the spermatozoa meet in the ampulla in time (Talbot *et al.*, 2003). After fertilization, only embryos which show a precise timing of transport are capable to implant (Pulkinnen, 1995). Thus, Akira *et al.* (1993) demonstrated that in rats superovulation treatment accelerates embryonic transport leading to impaired implantation. Both the oocyte and the embryo in the oviduct are transported by (a) ciliary beating of the tubal epithelial cells and (b) contraction of the oviductal smooth muscle (Croxatto, 2002; Halbert *et al.*, 1976, 1989). Ciliary transport is important but not essential for successful pregnancy as women with immotile cilia (Kartagener's syndrome) can in some cases become pregnant (McComb *et al.*, 1986). However, the precise mechanisms of regulation and modulation of gamete and embryo transport in the oviduct are still unclear. Especially in this aspect live cell imaging is a very valuable tool to gain more knowledge on the signal transduction pathways which are essential for precise transport and successful pregnancy.

Most of the knowledge on gamete transport, fertilization, and early ontogenesis have been gained by *in vitro* studies. However, to date especially the crosstalk between the oocyte/the early embryo and a functional oviductal epithelium has not yet been elucidated. This is due to the fact that the oviduct is difficult to investigate *in vivo*. As it is surrounded by mesosalpinx and is integrated in the bursa ovarica, it cannot be effectively investigated by ultrasonography or radiography. Although *in vitro* cell cultures are effective tools for obtaining essential information on the early events of fertilization and embryonic development in the oviduct (Bongso *et al.*, 1989; Thibodeaux *et al.*, 1992; Umezu *et al.*, 2003) they have several shortcomings. Thus, secretory, and ciliary function as well as hormone receptor activity are lost during oviductal *in vitro* culture (Bongso *et al.*, 1989). These functional damages are accompanied by morphological alterations (Bongso *et al.*, 1989). Live cell imaging for the first time provides the possibility to investigate sperm binding, sperm movement, oocyte transport, fertilization, and early embryonic development under near *in vivo* and *in situ* conditions. Additionally the interaction between the oocyte and the ampullar epithelium, the binding of sperm to the isthmic epithelium as well as the mechanisms of early embryo–maternal communication can be elucidated. Furthermore live cell imaging is a valuable tool to investigate the effects of drugs and hormones used in ART for therapeutic purposes.

The digital videomicroscopic system described here was established and optimized in our lab for *in situ* investigations of oviducts from mice, cows, and pigs. The following chapters will give important tips and tricks which are essential for successful investigation of a functional active oviductal epithelium. For this purpose, effective preparation of the oviduct and the handling of the system for precise qualitative and quantitative analyses in the oviduct are described.

2. Preparation of the Oviduct

For live cell imaging of oocytes, spermatozoa, and embryos in the mouse, the mouse is put under anesthesia, fixed on the front and hind legs and cleaned with 70% ethanol. The abdomen is vertically opened from the sternum to the pelvis as well as horizontally along the costal arch. To visualize the genital tract, the intestinum has to be put aside. The cranial ovarian ligament and the Lig. latum uteri are cut and the ovary, oviduct, and uterine horne of one side are laid in a Delta T-Dish (Bioptechs Inc, Butler, PA) covered with Sylgard gel (No. 184, Dow Corning, Wiesbaden, Germany) and filled with 1.5 ml warm (37 °C) Hepes buffer. A constant temperature of 37 °C is maintained by placing the Delta T-Dish on a heated Delta T stage holder (Fig. 21.1). The electrodes of the stage holder (Fig. 21.2) contact the Delta T-dish and maintain a constant temperature which is controlled by a thermostat (Bioptechs Inc., Butler, PA). It is essential to keep the temperature constant all the time, as even small changes in temperature induce alterations in ciliary function, secretion, and tubal receptivity to hormones or drugs.

As the oviduct *in situ* is very strongly curled, it is difficult to visualize every part of the oviduct. Therefore in most experiments it is much more advisable to kill the mouse by inhalation of an overdose of isoflurane, open the abdomen as described above and to excise the whole genital tract by cutting the cranial ovarian ligments and the cervix. In the following the mesosalpinx is removed using a stereomicroscope and the oviduct is longitudinally fixed in a Delta T-Dish by using insect needles (No.15, Ento Sphinx, Pardubice, Czech Republic). For qualitative analyses of the behavior of oocytes, spermatozoa, and embryos in the oviduct, we always use the closed oviduct. For quantitative analyses of particle transport speed (PTS) and ciliary beat frequency (CBF), it is advisable to open the oviduct longitudinally for precise tracking of the particles.

In the cow and in the pig, the oviducts are removed immediately after slaughter. After they have been freed from mesosalpinx, they are opened longitudinally and fixed with insect needles in a Delta T-Dish covered with Sylgard gel as described before (Fig. 21.3).

Our studies in mice and cows showed that the functional integrity of the oviduct is maintained up to 6 h after death—all the experiments should be

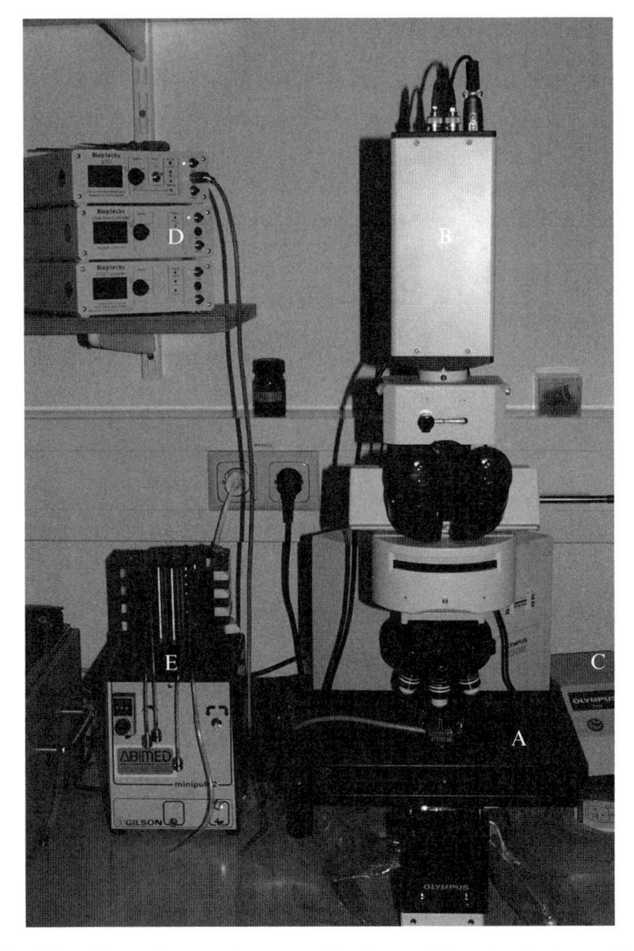

Figure 21.1 The videomicroscopic system for live cell imaging in the oviduct consists of (A) Delta T stage holder, (B) CCD camera, (C) light source, (D) thermostat, (E) machine for continuous application of drugs.

done within this time period. Ciliary beating as watched by light microscopy is generally visible up to 48–52 h after death. However, as shown by transmission electron microscopy, cellular alterations and degradation of cell organelles are starting from 6 h after death onwards. Thus, ciliary beating—which has often been described as a feature of cell viability in oviductal explants—does not reflect the integrity of the oviductal cell as it is maintained even when nearly all cell organelles are degraded. As shown by quantitative measurements before and after application of hormones, the receptivity of hormone receptors is generally maintained 6–12 h after death—this is dependant on the size of the receptor protein analyzed in the experiment.

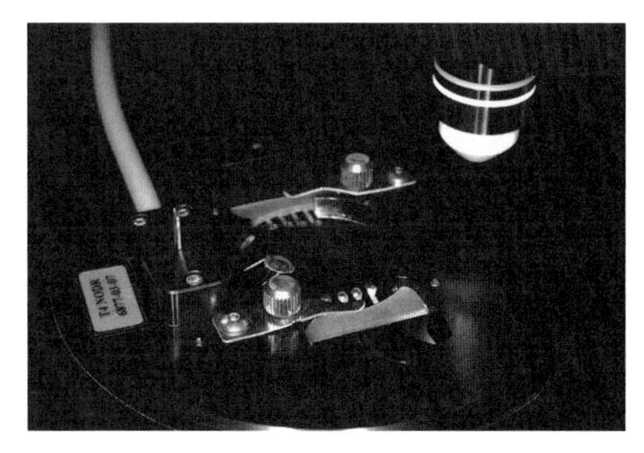

Figure 21.2 Electrodes which are in close contact to the Delta T dish maintain a constant temperature.

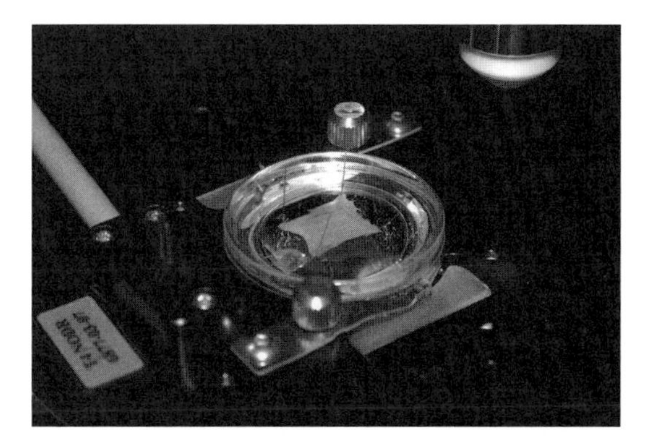

Figure 21.3 For quantitative analysis the bovine oviduct is opened longitudinally and is fixed with insect needles.

3. Live Cell Imaging: Qualitative Analysis

For live cell imaging, the fixed stage upright microscopes BX50WI or BX61WI (Olympus, Hamburg, Germany) is used (Fig. 21.1). The microscopes are equipped with four water immersion objectives with magnifications of 10, 20, 40, and 100 (UMPLFL N 10 × W/0.30, UMPLFL N 20 × W/0.50, UMPLFL N 40 × W/0.80, UMPLFL N 100 × W/0.70). The microscopes are additionally equipped with an Imago CCD camera with a 1280 × 960 pixel CCD chip (Till Photonics, Graefelfing, Germany). The normal image frequency is 20 Hz, 49 pictures per second provide good films. The interactions of cumulus–oocyte complexes and early embryos with the oviductal epithelium are monitored using the UMPLFL 10 × W/0.30

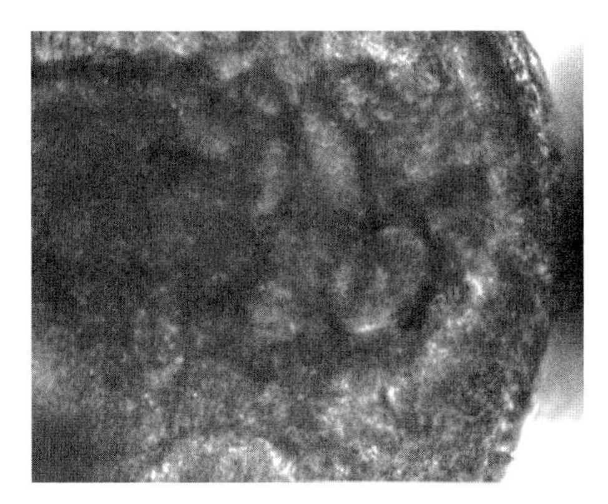

Figure 21.4 The interaction between a mouse embryo and the oviductal epithelium is viewed under near *in vivo* conditions.

objective (Fig. 21.4). For investigating the behavior of spermatozoa in the oviduct, the ×20 and ×40 objectives are appropriate. For analyzing ciliary beating and secretory activity of the oviductal lining, the UMPLFL N 100 × W/0.70 has to be used. As single animals show immotile cilia, ciliary activity in the oviduct should regularly be controlled before starting the experiments. In the mouse, oocytes, and early embryos can be watched in the oviduct up to 2 days after occurrence of the vaginal plug. In the cow, the early embryo can be observed in the oviduct up to 3 days after insemination. Movies demonstrating the behavior of the cumulus–oocyte complex, the spermatozoa and the early embryo in the bovine oviduct can be viewed at http://www.biolreprod.org/cgi/content/abstract/biolreprod.108.073874, see supplemental files (Kölle *et al.*, 2009).

4. Live Cell Imaging: Quantitative Analysis

For measuring the PTS in the oviduct 9 × 10^6 polystyrene beads with a diameter of 2.8 μm (Dynabeads, Invitrogen Dynal, Oslo, Norway) are added to the Hepes buffer in the Delta T dish (Fig. 21.5). To avoid that particles settle down, it is important to mix well. In the following, it is essential to wait for 3 min before starting measurement to avoid that movement of the fluid affects the measurement.

The basal PTS is measured eight times every 3 min followed by a minute-by-minute interval after adding drugs or hormones. Images are taken with the Olympus BX61WI upright light microscope (as described before) equipped with a 20× water immersion objective and a monochrome CMOS camera (EHD Imaging, Damme, Germany, or Sumix Corporation, Oceanside, CA).

Figure 21.5 The particle transport speed (PTS) is analyzed using small polystyrene particles (dynabeads), which are transported between the oviductal folds of the ampulla.

For visualizing and tracking the particles, the analysis software "Stream Pix" (Norpix Inc, Montreal, Canada) is used. At each time point, 200 images (640×512 pixels, 12 bit, 1/84 ms) are taken—the frame rate is 4932 images/s. The films then have to be converted from 12 to 8 bit grayscale so that the dynabeads can be tracked automatically by using the software Image Pro Plus (Media Cybernetics Inc, Silver Spring, USA). Only dynabeads that can be followed over a length of more than 10 frames are included in the calculation. Additionally only dynabeads the way of which does not deviate more than 15% from the direct connection between start and end point are tracked and included in the study.

It is important to know that in the isthmus of the mouse there is no ciliary transport because in this part of the oviduct there are mainly secretory cells throughout the cycle. The early embryo is physiologically transported in the isthmus mainly by smooth muscle contraction. Therefore dynabeads settle down and PTS cannot be measured in the isthmus. Thus, it only makes sense to investigate PTS in the murine ampulla. In the murine ampulla, the average PTS is 70–90 μm/s. In individuals, differences up to 20 μm/s occur which are not biologically relevant.

It is also important to know that some hormones such as progesterone can only be dissolved in buffer supplemented with ethanol. However, our studies showed that already 3% of ethanol induce an altered particle transport—the dynabeads are floating and show reduced speed. Our investigations revealed that concentrations of 0.06% ethanol can be used without altering the PTS. The mean PTS in the bovine ampulla is 150 μm/s. In individuals, differences up to 20 μm/s occur which are not biologically relevant (Kölle *et al.*, 2009, 2010).

The third important point for precise measuring is that the plain of focus must stay constant. In some cycle stages such as metestrus the contractility of the smooth muscle in the oviduct is very high—as a consequence the plain of documentation is always changing during one film sequence. In this case it helps to fix the oviduct with a narrow line of insect needles or—if this is not sufficient—to apply drugs inhibiting smooth muscle contraction.

For measuring the CBF, the general experimental setup described above is used. The only difference is that the images are taken without the application of dynabeads. For measuring the 40× water immersion objective (Olympus) is used. One thousand images in a resolution of 640 × 480 pixels are taken at a frame rate of 105 images/s. Ten ciliated cells per field of view are determined as "Area of interest" (AOI) for each time series (Fig. 21.6). In the following, the differences in brightness caused by ciliary beating are analyzed using the software Image Pro Plus (Media Cybernetics). CBF is then calculated by the Fourier transformation using the software Auto Signal (Systat Software GmbH, Erkath, Germany). Ciliary beat frequency in the oviduct is relatively similar in different species, it ranges from 18 to 20 Hz.

Statistical analysis is done using the software SPSS 17.0 (SPSS Inc, Chicago, USA). Multiple groups can be compared with the nonparametric Kruskal–Wallis test. If the result is $p < 0.05$, pairs of groups should be compared with the Mann–Whitney test. If $p < 0.05$ the results are rated significant, if $p < 0.01$ the results are rated as highly significant.

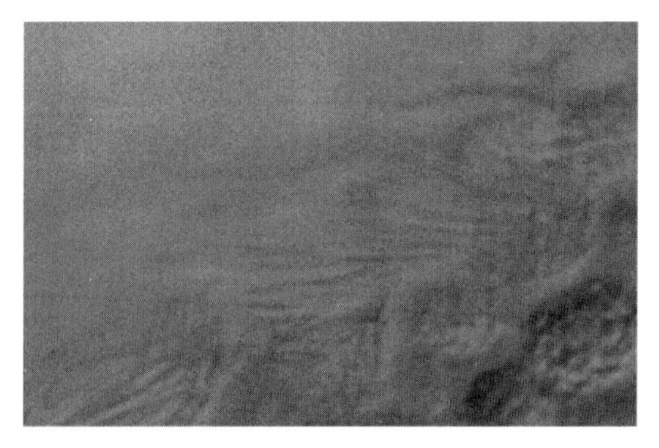

Figure 21.6 For the determination of ciliary beat frequency (CBF), ciliated cells in the oviduct are visualized and marked as "Area of Interest" (AOI).

ACKNOWLEDGMENTS

The research relevant to this chapter was supported by the German Research Foundation (DFG KO 1398/5-1) and by the Association for the Support of Research on Biotechnology (FBF).

REFERENCES

Akira, S., Sanbuissho, A., Lin, Y., and Araki, T. (1993). Acceleration of embryo transport in superovulated adult rats. *Life Sci.* **53,** 1243–1251.

Bongso, A., Chye, N. S., Sathananthan, H., Lian, N. P., Ruaff, M., and Ratnam, S. S. (1989). Establihsment of human ampullary cell cultures. *Hum. Reprod.* **5,** 486–495.

Croxatto, H. B. (2002). Physiology of gamete und embryo transport through the fallopian tube. *Reprod. Biomed. Online* **4,** 1160–1169.

Greve, T., and Callesen, H. (2001). Rendez-vous in the oviduct: Implications for superovulation and embryo transfer. *Reprod. Nutr. Dev.* **41,** 451–459.

Halbert, S. A., Tam, P. Y., and Blandau, R. J. (1976). Egg transport in the rabbit oviduct: The roles of cilia and muscle. *Science* **191,** 1052–1053.

Halbert, S. A., Becker, D. R., and Szal, S. E. (1989). Ovum transport in the rat oviductal ampulla in the absence of muscle contractility. *Biol. Reprod.* **40,** 1131–1136.

Kölle, S., Dubielzig, S., Reese, S., Wehrend, A., König, P., and Kummer, W. (2009). Ciliary transport, gamete interaction and effects of the early embryo in the oviduct: *Ex vivo* analyses using a new digital vidceomicroscopic system in the cow. *Biol. Reprod.* **81,** 267–274.

Kölle, S., Reese, S., and Kummer, W. (2010). New aspects of gamete transport, fertilization and embryonic development in the oviduct gained by means of live cell imaging. *Theriogenology* **73,** 786–795.

McComb, P., Langley, L., Villalon, L., and Verdugo, P. (1986). The oviductal cilia and Kartagener's syndrome. *Fertil. Steril.* **46,** 412–416.

Menezo, Y., and Guerin, P. (1997). The mammalian oviduct: Biochemistry and physiology. *Eur. J. Obstet. Gynecol. Reprod. Biol.* **73,** 99–104.

Pulkinnen, M. O. (1995). Oviductal function is critical for very early human life. *Ann. Med.* **27,** 307–310.

Talbot, P., Shur, B. D., and Myles, D. G. (2003). Cell adhesion and fertilization: Steps in oocyte transport, sperm-zona pellucida interactions, and sperm-egg fusion. *Biol. Reprod.* **68,** 1–9.

Thibodeaux, J. K., Meyers, M. W., Gouedeaux, L. L., Menezo, Y., Roussel, J. D., Broussard, J. R., and Godke, R. A. (1992). Evaluating an in viro culture system of bovine uterine and oviduct epithelial cells, for subsequent embryo co-culture. *Reprod. Fertil. Dev.* **4,** 573–583.

Umezu, T., Hanazono, M., Aizawa, S., and Tomooka, Y. (2003). Characterization of newly established colonal oviductal cell lines and differential hormonal regulation of gene expression. *In Vitro Cell. Dev. Biol.* **39,** 146–156.

CHAPTER TWENTY-TWO

Computational Quantification of Fluorescent Leukocyte Numbers in Zebrafish Embryos

Felix Ellett*,†,‡ *and* Graham J. Lieschke*,†,‡

Contents

1. Introduction 426
2. Computing LUs 427
   2.1. Imaging 427
   2.2. Image processing 428
   2.3. LU calculation 430
   2.4. Relationship between LUs and actual leukocyte numbers 430
3. Example of Applicability: Enumerating Leukocyte Populations Following Perturbation of Macrophage/Neutrophil Specification by *irf8* Misexpression 433
Acknowledgments 433
References 434

Abstract

Fluorescent transgenes with leukocyte-restricted expression have been essential to recent studies using zebrafish models of inflammation and the innate immune response. Many of the experiments performed using these models involve quantifying changes in the number of fluorescent leukocytes. Here, we describe a tool for deriving a quantitative variable proportional to fluorescent leukocyte numbers from single-plane fluorescent digital images of whole live embryos. The parameter, called "Leukocyte units," provides reliable values linearly proportional to actual leukocyte numbers in the range 50–400

* Australian Regenerative Medicine Institute, Monash University, Clayton, Victoria, Australia
† Cancer and Haematology Division, Walter and Eliza Hall Institute of Medical Research, Parkville, Victoria, Australia
‡ Department of Medical Biology, University of Melbourne, Parkville, Victoria, Australia

Methods in Enzymology, Volume 506
ISSN 0076-6879, DOI: 10.1016/B978-0-12-391856-7.00046-9

leukocytes/embryo, and its performance at higher leukocyte densities remains linear. Its usefulness is demonstrated by scoring changes in leukocyte numbers following perturbation of a characterized leukocyte specification pathway.

1. Introduction

Zebrafish are an excellent model for studying hematopoiesis and innate immunity (Ellett and Lieschke, 2010; Meeker and Trede, 2008; Meijer and Spaink, 2011; van der Sar *et al.*, 2004). The first zebrafish leukocyte precursors are specified from approximately 12–14 hours post-fertilization (hpf), during primitive hematopoiesis (Herbomel *et al.*, 1999; Lieschke *et al.*, 2002) and primitive macrophages exhibit active phagocytic and injury responses as early as 24 hpf (Herbomel *et al.*, 1999). Later, leukocytes arise from definitive hematopoietic stem cells that first arise from hemogenic endothelium in the ventral wall of the dorsal aorta (Bertrand *et al.*, 2010; Kissa and Herbomel, 2010; Lam *et al.*, 2010) and then go on to seed later sites of hematopoiesis, namely the caudal hematopoietic tissue and the pronephros (Murayama *et al.*, 2006).

The behavior of leukocytes in response to injury or infection can be visualized and compared *in vivo* using transgenes that drive fluorophore expression in leukocyte subpopulations. Such transgenic zebrafish have contributed to numerous observations about the behavior of leukocytes during inflammation (Cvejic *et al.*, 2008; d'Alencon *et al.*, 2010; Elks *et al.*, 2011; Ellett *et al.*, 2011; Gray *et al.*, 2011; Hall *et al.*, 2007; Lin *et al.*, 2009; Mathias *et al.*, 2006, 2007, 2009; Meijer *et al.*, 2008; Oehlers *et al.*, 2011; Redd *et al.*, 2006; Renshaw *et al.*, 2006; Yoo and Huttenlocher, 2011).

Leukocyte numbers may be altered following genetic manipulation of specification and differentiation pathways (Li *et al.*, 2010; Liongue *et al.*, 2009; Rhodes *et al.*, 2005), during chronic inflammatory responses (Mathias *et al.*, 2007), or in response to exogenous stimuli such as infection. Customarily, fluorescent leukocyte numbers have been quantified by direct counting, which, though accurate, is laborious and time consuming. A computational tool for quantifying fluorescent leukocyte numbers from digital images would simplify collecting such data and facilitate comparisons between and within experimental groups over time. This type of approach has been described for enumerating other fluorescent signals in zebrafish embryos that are difficult to enumerate by cell counting, for example, acridine orange-stained apoptotic cells (Tucker and Lardelli, 2007). Here, we describe the computation of a parameter that is proportional to leukocyte numbers, derived from high-contrast, single-plane fluorescent images acquired on a dissecting microscope processed using publically available software. We have called the parameter "Leukocyte units" (LUs).

2. Computing LUs

2.1. Imaging

2.1.1. Required materials

- Transgenic zebrafish with fluorophore expression restricted to leukocyte population of interest
- Fluorescence dissecting microscope with camera (e.g., Olympus SZX16 with DP71 camera or similar, operating DP Controller (Version 3.3.1.292) acquisition software)
- E3 zebrafish culture medium (5 m*M* NaCl, 0.17 m*M* KCl, 0.33 m*M* $CaCl_2$, 0.33 m*M* $MgSO_4$ in dH_2O) (Nusslein-Volhard and Dahm, 2002).
- 100× 1-phenyl-2-thiourea (PTU) stock solution (E3 with 0.3%, w/v, PTU; Sigma Aldrich)
- 25× tricaine stock solution (0.4%, w/v, tricaine (ethyl 3-aminobenzoate methanesulfonic acid) in E3, adjust to pH 7 with 1 *M* Tris–HCl (pH 9))

Additional disposables

- 90 mm Petri dishes
- Plastic transfer pipettes

2.1.2. Culturing of zebrafish embryos

- Raise embryos according to standard procedures (Westerfield, 2000), initially in E3 (without PTU), then transfer to 0.003% PTU in E3 at approximately 8–12 hpf to inhibit pigment formation.
- Avoid use of methylene blue as an antifungal agent as it can lead to brighter yolk autofluorescence.
- Dechorionate embryos at least 1–2 h prior to imaging. This allows embryos to straighten, increasing the in-focus region during imaging.

2.1.3. Fluorescence imaging

- Prewarm fluorescence light source for 10–15 min prior to imaging.
- Anesthetize embryos by adding 2 mL of working concentration tricaine in E3 per 90 mm dish of embryos. Allow 4–5 min for tricaine to take effect.
- If there are extraneous fluorescent spots on the culture dish, transfer to a fresh dish prior to imaging.
- Select a magnification that allows imaging of entire individual embryos (usually 32–40×). If a timecourse is intended, choose a lower magnification

that will allow subsequent imaging of older, larger embryos using the same magnification.

- Manipulate the embryo so that it can be imaged laterally, ideally so that the entire trunk is in the same focal plane.
- Ensure that the aperture is fully open. This increases the amount of light reaching the camera, allowing shorter exposure times to be used. If the fluorescence is very strong, decreasing the aperture can result in better in-focus images by increasing the depth of field. Since the embryos are not mounted during this procedure, short exposure times are necessary to reduce blurring caused by the movement of embryos.
- Using the appropriate fluorescence filter, bring into focus the fluorescent leukocytes in the trunk and tail of the embryo, selecting a focal plane by eye which provides an impression that a majority of fluorescent cells are in focus.
- Choose an exposure time that allows clear delineation of cells, while minimizing background fluorescence (Fig. 22.1A,i).
- If the camera software allows black-balance correction, this can be useful for removing autofluorescence.
- Capture image. We routinely use 1360 × 1024 RGB color tif format. Other formats can be used but may not work as smoothly in subsequent ImageJ processing steps described in Section 2.2.
- Repeat the process, using the same camera settings for all embryos and experimental groups.

2.2. Image processing

2.2.1. Required software

- ImageJ 1.42q or later (free download from http://rsbweb.nih.gov/ij/download.html)
- Microsoft Excel 12.2.8 or later

2.2.2. Image processing and measurements

- Open images using ImageJ. It is useful to approach this one experimental group at a time to reduce errors. We routinely use 5–20 images per experimental group.
- To convert RGB color images to binary, select: Process > Binary > Make Binary.
- This results in a binary image with a white background and black foreground (Fig. 22.1A,ii). Creating a shortcut for this command speeds up this process substantially.

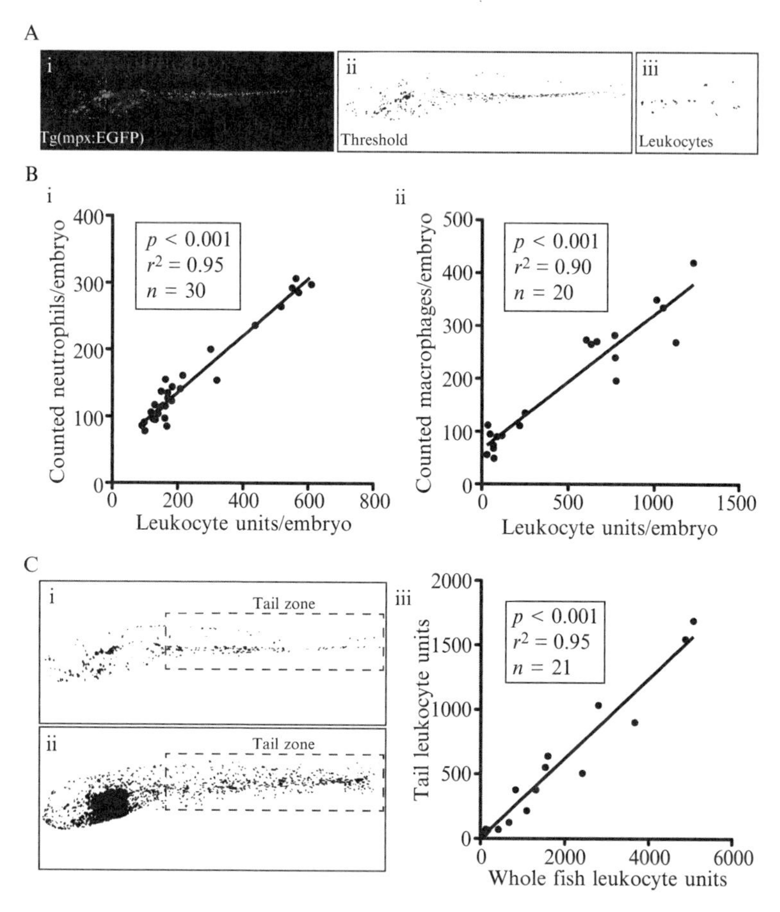

Figure 22.1 Steps in Leukocyte unit (LU) calculation and construction of standard curves. (A) Conversion of fluorescence images (i), to binary images (ii), allows measurement of fluorescence area in pixels. By dividing whole-body area by the average area of a single cell for that group (iii), the LU value, proportional to the number of leukocytes in each embryo, can be estimated. (B) Comparison of manually counted numbers of neutrophils (i) and macrophages (ii) to their respective calculated LU values, demonstrating linear correlation between the two values in the range 50–400 leukocytes/embryo. LUs are an overestimation of actual cell numbers in both examples. Neutrophil and macrophage values in (i) and (ii) were based on imaging green Tg(*mpx*:EGFP) and red photoconverted Tg(*mpeg1*:Gal4/UAS:Kaede) embryos, respectively, of age 2–6 dpf. (C) Comparison of whole-animal to tail zone only (dashed boxes) LUs over a range of embryos with low (i) and high (ii) amounts of fluorescent overlap shows that these values have a strongly linear correlation ($r^2 = 0.95$). This suggests that overlapping fluorescence does not skew estimations of cell numbers over the range tested. Linear regression analysis performed using GraphPad Prism.

- Once images have been made binary, the total foreground/fluorescent area can be measured. Prior to measuring, change the measurement settings using: Analyze > Set Measurements. Uncheck all boxes except "area" and "limit to threshold" and set decimal places to 3 (default).
- To measure total foreground area, select all using the "⌘A" shortcut, then select: Analyze > Measure. Alternatively, use the "⌘M" shortcut.
- Repeat for all images in the experimental group. This will produce a results window containing area measurements for each embryo.
- Cut and paste the results into an Excel spreadsheet for processing (see Section 2.3).
- To measure leukocyte size, zoom in (using "⌘+," to 200%) onto the cells on the tip of the tail (using "⌘+"). Note: ImageJ will zoom to the location of the mouse.
- Using the "wand (tracing) tool" on the toolbar, select an individual cell in the tail and measure ("⌘M") (Fig. 22.1A,iii).
- Repeat for five cells per embryo for five embryos per experimental group.
- Cut and paste these 25 measurements into an Excel spreadsheet for processing (see Section 2.3).

2.3. LU calculation

Once "total foreground" and "leukocyte size" measurements have been taken for each group and entered into an Excel spreadsheet, calculation of LU values for individuals in that group is simple.

- Divide the "total foreground" value of each embryo by the average "leukocyte size" for the group to which it belongs.
- If this assay is to be used frequently, it is easiest to set up an Excel template with calculations pre-entered so that data can be entered directly and LU values easily transformed into graphs or exported to separate graphing or statistics programs for further analysis.

2.4. Relationship between LUs and actual leukocyte numbers

Plotting of calculated LUs versus manually counted neutrophil and macrophage numbers demonstrates that the relationship between these values is approximately linear in the range 50–300 neutrophils/embryo and 50–400 macrophages/embryo (Fig. 22.1B). This shows that although LUs behave as a quantitative variable with an approximately linear proportionality over this range, the value itself is an overestimation of actual leukocyte numbers. By creating such a "standard curve" for each system, however, it is possible to calculate a better estimation of actual leukocyte numbers. For example, these plots for neutrophils and macrophages reveal a 2.3 and 3.9 times overestimation, respectively. If absolute leukocyte values are required, the variation between imaging systems, cell types and transgenes will require the calculation of standard curves for each experiment.

When the number of fluorescent neutrophils per embryo is large, such as in response to an infection, the number of overlapping and out of focus cells increase, potentially impacting on the proportionality between LUs and actual leukocyte numbers. Of particular concern in such embryos is the confluence of fluorescence and autofluorescence over the yolk, which was potentially a major artifact that would degrade the linear relationship between LUs and absolute leukocyte numbers. Whether this was an issue or not was assessed by comparing the relationship between LU values determined from whole embryo imaging with LU values derived just from the trunk/tail region of the same embryo, a region that is more in focus and that is free of yolk-related artifacts (Fig. 22.1C). The linear correlation observed between values generated for these two regions ($r^2 = 0.95$) suggests that these phenomena did not significantly corrupt the linearity of whole embryo LU values over the range of LUs tested. However, in the absence of extensive experience with high LU values, we suggest that using representative zonal LU values as a surrogate for whole embryo LUs should be considered in scenarios where very high leukocyte densities occur.

When the number of fluorescent leukocytes is low, or the transgene is not expressed highly in relation to background, background autofluorescence may interfere with the thresholding steps. This is due to the inability of the isodata algorithm used by ImageJ to clearly distinguish two peaks of pixel intensity within the image that correspond to background and foreground signals.

When leukocyte numbers in test groups are low, such as after cell ablation or interference with lineage specification, thresholding may be problematic. In this case, the control group should contain enough cells to enable accurate automatic thresholding, enabling an appropriate threshold setting generated for an image from the control group to be used to manually set the threshold for the corresponding test group.

When there are low fluorescence levels or high background fluorescence, manually setting the threshold may be required, although settings should be consistent between test and control groups. Comparison to a nontransgenic negative control may allow a lower threshold level to be set manually with greater accuracy. Again, limiting the analysis to a region with lower background (e.g., the tail rather than the head/yolk region) may allow better signal differentiation to be obtained.

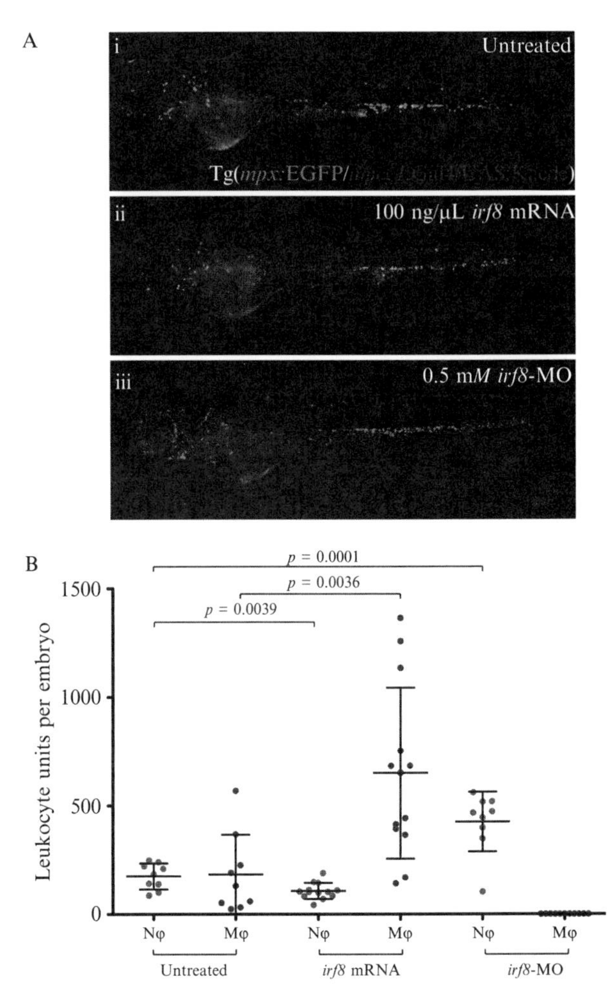

Figure 22.2 Using leukocyte units to document changes in leukocyte populations resulting from *irf8* misexpression. (A) Images demonstrating the altered balance of macrophage/neutrophil specification following *irf8* overexpression and knockdown in Tg(*mpeg1*:Gal4/UAS:Kaede/*mpx*:EGFP) embryos. In this compound transgenic line, macrophages fluoresce red from photoconverted Kaede with its lineage specificity conferred by the *mpeg1* promoter (Ellett *et al.*, 2011) and neutrophils fluoresce green from an EGFP transgene driven by the *mpx* promoter (Renshaw *et al.*, 2006). This arrangement allows for comparison not only between treatment groups but also between cell types. Compared to untreated groups (i), knockdown of *irf8* results in loss of macrophage specification and increased granulopoiesis in *irf8*-MO treated embryos (ii), while overexpression by *irf8* mRNA treatment leads to increased macrophage specification and reduced granulopoiesis (iii). (B) Graph displays LUs for each leukocyte-type and each treatment group. As has been described (Li *et al.*, 2010), *irf8* overexpression resulted in a small but significant decrease in neutrophil (Nφ) specification and a striking increase in macrophage (Mφ) specification. Conversely, *irf8* knockdown resulted in a complete loss of the macrophage lineage and a significant increase in

3. Example of Applicability: Enumerating Leukocyte Populations Following Perturbation of Macrophage/Neutrophil Specification by *irf8* Misexpression

During zebrafish primitive hematopoiesis, the specific cellular fate of myeloid precursors arising from the anterior lateral plate mesoderm (later termed the rostral blood island) is influenced by the activity of *irf8*, where expression of *irf8* drives specification of precursor cells toward a macrophage rather than a neutrophil fate (Li *et al.*, 2010).

To demonstrate and quantify these changes *in vivo* using LUs, Tg(*mpx*:EGFP/*mpeg1*:Gal4/UAS:Kaede) compound transgenic zebrafish embryos were injected at the one-cell stage with either *irf8*-MO (0.5 m*M*) or *irf8* mRNA (50 ng/μL). These embryos will develop into adults with neutrophils labeled in green (EGFP) and macrophages in red (photoconverted Kaede). Quantification of each leukocyte lineage was performed as described above and analyzed using GraphPad Prism (Fig. 22.2B). As previously reported (Li *et al.*, 2010), knockdown of *irf8* resulted in a complete loss of macrophages and an increase in neutrophil specification, whereas overexpression of *irf8* resulted in a striking increase in macrophage numbers with a corresponding small but significant decrease in neutrophil numbers (Fig. 22.2A). These changes were demonstrated by LU quantification and amenable to statistical analysis (Fig. 22.2B).

ACKNOWLEDGMENTS

The authors thank Ethan Scott, Herwig Baier, and Stephen Renshaw for transgenic lines; Li Li and Zilong Wen for *irf8* reagents; Mark Greer and Julian Cocks for animal care in the aquarium; Ben Kile, Alex Andrianopoulos, Joan Heath, and Luke Pase for advice. Microscopy used instruments in the Monash Micro Imaging facility and the Walter and Eliza Hall Institute of Medical Research. This work was supported by the National Institutes of Health (Grant R01 HL079545) and the National Health and Medical Research Council (Grant 637394; G. J. L.). F. E. was supported by an Australian Postgraduate Award and Walter and Eliza Hall Institute of Medical Research Edith Moffatt Scholarship. WEHI receives infrastructure support from the Commonwealth NHMRC Independent Research Institutes Infrastructure Support Scheme (361646) and a Victorian State Government Operational Infrastructure Support Scheme grant. The Australian Regenerative Medicine Institute is supported by grants from the State Government of Victoria and the Australian Government.

granulopoiesis. Bars represent mean ± standard deviation. *p*-Values are from a two-tailed *t*-test, $n \geq 9$ embryos/group. Distribution normality confirmed by Kolmogorov–Smirnov test (with Dallal–Wilkinson–Lilliefor *p* value) using GraphPad Prism. (See Color Insert.)

REFERENCES

Bertrand, J. Y., Chi, N. C., Santoso, B., Teng, S., Stainier, D. Y., and Traver, D. (2010). Haematopoietic stem cells derive directly from aortic endothelium during development. *Nature* **464,** 108–111.

Cvejic, A., Hall, C., Bak-Maier, M., Flores, M. V., Crosier, P., Redd, M. J., and Martin, P. (2008). Analysis of WASp function during the wound inflammatory response–Live-imaging studies in zebrafish larvae. *J. Cell Sci.* **121,** 3196–3206.

d'Alencon, C. A., Pena, O. A., Wittmann, C., Gallardo, V. E., Jones, R. A., Loosli, F., Liebel, U., Grabher, C., and Allende, M. L. (2010). A high-throughput chemically induced inflammation assay in zebrafish. *BMC Biol.* **8,** 151.

Elks, P. M., van Eeden, F. J., Dixon, G., Wang, X., Reyes-Aldasoro, C. C., Ingham, P. W., Whyte, M. K., Walmsley, S. R., and Renshaw, S. A. (2011). Activation of Hif-1alpha delays inflammation resolution by reducing neutrophil apoptosis and reverse migration in a zebrafish inflammation model. *Blood* **118,** 712–722.

Ellett, F., and Lieschke, G. J. (2010). Zebrafish as a model for vertebrate hematopoiesis. *Curr. Opin. Pharmacol.* **10,** 563–570.

Ellett, F., Pase, L., Hayman, J. W., Andrianopoulos, A., and Lieschke, G. J. (2011). mpeg1 promoter transgenes direct macrophage-lineage expression in zebrafish. *Blood* **117,** e49–e56.

Gray, C., Loynes, C. A., Whyte, M. K., Crossman, D. C., Renshaw, S. A., and Chico, T. J. (2011). Simultaneous intravital imaging of macrophage and neutrophil behaviour during inflammation using a novel transgenic zebrafish. *Thromb. Haemost.* **105,** 811–819.

Hall, C., Flores, M. V., Storm, T., Crosier, K., and Crosier, P. (2007). The zebrafish lysozyme C promoter drives myeloid-specific expression in transgenic fish. *BMC Dev. Biol.* **7,** 42.

Herbomel, P., Thisse, B., and Thisse, C. (1999). Ontogeny and behaviour of early macrophages in the zebrafish embryo. *Development* **126,** 3735–3745.

Kissa, K., and Herbomel, P. (2010). Blood stem cells emerge from aortic endothelium by a novel type of cell transition. *Nature* **464,** 112–115.

Lam, E. Y., Hall, C. J., Crosier, P. S., Crosier, K. E., and Flores, M. V. (2010). Live imaging of Runx1 expression in the dorsal aorta tracks the emergence of blood progenitors from endothelial cells. *Blood* **116,** 909–914.

Li, L., Jin, H., Xu, J., Shi, Y., and Wen, Z. (2010). Irf8 regulates macrophage versus neutrophil fate during zebrafish primitive myelopoiesis. *Blood* **117,** 1359–1369.

Lieschke, G. J., Oates, A. C., Paw, B. H., Thompson, M. A., Hall, N. E., Ward, A. C., Ho, R. K., Zon, L. I., and Layton, J. E. (2002). Zebrafish SPI-1 (PU.1) marks a site of myeloid development independent of primitive erythropoiesis: Implications for axial patterning. *Dev. Biol.* **246,** 274–295.

Lin, A., Loughman, J. A., Zinselmeyer, B. H., Miller, M. J., and Caparon, M. G. (2009). Streptolysin S inhibits neutrophil recruitment during the early stages of Streptococcus pyogenes infection. *Infect. Immun.* **77,** 5190–5201.

Liongue, C., Hall, C. J., O'Connell, B. A., Crosier, P., and Ward, A. C. (2009). Zebrafish granulocyte colony-stimulating factor receptor signaling promotes myelopoiesis and myeloid cell migration. *Blood* **113,** 2535–2546.

Mathias, J. R., Perrin, B. J., Liu, T. X., Kanki, J., Look, A. T., and Huttenlocher, A. (2006). Resolution of inflammation by retrograde chemotaxis of neutrophils in transgenic zebrafish. *J. Leukoc. Biol.* **80,** 1281–1288.

Mathias, J. R., Dodd, M. E., Walters, K. B., Rhodes, J., Kanki, J. P., Look, A. T., and Huttenlocher, A. (2007). Live imaging of chronic inflammation caused by mutation of zebrafish Hai1. *J. Cell Sci.* **120,** 3372–3383.

Mathias, J. R., Dodd, M. E., Walters, K. B., Yoo, S. K., Ranheim, E. A., and Huttenlocher, A. (2009). Characterization of zebrafish larval inflammatory macrophages. *Dev. Comp. Immunol.* **33,** 1212–1217.

Meeker, N. D., and Trede, N. S. (2008). Immunology and zebrafish: Spawning new models of human disease. *Dev. Comp. Immunol.* **32,** 745–757.

Meijer, A. H., and Spaink, H. P. (2011). Host-pathogen interactions made transparent with the zebrafish model. *Curr. Drug Targets* **12,** 1000–1017.

Meijer, A. H., van der Sar, A. M., Cunha, C., Lamers, G. E., Laplante, M. A., Kikuta, H., Bitter, W., Becker, T. S., and Spaink, H. P. (2008). Identification and real-time imaging of a myc-expressing neutrophil population involved in inflammation and mycobacterial granuloma formation in zebrafish. *Dev. Comp. Immunol.* **32,** 36–49.

Murayama, E., Kissa, K., Zapata, A., Mordelet, E., Briolat, V., Lin, H. F., Handin, R. I., and Herbomel, P. (2006). Tracing hematopoietic precursor migration to successive hematopoietic organs during zebrafish development. *Immunity* **25,** 963–975.

Nusslein-Volhard, C., and Dahm, R. (2002). Zebrafish—Practical Approach. pp. 237–281. Oxford University Press, New York.

Oehlers, S. H., Flores, M. V., Okuda, K. S., Hall, C. J., Crosier, K. E., and Crosier, P. S. (2011). A chemical enterocolitis model in zebrafish larvae that is dependent on microbiota and responsive to pharmacological agents. *Dev. Dyn.* **240,** 288–298.

Redd, M. J., Kelly, G., Dunn, G., Way, M., and Martin, P. (2006). Imaging macrophage chemotaxis *in vivo*: Studies of microtubule function in zebrafish wound inflammation. *Cell Motil. Cytoskeleton* **63,** 415–422.

Renshaw, S. A., Loynes, C. A., Trushell, D. M., Elworthy, S., Ingham, P. W., and Whyte, M. K. (2006). A transgenic zebrafish model of neutrophilic inflammation. *Blood* **108,** 3976–3978.

Rhodes, J., Hagen, A., Hsu, K., Deng, M., Liu, T. X., Look, A. T., and Kanki, J. P. (2005). Interplay of pu.1 and gata1 determines myelo-erythroid progenitor cell fate in zebrafish. *Dev. Cell* **8,** 97–108.

Tucker, B., and Lardelli, M. (2007). A rapid apoptosis assay measuring relative acridine orange fluorescence in zebrafish embryos. *Zebrafish* **4,** 113–116.

van der Sar, A. M., Appelmelk, B. J., Vandenbroucke-Grauls, C. M., and Bitter, W. (2004). A star with stripes: Zebrafish as an infection model. *Trends Microbiol.* **12,** 451–457.

Westerfield, M. (2000). The Zebrafish Book. A Guide for the Laboratory Use of Zebrafish (Danio rerio). 4th edn University of Oregon Press, Eugene.

Yoo, S. K., and Huttenlocher, A. (2011). Spatiotemporal photolabeling of neutrophil trafficking during inflammation in live zebrafish. *J. Leukoc. Biol.* **89,** 661–667.

CHAPTER TWENTY-THREE

Four-Dimensional Tracking of Lymphocyte Migration and Interactions in Lymph Nodes by Two-Photon Microscopy

Masahiro Kitano *and* Takaharu Okada

Contents

1. Introduction 438
2. Sample Preparation 439
   2.1. Labeling of target cells 439
   2.2. Adoptive transfer and immunization 440
3. Imaging Preparations and Data Acquisition 444
   3.1. Lymph node explants for live imaging 444
   3.2. Image data acquisition 445
   3.3. Postimaging samples 447
4. Data Analysis 447
   4.1. General workflow of cell tracking analysis 447
   4.2. Analysis of pregerminal center B cell entry to the germinal center clusters 451
   4.3. Analysis of B cell–T cell interactions 452
Acknowledgments 452
References 453

Abstract

Two-photon microscopy of live tissue imaging provides insightful information about the four-dimensional dynamics of cell behavior and has contributed to the discoveries of new biological mechanisms including those in the immunology field. In recent years, it has become easier for many researchers to perform the tissue imaging experiments, due to the refinement of the commercially available microscope systems. However, it is still crucial for the efficient visualization of biological events to optimize the sample preparation by using the best available reagents. Further, it is equally important to elicit key information from

Research Unit for Immunodynamics, RIKEN, Research Center for Allergy and Immunology, Yokohama, Japan

Methods in Enzymology, Volume 506
ISSN 0076-6879, DOI: 10.1016/B978-0-12-391856-7.00047-0

the large quantity of imaging data, whose handling is often complex and laborious. Here, by taking as an example the two-photon analysis of lymphocyte migration and interaction in explanted lymph nodes, we describe the key points that need to be considered for the successful experimental design, sample preparation, image acquisition, and data analysis.

1. Introduction

Until a decade ago, dynamic tissue imaging in the immunological studies was limited to intravital microscopy of leukocyte locomotion promoted by blood or lymph flow. More autonomous behavior of immune cells in tissue parenchyma was studied mostly through evaluation of cell localization changes in tissue samples fixed at various time points of immunological events. Even though such studies have constructed a large part of the current knowledge about dynamics of the immune system, direct observation of immune cell migration *in vivo* is still important for the understanding of precise mechanisms. Recent advances in time-lapse imaging techniques, especially, since two-photon excitation microscopy was introduced to the analysis of lymphocyte migration in intact lymph nodes (Miller *et al.*, 2002), have made it possible to study the four-dimensional immune cell behavior in intact tissues. With this powerful method, on one hand, we can confirm and extend currently accepted notions; for example, direct interactions of cognate antigen-engaged B cells and T cells in lymph nodes, whose location had been suggested by the snapshot images (Garside *et al.*, 1998), were demonstrated to occur with fascinating dynamics during antibody responses by the dynamic two-photon imaging of antigen-draining lymph nodes (Okada *et al.*, 2005). On the other hand, this new technique has also made contradictory findings to the previous understandings; for example, the integrin molecules, which had been thought to play crucial roles for cell motility, were shown to be largely dispensable for leukocyte interstitial migration (Lämmermann *et al.*, 2008). The unexpected findings obtained through two-photon imaging have lead to the propositions of new biological mechanisms, highlighting the importance of this technique.

Two-photon microscopes used to be only available in limited laboratories. Many of them were custom-built, and their use required expertise unfamiliar to most of biologists. Now that the commercially available systems have been improved for its suitability to deep-tissue imaging and its user-friendliness, the number of researchers using two-photon live imaging is increasing. In order to take full advantage of the improved two-photon microscopes, however, researchers need to carefully design the sample preparation and optics settings suitable to it. In this chapter, we show a series of experimental procedures for imaging of activated B cell migration and B cell–T cell interaction in intact lymph nodes during the germinal center formation process (Kitano *et al.*, 2011) and describe the important points in the imaging procedure, as well as in the sample preparation several days prior to the data acquisition. Equally important as data acquisition is data analysis. Thanks to the recent advances of four-dimensional imaging analysis software, the accuracy of automated cell tracking has been improved. However, manual processing of automated tracking data is still required for verification of tracks and for quantification of events like cell interactions and entry into tissue subdomains. Here, we describe our methods of data analysis using the Imaris (Bitplane) software. Several preceding technical reviews including the recent article published as a chapter of Methods in Enzymology (Zinselmeyer *et al.*, 2009) have nicely described the imaging of immune cell dynamics using two-photon microscopy. We focus on our specific design of the sample preparation, data acquisition, and data analysis without overlapping too much with the contents of the preceding reviews, and aim to extract technical tips that may be generally useful for analysis of lymphocyte dynamics in the lymph node.

2. Sample Preparation

2.1. Labeling of target cells

Two kinds of signals can be detected by two-photon microscopy, fluorescence and emission derived from second harmonic generation (SHG), the nonlinear optical process that occurs when photons interact with noncentrosymmetrical materials like collagen fibers. SHG gives emissions with the half wavelength of incident light and can provide us tissue structural information without any treatment to the sample. A recent report also shows the utilization of $BaTiO_3$ nanocrystals for detecting the probe-incorporating cells as well as labeling of endogenous molecules *in vivo* (Pantazis *et al.*, 2010). In order to visualize dynamics of various immune cells in complex tissue structures, it would be best to detect signals from multiple fluorophores in combination with SHG signal. Labeling of cells with fluorophores is achieved by utilization of chemically synthesized fluorescent dyes and/or exogenous expression of fluorescent proteins.

In this chapter, we mainly describe the procedures for monitoring the adoptively transferred lymphocytes in the lymph nodes of recipient mice. Visualization of endogenous immune cells by using transgenic animals that express fluorescent proteins in the cell type-specific manner has been described elsewhere (Zinselmeyer *et al.*, 2009). To visualize transferred cells in recipient mice, donor cells are labeled with appropriate fluorophores prior to the transfer. The standard method for the labeling of donor cells with commercially available fluorescent dyes was previously described (Zinselmeyer *et al.*, 2009). In many experiments, we label cells with 5-(and-6)-carboxyfluorescein diacetate, succinimidyl ester (CFSE, Invitrogen), and/or 5-(and-6)-(((4-chloromethyl)benzoyl)amino)tetramethylrhodamine (CellTracker™ Orange CMTMR, Invitrogen). Due to the efficient excitation by the two-photon laser at the same wavelength (800–950 nm) and enough emission signals, these dyes are quite useful for monitoring the behavior of adoptively transferred cells. However, they are not suitable for the tracking of proliferating cells, because they become diluted through cell divisions. This problem can be circumvented by using various—cyan, green, yellow, and red—fluorescent protein-expressing cells. A simple strategy is to purify and transfer cells of interest from transgenic mice constitutively expressing fluorescent proteins. We use ubiquitin-GFP (JAX: 004353) and β-actin-CFP (JAX: 004218) transgenic lines crossed with other transgenic models according to the purpose of each experiment. Lymphocytes from heterozygous ubiquitin-GFP mice usually give enough signals for two-photon microscopy regardless of their activation/differentiation status. Fluorescence of β-actin-CFP lymphocytes is generally less intense, and it is recommended to generate homozygous animals. In addition, CFP expression levels in these lymphocytes change during their activation and proliferation. Naive T cells express detectable amounts of CFP, and antigen-engaged T cells increase CFP expression during their proliferation. Naive B cells also express detectable CFP, but antigen-engaged B cells seem to decrease CFP expression transiently and restore it during their proliferation.

We routinely visualize SHG, CFP, GFP (or CFSE), and CMTMR signals. By using the optics setting shown in Fig. 23.1A, simultaneous identification of cells with these fluorophores can be satisfactorily achieved. By using a different filter set, even GFP and YFP signals can be simultaneously visualized (Fig. 23.1B). This filter set is most likely useful for tracking of GFP-reporter expression in YFP-expressing cells. A drawback of this approach is that GFP signal intensities would be reduced because of the narrow range of the band-pass filter.

2.2. Adoptive transfer and immunization

For visualizing the behavior of transferred lymphocytes during the immune response, it is important to optimize the number of antigen-specific cells to be transferred. In the case of monitoring lymphocytes before proliferation in lymph nodes, transferring 5×10^5–2×10^6 cells per recipient should result in the appropriate cell density in the imaging field. If anti-CD62L antibody (clone: Mel-14) is administered several hours after the transfer to restrict the subsequent lymphocyte homing to lymph nodes (Mempel *et al.*, 2004), it is recommended to transfer at least 5×10^6 cells per recipient. If imaging is to be conducted at 2–3 days postimmunization, the number of antigen-specific cells transferred should be reduced to 1×10^4–1×10^5 cells for cell tracking analysis of individual lymphocytes. When a tissue region of interest is expected to be packed

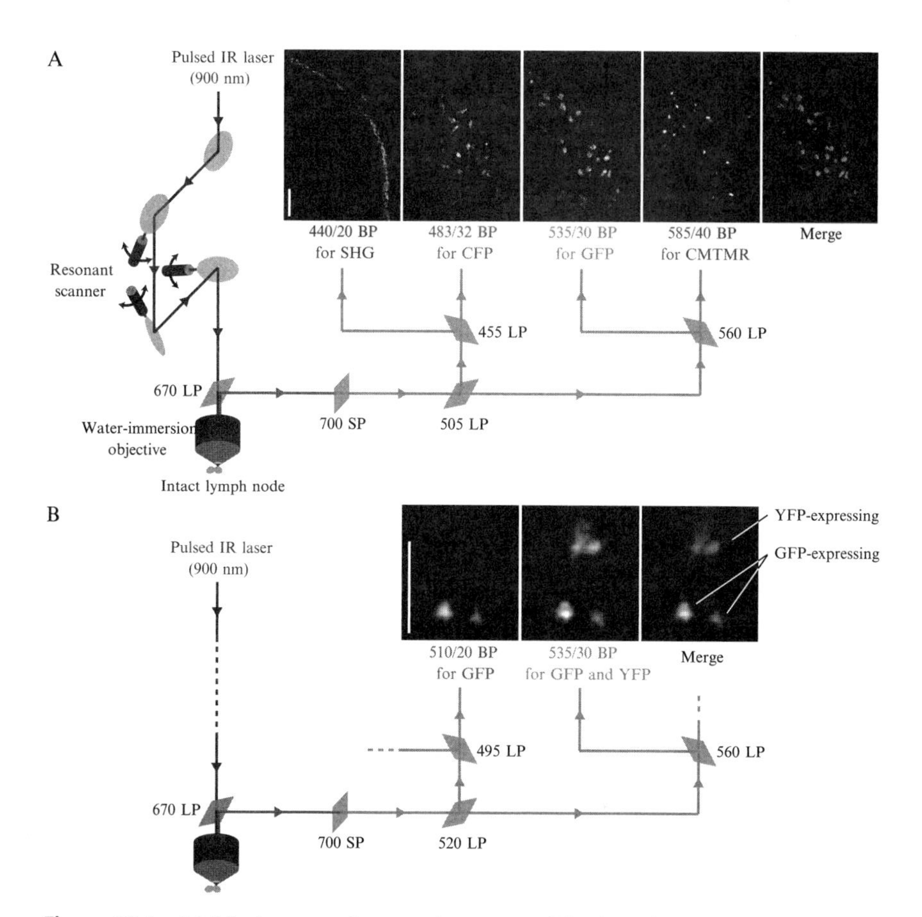

Figure 23.1 Multicolor two-photon microscopy. (A) The schematic diagram shows one of the most frequently used configurations of our two-photon microscope. The combination of 900 nm laser, dichroic mirrors, and band-pass filters shown here enables simultaneous observation and discrimination of SHG signals and fluorescence signals from CFP, GFP, and CMTMR. Scale bar: 50 μm. (B) An optical filter selection to visualize GFP and YFP fluorescence. Scale bar: 30 μm. (For color version of this figure, the reader is referred to the Web version of this chapter.)

with the transferred antigen-specific cells (like germinal centers), cotransfer of the antigen-specific cells expressing fluorescent protein (1×10^3–1×10^4) together with a larger number of nonfluorescent cells with the same antigen specificity (1×10^4–1×10^5) is a useful strategy to be able to track individual cells in the region (Allen *et al.*, 2007).

The other important factor is the method of immunization. Some adjuvant like complete Freund's adjuvant (CFA) induces massive swelling of lymph nodes including thickening of B cell regions. This makes it difficult to visualize deep parts of the B cell follicle where germinal centers are formed and the T cell zone located under the B cell region. When possible, other adjuvant like alum, which induces less swelling, are preferable for imaging. In order to minimize the lymph node swelling, a recent study also developed a protocol to visualize secondary germinal centers induced in lymph nodes by subcutaneous injection of antigen without adding adjuvant (Victora *et al.*, 2010).

The followings are the adoptive transfer and immunization procedures we use for the imaging of pregerminal center B cell migration near the nascent germinal center cluster. In this experiment, it is required to visualize the germinal center B cell cluster and pregerminal center B cells simultaneously in different colors. As shown in Fig. 23.2A, CFP^{+} B cells and GFP^{+} B cells, both of which have the same transgenic B cell receptor (BCR) against the model-protein antigen hen egg lysozyme (HEL), are sequentially exposed to antigen stimulation in recipient mice. Then we perform imaging at the time point when CFP-expressing B cells have just formed germinal center clusters while majority of GFP-expressing cells are still in the outer follicle region (Fig. 23.2B).

Day −2.5: *1st adoptive cell transfer.* B cells are purified from the double-transgenic mice harboring the MD4 (HEL-specific BCR transgenic, JAX: 002595) and homozygous β-actin-CFP transgenes by using CD43-microbeads (Miltenyi Biotech) according to the manufacture's guide. To increase the reproducibility of the time course of the germinal center response, we cotransfer cognate-antigen-specific helper T cells, in this case, nonfluorescent OT-II (OVA-specific TCR transgenic) T cells, which need not be purified. Check the OT-II cells by flow cytometry to determine the density of TCR $V\alpha2^{+}$ $V\beta5^{+}$ T cells. CFP^{+} MD4 B cells (1×10^{5} per head) and OT-II $CD4^{+}$ $V\alpha2^{+}$ $V\beta5^{+}$ T cells (1×10^{5} per head) are injected to wild-type C57BL/6 (B6) mice intravenously.

Day −1.5: *Immunization.* HEL-OVA (proteins chemically conjugated and purified in our laboratory, see Allen *et al.*, 2007) is prepared at 1 mg/ml in PBS. Alum (Imject Alum, Thermo Scientific) is added dropwise to the HEL-OVA solution (1:1 volume ratio) while vortex-mixing, and the resultant emulsion is continued to be mixed with a rotator at least for 30 min. The emulsion is subcutaneously injected to the mice that have received CFP^{+} B cells and helper T cells on Day −2.5. Typically, we perform injections in the footpad, base of tail, upper flank, and scruff to induce germinal centers in popliteal, inguinal, axillary, brachial, and facial lymph nodes.

Day 0: *2nd adoptive cell transfer.* B cells are purified from MD4 ubiquitin-GFP double-transgenic mice and transferred (1×10^{5} per head) to the mice immunized on Day −1.5. It is also recommended to cotransfer a large number of ($\sim 1 \times 10^{7}$) polyclonal B cells that have been purified from wild-type B6 mice and labeled with CMTMR. Their visualization helps demarcate the outer follicle region (Fig. 23.2B). In addition, the behavior of polyclonal B cells will serve as an index to whether lymph node samples are kept in a good condition during imaging.

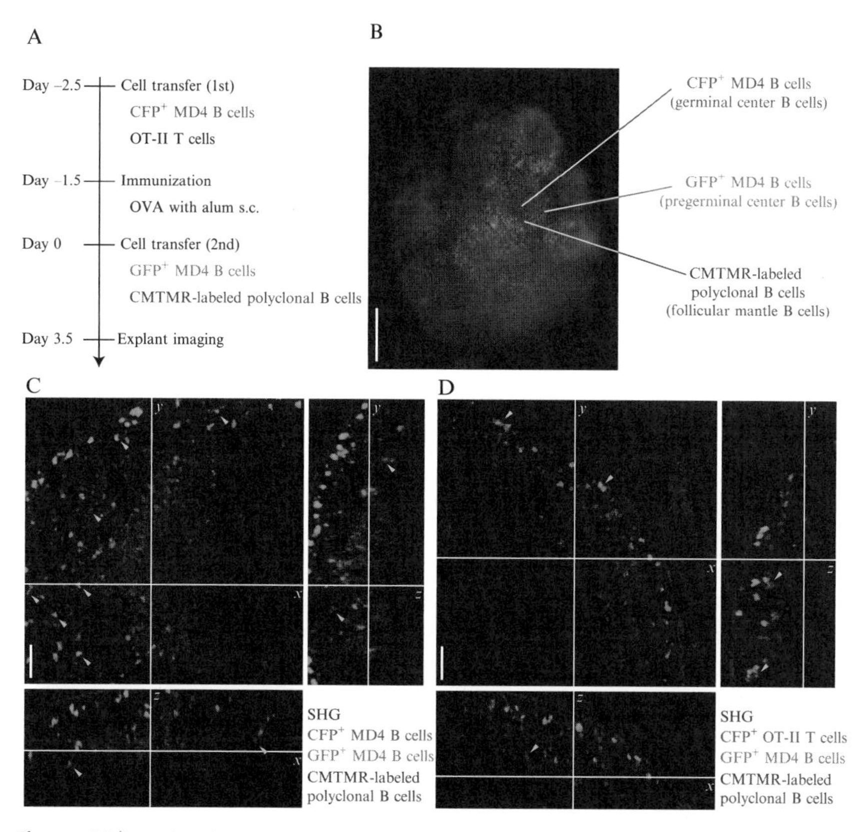

Figure 23.2 Visualization of activated B cells near the germinal center and their contacts with helper T cells. (A) A time line of the experiment to visualize germinal center B cells and pregerminal center B cells in the same lymph node. (B) A wide-field fluorescent image of an intact antigen-draining lymph node as prepared in (A). CFP^+ MD4 B cells have already formed the germinal center cluster, while majority of GFP^+ MD4 B cells are still in the outer follicle region. Magnification: 4×. Scale bar: 500 μm. (C) Two-photon imaging of B cell migration to the germinal center in the intact lymph node. *xy*-, *xz*-, and *yz*-cross sections are shown. Arrowheads indicate examples of cells with migratory appearance. Scale bar: 50 μm. (D) Two-photon imaging of B cell–T cell interaction in the intact lymph node 3 days after immunization. *xy*-, *xz*-, and *yz*-cross sections are shown. Arrowheads highlight MD4 B cells contacting OT-II T cells. Scale bar: 50 μm. (For color version of this figure, the reader is referred to the Web version of this chapter.)

Day 3–3.5: *Two-photon imaging*. See the next section.

Compared to the above protocol for imaging of pregerminal center B cells, preparation for imaging of B cell–T cell interactions is relatively simple. In brief, purified GFP^+ MD4 B cells (1×10^5 per head) and purified CFP^+ OT-II $CD4^+$ T cells (1×10^5 per head) are cotransferred to recipient mice. The method for $CD4^+$ T cell purification using magnetic beads has been introduced by video presentation (Matheu and Cahalan, 2007). CMTMR-labeled polyclonal B cells (1×10^7 per head) are simultaneously transferred. One day after the transfer, alum-absorbed HEL-OVA is injected subcutaneously as described above and wait for 1–3 days.

3. Imaging Preparations and Data Acquisition

3.1. Lymph node explants for live imaging

There are two preparation types of live tissue samples visualized under the two-photon microscope. One is the intravital imaging preparation, which enables monitoring of cells in the surgically exposed tissue in living mice under anesthesia. The intravital imaging procedures of peripheral lymph nodes have been described elsewhere (Celli *et al.*, 2008; Mempel, 2010; Zinselmeyer *et al.*, 2009). The other is the explant imaging preparation using the perfusion system. In this chapter, we focus on the two-photon imaging of explanted lymph nodes. The advantages of this method are the high reproducibility of imaging conditions and the capability to reduce the number of animals used. In the explant imaging experiments, we can screen many lymph nodes from each animal under conventional fluorescent microscopes prior to two-photon imaging (Fig. 23.2B). This preview process helps find lymph nodes that are promising for excellent visualization of the lymph node anatomy and cells in the deep area. Several studies have shown that lymphocyte motility and interaction dynamics observed by explants imaging are comparable to those from intravital imaging (Allen *et al.*, 2007; Grigorova *et al.*, 2009; Miller *et al.*, 2003, 2002). Thus, it would be reasonable to perform explant imaging as a model system to analyze lymphocyte motility and interactions, unless the imaging targets are the events known to be dependent on blood flow and/or lymphatic flow. In order to interpret results from explants imaging, it is important to address whether cell migration visualized in lymph node explants is consistent with cell localization changes that happen *in vivo*. For example, migration of antigen-engaged B cells in the lymph node explants result in their follicular exclusion like in the *in vivo* situation (Okada *et al.*, 2005). In the analysis of pregerminal center B cell dynamics in lymph node explants, which are prepared using the protocol described above, we found that entry of pregerminal center B cells into the nascent germinal center area was significantly more efficient than egress of these B cells from the germinal center, suggesting that pregerminal center B cells in the lymph node explants are clustering into the germinal center like in *in vivo* lymph nodes. Our explant imaging setups are almost identical to those previously described in detail (Zinselmeyer *et al.*, 2009). The procedures for collecting peripheral lymph nodes and mounting them on plastic cover slips have been demonstrated on video presentation (Matheu *et al.*, 2007).

3.2. Image data acquisition

The parameters important for the two-photon acquisition are following: (1) photomultiplier tube (PMT) sensitivity, (2) laser wavelength, (3) laser intensity, (4) scan averaging, (5) zooming factor, (6) *z*-interval, (7) time interval, and (8) image bit number.

(1) *PMT sensitivity*. Conventional PMTs allow us to adjust three parameters, gain, offset, and gamma. It is safe to leave the gamma value as default, although this parameter can be useful to avoid signal saturation when a great difference in intensity exists among the target populations. The offset value should be first increased or decreased from zero to the extent enough for canceling the background signal, which is homogenously observed throughout the field. Then we adjust the most important parameter, the PMT gain. In order to minimize the laser illumination to the specimen, we set the PMT gain maximum below the level where influence of shot noise becomes prominent. It is recommended to set the values to each PMT because the performance of PMTs varies among manufacturing lots.
(2) *Laser wavelength*. We routinely excite CFP, GFP, and CMTMR efficiently with 900 nm laser as described above (Fig. 23.1A). Although CFP signals may be increased by using shorter wavelength like 850, 900 nm excitation produces less background signals at least in our hands. By using the optics setting in Fig. 23.1A, we can achieve good separation of SHG and CFP signals (Fig. 23.1A), which is difficult at longer wavelength longer than 920 nm. Using shorter wavelength like 800 nm, we more often experience the phototoxicity problem, that is, cells stop moving during the data acquisition.
(3) *Laser intensity*. It is difficult to know what maximum laser power one can use without disrupting important function of cells in tissues. However, it is always recommended to make sure that motility of naive B cells or T cells in the imaging volume is comparable to that reported previously. We typically use 20–80 mW (after the objective lens) at 900 nm. If a fixed amount of the laser power is used, tissue regions close to the surface are exposed to more intense laser than deeper regions. To compensate this, the laser intensity can be automatically adjusted by the electro-optical modulator (EOM) according to *z* focus positions. However, we usually do not use this adjustment when we use the resonant scanners (see below), as the EOM adjustment slows down the scan rate. When this laser adjustment is used, one should carefully watch naive lymphocyte motility in deep regions, which may be exposed to too much laser power.
(4) *Scan averaging*. While most of the two-photon microscopes adopt the conventional galvano mirror-based scanner systems, there are several commercially available fast-scanning systems enabling near video-rate

data acquisition. The system we are currently using is the resonant mirror-based (Leica SP5), whose scanning rate (~50 frames/s at 256 × 256 pixels) is much faster than the conventional system (~3 frames/s). However, the signal-to-noise ratio in images obtained from single scanning with the resonant scanner is usually low, and it is required to obtain averaged images from at least eight scans. Even then, we can achieve 1.5–2 times faster image acquisition rate than that with the conventional galvano scanner systems.

(5) *Zooming factor.* We routinely scan the *xy* plane composed of 256 × 256 pixels at a resolution of 1.2–1.7 μm/pixel. This resolution seems to be enough for the quantification of cell motility and intercellular interaction. Increasing the zooming factor would enable the visualization of more detailed structures, because the optical resolution of GFP signals at 510 nm accomplished by our objective lens (Leica 20×, NA = 1.0) is around 0.3 μm in theory. However, as the *z*-resolution of two-photon microscopy is about 1.5 μm or worse (Scherschel and Rubart, 2008), higher *xy* resolution may not significantly improve the three-dimensionally reconstructed image. Further, the amount of illumination per unit area increases when the zooming factor is heightened. Thus, the laser power or scan averaging should be adjusted according to the zooming factor.

(6) *z-Interval.* As the *z*-resolution in our two-photon system is ~1.5 μm as described above, we set the *z*-interval to 2–3 μm in most of experiments. Because the diameter of lymphocytes is 4–10 μm (Zinselmeyer *et al.*, 2009), scanning planes hardly miss cells at the *z*-interval.

(7) *Time interval.* To track the lymphocyte migration behavior, we usually acquire the image stacks every 15–30 s. The interval longer than 30 s would compromise tracking cells faster than 10 μm/min.

(8) *Image bit number.* We use the 8-bit mode for most of the data acquisition. The higher bit numbers (10- and 12-bit modes are available in our system) tremendously increase the data volume, making the data handling cumbersome. If images need to be analyzed with a finer signal intensity scale by using the higher bit number, it is realistically required to reduce the pixel number in each *xy* plane, the number of *xy* planes in each *z*-stack, and/or the time resolution.

With the microscope settings described above, images of the lymph node samples are acquired. As shown in Fig. 23.2C, the germinal center cluster of CFP^+ MD4 B cells, GFP^+ pregerminal center MD4 B cells, naive B cells labeled with CMTMR, and the lymph node capsule structure emitting SHG signals are simultaneously monitored. The important point is to select the field covering the structure of interest (in this case, the nascent germinal center and the neighboring outer follicle region). For this purpose, it is recommended to start with the lowest digital zooming factor. The algorism to define cell centroids still works well in the low resolution (1.7 μm/pixel, Fig. 23.2C). Another point that requires the attention is the morphology of the cells. If naive lymphocytes are uniformly round-shaped, this suggests that the preparation had some problems. Migratory cell shapes (e.g., irregular shapes with cellular processes like uropods, as shown in Fig. 23.2C, arrows) are a good sign for the successful preparation. The cell shape difference should be more apparent with higher zooming factor, around 1.2 μm/pixel.

The intact lymph node samples prepared for the imaging of B cell–T cell interactions can be visualized as shown in Fig. 23.2D. It is apparent from these images that GFP^+ MD4 B cells are in the outer follicle region and CFP^+ OT-II cells are in B cell follicles where CMTMR-labeled naive B cells are. At the same time, we can see that some of MD4 B cells are in contact with OT-II T cells (Fig. 23.2D, arrowheads).

3.3. Postimaging samples

It is recommended to keep the tissue samples on ice after imaging and analyze them by flow cytometry or immunofluorescence staining of their cryosections. Data from these analyses are often helpful for the interpretation of imaging data. By flow cytometric analysis, we can obtain the information about the number of cells with each fluorophore used for imaging and the activation state of individual cells. Immunofluorescence staining of the cryosections can give us information about the precise anatomy of target cell locations in the tissue. After explants imaging, we detach lymph nodes from plastic cover slips with a razor blade and mash them for flow cytometry staining or fix them in PBS containing 3% paraformaldehyde at 4 °C for 1 h for fixation of soluble fluorescent proteins to freeze them in the OCT compound (Sakura).

4. Data Analysis

4.1. General workflow of cell tracking analysis

The above procedure from the cell transfer through the image acquisition can be completed within a week. However, it takes a lot of time and effort to accumulate data of verified cell tracks from acquired images. Further, researchers need to come up with good strategies to quantify differences in cell dynamics between control and experimental cells. Quantification of cell motility or displacement is relatively straightforward (Zinselmeyer *et al.*, 2009), whereas enumeration of entry/egress frequencies from one microenvironment to another requires criteria for defining these events.

Visualization of clear boundaries between different microenvironments is valuable to define the criteria of entry/egress and to sort cell tracks based on them. Here, we describe the procedure of data analysis with the tracking tool in the Imaris Ver. 5 software (Bitplane) by taking as examples the two types of imaging results shown above. Imaris is one of the most popular software along with Volocity (PerkinElmer) that enables automated cell tracking from four-dimensional data, and currently, the latest Ver. 7 of Imaris is available with several new functions. Because the basic method is similar between Ver. 5 and Ver. 7, we mainly describe the method using Imaris Ver. 5 and also introduce improvements newly implemented in Ver. 7 to facilitate multicolor cell tracking. We choose not to use Ver. 6 because it lacked the capability to correct cell tracking errors. The software modules needed for the analysis we describe below are the Imaris basic module, the MeasurementPro module, and the ImarisTrack module.

The workflow is: (1) *importing the image data in Imaris*, (2) *cell tracking*, (3) *verification of created tracks*, and (4) *extracting the parameters of interest*. The time-consuming processes are (3) and (4), because these processes still contain manual operations due to difficulties in complete automation. We describe our procedure to minimize the chance of biased data extraction that is potentially caused by the manual processing of cell tracks.

(1) *Importing the image data in Imaris*. Imaris Ver. 7 can directly read the image formats of the major microscope systems (e.g., .lif, .oib, and .lsm for Leica, Olympus, and Zeiss, respectively), whereas files should be exported in the tiff format to open in Imaris Ver. 5. Most of the acquisition software can export a series of tiff files with file names corresponding to detection channels, *z*-positions, and time points, which can be automatically recognized or manually assigned in Imaris.
(2) *Cell tracking*. Clusters of fluorescent signals are recognized as cell centroid spots, which are used for tracking analysis. By selecting "Spots" in the "Surpass" menu, we can create the cell centroid spots based on two parameters; the minimum diameter of cells and the minimum threshold of signal intensity. We usually enter 6 μm for the minimum diameter of naive lymphocytes and 8–10 μm for activated lymphocytes. Cells having elongated shapes or contacting with other cells that emit signals in the same channel are often assigned with false spots, which should be deleted later as described in the next paragraph. The minimum threshold of signal intensity is determined such that every cluster of relevant signals is assigned with a spot. As shown in Fig. 23.3A, however, by lowering the intensity threshold to detect GFP^+ cells in the deep region, CFP^+ cells near the tissue surface are also assigned with spots. In Ver. 5, we manually remove these false spots on bleedthrough signals. In Ver. 7, the maximum intensity threshold in the CFP channel can be additionally applied to exclude cells emitting signals in the both channels (Fig. 23.3B).

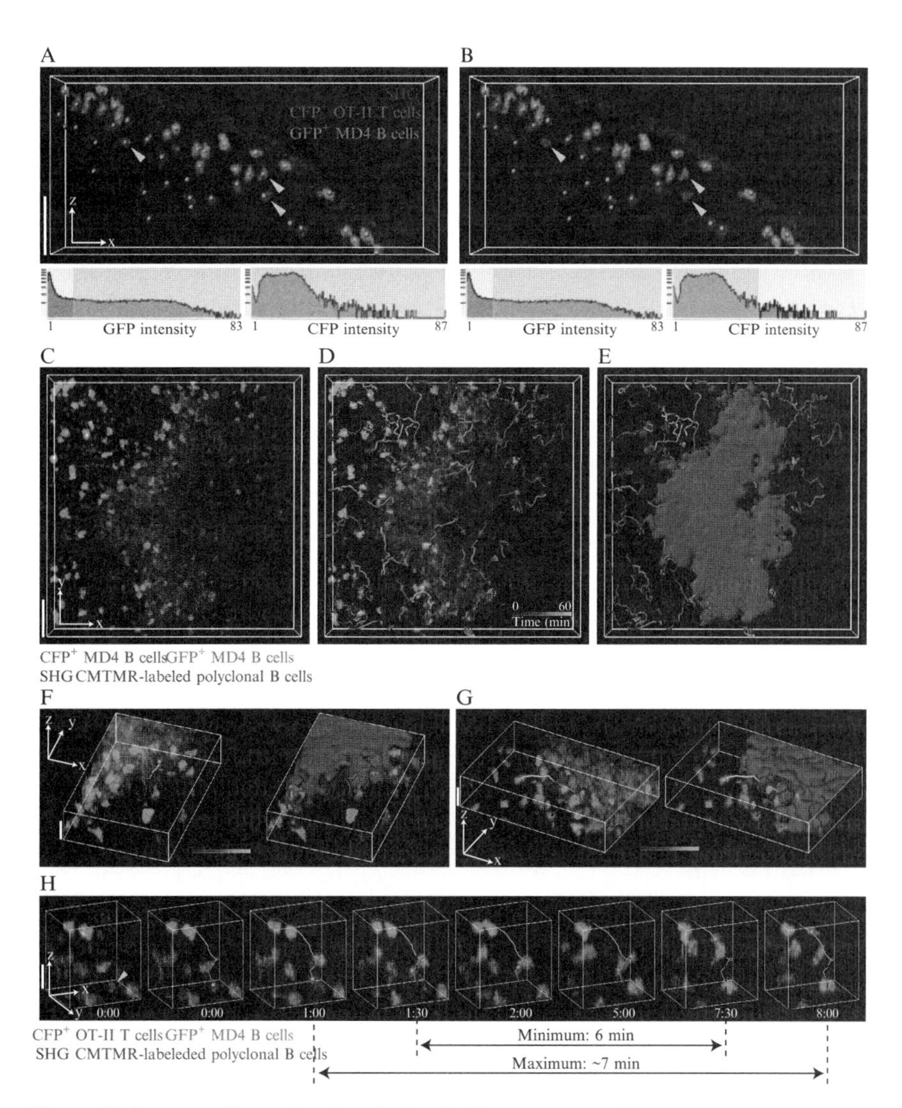

Figure 23.3 Four-dimensional analysis of cell migration and interaction. (A,B) Automated spot assignments to GFP^+ cells in the presence of CFP^+ cells. In (A), the low minimum intensity threshold in the GFP channel to assign spots to weak GFP signals in the deep region causes spot assignments also to strong CFP signals (arrowheads). By combining the maximum intensity threshold in the CFP channel to the minimum threshold in the GFP channel, successfully remove spots from CFP signals as shown in (B). The images are the three-dimensional projection of *xz* images. Scale bar: 50 μm. Image depth in the *y* direction: 25 μm. In the histograms, pixel counts are plotted against the signal intensities in the GFP and CFP channels. The signal intensity ranges used for cell detection are shown in yellow. (C–G) Analysis of pregerminal center B cell migration to the germinal center cluster. (C) The three-dimensional projection of *xy* images clipped from the data shown in Fig. 23.2C. Scale bar: 50 μm. Image depth:

After creating spots, clicking "Create Track" in the "Tracking" tab will launch the cell tracking wizard. Here, we can select a tracking algorism from "Brownian Motion" and "Autoregressive Motion." The former is to cover any kinds of random movements, while the latter would be suitable for tracking more rectilinear movements like naive T cell migration. An essential parameter for the tracking is the "Maximum Distance" of cell movements between two adjacent time points. The large value of this parameter enables tracking of fast moving cells, while substantially increasing the possibility of erroneous track connections between adjacent cell spots. We usually enter 15 μm for T cells and 10 μm for B cells for data from every 30 s recordings.

(3) *Verification of tracks.* This is one of the most time-consuming processes. The created tracks are shown in the three-dimensional projection and can be sorted in folders. We highlight tracks one by one and check the spot assignment at each time point and connections of spots. False spots are deleted to skip time points containing them. Erroneous connections in tracks are also deleted, and disconnected tracks that are obviously of the same cell are connected by using the "Edit Track" function. Tracks that have been verified or corrected are sorted in a folder. If the imaging field contains a large number of cells, it may not be realistic to repeat this procedure for all the tracks. One way to reduce the number of tracks to be used, without bias and without reducing the image stack volume or time points, is to check only tracks that have spots at the midpoint of the imaging duration. For analyzing migration across the border between different tissue domains, it may also help to reduce the image stack volume and to focus on the region containing the border as described below. When two (or more) cell tracks in the same channel intersect, it is often difficult to determine connections between tracks before and after crossing. In such cases, we remove the tracks before or after crossing from the analysis. We also remove short tracks, for example, tracks covering less than 5 min. In some cases, tracks of cells that have divided during imaging may be removed to analyze cell

30 μm. (D) The verified migration tracks of GFP-expressing MD4 B cells that appear at the midpoint of the imaging duration (60 min). (E) The "Isosurface" of the germinal center has been created by three-dimensional rendering of CFP signals. (F, G) The migration track of a GFP^+ MD4 B cell accessing the germinal center border to enter (F) or stay outside (G) the cluster. The germinal center clusters are shown as the CFP fluorescence images (left) and isosurface images (right). Scale bars: 20 μm. (H) Time-lapse images capturing a T cell contacting a B cell. The track of T cell migration throughout the time points is overlaid on each time point image. The arrowhead in the far left panel indicates the tracked T cell. Elapsed time is presented as m:s. Scale bar: 20 μm. The possible range of the contact duration is shown at the bottom. (For interpretation of the references to color in this figure legend, the reader is referred to the Web version of this chapter.)

motility. The criteria for selecting tracks need to be determined based on the aim of the analysis.

(4) *Extracting the parameters of interest.* After the track verification, we can export spatial coordinates data as an Excel (Microsoft) file or a CSV file by clicking "Statistics" with the verified track folder being selected. The subsequent procedure would be conducted with Excel to compare the cell motility parameters, for example, velocity, displacement, turning angle, and arrest coefficient (Sumen *et al.*, 2004; Zinselmeyer *et al.*, 2009). However, further analysis with Imaris is suited to investigate cell migration between different tissue microstructures and intercellular interactions. In the next sections, our manual quantification methods to analyze these types of events are described.

4.2. Analysis of pregerminal center B cell entry to the germinal center clusters

As shown in Fig. 23.2C, the germinal center is clearly visualized by a cluster of CFP^+ cells surrounded by polyclonal (and mostly naïve) B cells labeled with CMTMR. Although the germinal center can be discernable solely based on its paucity of naive B cells (Green *et al.*, 2011; Qi *et al.*, 2008), the CFP^+ cell cluster enables us to visualize the more precise surface of the germinal center. When possible, by focusing on a part of the germinal center surface that is nearly perpendicular to xy planes, the eye detection of migration events across the surface becomes easier, because the xy spatial resolution is higher than that in the z direction. For example, we can crop the imaging data shown in Fig. 23.2C to generate the image volume shown in Fig. 23.3C. In this volume, GFP^+ pregerminal center MD4 B cells are tracked, and the verified tracks of the cells that appear in the analysis volume at the midpoint of the imaging period are shown in Fig. 23.3D. Then we evaluate the accessibility of tracked cells to the germinal center structure. Creating three-dimensionally rendered "Isosurface" on Imaris is helpful for the precise demarcation of the surface (Fig. 23.3E). The tracked cells are first categorized into the following four groups. (1) Cells that were outside the germinal center and made no contacts with the germinal center surface. (2) Cells that made contacts with the surface from the outside of the germinal center. (3) Cells that were in the germinal center and made no contacts with the surface. (4) Cells that made contacts with the surface from the inside. The cells in the category of (2) are further separated into three subgroups: (a) Cells that crossed the surface and migrated into the germinal center at least 10 μm away from the nearest surface (Fig. 23.3F). (b) Cells that moved back to the outside of the germinal center at least 10 μm away from the surface (Fig. 23.3G). (c) Cells that stayed within 10 μm of the surface until the end points of their verified tracks. Spheres with the diameter of 20 μm, whose center is positioned at the cell centroid, are used to determine whether or not cells are 10 μm away from the surface of germinal center. By defining the number of these events as n_a, n_b, and n_c, respectively, $n_a/(n_a + n_b)$, $n_a/(n_a + n_b + n_c)$, and/or $(n_a + n_c)/(n_a + n_b + n_c)$ can be presented as the relative entry frequency to compare between control and experimental cells. The similar enumeration and calculation can be performed for the cells in the category of (4) to obtain the relative egress frequency.

4.3. Analysis of B cell–T cell interactions

Lengths of B cell–T cell contacts are measured along verified tracks of either B cells or T cells. Tracks along which the contact analysis has been completed are stored in a folder to avoid analyzing the same contact more than once. We usually use verified tracks of T cells rather than B cells for the contact analysis because T cells reduce their motility during their antigen-specific conjugation with B cells. By using verified tracks of T cells for the analysis, we can also obtain the motility information of T cells that are or are not in contacts with antigen-specific B cells. To measure the contact length, each T cell track is highlighted, and the track identity number assigned by Imaris is recorded in an Excel spreadsheet. Then the contacts of B cells with the tracked T cell, including those mediated through cell processes, are carefully examined. The time points at which each contact along the track is first and last observed are recorded in the Excel file. Some contacts, especially when they are long and dynamic, are unable to be tracked for the entire contact period because cells may enter the field as a conjugate and/or leave the field as a conjugate. We usually include these incompletely tracked interactions for the quantification as underestimated contact time. When more than two B cells are in contact with a tracked T cell at the same time, demarcation between the B cells may be difficult. In this case too, we record the shortest possible contact time for each B cell. Even for contacts that are completely tracked, the resolution of contact time information is twice the time interval of the image acquisition (Fig. 23.3H). Again, we use the shortest possible contact time with the resolution. It is important to note that this analysis is based on an assumption that the frequency of detachment plus reattachment (or attachment plus redetachment) to occur within acquisition intervals is negligible. After recording the length of contacts, we perform binning of the contact length to generate a histogram showing frequencies of contacts in each bin of the contact length. Contributions of incompletely tracked contacts to frequencies in each bin can be shown in the form of stacked bar graph (Allen *et al.*, 2007; Kitano *et al.*, 2011; Okada *et al.*, 2005).

ACKNOWLEDGMENTS

We thank S. Moriyama and Y. Wakabayashi for helpful discussions. M. K. is a RIKEN Special Postdoctoral Researcher. This work was supported by the Ministry of Education, Culture, Sports, Science and Technology of Japan (M. K. and T. O.), the Uehara Memorial

Foundation (T. O.), the Sumitomo Foundation (T. O.), and the Mochida Memorial Foundation for Medical and Pharmaceutical Research (T. O.).

REFERENCES

Allen, C. D. C., Okada, T., Tang, H. L., and Cyster, J. G. (2007). Imaging of germinal center selection events during affinity maturation. *Science* **315,** 528–531.

Celli, S., Breart, B., and Bousso, P. (2008). Intravital two-photon imaging of natural killer cells and dendritic cells in lymph nodes. *Methods Mol. Biol.* **415,** 119–126.

Garside, P., Ingulli, E., Merica, R. R., Johnson, J. G., Noelle, R. J., and Jenkins, M. K. (1998). Visualization of specific B and T lymphocyte interactions in the lymph node. *Science* **281,** 96–99.

Green, J. A., Suzuki, K., Cho, B., Willison, L. D., Palmer, D., Allen, C. D., Schmidt, T. H., Xu, Y., Proia, R. L., Coughlin, S. R., and Cyster, J. G. (2011). The sphingosine 1-phosphate receptor S1P(2) maintains the homeostasis of germinal center B cells and promotes niche confinement. *Nat. Immunol.* **12,** 672–680.

Grigorova, I. L., Schwab, S. R., Phan, T. G., Pham, T. H. M., Okada, T., and Cyster, J. G. (2009). Cortical sinus probing, S1P1-dependent entry and flow-based capture of egressing T cells. *Nat. Immunol.* **10,** 58–65.

Kitano, M., Moriyama, S., Ando, Y., Hikida, M., Mori, Y., Kurosaki, T., and Okada, T. (2011). Bcl6 protein expression shapes pre-germinal center B cell dynamics and follicular helper T cell heterogeneity. *Immunity* **34,** 961–972.

Lämmermann, T., Bader, B. L., Monkley, S. J., Worbs, T., Wedlich-Söldner, R., Hirsch, K., Keller, M., Förster, R., Critchley, D. R., Fässler, R., and Sixt, M. (2008). Rapid leukocyte migration by integrin-independent flowing and squeezing. *Nature* **453,** 51–55.

Matheu, M. P., and Cahalan, M. D. (2007). Isolation of CD4+ T cells from mouse lymph nodes using Miltenyi MACS purification. *J. Vis. Exp.* **9,** 409.

Matheu, M. P., Parker, I., and Cahalan, M. D. (2007). Dissection and 2-photon imaging of peripheral lymph nodes in mice. *J. Vis. Exp.* **7,** 265.

Mempel, T. R. (2010). Single-cell analysis of cytotoxic T cell function by intravital multiphoton microscopy. *Methods Mol. Biol.* **616,** 181–192.

Mempel, T., Henrickson, S., and Von Andrian, U. (2004). T-cell priming by dendritic cells in lymph nodes occurs in three distinct phases. *Nature* **427,** 154–159.

Miller, M. J., Wei, S. H., Parker, I., and Cahalan, M. D. (2002). Two-photon imaging of lymphocyte motility and antigen response in intact lymph node. *Science* **296,** 1869–1873.

Miller, M. J., Wei, S. H., Cahalan, M. D., and Parker, I. (2003). Autonomous T cell trafficking examined *in vivo* with intravital two-photon microscopy. *Proc. Natl. Acad. Sci. USA* **100,** 2604–2609.

Okada, T., Miller, M. J., Parker, I., Krummel, M. F., Neighbors, M., Hartley, S. B., O'Garra, A., Cahalan, M. D., and Cyster, J. G. (2005). Antigen-engaged B cells undergo chemotaxis toward the T zone and form motile conjugates with helper T cells. *PLoS Biol.* **3,** e150.

Pantazis, P., Maloney, J., Wu, D., and Fraser, S. E. (2010). Second harmonic generating (SHG) nanoprobes for *in vivo* imaging. *Proc. Natl. Acad. Sci. USA* **107,** 14535–14540.

Qi, H., Cannons, J. L., Klauschen, F., Schwartzberg, P. L., and Germain, R. N. (2008). SAP-controlled T-B cell interactions underlie germinal centre formation. *Nature* **455,** 764–769.

Scherschel, J. A., and Rubart, M. (2008). Cardiovascular imaging using two-photon microscopy. *Microsc. Microanal.* **14,** 492–506.

Sumen, C., Mempel, T. R., Mazo, I. B., and von Andrian, U. H. (2004). Intravital microscopy: Visualizing immunity in context. *Immunity* **21,** 315–329.

Victora, G. D., Schwickert, T. A., Fooksman, D. R., Kamphorst, A. O., Meyer-Hermann, M., Dustin, M. L., and Nussenzweig, M. C. (2010). Germinal center dynamics revealed by multiphoton microscopy with a photoactivatable fluorescent reporter. *Cell* **143,** 592–605.

Zinselmeyer, B. H., Dempster, J., Wokosin, D. L., Cannon, J. J., Pless, R., Parker, I., and Miller, M. J. (2009). Chapter 16. Two-photon microscopy and multidimensional analysis of cell dynamics. *Methods Enzymol.* **461,** 349–378.

Author Index

Note: Page numbers followed by "*f*" indicate figures, and "*t*" indicate tables.

A

Abad, J. L., 95–97
Abdulla, A., 366
Abe, T., 365–366, 367*t*, 368–369, 370*t*, 371–372
Abril, M., 408, 409
Abromaitis, S., 103–104
Acab, A., 357
Achtman, M., 94
Ackermann, M., 397
Adams, S. R., 337, 365–366
Adjei, A. A., 177–178
Aguilar, B., 95–97
Aho, R. J., 363–365, 366–368
Aizawa, S., 365–366, 367*t*, 368–369, 370*t*, 371–372, 416
Akaki, M., 24, 25
Akira, S., 416
Akiyama, H., 159*t*, 368–369
Alberius, P., 250
Alberta, J. A., 313
Alcor, D., 228, 230–231, 232, 234, 235
Alexander, S., 176, 201
Allan, C., 344–345
Allen, C. D. C., 442, 444, 451–452
Allende, M. L., 139–140, 426
Allen, E. D., 184, 185
Almholt, K., 180, 185
Alsberg, E., 257
Alsteens, D., 1–17
Altmann, S., 158
Altschuler, S. J., 352–355
Al-Tuwaijri, M., 199, 211
Alvarez-Buylla, A., 312
Amino, R., 20–23, 24–25, 82, 292–293, 302
Amoh, Y., 201, 203, 205, 220
Anderson, C. J., 177–178
Anderson, D. J., 313
Ando, D. M., 331–360
Andoh, A., 185
Ando, T., 6–7
Ando, Y., 438–439, 452
Andreasen, P. A., 185
Andre, G., 13, 14
Andresen, V., 201
Andreu, N., 103
Andrews, P. D., 181
Andrianopoulos, A., 141*f*, 426, 432*f*
Ang, M. K., 177–178
Angot, E., 158
Annovazzi, L., 313
Antosiewicz-Bourget, J., 357
Aonuma, H., 408–409
Appelmelk, B. J., 426
Applebee, C. J., 227, 239
Arai, T., 159*t*
Arakawa, Y., 397
Araki, T., 416
Araki, Y., 185
Argon, Y., 94
Arhel, N., 292–293
Arif, B. M., 163*t*
Armoogum, D. A., 227, 239
Arnold, S. J., 362
Arnold, W. D., 180
Aroian, R. V., 138*t*
Arrasate, M., 158, 333, 338, 340–341
Arteaga, C. L., 241
Artus, J., 363, 366–368, 368*f*
Aslam, R., 54
Assfalg-Machleidt, I., 176
Astudillo, A., 176, 177–178, 185
Athaide, C., 325
Atkins, D., 241
Atkinson, S. J., 36
Atsuta, R., 180–181
Aubin, J. E., 250
Aubry, L., 357
Auerswald, E. A., 176
Augustin, N., 180, 185
Auld, D. S., 332–333
Aumailley, M., 179
Austin, A. P., 364*t*, 368–369
Austin, C. P., 332–333
Axelrod, D., 274

B

Baba-Aissa, L., 291–309
Babcock, H. P., 64–65, 65*t*
Babister, J. C., 256
Bacallao, R. L., 36
Bachmann, M. H., 119, 129–130
Bachoo, R. M., 313
Bader, B. L., 438
Baek, S. J., 119
Baena-Lopez, L. A., 409
Baggio, L., 52–53

Bai, X. T., 138*t*
Bajenoff, M., 266
Baker, L., 177–178
Baker, N. E., 407–408, 409
Bak-Maier, M., 426
Balagopalan, L., 266
Balbin, M., 176, 177–178, 185
Baldacci, P., 21–23, 24
Baldock, R., 372
Baldwin, T. M., 285–286
Balla, K. M., 138*t*
Bamba, S., 185
Bamba, T., 185
Bancroft, G. J., 103
Banker, G., 393
Bannister, L. H., 82
Banuelos, K., 138*t*
Bao, Z., 296–297, 298, 301
Baranov, E., 199–200, 211
Baranowski, B., 311–329
Barbero, R. J., 13
Bareil, C., 118–119
Barker, N., 384
Barlow, W. E., 178
Barmada, S. J., 338
Barnard, E., 163*t*
Baron, M. H., 363–365, 364*t*, 366–368
Barr, V. A., 266
Barry, J. D., 21–23
Bartel, F. O., 334
Baruch, A., 180
Basler, K., 408, 409
Bastin, M. E., 372
Batchelor, P. G., 347
Bates, M., 78–79, 376
Batista, F. D., 267, 286
Baumeister, W., 6–7
Baumgartel, V., 65*t*
Bavari, S., 94
Bavister, B. D., 37
Bayani, J., 314, 315, 317, 319
Bayati, B., 65*t*
Beal, M. F., 333
Beaulieu, J. M., 124–125
Becker, D. R., 416
Becker, T. S., 138*t*, 426
Beck, H. P., 88
Begbie, J., 364*t*, 365–366, 368–369
Behrendt, M., 118–119
Behrendt, N., 176
Behringer, R. R., 364*t*, 368–372
Bekar, L., 37
Bekris, L. M., 159*t*
Béland, M., 103–104
Belayev, L., 124–125
Bellosta, P., 408, 409
Bellows, D. S., 314
Belousov, V. V., 136–137, 148
Benchoua, A., 313, 319
Bennis, R. A., 384
Benns, S., 23–24
Bercury, K., 158–159
Berezovskaya, O., 207
Berger, K. L., 65*t*
Berget, P. B., 180–181
Bergstrom, A. L., 158
Berland, K. M., 376–377
Berman, J. N., 138*t*
Berman, T., 185
Bernad, A., 95–97
Bernstein, M., 314, 315, 317, 319
Berón, W., 94
Bertling, E., 391–406
Bertrand, J. Y., 138*t*, 426
Bestor, A., 94, 103–104
Betzig, E., 78–79, 199, 376
Beuzón, C., 94, 110
Beveridge, T. J., 4
Bhargava, V., 185
Bialecka, M., 363–365
Biechele, S., 364*t*
Biella, G., 313, 317, 319
Bilder, D., 408–409
Binda, E., 314
Bingen, E., 94
Bird, T. D., 159*t*
Birkhead, J. R., 180
Birkhoff, G., 365–366
Birkmann, E., 158–159
Bissell, M. J., 184, 186–187, 408–409
Bito, H., 397
Bitter, W., 138*t*, 426
Blagborough, A. M., 20
Blandau, R. J., 416
Blasi, F., 177–178
Blau, H. M., 158–159
Bloch, R. J., 392–393, 397
Blum, G., 176, 180, 188
Bobek, V., 206–207
Bodner, R. A., 158
Boehm, J., 398
Boekel, J., 36, 37, 41*f*, 46*f*, 49*f*, 50, 51, 53–54, 53*f*, 56, 58–59
Boes, M., 267
Bogdanova, E. A., 202, 366
Bogenrieder, T., 185
Boguslawski, G., 95–97
Bogyo, M., 176, 178–179, 180, 185, 188
Boleti, H., 83
Bolhassani, A., 20
Bollenbach, T., 370*t*, 371
Bonacorsi, S., 94
Bonadio, J., 257
Bonato, M. S., 241
Bongso, A., 416
Bonhoeffer, T., 392–393

Bonifacino, J. S., 78–79, 199, 376
Bonn, V. E., 314
Boring, L., 128–129
Bortner, C. D., 294
Bosco, D. A., 159*t*
Bourne, J. N., 392
Bousso, P., 444
Boutin, A., 43–44
Boutros, M., 320
Bouvet, M., 199–200, 203, 204, 205–206, 211, 212, 213
Boyd, A. E., 365–366
Boyd, P. J., 185
Bradshaw, C. R., 352–355
Brady, J. D., 398
Brameshuber, M., 273
Brandenburg, B., 64, 65*t*, 66, 76
Brauchle, C., 64, 65*t*
Bray, D., 267
Brazil, D. P., 227, 228
Breart, B., 444
Bremer, C., 179
Brennan, C., 313
Briddon, S. J., 163*t*, 165
Briggs, B. J., 94, 110
Briggs, J. A., 65*t*
Briolat, V., 426
Britanova, O. V., 366
Brivanlou, A. H., 363
Broccoli, V., 357
Brockhoff, G., 185–186
Brodoefel, H., 180, 185
Broitman-Maduro, G., 94, 101–103
Bromme, D., 180
Bronner-Fraser, M., 314, 363
Brooks, P. C., 179, 184
Brouillet, E., 333
Broussard, J. R., 416
Brown, C. M., 110
Brown, E. B., 37, 52–53, 202
Brugge, J. S., 184, 186, 187–188
Bruijn, L. I., 159*t*
Brumby, A. M., 408–409
Bryant, P. J., 408–409
Brzostowski, J. A., 179, 265–290
Bucci, C., 94, 96*t*, 110
Buchholz, F., 352–355
Buckles, E. L., 43–44
Bugge, T. H., 176
Buning, H., 64, 65*t*
Burckhardt, C. J., 64, 65*t*, 78–79
Bureau, I., 371
Burel, J. M., 344–345
Burgdorf, S., 176
Burgle, M., 176
Burgos, J. S., 199–200
Burland, V., 43–44
Burlingame, A., 180
Bustamante, C., 4–5
Busto, R., 124–125
Butt, H. J., 6–7
Byers, H. R., 397
Byrne, M. H., 180

C

Cagan, R. L., 408–409
Caglic, D., 176, 188
Cahalan, M. D., 438, 443–444, 452
Cahill, M. E., 392
Cai, X., 392–393, 397
Caldera, V., 313
Calderwood, D. A., 4–5
Callamaras, N., 41–42
Calleja, V., 225–246
Callesen, H., 416
Cambier, J., 266
Campbell, R. B., 37, 52–53, 202
Campbell, R. E., 96*t*, 199–200, 202
Campo, E., 185
Campos, S., 40
Cancel, G., 159*t*
Canivet, A., 291–309
Cannon, J. J., 438–440, 444, 446, 447–448, 451
Cannons, J. L., 451–452
Cao, C., 110
Caparon, M. G., 426
Capito, R. M., 256
Caplan, A. I., 247–263
Caras, I., 185
Cardesa, A., 185
Carmeliet, P., 37, 52–53, 202
Carmel, L., 325
Caron, E., 94, 110
Carpenter, 353*t*
Carpenter, A. E., 315, 321–322, 372
Carrasco, Y. R., 267
Carromeu, C., 357
Casas-Tinto, S., 409
Casuso, I., 6–7, 14
Cates, S., 340
Cattaneo, E., 313, 317, 319
Cavallo-Medved, D., 176, 179, 180, 183, 186–188
Cavenee, W. K., 313
Cavicchi, S., 409
Celli, S., 82, 444
Chalfie, M., 199
Chambers, A. F., 200–201
Chang, J. H., 315
Chao, L. L., 25
Chapman, H. A. Jr., 185
Charlton, C. A., 158–159
Chau, I., 177–178
Cheezum, M. K., 74–75
Chehade, K., 180

Chen, A., 179
Chen, B., 392–393
Chen, C., 65*t*
Chen, F. H., 248–249
Chen, G., 357
Cheng, D., 159*t*
Cheng, L., 199
Chen, J., 248
Chen, M. R., 384
Chen, N., 184
Chen, S. G., 159*t*
Chen, Y., 138*t*, 202
Chen, Z., 138*t*
Chepurnykh, T. V., 202, 366
Cheung, G. Y. C., 103
Chiantia, S., 273
Chico, T. J., 138*t*, 426
Chien, A., 138*t*
Chien, D. S., 177–178
Chim, K., 103
Chi, N. C., 426
Chinenov, Y., 158–160
Chin, L., 313
Chippendale, G. R., 43–44
Chishima, T., 199, 211
Cho, B., 451–452
Choi, A. O., 118–119
Choong, F. X., 35–62
Choong, P. F., 177
Chow, A. B., 186
Chow, L. H., 180
Christian, M., 177–178
Christopher, N., 314
Chudakov, D. M., 90, 91, 202, 366, 384
Chu, J., 160, 163*t*
Chung, J. W., 94
Chung, T. D., 322–323, 332–333, 356
Chye, N. S., 416
Ciccarone, V. C., 340
Ciceron, L., 29
Cidlowski, J. A., 294
Cipelletti, B., 314
Clarke, C., 315
Clarke, I. D., 314, 315, 317, 319
Clevers, H., 384
Close, D. M., 119
Cody, C. W., 199–200
Cody, S. H., 138*t*, 199
Cohen, J., 23–24
Colgan, S., 185
Coller, K. E., 65*t*
Collinet, C., 352–355
Collins, N. L., 184
Collins, V. P., 313
Colombo, M. I., 94
Colville, K., 10–11
Conchello, J. A., 372, 373
Condeelis, J. S., 199, 200–201
Conde, F., 333
Condron, M. M., 138*t*
Constantine-Paton, M., 357
Contag, C. H., 119, 129–130
Contag, P. R., 128–129
Conti, L., 313, 317, 319
Coombes, J. L., 20
Cooper, J. D., 65*t*
Coppi, A., 21–24
Coquoz, O., 128–129
Cordeau, P. Jr., 115–133
Cordero, J. B., 408–409
Cormack, B., 199–200
Cormier, M. J., 199
Coughlin, S. R., 451–452
Coussens, L. M., 176, 177–178, 185
Cowman, A., 83
Cozens, B. A., 159*t*
Creutzburg, S., 176
Critchley, D. R., 438
Crosier, K. E., 138*t*, 139–140, 426
Crosier, P. S., 138*t*, 139–140, 426
Cross, A. S., 94
Crossman, D. C., 138*t*, 426
Crowhurst, M. O., 139–140
Crowley, J., 177–178
Croxatto, H. B., 416
Csiszar, K., 186
Cubitt, A. B., 199–200, 365–366
Cuervo, A. M., 333
Cui, Z. Q., 163*t*
Cunha, C., 138*t*, 426
Cunningham, D., 177–178
Cureton, D. K., 65
Curtis, J. M., 285–286
Curtiss, R. III., 103–104
Cusimano, M. D., 314
Cvejic, A., 426
Cyr, R., 294, 301–302, 305, 306–307
Cyster, J. G., 267, 438, 442, 444, 451–452

D

Dagher, P. C., 36
Dague, E., 9–10, 11
Dahlstrand, J., 313
Dahm, R., 427
Daigle, F., 103–104
Dalby, B., 340
d'Alencon, C. A., 139–140, 426
Dalkara, T., 118–119
Dal Porto, J. M., 266
Daniel, R. A., 4
Dano, K., 176, 185
Danovi, D., 311–329
DaRocha, W. D., 20
Darula, Z., 180
Datsenko, K. A., 39

Datta, N. S., 184
Daub, A., 331–360
Dautry-Varsat, A., 83
Davare, M. A., 398
Davey, A., 265–290
Davidson, A., 138*t*
Davidson, M. W., 78–79, 199, 371–372, 376
Davies, S. W., 159*t*
Dawson, T. M., 159*t*
Dawson, V. L., 159*t*
Day, P. M., 65, 65*t*, 67–68
De, A., 199–200
Dean, G., 94, 101–103
de Beco, S., 409
Debnath, J., 184, 187–188
de Boer, S., 363–365
DeClerck, Y. A., 176, 179, 185
Deed, S., 82
DeFranco, A. L., 266
Deghorain, M., 14
Deguchi, T., 334
Dei Tos, A. P., 240–241
de la Cova, C., 408, 409
Delagrave, S., 199–200
de Louvois, J., 94
Demchik, L. L., 180
Dempster, J., 438–440, 444, 446, 447–448, 451
Deng, M., 138*t*, 426
Deng, X., 163*t*, 165, 171–172
Denk, W., 37, 52–53, 202, 376–377
Dennis, J. E., 247–263
de Paula, F. J., 256
DePinho, R. A., 313
Depoil, D., 286
Derkach, V. A., 398
Deryugina, E., 176
de Sousa Lopes, S. M., 363–365
Desplats, P., 158
Deussing, J., 180, 185
De Vitis, S., 314
De Vos, W. H., 294, 301, 306–307
Diamandis, P., 314
Diaspro, A., 293
Diaz, R., 177–178
Di Cello, F., 94, 101–103, 106–107
Dickinson, M. E., 201, 362, 363, 364*t*, 365–368
Dickstein, D. L., 393, 398
Dieguez-Hurtado, R., 364*t*, 366
Diglio, C. A., 185
Dimeco, F., 314
Ding, S., 314–315
Di Paolo, N. C., 267
Dirks, P. B., 314, 315, 317, 319
Dirnagl, U., 118–119
Dixit, R., 294, 301–302, 305, 306–307
Dixon, G., 426
Dobler, M. A., 185–186
Dodd, M. E., 426
Doetsch, F., 312–313
Doktycz, M. J., 10
Domenech, O., 11–13, 14
Dong, C. Y., 376–377
Donin, N. M., 314
Donnenberg, M. S., 37
Dooley, K., 138*t*
Dorscheid, D. R., 180–181
Dosescu, J., 176, 180, 183, 185–188
Douglass, A. D., 282
Downing, K. H., 6–7
Draft, R. W., 384
Dragavon, J., 291–309
Drecktrah, D., 94
Drucker, D. J., 52–53
Dubertret, B., 363
Dubielzig, S., 419–420, 421–422
Duce, S., 372
Dudek, H., 340
Dufort, D., 363–365, 366–368
Dufrêne, Y. F., 4–6
Dugich-Djordjevic, M. M., 312–313
Dunaevsky, A., 392–393
Dunn, G., 137, 426
Dunn, K. W., 36, 37, 39–40, 52–53
Dupont, A., 65*t*
Dupre, D. J., 165
Dupres, V., 9–10, 11–13
Dustin, M. L., 441–442
Duyckaerts, C., 159*t*
Dvorak, J. A., 24, 25, 82
Dyba, M., 78–79
Dykstra, K. M., 180–181

E

Eakin, G. S., 363–365, 364*t*
Ebert, A. D., 357
Eckenrode, V. K., 199
Edwards, D. R., 177–178
Egeblad, M., 177, 186
Egen, J. G., 20–21, 24, 25, 267
Eggan, K., 365–366
Egli, D., 365–366
Egner, A., 78–79
Ehlenberger, D. B., 393, 398
Ehlers, M. D., 398
Elkington, P. T., 103
Elks, P. M., 426
Ellenberg, J., 320, 372
Ellerbroek, S. M., 176, 180
Ellett, F., 136–137, 140–141, 141*f*, 425–435
Elliott, S. J., 94, 101–103, 106–107
Ellis-Davies, G. C., 392, 397
Ellis, I., 240–241
Ellisman, M. H., 337
Ellis, P., 312

Elmendorf, H. G., 83
Elmquist, J. K., 52–53
Elso, C. M., 285–286
Elworthy, S., 139–140, 426, 432*f*
Emergy, D. L., 21–23
Emmert-Buck, M. R., 185
Endress, T., 64, 65*t*
Engel, A., 6–7
Engelholm, L. H., 176
Engelke, M., 65*t*
English, D., 95–97
Enninga, J., 292–293
Erler, J. T., 186
Ermakova, G. V., 202, 366, 384
Errington, J., 4
Ethier, N., 165
Evans, E. A., 4–5
Ewers, H., 63–81

F

Fagan, B. M., 312
Faix, J., 397
Falco, C. N., 180–181
Falk, A., 313, 314, 315, 317
Falkow, S., 96*t*, 199–200
Fan, J. Y., 163*t*
Fantner, G. E., 13
Farina, K. L., 200–201
Farrell, M. J., 138*t*
Farre, X., 185
Fässler, R., 438
Faucher, S., 103–104
Fava, E., 352–355
Feng, Y., 138*t*, 139–140
Fenton, T., 313
Fergusen, D. J., 88
Fernandez, E., 185
Fernandez, G., 124–125
Fernandez, P. L., 185
Ferrer-Vaquer, A., 362, 362*f*, 363–365, 366–368, 377*f*, 379–380, 381
Filipe-Santos, O., 25
Filippov, S., 185
Fine, H. A., 314
Fingleton, B., 177–178, 179, 180–181
Finkbeiner, S., 158, 331–360
Fink, D., 365–366
Fink, K., 267
Finlay, B. B., 101–103, 110
Finne, J., 106–107
Fiocco, R., 314
Firth, A. L., 54
Fischer, B., 320
Fischer, M., 392–393
Fish, E. M., 408–409
Fleire, S. J., 267, 286
Fletcher, T., 103
Flores, M. V., 138*t*, 139–140, 426
Folarin, A. A., 311–329
Fong, S. F., 186
Fonovic, M., 178–179, 180
Fooksman, D. R., 441–442
Foote, S. J., 285–286
Fordyce, C. B., 288
Forget, B. G., 357
Foroni, C., 314
Förster, R., 438
Forstner, M. B., 267
Fortin, D. A., 398
Fowler, R. E., 82
Fradkov, A. F., 136–137, 202, 366, 384
Francius, G., 11–13, 14
Frane, J. L., 357
Franetich, J. F., 21–23
Frankenberg, S., 363, 366–368, 368*f*
Frankland, S., 82
Fraser, A., 372
Fraser, S. E., 201, 363, 366–368, 370*t*, 371, 439
Fraser, S. T., 363–365, 364*t*, 366–368
French, D. L., 176
Frevert, U., 20–23, 24–25, 82
Friedl, P., 176, 201
Fries, E., 64
Friesen, J., 29
Frigault, M. M., 110
Friman, O., 315
Frings, S., 163*t*
Frischknecht, F., 19–33, 82, 292–293
Froldi, F., 409
Fuchs, F., 320
Fuchs, S. Y., 177
Fu, D., 357
Fueyo, A., 176, 177–178, 185
Fu, J., 199
Fujimori, T., 365–366, 367*t*, 368–369, 370*t*, 371–372
Fujiwara, K., 397
Fujiwara, T. K., 209, 267–268
Fujiyama, Y., 185
Fukami, K., 199, 202, 203
Fukuda, M., 392
Fukumura, D., 37, 52–53, 202
Fuller, S. A., 99–101
Funke, S. A., 158–159
Fu, Q., 94
Furnari, F. B., 313
Furuyashiki, T., 397
Fu, Y. F., 138*t*

G

Gabay, L., 313
Gadea, B. B., 185
Gadella, T. W., 233
Gad, M., 10

Gage, F. H., 357
Gaines, P, 357
Galarneau, A., 158–159
Galetin, A., 332–333
Gallagher, P. G., 366–368
Gallagher, S. R., 99–101
Gallant, P., 408, 409
Gallardo, V. E., 139–140, 426
Galli, R., 314
Galy, V., 291–309
Gambhir, S. S., 199–200
Ganesan, S. K., 95–97
Gao, J., 254
Gao, L., 28
Garcia, J. G. N., 95–97
Garcia, M. D., 362
Garfall, A. L., 185
Garoia, F., 409
Garred, P., 176
Garside, P., 438
Gasser, D. L., 186
Gauld, S. B., 266
Gautam, S., 179, 180–181
Gayraud, B., 179
Gazzola, M., 65*t*
Gemski, P., 94
Gerlich, D. W., 320
Germain, R. N., 266, 267, 451–452
Gertsenstein, M., 364*t*
Geschwind, D. H., 314
Ghibaudo, M., 26–27
Ghosh, A., 340
Ghosh, I., 158–160
Giaccia, A., 186
Giepmans, B. N. G., 96*t*, 199–200, 202, 337
Gilberger, T. W., 82
Ginsberg, M. D., 124–125
Giovannini, D., 24–25
Giraudo, E., 180
Gitai, Z., 4
Gleave, J. A., 249
Glinskii, A. B., 206–207
Glinsky, G. V., 207
Glode, M. P., 94
Gocheva, V., 185
Goderie, S. K., 312–313
Godke, R. A., 416
Goeppert-Mayer, M., 37
Goffredo, D., 313, 319
Goh, L. M., 23–24
Goldman, J. P., 267
Goldring, M. B., 180
Goldstein, S. A., 257
Golland, P., 315, 316*f*, 321–322
Gonçalves, S. A., 157–174
González, M. A., 95–97
Goodison, S., 208
Goo, J. H., 158–159
Gorba, T., 313, 317, 319
Gordh, M., 250
Gorodnicheva, T. V., 366
Gortz, P., 158–159
Gorvel, J.-P., 101–103, 110
Goslin, K., 393
Goswami, S., 200–201
Gotschlich, E. C., 94
Gouedeaux, L. L., 416
Gouon-Evans, V., 185–186
Gourfinkel-An, I., 159*t*
Gousset, K., 292–293
Gowing, G., 124–125
Grabher, C., 136–137, 139–140, 426
Graham, J., 314
Grass, D. S., 128–129
Gravel, M., 118–119, 121, 126–127, 128–129, 131–132
Gray, C., 138*t*, 426
Gray, N. W., 371
Graziotto, J., 357
Greber, U. F., 64, 65*t*, 78–79
Greenbaum, D. C., 180
Greenberg, M. E., 340
Green, D. M., 43–44
Green, J. A., 451–452
Green, J. M., 138*t*
Greger, K., 375–376
Greve, T., 416
Grifoni, D., 409
Grigorova, I. L., 267, 444
Griswold, S. L., 371–372
Gritti, A., 314
Gronthos, S., 250
Groom, A. C., 200–201
Groshen, S., 177–178
Gross, L. A., 365–366
Gross, R., 101–103
Grotewold, L., 313
Groves, J. T., 267
Grüring, C., 81–92
Gudjonsson, T., 184
Gueirard, P., 21–23
Guerin, P., 416
Guertin, D. A., 315
Guignot, J., 94, 110
Guilford, W. H., 74–75
Guo, L., 288
Gustafsson, M. G., 78–79
Gutierrez, M. G., 94
Gutschow, P. W., 138*t*
Guy, J., 199–200
Gyorvary, E. S., 6–7

H

Haas, A., 101–103
Haas, T. L., 185

Haataja, S., 106–107
Habermann, B., 352–355
Hackett, J., 43–44
Hackett, T. L., 180–181
Hadjantonakis, A. K., 361–389
Hafalla, J. C., 82
Hagemann, A. I., 163*t*, 171–172
Hagen, A., 138*t*, 426
Hager, J. H., 180
Hahn-Dantona, E., 176
Hahn, R. E., 119
Hahn, W. C., 313
Halbert, S. A., 416
Haldar, K., 82, 83
Halket, S., 94
Hall, C. J., 138*t*, 139–140, 426
Hallek, M., 64, 65*t*
Hall, N. E., 426
Hama, H., 199, 202, 203
Hamilton-Dutoit, S., 185
Hamilton, N., 372
Hammarstrom, P., 158
Hanahan, D., 180, 185
Hanazono, M., 416
Handin, R. I., 138*t*, 426
Handman, E., 285–286
Hanna, J., 268–269, 274, 281–282, 288
Hansell, E., 185
Hansen, C., 158
Hansen-Wester, I., 94, 103–104
Hansma, P. K., 5–7
Hanson, L. A., 94
Hanssen, E., 82, 83
Hantraye, P., 333
Han, X., 185
Hanyu, A., 199, 202, 203
Hariharan, I. K., 408
Harper, I. S., 181
Harris, A., 340
Harris, K. M., 392, 398
Harris, M. C., 94
Hartley, S. B., 438, 444, 452
Harvey, D., 94
Harvey, K. A., 95–97
Harvill, E. T., 94
Harwood, N. E., 286
Hasegawa, M., 159*t*
Hasegawa, S., 199, 211
Hashimoto, R., 52–53
Hastings, J., 199–200
Hatta, K., 369–371
Haugland, R. P., 83
Hauswirth, W. W., 199–200
Hautefort, I., 36, 37, 39, 41*f*, 46*f*, 50, 51, 56, 59
Hawkes, D. J., 347
Hawkins, C., 314
Hawley, R. G., 199
Hawtin, R. E., 199–200
Hayashi, K., 199, 204, 207–208, 209
Hayashi, S., 312
Hayashi-Takagi, A., 392
Hayashi-Takanaka, Y., 364*t*
Hayes, J. A., 94
Hayman, J. W., 141*f*, 426, 432*f*
Hayrapetian, L., 180
Hazel, T. G., 312–313
Hazelwood, K. L., 366, 371–372
Hazenbos, W. L. W., 101–103
Head, R., 314, 315, 317, 319
Heaton, N. S., 65*t*
Hebert, T. E., 165
Hedlund, E., 357
Hegge, S., 29–30
Heiber, A., 82
Heim, R., 199–200, 365–366
Heinrich, D., 27
Held, M., 320
Helenius, A., 64, 65, 65*t*, 67–68, 76, 77–78
Hellmann, J. K., 20–21, 25–27
Hell, S. W., 4, 78–79, 375–376
Helmchen, F., 52–53, 376–377
Helmuth, J. A., 65*t*, 78–79
Helton, T. D., 398
Hemmati, H. D., 314
Henke, F., 158–159
Henkelman, R. M., 314
Henn, C., 6–7
Henrickson, S. E., 267, 440–441
Henriet, P., 179
Henschel, R., 352–355
Hensel, M., 94, 103–104
Herbomel, P., 137, 426
Herlyn, M., 185
Hermetter, A., 273
Hernandez, P. P., 139–140
Herrera, F., 157–174
Herroon, M. K., 176
Hertzberg, R. P., 332–333
Hess, D., 227, 228
Hess, H. F., 78–79, 199, 376
Heupel, W.-M., 201
Heussler, V. T., 82
Hewitt, D., 128–129
He, Z., 366–368
Hide, T., 314
Higuchi, T., 334
Hihara, J., 208
Hikida, M., 438–439, 452
Hill, D. L., 347
Hill, M. M., 227, 228
Hill, P. J., 94, 103
Hines, L. M., 186
Hinterdorfer, P., 6
Hinton, J. C. D., 39
Hirschberg, K., 96*t*
Hirschberg, M., 201

Hirsch, K., 438
Hirsh, D., 295–296
Hirvonen, H., 106–107
Hishikawa, N., 159*t*
Hodgson, J. G., 177–178
Hodgson, J. J., 163*t*
Hoebe, R. A., 294
Hoeppner, D. J., 325
Hoffman, R. M., 195–224
Hoffman, S. L., 23–24
Hof, P. R., 393, 398
Hofsteenge, J., 227, 228
Hojilla, C. V., 177
Holden, D. W., 94, 110
Holden, M., 347
Hollenberg, A. N., 52–53
Holliday, N. D., 163*t*, 165
Holt, D. E., 94
Holtzer, L., 284
Hong, J., 179
Hong, N. W., 364*t*, 368–369
Hong, Y.-K., 95–97
Honigman, A., 249
Honkura, N., 397
Honma, T., 248
Honore, C., 176
Hoover, T. A., 94
Hopkins, J. M., 82
Hörber, J. K. H., 5–6
Ho, R. K., 426
Horvitz, H. R., 295–296
Horwitz, E. M., 254
Hotary, K. B., 184
Hotulainen, P., 391–406
Houseweart, M. K., 159*t*
Houston, J. B., 332–333
Howe, D., 94
Hoyer-Hansen, G., 177–178, 185
Hoyer-Hansen, M., 176
Hsu, K., 138*t*, 426
Huang, A. Y., 267
Huang, C. C., 368–369
Huang, S.-H., 94
Huber, W., 320
Hu, C. D., 158–160, 163*t*, 165, 171–172
Hudson, L. G., 176
Huganir, R., 398
Huisken, J. 375–376
Huling, S., 185
Hulit, J., 366
Humphrey, J. S., 177–178
Humphreys, P., 314, 315
Humphris, A. D. L., 6–7
Hunter, K. E., 185
Hunt, T., 54
Hurlstone, A., 139–140
Hurrell, J. G., 99–101
Huttenlocher, A., 138*t*, 426
Hutton, S., 312
Hu, Y., 138*t*

I

Iadecola, C., 118–119
Iannacone, M., 267
Ichisaka, T., 357
Igaki, T., 407–413
Iglesias, P. A., 376–377
Ikai, A., 7, 10
Imai, J. H., 371–372
Imamura, T., 199, 202, 203
Ingham, P. W., 139–140, 426, 432*f*
Inglese, J., 332–333
Ingulli, E., 438
Ingvarsen, S., 176
Inoue, K. I., 365–366, 367*t*, 368–369, 370*t*, 371–372
Inoué, S., 307
Inoué, T., 307
Iris, B., 249
Isacson, O., 357
Isern, J., 363–365, 364*t*, 366–368
Ishigaki, S., 159*t*
Itabashi, H., 94
Ivanchenko, S., 65*t*
Ivy, P., 177–178
Ivy, S. P., 177–178

J

Jach, G., 163*t*
Jacobsen, V., 65*t*
Jaenisch, R., 180, 357
Jain, R. K., 37, 52–53, 202
Jakobs, S., 78–79
Jakobsson, E., 106–107
Jamieson, L. G., 314
Jang, C. W., 364*t*, 368–369, 371–372
Jedeszko, C., 176, 178, 180, 183, 186–188
Jena, B. P., 5–6
Jenkins, M. K., 438
Jensen, J. B., 87
Jeong, H., 357
Jericho, M. H., 11, 13
Jessen, J. R., 138*t*
Jessen, T., 103
Jiang, P., 199–200, 201, 202–203, 204–205, 206, 207, 208–209, 210–211, 212, 220
Jiang, Y., 248
Ji, N., 294, 301–302, 305
Jin, H., 426, 432*f*, 433
Jin, T., 179, 269
Jin, Y., 138*t*
Johe, K. K., 312–313
Johnsen, M., 185
Johnson, D. E., 43–44
Johnson, J. G., 438

Johnson, R. L., 332–333
Johnsson, N., 158–159
Johnston, L. A., 407–408, 409
Jokilammi, A., 106–107
Jones, B. D., 43–44
Jones, E. A., 364*t*
Jones, J. G., 200–201
Jones, K. A., 392
Jones, R. A., 139–140, 426
Jones, T. R., 315, 316*f*, 321–322
Jonsdottir, G. A., 357
Josefsson, L., 57*f*, 59
Joyce, J. A., 180, 185
Joyner, A. L., 363–365, 369–371
Julien, J. P., 118–119, 121–122, 124–125
Junt, T., 267
Jurgensen, H. J., 176
Ju, W., 398

K

Kader, A., 57*f*, 59
Kadurugamuwa, J. L., 128–129
Kaech, S., 392–393
Kagan, J., 94, 110
Kailas, L., 10, 11, 13
Kajava, A. V., 384
Kajiya, F., 52–53
Kalajzic, I., 249
Kalajzic, Z., 250–251
Källskog, O., 49*f*, 50, 51, 53–54, 53*f*, 58–59
Kalra, V. K., 95–97
Kamphorst, A. O., 441–442
Kanagawa, O., 370*t*, 371–372
Kanaya, F., 207–208
Kanda, T., 366–368
Kang, H. C., 83, 384
Kang, I. H., 315, 316*f*, 321–322
Kang, J. J., 52–53, 95–97
Kanki, J. P., 138*t*, 426
Karasawa, S., 370*t*, 371–372
Karlyshev, A. V., 103
Kartenbeck, J., 64
Kasai, H., 392–393, 397
Kasai, R. S., 267–268
Kasas, S., 7
Kashiwagi, S., 199, 202, 203
Kass, L., 186
Kato, S., 159*t*
Katsuoka, K., 201, 205, 220
Kaushal, S., 206, 213
Kawahara, A., 138*t*
Kawai, K., 208
Kawakami, K., 138*t*
Kaye, J., 331–360
Kebaier, C., 21
Keller, A. F., 118–119, 126–127, 131–132
Keller, M., 438
Keller, P. J., 375–376
Kelly, G., 137, 426
Kelly, K. J., 36
Kelly, O. G., 364*t*
Kenny, P. A., 184
Kenski, D., 352–355
Kenworthy, A. K., 96*t*
Keppler, D., 185
Keren, K., 180
Kerppola, T. K., 158–160, 163*t*, 165–166, 171–172
Khankaldyyan, V., 199–200
Khan, N. A., 94, 97
Khazaie, K., 179
Kho, A. T., 313
Khokha, R., 177
Kienberger, F., 10–11
Kikuchi, T., 120
Kikuta, H., 138*t*, 426
Kimble, N., 295–296
Kim, J., 176
Kim, K. J., 94, 101–103, 106–107
Kim, K. S., 94, 101–103, 106–107
Kimmel, C. B., 140–141
Kim, S. B., 160–161, 251, 384
Kimura, H., 206, 212, 213, 217–218, 364*t*
Kim, Y. H., 94, 250
Kinder, S. J., 369–371
Kirchhausen, T., 65
Kirschenhofer, A., 176
Kishimoto, H., 209
Kissa, K., 426
Kitaguchi, T., 138*t*
Kitano, M., 437–454
Kiyonari, H., 365–366, 367*t*, 368–369, 370*t*, 371–372
Kjoller, L., 176
Klar, T. A., 78–79
Klauschen, F., 451–452
Klein, A. M., 384
Klein, C., 357
Kleinfeld, D., 52–53
Klein, J. P., 355
Klinger, M., 24
Klonis, N., 82, 83
Klotzsch, E., 65*t*
Klucken, J., 158–159
Knochel, W., 369–371
Knodler, L. A., 94, 103–104
Knuechel, R., 185–186
Knutti, D., 392–393
Koblinski, J. E., 176
Koc, O. N., 249
Kodama, Y., 160, 163*t*, 171–172
Kodera, N., 6–7
Koester, H. J., 294, 301–302, 305
Koh, C. J., 95–97
Kohn, D. B., 199–200

Kohn, E. C., 179, 185
Kohwi, M., 312
Kolf, C. M., 254
Kölle, S., 415–423
Kolostova, K., 206–207
Kondoh, G., 363–365, 364*t*, 367*t*, 368–369
Kondo, J., 267–268
Kong, A., 227, 242, 244, 244*f*, 245
König, K., 294, 301–302, 305
König, P., 419–420, 421–422
Kopec, C. D., 398
Kopitz, C., 176
Korb, E., 338
Korets, L., 177–178
Korja, M., 106–107
Korkotian, E., 397
Kornblum, H. I., 314
Korovin, M., 180, 185
Koskinen, M., 391–406
Kotliarova, S., 314
Kotliarov, Y., 314
Koumoutsakos, P., 65, 65*t*, 74–75, 77–78
Koundinya, M., 186
Kovala, A. T., 95–97
Kovalski, K., 176
Koyama-Honda, I., 267–268
Krainc, D., 357
Krane, S. M., 180
Krausslich, H. G., 65*t*
Krell, P. J., 163*t*
Kremers, G. J., 371–372
Kreutz, M., 185–186
Kriegstein, A., 312
Krishna, S., 82
Kriz, J., 115–133
Krohn, K. A., 178
Krol, J., 176
Kroon-Veenboer, C., 384
Krueger, A., 82
Krueger, P. D., 268–269, 274, 281–282, 288
Kruger, A., 176, 180, 185
Krummel, M. F., 438, 444, 452
Krumrine, J., 180
Kruse, F., 82
Krzic, U., 375–376
Kudryashev, M., 29–30
Kukura, P., 65*t*
Kulakauskas, S., 13
Kulesa, P. M., 369–371
Kumar, B., 285–286
Kummer, W., 419–420, 421–422
Kuntz, I. D., 180
Kunz-Schughart, L. A., 185–186
Kurokawa, H., 199, 202, 203
Kurosaki, T., 438–439, 452
Kusécek, B., 94
Kusumi, A., 267–268
Kutok, J. L., 138*t*
Kuwahara, T., 185
Kwaan, M. R., 313
Kwiatkowski, T. J. Jr., 159*t*
Kwon, G. S., 363–368, 364*t*

L

LaBaer, J., 186
Labas, Y. A., 202
Lackan, C. S., 363–366, 364*t*, 367*t*, 368–369
Lacoste, J., 110
Laguerre, M., 228, 230–231, 232, 234, 235
Lajoie, P., 159–160
Lakadamyali, M., 64–65, 65*t*
Lakins, J. N., 186
Lalancette-Hebert, M., 118–120, 122, 124–125, 126–127, 128–130
Lamb, D. C., 65*t*
Lamberti, Y. A., 94
Lamers, G. E., 138*t*, 426
Lam, E. Y., 426
Lämmermann, T., 438
Lampe, M., 65*t*
Lamprecht, M. R., 315
Lange-Asschenfeldt, C., 158–159
Langley, L., 416
Lang, P. A., 267
Lanner, F., 365–366
Laplante, M. A., 138*t*, 426
Lappalainen, P., 397
Lardelli, M., 426
Larijani, B., 225–246
Larina, I. V., 365–366
Laskay, T., 24
Laszlo, L., 364*t*
Lau, A., 288
Laug, W. E., 176, 199–200
Laveran, A., 82
Lavis, L. D., 337
Lawson, N. D., 138*t*
Layton, J. E., 139–140, 426
Lazareff, J. A., 314
Le Blanc, M., 177–178
Leboucher, P., 225–246
Leclerc, A. L., 159*t*
le Digabel, J., 26–27
Lee, C. E., 52–53
Lee, E. H., 184
Lee, G. Y., 184
Lee, H., 200–201, 357
Lee, J., 314
Leek, R., 227, 242, 244, 244*f*, 245
Lee, M., 314, 315, 317, 319
Lee, S. J., 95–97, 158
Lee, T., 409, 411
Lee, Z., 247–263
Lehmann, M. J., 65, 65*t*
Le, H. Y., 138*t*

Leibovich, S. J., 136
Leighton, T. J., 11, 13
Lendahl, U., 312, 313
Lendvai, B., 392–393
Leppla, S. H., 103
Levental, K. R., 186
Levitt, D. L., 325
Li, A., 314
Lian, N. P., 416
Liao, D., 398
Liao, L., 159*t*
Li, B., 398
Libchaber, A., 363
Lichtman, J. W., 372, 373, 384
Liebel, U., 139–140, 358, 426
Lieschke, G. J., 135–156, 425–435
Ligon, K. L., 313
Li, J. F., 200–201
Li, L., 199, 201, 204–205, 211, 220, 365–366, 368–369, 426, 432*f*, 433
Lilie, H., 65, 65*t*, 77–78
Lim, W. T., 177–178
Lin, A., 426
Lindgren, M., 158
Lindquist, R. A., 315–316, 321–322
Lindwasser, O. W., 78–79, 199, 376
Linebaugh, B. E., 176
Lin, E. Y., 185–186, 200–201
Ling, E. K., 314
Lin, H. F., 138*t*, 426
Lin, J. H. C., 37
Linkert, M., 344–345
Lin, S., 138*t*
Lin, Y., 416
Liongue, C., 426
Liotta, L. A., 179, 185
Liou, S. R., 43–44
Lippincott-Schwartz, J., 78–79, 82, 94, 96*t*, 199, 376
Li, Q., 176, 185–186, 288
Li, S., 180
Liu, D. W., 140–141
Liu, H., 163*t*, 165, 171–172
Liu, T. X., 138*t*, 426
Liu, W., 265–290
Livet, J., 384
Li, W. P., 177–178, 408, 409
Li, X.-M., 199, 207, 211
Li, Z. Y., 384
Llano, O., 397
Lockatell, C. V., 43–44
Lodge, R., 96*t*
Lo, E. H., 118–119
Logan, D. J., 315
Loh, L. N., 93–113
Loimaranta, V., 106–107
Lombardia, L., 409
Long, J. Z., 363–366, 364*t*, 367*t*, 368–369
Long, M. W., 184
Long, Q., 138*t*
Look, A. T., 136–137, 138*t*, 426
Loo, L. H., 352–355
Loosli, F., 139–140, 426
Lopez-Gay, J. M., 409
Lopez, M. E., 52–53
Lopez-Otin, C., 176, 177–178, 185
Loranger, B., 344–345
Lorson, C. L., 357
Loughman, J. A., 426
Louis, D. N., 313
Love, Z., 249, 250, 254
Lowell, S., 313, 319
Loynes, C. A., 138*t*, 139–140, 426, 432*f*
Ludwig, A., 391–406
Lu, F., 180
Lugo-Villarino, G., 138*t*
Luiken, G., 211
Lu, J., 384
Luker, G. D., 119, 129–130
Luker, K. E., 119, 129–130, 366
Lukyanov, K. A., 90, 136–137, 202, 366, 384
Lukyanov, S. A., 90, 136–137, 202, 366, 384
Lu, N., 202
Lund, L. R., 185
Luo, L., 365–366, 368–369, 409, 411
Lu, W. Y., 398
Lu, Y., 186
Lynch, C. C., 179, 180–181

M

Ma, C., 94, 103–104
Macagno, J. P., 408–409
Macarron, R., 332–333
Macdonald, D., 344–345
MacDonald, I. C., 200–201
MacDonald, J. F., 398
Machleidt, W., 176
Macosko, J. C., 4–5
Madri, J. A., 185
Madsen, C. A., 176
Madsen, D. H., 176
Magdolen, U., 176
Magdolen, V., 176
Magness, S. T., 312
Mahmoudi, N., 29
Maitra, B., 248
Majewska, A., 41–42, 392–393
Malinow, R., 398
Maloney, J., 439
Mamarbachi, A. M., 165
Mangano, M., 363–365, 364*t*
Man, H. Y., 398
Mankoff, D. A., 178
Månsson, L. E., 36, 37, 41*f*, 46*f*, 49*f*, 50, 51, 53–54, 53*f*, 56, 59

Marchbank, K. L., 286
Marchetto, M. C., 357
Marcus, J. N., 52–53
Marijanovic, I., 248–249
Markelov, M. L., 202
Marks, C. B., 65, 65*t*
Marsh, R. J., 227, 239
Marten, K., 179
Martin, B., 24, 82
Martinez-Corral, I., 364*t*, 366
Martinez, M. D., 364*t*, 366
Martinez-Salas, E., 122–123
Martin, F. A., 407–408, 409
Martin, J., 364*t*, 366
Martin, O. C., 83
Martin, P., 136, 137, 139–140, 426
Maruvada, R., 94
Masliah, E., 158, 333, 338
Mason, C., 392–393
Massberg, S., 267
Massol, R. H., 65
Masterman-Smith, M., 314
Masters, B. R., 376–377
Masters, T. A., 227, 239
Matheu, M. P., 443–444
Mathias, J. R., 138*t*, 426
Matias, V. R. F., 4
Matley, M., 185
Matos, J. E., 159–160
Matrisian, L. M., 177–178, 179, 180–181, 185
Matsuguchi, T., 120
Matsuzaki, M., 392–393, 397
Mattheyses, A. L., 274
Mattingly, R. R., 175–195
Mattis, V. B., 357
Matus, A., 392–393, 397
Matuschewski, K., 23–24
Matz, M. V., 202, 384
Mauney, J. R., 254
Mayhew, G. F., 43–44
Maysinger, D., 118–119
Mazier, D., 29
Mazo, I. B., 451
McCall, M., 23–24
McCarthy, J., 96*t*
McComb, P., 416
McCracken, G. H. Jr., 94
McCulloch, F., 52–53
McDole, K., 376–377
McFadden, G., 83
McFerran, N. V., 163*t*
McGlynn, P., 95–97
McIntyre, J. O., 177–178, 179, 180–181
McKay, R. D., 312–313, 325
McKerrow, J. H., 185
McMahon, A., 312
McPhee, D. O., 138*t*
Meckel, T., 179, 268–269, 274, 284, 288
Medzhitov, R. (2008), 136
Medzihradszky, K. F., 180
Meeker, N. D., 426
Megias, D., 364*t*, 366
Meijer, A. H., 138*t*, 426
Melander, M. C., 176
Melican, K., 36, 37, 41*f*, 46*f*, 49*f*, 50, 51, 53–54, 53*f*, 56, 57*f*, 59
Mellai, M., 313
Mellits, K. H., 94, 103
Mempel, T. R., 267, 440–441, 444, 451
Menard, R., 20, 24–25, 82
Menezo, Y., 416
Meng, A., 138*t*
Meng, L. F., 384
Mercer, A., 94
Mercer, J., 76
Méresse, S., 101–103, 110
Merica, R. R., 438
Merkle, F. T., 312
Merrell, K. T., 266
Merzlyak, E. M., 202, 366
Meyer-Hermann, M., 441–442
Meyer-Morse, N., 180
Meyers, M. W., 416
Michnick, S. W., 158–159
Mikkelsen, T., 180, 185
Mikos, A. G., 256
Miles, M. J., 6–7
Miller-Fleming, L., 158, 159–160, 165–166, 168, 171
Miller, J. L., 325, 333, 338
Miller, L. H., 82
Miller, M. J., 25–26, 426, 438–440, 444, 446, 447–448, 451, 452
Mills, D., 266
Mills, K. R., 184
Milne, J. L. S., 4
Min, J. J., 199–200
Mione, M., 139–140
Miquel, R., 185
Mironov, S., 41–42
Mitchell, G. H., 82
Mitchison, T. J., 136–137
Mitra, P. P., 52–53
Mitra, S., 158, 333, 338
Miura, M., 408–409
Miwa, Y., 370*t*, 371–372
Miyagi, Y., 199
Miyamichi, K., 365–366, 368–369
Miyata, T., 199, 202, 203
Miyawaki, A., 370*t*, 371–372, 408–409, 411, 411*f*
Miyoshi, H., 199, 202, 203
Mller, M., 41–42
Moats, R. A., 199–200
Mobley, H. L., 37, 43–44
Modi, K., 128–129
Moeschberger, M. L., 355

Moffat, J., 315
Mohammed, F. F., 177
Moin, K., 175–195
Molitoris, B. A., 35–62, 408–409
Momiyama, M., 212, 217–218
Monetti, C., 364*t*
Monkley, S. J., 438
Monsky, W. L., 202
Monteiro, R. M., 363–365
Mooi, F. R., 101–103
Moore, J., 344–345
Moore, R., 366–368
Moore, W., 344–345
Moossa, A. R., 199–200, 201, 203, 204–205, 208–209, 210, 211, 212, 220
Moquin, A., 118–119
Morata, G., 407–408, 409
Mordelet, E., 426
Moreno, E., 407–408, 409
Morimoto, R. I., 333
Morimura, T., 199, 202, 203
Morin, J., 199–200
Mori, Y., 438–439, 452
Moriyama, S., 438–439, 452
Morris, T. J., 288
Morris, V. L., 200–201
Moseman, E. A., 267
Moskowitz, M. A., 118–119
Moss, L. G., 199
Mota, M. M., 29, 82
Mothes, W., 65, 65*t*
Moxon, E. R., 94
Moyer, A., 176
Muchowski, P. J., 333
Muehlenweg, B., 176
Mueller, M. S., 352–355
Mukasa, A., 313
Mulholland, K., 94
Muller, B., 65*t*
Muller, C., 65*t*
Müller, D. J., 6–7
Muller, T., 312–313
Mullins, S. R., 176, 180, 183, 185–188
Mummery, C. L., 363–365
Munno, D. M., 325
Munoz, J., 185
Münter, S., 24–25, 26–27
Muotri, A. R., 257
Murakoshi, H., 267–268
Murase, K., 267–268
Murayama, E., 426
Murphy, C. S., 366, 371–372
Murray, M. J., 369–371
Murray, R., 185
Musikacharoen, T., 120
Muthuswamy, S. K., 184, 186, 187–188
Mu, Y., 257
Muzumdar, M. D., 365–366, 368–369
Muzycka, N., 199–200
Mwakingwe, A., 30–31
Myles, D. G., 416

N

Nadal, A., 185
Nadesapillai, A. P., 177
Nagase, H., 176
Nagel, W. E., 352–355
Nagy, A., 4–5, 364*t*, 369–371, 379–380, 380*f*, 381
Nakada, C., 267–268
Nakahara, H., 392–393
Nakamoto, H., 52–53
Nakano, I., 314
Nakao, K., 365–366, 367*t*, 368–369, 370*t*, 371–372
Nakazawa, T., 256
Nam, J. M., 186
Narita, M., 257
Narumiya, S., 397
Natarajan, R., 23–24
Naumov, G. N., 200–201
Nedergaard, M., 37
Neighbors, M., 438, 444, 452
Nelson, J., 163*t*
Nelson, K., 180
Nelson, M. D. Jr., 199–200
Nestor, M. W., 392–393, 397
Neto-Silva, R. M., 409
Neuman, K. C., 4–5
Neumann, B., 320, 357–358
Neves, C., 344–345
Newman, S., 94, 101–103
Ng, M. R., 186
Nguyen, A. V., 185–186
Nguyen, Q., 41–42
Nguyen, T. M., 312
Nichols, B. J., 96*t*
Nicoziani, P., 96*t*
Nie, J., 357
Nielsen, B. S., 180, 185
Nielsen, G., 202
Niethammer, P., 136–137, 148
Nishigaki, T., 294, 301–302
Nishikawa, Y., 20
Niu, W. P., 384
Niwa, J., 159*t*
Noelle, R. J., 438
Noguchi, J., 392–393, 397
Noireaux, V., 363
Nolan, G. P., 332–333
Norris, D. J., 363
Nowell, C. J., 135–156
Nowotschin, S., 361–389
Nozaki, N., 364*t*
Nozaki, S., 177–178
Nussenzweig, M. C., 441–442

Nussenzweig, R. S., 82
Nussenzweig, V., 82
Nusslein-Volhard, C., 138*t*, 427
Nutt, C. L., 313
Nygaard, S., 398

O

Oates, A. C., 139–140, 426
O'Connell, B. A., 426
Oehlers, S. H., 426
Offersen, B. V., 185
O'Garra, A., 438, 444, 452
Ogasawara, Y., 52–53
Ogawa, M., 199, 202, 203, 334
Ohki, E. C., 340
Ohnuki, M., 357
Ohsawa, S., 407–413
Ojcius, D. M., 83
Okada, T., 267, 437–454
Okuda, K. S., 426
Oldenburg, K. R., 322–323, 332–333, 356
Olenych, S., 78–79, 199, 376
Olivari, F. A., 139–140
Olive, M. B., 176, 178, 180, 183, 186–188
Ollikka, P., 106–107
Olma, M. H., 320
Olmeda, D., 364*t*, 366
Omura, T., 369–371
Ong, C. J., 365–366
Onnebo, S. M., 138*t*
Onodera, Y., 186
Oray, S., 392–393
Orfanelli, U., 314
Ørskov, F., 94
Ørskov, I., 94
Ortega, S., 364*t*, 366
Osawa, H., 199, 202, 203
Ossowski, L., 176
Ostroff, L., 398
O'Sullivan, F., 178
Oswald, F., 369–371
Outeiro, T. F., 157–174
Overall, C. M., 176, 177–178, 180, 185
Overgaard, J., 185
Overton, J. M., 52–53
Owen, T. A., 248–249

P

Pagano, R. E., 83
Pagliarini, R. A., 408–409
Palacin, A., 185
Palavalli, L. H., 177–178
Palmer, A. E., 96*t*, 199–200, 202
Palmer, D., 451–452
Palm, F., 49*f*, 50, 51, 53–54, 53*f*
Panchision, D. M., 325
Pantazis, P., 370*t*, 371, 439
Panthier, J. J., 366–368
Papaioannou, V. E., 363–365, 364*t*, 366–368, 367*t*
Papallo, A., 315, 316*f*, 321–322
Papazoglou, A., 180, 185
Parish, T., 103
Park, C. C., 186
Park, E. C., 202
Parker, I., 41–42, 438–440, 444, 446, 447–448, 451, 452
Parkin, G., 313, 317
Park, J. K., 227, 228, 230–231, 232, 234, 235, 314, 357
Park, M., 398
Park, W. J., 158–159
Parrado, S. G., 176
Parthasarathy, R., 267
Pase, L., 135–156, 426, 432*f*
Pastorino, S., 314
Pastor-Pareja, J. C., 408–409
Pastrana, E., 312–313
Patterson, A., 344–345
Patterson, G. H., 78–79, 199, 376
Patterson, S. S., 119
Pau, G., 320
Paw, B. H., 138*t*, 426
Pease, S., 370*t*, 371
Pehling, G., 397
Peiro, N., 185
Pelkmans, L., 357–358
Pena, O. A., 139–140, 426
Penberthy, W. T., 138*t*
Pendas, A. M., 176, 177–178, 185
Penman, S., 199–200, 201, 205, 209
Penzes, P., 392
Peppler, M., 94, 101–103
Pereira, R. F., 249
Perez Vidakovics, M. L., 94
Peri, F., 138*t*
Perret, E., 291–309
Perrier, A. L, 357
Perrin, B. J., 138*t*, 426
Perry, G., 159*t*
Peschanski, M., 357
Pescher, P., 25
Pesch, M., 163*t*
Pession, A., 409
Peter, M., 320
Peters, C., 180, 185
Petersen, O. W., 184
Peters, N. C., 20–21, 24, 25
Peti-Peterdi, J., 52–53
Pevny, L., 312
Pfeffer, M., 365–366
Phair, R. D., 96*t*
Pham, T. H. M., 444
Phaneuf, D., 118–120, 122, 126–127, 128–129
Phan, T. G., 267, 444

Phillips, G. N. Jr., 199
Piatkevich, K. D., 366, 371–372, 384
Pickering, J. G., 180
Piepenburg, O., 163*t*, 171–172
Pieper, S., 372
Pierce, S. K., 265–290
Piliszek, A., 363–365, 366–368, 368*f*
Pinterova, D., 206–207
Pires, S. F., 20
Pirity, M. K., 363–365, 364*t*, 367*t*, 368–369
Piston, D. W., 179, 180–181, 371–372
Pitiot, A. S., 176, 177–178, 185
Plachta, N., 370*t*, 371
Planchon, T. A., 28
Platani, M., 181
Platoshyn, O., 54
Pleiss, M. A., 338
Pless, R., 438–440, 444, 446, 447–448, 451
Pletnev, V. Z., 366
Plochberger, B., 273
Ploemen, I. H., 30–31
Plomp, M., 11, 13
Plunkett, G., 43–44
Plusa, B., 363, 366–368, 368*f*
Pluschke, G., 94
Poche, R. A., 365–366
Podgorski, I., 176, 177, 179, 187–188
Pohl, A., 94
Polin, R. A., 94
Polishchuk, R. S., 96*t*
Pollard, J. W., 185–186, 200–201
Pollard, S. M., 311–329
Pollok, J. M., 82
Pomerantsev, A. P., 103
Pope, A. J., 332–333
Popovtzer, A., 241
Poss, K. D., 138*t*
Potter, M., 199–200
Pradel, G., 82
Praitis, V., 298
Prasadarao, N. V., 94
Prasher, D. C., 199–200
Prendergast, F. G., 199–200
Price, P. J., 340
Primeau, M., 158–159
Proença, M. J., 39
Proia, R. L., 451–452
Prudencio, M., 29, 30–31
Pruszak, J., 357
Pugh-Bernard, A. E., 266
Pulkinnen, M. O., 416
Punturieri, A., 185
Purchio, A. F., 128–129
Purchio, T., 128–129
Purow, B. W., 314
Putcha, P., 158, 159–160, 165–166
Pylypenko, O., 110
Pypaert, M., 65, 65*t*

Q

Qazi, S. N. A., 94, 103
Qian, X., 312–313, 325
Qi, H., 267, 451–452
Quigley, J. P., 176
Quinlan, G. A., 369–371

R

Rabinovitch, M., 94
Racz, B., 397
Radisky, D., 408–409
Ragan, M. A., 372
Rajamaki, M. L., 65, 65*t*, 67–68
Rajewsky, K., 266
Rak, A., 110
Ramboz, S., 128–129
Ramos-DeSimone, N., 176
Ramu, S., 95–97
Randall, G., 65*t*
Ranheim, E. A., 426
Rank, F., 184, 185
Rankl, C., 10–11
Rao, E. J., 338
Rao, M., 312
Rappoport, J. Z., 274
Rasch, M. G., 186
Rasko, D., 43–44
Ratain, M. J., 177–178
Ratcliffe, E. C., 4, 10, 11, 13
Ratnam, S. S., 416
Ravin, R., 325
Ray, P., 199–200
Reddi, A. H., 256
Redd, M. J., 137, 426
Reddy, M. A., 94
Reddy, V. Y., 185
Redford, P., 43–44
Rees, C. E. D., 94, 103
Reese, S., 419–420, 421–422
Regen, T., 82
Reginato, M. J., 184
Reiffen, K. A., 241
Reinheckel, T., 180, 185
Reitano, E., 313, 317, 319
Remington, S. J., 94
Renia, L., 29
Renn, A., 65*t*
Rennenberg, A., 82
Renshaw, S. A., 138*t*, 139–140, 426, 432*f*
Reth, M., 266
Retzlaff, S., 82
Reyes-Aldasoro, C. C., 426
Reynaud, E. G., 375–376
Reynolds, B. A., 312–313
Reynoso, J., 204–206, 212, 213
Rhee, J. M., 363–365, 364*t*, 367*t*, 368–369
Rhee, J. S., 177–178

Rhen, M., 58–59
Rhiner, C., 409
Rhodes, J., 138*t*, 426
Ribot, W. J., 94
Rice, B., 128–129
Rice, M. J., 180
Richardson, H. E., 408–409
Richer, M., 165
Richter, D., 41–42
Richter-Dahlfors, A., 35–62
Richter, K., 163*t*
Ried, M. U., 64, 65*t*
Riese, H. H., 95–97
Riesner, D., 158–159
Rietze, R. L., 312–313
Rigg, A., 177–178
Ringler, P., 6–7
Rink, J. C., 352–355
Ripoll, J., 103
Ripoll, P., 407–408, 409
Ripp, S. A., 119
Ritchie, K., 267–268
Ritter, U., 24
Rivera, C., 397
Rivera, J., 200–201
Robbins, J. B., 94
Roberts, L. J., 285–286
Robertson, B. D., 103
Robertson, E. J., 362
Robertson, J., 118–119
Roberts, T. H., 96*t*
Robey, E. A., 20
Robida, A. M., 171–172
Robinson, C. G., 398
Robitaille, M., 165
Roche, P. M., 366
Rodrigues, C. D., 29
Rodriguez, A., 29, 82, 393, 398
Rodriguez, M. E., 94
Roesch, P., 43–44
Roestenberg, M., 23–24
Roizman, P., 241
Romer, J., 185
Romo-Fewell, O., 138*t*
Ronnov-Jessen, L., 184
Rosains, J., 365–366
Rose, F. F. Jr., 357
Rosen, C. J., 256
Rose, R. H., 163*t*, 165
Roshy, S., 176
Rosol, M., 199–200
Rossant, J., 362, 365–366
Rossi, F., 158–159
Rothberg, J. M., 175–195
Roussel, J. D., 416
Roux, P., 291–309
Rowitch, D. H., 313
Roy, C., 94, 110
Roy, S., 184
Ruaff, M., 416
Rubart, M., 446
Rubin, G. M., 409
Rubin, L. L., 313
Rubinstein, E., 21–23
Rubinstein, L., 177–178
Rudy, D. L., 176, 180, 183, 186–188
Rueden, C. T., 344–345
Ruotti, V., 357
Russell, R. G., 185–186, 314, 315, 317, 319
Rust, M. J., 64–65, 65*t*, 78–79, 376
Ruthel, G., 94
Ruttimann, U. E., 347
Ryabova, A. V., 366
Rydén-Aulin, M., 58–59
Rzomp, K. A., 94, 110

S

Sabass, B., 24–25, 27
Sabatini, D. M., 315, 316*f*, 321–322
Sabeh, F., 179, 186
Sacher, A. G., 314
Sadoff, J., 94
Saetzler, R. K., 301–302, 305, 306–307
Sahr, K. E., 363–365, 364*t*
Saik, J. E., 365–366
Saint, R., 369–371
Sajja, K. C., 371–372
Sakaue-Sawano, A., 199, 202, 203
Saka, Y., 163*t*, 171–172
Sakthianandeswaren, A., 285–286
Salgado, J. M., 398
Samelson, L. E., 266
Sameni, M., 175–195
Samuels, Y., 177–178
Samusik, N., 352–355
Sanbuissho, A., 416
Sandoghdar, V., 65*t*
Sandoval, R. M., 35–62
Sanes, J. R., 384
Sano, H., 163*t*
Santi, P. A., 375–376
Santoriello, C., 139–140
Santoso, B., 426
Sanzone, S., 313, 317, 319
Saper, C. B., 52–53
Saraf, A., 256
Sargent, D., 177–178
Sasaki, H., 369–371
Sastalla, I., 103
Sathananthan, H., 416
Sato, M., 160–161
Sato, T., 384
Savitsky, A. P., 202
Savitt, J. M., 159*t*
Sayler, G. S., 119

Sbalzarini, I. F., 65, 65t, 74–75, 77–79
Scannevin, R. H., 398
Schabert, F. A., 6–7
Schacht, V., 176
Schaible, U., 103
Schelhaas, M., 63–81
Scherer, R. L., 180–181
Scherschel, J. A., 446
Scheuring, S., 6–7, 14
Schiffer, D., 313
Schiller, J. T., 65, 65t, 67–68
Schitter, G., 6–7
Schlaepfer, D. D., 94
Schmeichel, K. L., 184, 186
Schmidt, A. D., 375–376
Schmidt, B. T., 366
Schmidt, E. E., 200–201
Schmidt, J., 41–42
Schmidt, T. H., 284, 451–452
Schmitt, M., 176
Schmitz, M. H., 320
Schneider, B., 101–103
Schneider, J. E., 372
Scholtes, L. D., 94, 110
Schulten, K., 4–5
Schulz, V., 366–368
Schutz, G. J., 273
Schwab, S. R., 444
Schwartzberg, P. L., 451–452
Schwartz, S. L., 110
Schweitzer, E. S., 158, 333, 338
Schwickert, T. A., 441–442
Schwille, P., 273
Scidmore, M. A., 94, 110
Scott, M. L., 365–366
Seed, B., 202
Segall, J. E., 199, 200–201, 366
Segal, M. R., 158, 333, 338, 397
Segura, I., 95–97
Seibler, P., 357
Seidman, C. E., 103
Seisenberger, G., 64, 65t
Selig, M., 202
Selinummi, J. 353t
Serbedzija, G. N., 363
Sergeev, M., 65t
Serrano, A., 95–97
Seton-Rogers, S. E., 186
Severinghaus, J. W., 55–56
Seymour, L., 177–178
Shaby, B. A., 333, 338
Shakhbazov, K. S., 136–137
Shaner, N. C., 96t, 103, 199–200, 202
Shapiro, S. D., 176, 177–178, 185
Sharma, P., 331–360
Shattuck, D. W., 372
Shaw, K. R., 184, 187–188
Shayakhmetov, D. M., 267
Shcheglov, A. S., 366
Shcherbo, D., 202, 366
Sheen, J., 103
Shemiakina, I. I., 366
Shen, Q., 312–313
Sherer, N. M., 65, 65t
Sherman, E., 266
Sheyn, D., 250
Shibata, M., 6–7
Shichinohe, H., 249
Shidlovskiy, K. M., 366
Shi, G. P., 185
Shih, C. M., 25
Shimada, H., 94, 199, 211
Shimizu-Hirota, R., 179, 186
Shimomura, O., 199
Shin, S., 94
Shinya, T., 163t
Shioi, G., 365–366, 367t, 368–369, 370t, 371–372
Shirai, Y. M., 267–268
Shiroishi, T., 82
Shi, S. H., 371–372
Shi, Y., 314–315, 426, 432f, 433
Shorte, S., 82, 291–309
Shpitzer, T., 241
Shree, T., 185
Shu, A. C., 13
Shur, B. D., 416
Shyu, Y. J., 163t, 165, 171–172
Siddiqui, R., 97
Sidjanski, S., 21–23
Siebert, P., 384
Siedlak, S. L., 159t
Sigurdson, C., 158
Silva, C. M., 199–200
Silva-Vargas, V., 312–313
Silvie, O., 21–23, 29
Silzle, T., 185–186
Simard, A., 124–125
Simbuerger, E., 201
Simeonov, A., 332–333
Simons, B. D., 384
Simons, K., 64
Simon, S. M., 274
Simpson, P., 408
Simunovic, F., 357
Singer, M., 19–33
Singhera, G. K., 180–181
Singh, S. K., 314
Sipley, J., 176
Sipos, A., 52–53
Sissons, S., 177–178
Sixt, M., 438
Skibinski, G., 338
Sklyar, O., 320
Skourides, P., 363
Sledge, G. W. Jr., 177–178
Sloane, B. F., 175–195

Slukvin, I. I., 357
Smirnov, S., 397
Smith, A. E., 65, 65*t*, 77–78
Smith, A. G., 313, 314, 315, 317, 319
Smith, B. A., 207
Smith, J. C., 163*t*, 171–172
Smith, M. A., 159*t*
Smith, S., 128–129
Smit, J. M., 65*t*
Smuga-Otto, K., 357
Smyth, G. K., 285–286
Snapp, E. L., 159–160
Snijder, B., 357–358
Snippert, H. J., 384
Soderling, T. R., 398
Sohn, H. W., 265–290
Solchaga, L. A., 248
Soldini, D., 409
Soldner, F., 357
Solovieva, E. A., 202, 366
So, P. T., 376–377
Sorensen, P. H., 365–366
Sotomayor, M., 4–5
Soucy, G., 118–120, 122, 126–127, 128–129
Sougrat, R., 78–79, 82, 199, 376
Souslova, E. A., 366
Spaink, H. P., 138*t*, 426
Spector, M., 256
Spielmann, T., 81–92
Spitsbergen, J. M., 138*t*
Spriggs, D., 177–178
Spyropoulos, D. D., 334
Squire, J. A., 314, 315, 317, 319
Squirrell, J. M., 37
Srinivas, S., 364*t*, 365–366, 368–369, 382
Srivastava, T., 398
Staats, J., 334
Stacey, A., 180
Stachura, D. L., 138*t*
Stack, M. S., 176, 180
Stahlberg, H., 6–7
Stainier, D. Y., 426
Stanley, E. F., 288
Stanley, E. R., 200–201
Stark, D. A., 369–371
Staroverov, D. B., 136–137
Steele-Mortimer, O., 94, 101–104, 110
Stefanatos, R. K., 408–409
Stegh, A., 313
Steinbach, P. A., 96*t*, 103, 199–200, 202
Steinert, A. F., 256
Stein, J. V., 136
Stein, O., 6–7
Stelzer, E. H., 375–376
Stern, E. A., 392–393
Stern, J. H., 312–313
Stern, Y., 241
Stevens, J. K., 392
Stewart, D., 177–178
Stewart, K., 250
Stewart, M. D., 364*t*, 368–369
Stewart, M. J., 29
Stewart, R., 357
Sticco, C., 344–345
Stiles, C. D., 313
Stins, M. F., 94, 101–103, 106–107
Stockner, T., 273
Stommel, J. M., 313
Stone, M. R., 392–393, 397
Storm, T., 139–140, 426
Stoter, M., 352–355
Stover, M. L., 250–251
Strathdee, K. E., 408–409
Stricker, S., 314, 315, 317, 319
Strickler, J. H., 37, 202
Stroud, D., 43–44
Struhl, K., 103
Strukova, L., 366
Sturm, A., 82
Su, 163*t*
Subach, O. M., 366
Subramaniam, S., 4
Sugimoto, N., 207–208
Sugimura, K., 407–413
Sugiura, T., 202
Sullivan, C. J., 10–11
Sullivan, J., 325
Sullivan, K. F., 366–368
Sulston, J. E., 295–296
Sumen, C., 451
Sun, F.-X., 199–200, 211
Sun, M., 25, 27
Sun, Y., 313, 317, 319
Su, Q., 314
Surfus, J. C., 138*t*
Sur, M., 392–393
Sutton, A., 94
Sutton, T., 36, 37, 39–40, 52–53
Suzuki, K. G., 267–268, 451–452
Svendsen, C. N., 357
Svoboda, K., 37, 371, 392–393
Swarup, V., 118–119
Swedlow, J. R., 181
Swift, J. L., 110
Szal, S. E., 416

T

Taheri, T., 20
Takahashi, K., 357
Takano, T., 37
Takeda, J., 363–365, 364*t*, 367*t*, 368–369
Takehara, K., 52–53
Takemoto-Kimura, S., 397
Takino, K., 407–413
Taksir, T., 202

Talbot, P., 416
Talman, A. M., 20
Tam, P. P., 362, 369–371
Tam, P. Y., 416
Tanabe, K., 357
Tanaka, J., 370*t*, 371–372
Tanaka, N., 209
Tang, H. L., 442, 444, 452
Tang, Y. L., 248
Tanhuanpää, K., 397
Tanner, G. A., 36, 37, 41*f*, 46*f*, 49*f*, 50, 51, 53–54, 53*f*, 56, 57*f*, 59
Tan, P., 384
Tan, S. B., 177–178
Tao, H., 160–161
Taranova, O., 312
Tarin, D., 208
Tarkowska, A., 344–345
Tashiro, A., 392–393
Tasic, B., 365–366, 368–369
Tavares, J., 21–23
Tegtmeier, C. L., 241
Temple, S., 312–313
Teng, S., 426
Tenreiro, S., 158, 159–160, 165–166, 168, 171
Terasaki, M., 314
Terskikh, A. V., 136–137, 384
Tester, A. M., 176, 177–178, 185
Tetzlaff, J. E., 158, 159–160, 165–166
Tetzlaff, W., 312–313
Thelen, M., 136
Thevenaz, P., 347
Thiberge, S., 20–23, 24–25, 82, 292–293
Thibodeaux, J. K., 416
Thisse, B., 137, 426
Thisse, C., 137, 426
Thompson, M. A., 426
Thompson, S. M., 392–393, 397
Thomsen, P., 96*t*
Thomson, J. A., 357
Thomson, N. H., 8–9, 11, 13
Thomson, R. J., 285–286
Tian, G. F., 37, 363–365, 366–368
Tian, S., 357
Tilkins, M. L., 340
Tilley, L., 82, 83
Timson, D. J., 163*t*
Tinevez, J.-Y., 291–309
Ting, L. M., 30–31
Tinkle, C. L., 185
Tinsley, J., 314, 315
Toba, S., 180
Toh, B.-H., 101–103, 110
Tolar, P., 267, 268–269, 274, 281–282, 286, 288
Toma, C., 248
Toma, I., 52–53
Tombe, T., 365–366
Tome, Y., 207–208
Tomita, K., 199, 201, 203, 204, 207–209, 210, 212, 217–218
Tomlinson, S., 313
Tomoda, K., 357
Tomooka, Y., 416
Tompkins, N., 10–11
Tomura, M., 370*t*, 371–372
Toselli, M., 313, 317, 319
Touhami, A., 11, 13
Trager, W., 85, 87
Tran Cao, H. S., 206, 213
Traver, D., 138*t*, 426
Treanor, B. L., 286
Trede, N. S., 426
Trichas, G., 364*t*, 365–366, 368–369
Trifillis, A. L., 43–44
Trimble, W. S., 398
Trotta, V., 409
Trudeau, L. E., 158–159
Trudgett, A., 163*t*
Trushell, D. M., 139–140, 426, 432*f*
Tsai, F. Y., 180
Tsang, T. E., 369–371
Tsien, R. Y., 96*t*, 103, 199–200, 202, 337, 365–366
Tsigelny, I., 158
Tsuang, D. W., 159*t*
Tsuchiya, H., 199, 201, 203, 204, 207–209, 210, 212, 217–218
Tsujii, H., 369–371
Tsuji, K., 203
Tsukamoto, A., 199
Tsuzuki, Y., 37, 52–53, 202
Tsvetkov, A. S., 338
Tuck, D., 366–368
Tucker, B., 426
Tung, C. H., 179
Turmaine, M., 159*t*
Turner, R. D., 4, 8–9, 11, 13
Tybulewicz, V. L., 286
Tyers, M., 314
Tyler, D. M., 408, 409

U

Udan, R. S., 362
Uhrig, J. F., 163*t*
Ulanovski, D., 241
Umezu, T., 416
Ungefehr, J., 82
Unser, M., 347
Urata, Y., 209
Urquidi, V., 208

V

Valdivia, R. H., 96*t*, 199–200
Valente, G., 313
Vale, R. D., 282

Vallance, B. A., 94, 103–104
van den Berg, B. M., 101–103
van den Born, M., 384
Vandenbroucke-Grauls, C. M., 426
Vanderberg, J. P., 20–23, 24–25, 29, 82
van der Ende-Metselaar, H., 65*t*
van der Flier, L. G., 384
van der Sar, A. M., 138*t*, 426
van der Schaar, H. M., 65*t*
van de Sand, C., 82
van Deurs, B., 96*t*
van Eeden, F. J., 426
van Es, J. H., 384
van Furth, R., 101–103
van Kempen, L. C., 185
VanLeeuwen, J.-E., 392
van Munster, E. B., 233
van Rheenen, J., 176, 384
van Rooijen, N., 267
VanSaun, M. N., 180–181
van't Wout, J. W., 101–103
van Zandbergen, G., 24
Varma, S., 138*t*
Varshavsky, A., 158–159
Vasiljeva, O., 180, 185
Venkataraman, S., 10–11
Verbelen, C., 11–13
Verdugo, P., 416
Verkhusha, V. V., 202, 366, 371–372, 384
Vermot, J., 307
Vescovi, A., 314
Vickers, R., 314, 315
Victora, G. D., 441–442
Victor, B. C., 175–195
Vidal, M., 408–409
Villadsen, R., 184
Villalon, L., 416
Viotti, M., 363–366, 364*t*
Visnjic, D., 250
Vodyanik, M. A., 357
Vojnovic, B., 228, 230–231, 232, 234, 235
von Andrian, U. H., 176, 440–441, 451
Vonderheit, A., 67–68
von Recum, H., 251
Voza, T., 21

W

Wacker, S. A., 369–371
Wada, M., 160, 163*t*
Wadsworth, S. J., 180–181
Wagner, M., 294, 301–302, 305
Wahl, G. M., 201, 209, 210, 366–368
Wakamiya, M., 369–371
Wakitani, S., 254
Walker, C., 140–141
Walker, W. F., 74–75
Wallbank, R., 313
Walmsley, S. R., 426
Walters, K. B., 138*t*, 426
Walter, T., 320, 372
Wandinger-Ness, A., 110
Wang, E., 40
Wang, H. W., 138*t*, 185
Wang, J.-W., 159*t*, 199–200
Wang, K. M., 11–13
Wang, X., 37, 199–200, 371–372, 426
Wang, Y., 94, 200–201
Wang, Y. P., 163*t*
Wang, Y. T., 398
Wanli Liu., 265–290
Wanner, B. L., 39
Ward, A. C., 138*t*, 139–140, 426
Ward, R. J., 314
Ward, T. H., 93–113
Ward, W. W., 199–200
Warren, J. W., 37, 43–44
Warren, R. L., 94, 185
Wartel, M., 25, 27
Wass, C. A., 94
Watanabe, S., 392
Waters, C. W., 201
Watson, D. K., 334
Way, M., 137, 426
Wearne, S. L., 393, 398
Weaver, V. M., 186
Webb, W. W., 37, 202
Weber, M., 267, 286
Wedlich-Söldner, R., 438
Weghuber, J., 273
Wehrend, A., 419–420, 421–422
Weigelt, B., 186–187
Wei, H. P., 163*t*
Weimer, R. M., 371
Weinberg, K. I., 179
Weinberg, R. J., 397
Weingart, C. L., 94, 101–103
Weinstein, B. M., 138*t*
Weisenhorn, A. L., 5–6
Wei, S. H., 25–26, 438, 444
Weiss, A. A., 94, 101–103
Weiss, J., 313
Weissleder, R., 179
Weissman, I., 384
Weissman, S. J., 37
Weissman, T. A., 384
Weiss, S. J., 179, 184, 185, 186, 312–313
Wei, W., 398
Welch, R. A., 43–44
Wells, K. S., 179, 180–181
Welstler, V. M., 199–200
Welter, J., 247–263
Weng, Y. C., 118–120, 121, 122, 124–125, 126–127, 128–130
Weninger, W., 186
Wen, Z., 426, 432*f*, 433
Werb, Z., 177, 185

Wernig, M., 357
Westerfield, M., 140–141, 427
West, J. L., 365–366
Wheeler, D. B., 315, 316*f*, 321–322
Whelan, S. P., 65
Whitehouse, W. C., 82
White, J. G., 37
Whittaker, G. R., 94, 110
Whyte, M. K., 138*t*, 139–140, 426, 432*f*
Wiedenmann, J., 369–371
Wienands, J., 266
Wildenhain, J., 314
Wiles, S., 103
Willatt, L., 313, 317
Willbold, D., 158–159
Willey, S., 364*t*
Williams, T. D., 52–53
Willison, L. D., 451–452
Wilschut, J., 65*t*
Wilson, S. M., 200–201
Winston, S. E., 99–101
Winter, S., 227, 242, 244, 244*f*, 245
Winther, H., 241
Wiseman, P. W., 65*t*
Wittamer, V., 138*t*
Wittbrodt, J., 375–376
Wittmann, C., 139–140, 426
Wohrer, S., 365–366
Wokosin, D. L., 37, 438–440, 444, 446, 447–448, 451
Wolf, K., 176
Wong, J. S., 338
Woods, D. F., 408–409
Woolfrey, K. M., 392
Worbs, T., 438
Wren, B. W., 103
Wrobel, C. N., 184, 187–188
Wu, B., 366
Wu, C. C., 13
Wu, D., 439
Wu, H., 180
Wu, J. Y., 338
Wu, L. F., 352–355
Wu, W., 54
Wu, Y. I., 180
Wu, Z. X., 384
Wyckoff, J. B., 200–201

X

Xavier, R., 202
Xenopoulos, P., 361–389
Xia, M. H., 332–333
Xiong, Y., 376–377
Xu, J., 426, 432*f*, 433
Xu, L., 37, 52–53, 202
Xu, M., 199, 201, 203, 208–209, 210, 212
Xu, T., 408–409, 411, 411*f*
Xu, Y., 314–315, 451–452

Y

Yamada, K. M., 176
Yamagata, K., 364*t*
Yamamoto, H., 52–53
Yamamoto, N., 199, 201, 203, 204, 208–209, 210, 212
Yamamoto, T., 52–53
Yamanaka, S., 357
Yamanaka, Y., 365–366
Yamaoka, H., 199
Yamashita, H., 6–7
Yamauchi, K., 199, 201, 203, 204, 207–208, 209, 210, 212
Yamauchi, L. M., 21–23
Yamauchi, M., 186
Yang, F., 199
Yang, L., 11
Yang, M. M., 199–200, 201, 202–203, 204–206, 207, 208–209, 210–211, 212, 213, 219–220
Yang, S., 384
Yan, S., 176
Yasuda, R., 37
Yasui, H., 185
Yasumatsu, N., 392–393
Yates, B. P., 180–181
Yeo, G. W., 357
Ying, Q. L., 313, 317, 319
Yiu, G., 41–42
Yoo, J., 95–97
Yoon, R., 65*t*
Yoo, S. K., 138*t*, 426
Yoshida, N., 370*t*, 371–372
Yoshikai, Y., 120
Yoshikawa, K., 314, 315, 317, 319
Youssef, S., 27
Youvan, D. C., 199–200
Yuan, F., 202
Yuan, J. X. J., 54
Yu, C. E., 159*t*
Yu, D., 357
Yue, F., 251
Yu, H., 186
Yu, J., 357
Yuste, R., 41–42, 392–393
Yu, W., 176

Z

Zapata, A., 426
Zaraisky, A. G., 202, 366, 384
Zdolsek, J. M., 294, 301–302, 305, 306–307
Zelmer, A., 103
Zervas, M., 363–365, 369–371
Zhang, F., 65
Zhang, H., 20
Zhang, J. H., 322–323, 332–333, 356
Zhang, P., 364*t*
Zhang, Q. Y., 185

Zhang, W., 314
Zhang, X. E., 163*t*
Zhang, Y., 138*t*
Zhang, Z. P., 160, 163*t*
Zhao, J. P., 357
Zhao, M., 209
Zhao, W., 124–125
Zhao, X., 384
Zheng, W., 332–333
Zheng, Y., 160, 163*t*, 376–377
Zhong, Z. D., 179
Zhou, Y. F., 163*t*
Zhuang, X., 64–65, 65*t*, 66, 76, 78–79, 376
Zhu, H., 138*t*
Zhu, J., 138*t*
Zhu, K. Y., 138*t*
Zhu, L., 128–129
Zhuo, N., 408, 409
Zhu, Q., 121–122
Zhu, X., 159*t*
Zimmerman, L. B., 312
Zimmermann, B., 201
Zinselmeyer, B. H., 426, 438–440, 444, 446, 447–448, 451
Ziosi, M., 409
Zolotukhin, S., 199–200
Zon, L. I., 138*t*, 426
Zwijsen, A., 363–365

Subject Index

Note: Page numbers followed by "*f*" indicate figures, and "*t*" indicate tables.

A

ABPs. *See* Activity-based probes
Acousto-optical tunable filter (AOTF)
  and control box, 276
  and gas lasers, 275–276, 275*f*
Activity-based probes (ABPs), 180
AFM. *See* Atomic force microscopy
AGC kinases, 227
Ag-presenting cells (APCs)
  in vitro, 267
  in vivo, 267
  replacement, 267
Alkaline phosphatase (ALP), 248–249
ALP. *See* Alkaline phosphatase
Angiogenesis
  dual-color cancer cells expression, 203
  red fluorescent protein, 202–203
  two-photon excitation and emission spectra, 201–202
Anticoagulant therapy, 53–54
AOTF. *See* Acousto-optical tunable filter
APCs. *See* Ag-presenting cells
ART. *See* Assisted reproduction
Aspergillus fumigatus, 9–10, 9*f*
Assisted reproduction (ART), 414
Atomic force microscopy (AFM)
  dynamic process imaging
    A. fumigatus conidia germination, 11
    HS-AFM, 13
    mechanical immobilization, 11–13
    Mycobacterium tuberculosis cells, 11–13
  HS-AFM, 14
  imaging, underlying layers, 14
  microbial cell substructures elucidation
    DNA transfer, *E.coli*, 13
    peptidoglycan, 13
  nanoscale surface exploration, microbes, 4
  parts and setup, 5*f*
  principle, 5–6
  tips, 6

B

B cell receptors (BCRs) signaling
  Ag binds, 267–268
  APCs, 267
  biochemical approaches, 266
  cell preparation
    B cell lines, 269
    human/mouse primary B cells, 268–269
    labeling B cell surface receptors, 269–270
  image acquisition and analysis
    cluster dynamics, 282–286
    oligomerization, 281–282
    signaling, 286–288
  MIRR, 268
  oligomers, 267–268
  planar lipid bilayers, Ags
    glass items cleans, 270–271
    PLBs, 272–274
    SUVs preparation, 271–272
  PLBs, 267
  role, 266
  TIRF microscope design
    ancillary equipment, 279–280
    excitation light source, 274–276
    filters, mirrors and lenses, 277–279
    illuminator port, 276–277
    image acquisition hardware and software, 280–281
B cell–T cell interactions, 450
BiFC. *See* Bimolecular fluorescence complementation
Bimolecular fluorescence complementation (BiFC)
  creation, platform
    interaction, 165–166
    multicolor, 167
    PCR products, 166–167
    plasmids and Venus halves, 166–167, 166*f*
  fluorescent protein reporters, 162–165, 163*t*
  fundamental, 159–160, 160*f*
  GFP, 165
  incubation, cells, 162
  luciferase, 160–161
  multicolor, limiting, 162–165
  optimization
    background fluorescence, 166*f*, 171
    interaction, 170
    signal, 171
    transfection efficiency, 170
    Venus protein, 171–172
    versions, protein of interest, 170–171
  PPI pairs, 160
  testing
    determination, qualitative changes, 168
    immunoblotting, 168–169
    a priori, 167–168
    transfected cells, 166*f*, 168

BLI imaging
  animal, 256–257
  in vivo osteogenesis assays, 257
  MSCs aggregation, 260*f*
  use, 254
Bodipy-TR-C5-ceramide
  P. falciparum in vitro culture, 86
  Plasmodium berghei, 86
  red blood cells, 83, 84*f*
Brain response to ischemic injury
  bioluminescence imaging, stroke research, 119
  cerebral ischemia
    material/equipment, 124
    surgical procedures, 124–125
  description, stroke, 118–119
  dual reporter mouse models, 119–123
  in vivo bioluminescence imaging, 125–132

C

Caenorhabditis elegans (C. elegans)
  based phototoxicity measurements, 297
  embryogenesis
    cell divisions, 296
    environmental control, 301
    number of nuclei, 299–300
    "phototoxicity threshold", 299–300, 300*f*
    sensitive, 296
    sigmoidal relationship, 299–300
  imaging and phototoxicity, 298–299
  occurs, 295–296
  strain, 298
  transparent, 295–296
Cataclysmic phototoxicity, 302
CBF. *See* Ciliary beat frequency
Cell culture and transfection, HBMEC
  approaches, 97–98
  DNA–lipid complex, 97
  growth medium, 97
  jetPRIME^{TM}, 97
  materials, 95*t*
  microscopy scoring
    materials, 98*t*
    methods, 99*t*
  Western blots
    materials, 99*t*
    methods, 101*t*
Cell tracking analysis
  extraction, parameters, 449
  four-dimensional analysis, 447*f*
  import, image data, 446
  quantification, cell motility, 445–446
  signal intensity, 446
  verification, tracks, 448
  workflow, 446–449
Cell trafficking
  blood vessels, 203–204
  circulating cells and gene transfer, 207–208
  clonality, 209
  dormant cancer cells, 208
  high to low-metastatic osteosarcoma cells, 208
  lymphatic vessels, 204
  surgical navigation, 209–210
  tumor–host interaction, 204–205
  tumor microenvironment, 206–207
Cellular and subcellular labelling
  biomarker, neuronal HCS, 337–338, 338*t*
  cytosolic FP, 339–340
  fluorescent method, 337
  multiplexing fluorescent proteins, 338–339, 339*t*
  neuron lifetimes, FPs advantages *vs.* disadvantages, 338, 339*t*
  signaling peptides, 340
Cellular markers, 76–77
CFA. *See* Complete Freund's adjuvant
CFP. *See* Cyan fluorescent protein
Chamber imaging systems
  INDEC FluorVivo, 218
  Olympus OV100, 218–220
  UVP iBox, 218–219
Ciliary beat frequency (CBF)
  determination, 420*f*
  measurement, 420
Collection and culturing, early mouse embryos
  KSOM microdrop, 379, 380*f*
  postimplantation, 380–381
  preimplantation, 379
  Roller conditions, 381–382
Complete Freund's adjuvant (CFA), 439–440
Computational quantification, fluorescent leukocytes
  behavior, 424
  genetic manipulation, 424
  LUs
    and actual leukocyte numbers, 428–430
    calculation, 428
    image process, 426–428
    imaging, 425–426
  macrophage/neutrophil, *irf8* misexpression, 431
  zebrafish, 424
Cox proportional hazards (CPH)
  compare cell populations, 355–356
  hazard ratio, 356–357
  splines and survival, R packages, 356
CPH. *See* Cox proportional hazards
Cryo-electron microscopy, 4
Cumulative phototoxicity, 301–302
Cyan fluorescent protein (CFP), 206

D

Data acquisition and analysis, single particle tracking
  cellular markers
    fluorescent virions, overlap, 77

functional analysis, 76
time-lapse acquisition, 77
microscopy
cell tracking, 73–74
confocal techniques, 73
epifluorescence, 73
TIRFM, 73
tracking virus particles
algorithms, 75–76
entry process, 74
identification, 74–75
Simian Virus 40, 76
single particle, 74
and troubleshooting steps, 75
viral trajectory
diffusion coefficient, 77
parameters, 77, 78*f*
square displacement plot, 77
transition point, 78
Dendritic spine dynamics, 3D analysis
cLTP at DIV 21, 396
excitatory synapses, 390
imaging conditions
culture and reliability, live-cell, 394*f*
hippocampal neurons, DIV 10–21, 394
TCS SP5 confocal microscope, 393–394
Latrunculin B treatment at DIV 14, 395
morphology categorization, 390, 390*f*
motility
Bitplane's Imaris software, 401
channel volume, 401–402
colocalization, 401
and morphing, head, 390–391
noise and movement, specimen, 400–401
percentage, displaced voxels, 402
subsequent voxel, 399–402, 400*f*
NeuronStudio software, 391
primary cultures and transient transfections, 391–393
quantitative methods
latrunculin B/32 cLTP, 398–399
NeuronStudio sofware, 398
Rayburst algorithm, morphology, 396, 397*f*
1,4-Diazabicyclo[2.2.2] octane (DABCO)
Mowiol contains, 235, 237, 243
photobleaching, 229
Dichroic beam splitters, 278
Diffuse luminescent imaging tomography (DLIT), 129–130
1,2-Dioleoyl-sn-glycero-3-phosphoethanolamine-cap-biotin (DOPE-cap-biotin), 271
Disease-free survival, 245
DLIT. *See* Diffuse luminescent imaging tomography
DMEM. *See* Dulbecco's Modified Eagle Medium
dMyc. *See* Drosophila myc
DOPE-cap-biotin. *See* 1,2-Dioleoyl-sn-glycero-3-phosphoethanolamine-cap-biotin
Drosophila imaginal discs
apico-basal polarity, 406–407
cell competition, 405–406
dmyc, 406
Eiger–JNK signals, 406–407
live imaging
chambers, 407–408
culture, 408
polarized and nonpolarized cells, epithelium, 409
"losers", 405–406
Minute mutants, 406
spatiotemporal analysis, 407
"winners", 405–406
Drosophila myc (dMyc), 406
Dual reporter mouse models
advantages, multireporter approach
expression, 122–123
luciferase light emission, 122–123, 123*f*
protocols, 123
GAP-43/luc/gfp, 120–122
genotyping, 122
TLR2/luc/gfp, 119–120
Dulbecco's Modified Eagle Medium (DMEM), 252, 255, 257

E

EGFR. *See* Epidermal growth factor receptor
Eiger–JNK signals, 406–407
Electro-optical modulator (EOM), 443
Emission filters, 278–279
EOM. *See* Electro-optical modulator
Epidermal growth factor receptor (EGFR)
activation status, 242, 244*f*
expression, 244
pathway, 241
phosphorylation levels, 240
role, 241
two-site FRET assay, phosphorylation, 242–243
Escherichia coli K1 (*E. coli* K1)
colonization, 94
fluorescent, 101–107
HBMEC
cell culture and transfection, 95–101
identification, 94
live imaging, 107–110
Excitation/cleanup filters, 278

F

FAP. *See* Fluorogen-activating protein
FITC. *See* Fluorescein isothiocyanate

FLIM. *See* Fluorescence lifetime imaging microscopy
Fluorescein isothiocyanate (FITC), 101–103
Fluorescence lifetime imaging microscopy (FLIM)
  frequency-domain
    acousto-optic modulator (AOM) modulates, 233
    emission and excitation modulation, 233
    FRET analysis, 233–234
    and high-throughput process, 242
    phase-dependent image, 233
  time domain
    FRET analysis, 232–233
    two-photon, 231–232
Fluorescent *E. coli* K1
  effect, potein production, 105*f*
  protein-expressing
    bacterial growth kinetics, 104*t*
    gentamicin protection assay, 106*t*
    K1 capsule, 106–107
    materials, 104*t*
    transformation, 103
    variations, 103
  surface labeling, FITC
    materials, 102*t*
    methods, 102*t*
Fluorescent labeling
  approaches, 66
  chemical labeling
    aggregation, 69–70
    description, 67–68
    HPV-16 virions, 70–71
    labeling efficiency, analysis, 69
    purification, 69
    reaction buffer, 69
    reactive dye, 68–69
    reagents, 67–68, 68*t*
    virus, 68
  description, 67, 67*f*
  experimental challenge, 66–67
  lipid tracers
    description, 71
    dialkylcarbocyanines, 71, 71*t*
    influenza A virus (strain X31), 72–73
    reactive dye, 72
    virus, 72
Fluorescent proteins (FPs)
  development, 362, 362*f*
  fluorophores, 363
  genetically encoded FPs, live imaging
    labeling and tracking, 366–369
    multispectral, 363–366
    photomodulatable, 369–372
  live cell imaging tools
    confocal-based microscopes, 377
    environmental chamber, 377, 377*f*
    light sheet-based fluorescence microscopes, 375–376
    point laser scanning, 373–374
    slit scanning, 374
    spinning disk-nipkow type, 374–375
    super resolution fluorescence microscopy, 376
    two-and multiphoton microscopy, 376
    types, confocal microscope, 372, 373*t*
  live imaging *ex utero*, 362–363
  methodology
    collecting and culturing, 379–382
    imaging, 382–383
    microscope setup, 378
Fluorogen-activating protein (FAP)
  fluorophores, 180–181
  MMP activity, 180–181
Fluorophore-conjugated dextrans, 40*t*
FluorVivo small-animal imaging system, 213
Förster resonance energy transfer (FRET)
  AGC kinases, 227
  EGFR level measurement
    FLIM, 240
    frequency-domain FLIM and high-throughput process, 242
    molecular targets, cancer therapy, 240–241
    pathway, 241
    phosphorylation, 242–243
    preparation, 241–242
    role, 241
    statistical analysis, 243–245
    tumor micro array preparation, 243–245
  monitor molecular interactions, 226–227
  protein–protein interactions and conformation dynamics
    elemental aspects, 229–234
    heterodimerisation, PKB and PDK1, 236–238
    materials, 227–229
    PDK1, live cells, 239
    protein kinase B (PKB/Akt), 234–236
  signal transduction inhibitors, treatment, 227
FPs. *See* Fluorescent proteins
FPs and live animals
  angiogenesis
    dual-color cancer cells, 203
    red fluorescent protein, 202–203
    two-photon imaging with fluorescent proteins, 201–202
  cell trafficking (*see* Cell trafficking)
  chamber imaging systems (*see* Chamber imaging systems)
  histological techniques
    antiangiogenetic agents, 219
    immunohistochemical staining, 220
    tumor tissue, 219
  in vivo imaging, 199–200
  metastasis *in vivo*, 200–201

methods
GFP retrovirus production, 210
GFP/RFP gene transduction, cancer cells, 211
histone H2B-GFP vector, 211
imaging apparatus, 211–213
RFP retroviral vector production, 210–211
microscopy, 216–217
Olympus MVX10 MacroView, 217–218
stereomicroscopy, 217
technical details
GFP retrovirus production, 214–215
RFP/GFP gene transduction, tumor cell lines, 215
RFP retrovirus production, 213–214
spontaneous metastasis model, 216
two-color experimental metastasis model, 215
transgenic nude mouse
CFP, 206
GFP, 205
RFP, 205–206
FRET. *See* Förster resonance energy transfer
Functional imaging, proteolysis
analysis
cell migration and cell–cell interaction, 181–183, 182*f*
3D and 4D quantitation, 183
live-cell imaging, 181–183
PET, 178
probes, 178–181

G

GAP-43/luc/gfp mouse model
characteristics, 121
fragments, 121–122
plasmids and vectors, 120*t*
GFP. *See* Green fluorescent protein
Glioblastoma neural stem cell (GNS)
and human NS screening, 315–320
live cell imaging, 315
NS cell lines, 313
self-renewal, 314
GNS. *See* Glioblastoma neural stem cell
Green fluorescent protein (GFP)
β-barrel structure, 165
C-terminal fragments, 160, 167
dual-color cancer cells express, 203
expressing tumor blood vessel, 219–220
expression
cell sorting, 250
dual promoter multireporter gene construct, 252*f*, 253*f*
OC and Col1α1 promoter, 250–251
histone H2B-GFP vector, 211
in vivo imaging, 199–200
N-terminal fragments, 165
retrovirus, 210, 214–215
and RFP gene transduction, cancer cells, 211, 215
transgenic nude mouse, 205

H

Hank's balanced salt solution (HBSS), 273, 286
HBMEC. *See* Human brain microvascular endothelial cells
HBSS. *See* Hank's balanced salt solution
HCS. *See* High content cell screening
Head and neck squamous cell carcinoma (HNSCC)
patients, 240
tumor arrays, 242, 243
Heterodimerisation
PDK1, live cells
cell seeding and transfection, 239
data analysis, 239
PKB activation, PDK1
cell seeding and transfection, 237
data analysis, 238
stimulation, fixation and mounting, 237–238
High content cell screening (HCS)
phenotypic information, 314
stem cell biology, 314–315
High-speed AFMs (HS-AFMs), 6–7, 14
HNSCC. *See* Head and neck squamous cell carcinoma
Hours post-fertilization (hpf), 424
Human brain microvascular endothelial cells (HBMEC)
cell culture and transfection
approaches, 97–98
DNA–lipid complex, 97
growth medium, 97
jetPRIME^{TM}, 97
materials, 95*t*
microscopy scoring, 98–99
Western blots, 99–101
E. coli K1 interaction
live imaging, 107*t*
materials, 107*t*
Rab GTPases, 110
time-lapse imaging, 108*t*
Human MSCs (hMSCs)
chondrogenesis assays, *in vitro*, 258–260
in vitro osteogenic induction and col1α1 promoter activity
Col2.3 promoter expression, 255
extracellular matrix mineralization, 256
RT-PCR, 255–256
transduction, 254
in vivo osteogenesis assays
encapsulation, 257
histology, 258

Human MSCs (hMSCs) (*cont.*)
imaging, 257
segmental bone defect model, 257
Human OC (hOC) promoter, 249
Human papillomavirus type 16 (HPV-16) virions labeling, 70–71
HyPer, H_2O_2 levels determination
concentrations, 148
embryos, 151
filters, 148
genetic-encoded sensors, 148
in vivo visualization, leukocyte behavior, 152*f*, 153
MetaMorph journal, 148–153, 149*f*, 151*f*
ratiometric properties, 153
signal, 152

I

IHC. *See* Immunohistochemistry
Image acquisition
data storage and retrieval system
cell ID and structure, 346, 346*f*
hierarchical folder structure, 345
open source management programs, 345
Pipeline Pilot, 345–346
wild-type (WT) cells, 346
plate management
bar-coding devices, 343
liquid-handling workstations, 342
transporting robots, 342
Image analysis pipeline
algorithm, replicate wells, 352–353
different population levels, 352–353
modularity
cell tracking, 351
feature analysis, 352
preprocessing workflow, 347, 348*f*
registration, 348
segmentation, 348, 350*f*
multiple levels, 352–353
Imaging early mouse embryos
embryonic drift or movement, 382–383
fluctuations, gas flow and temperature, 383
risk, photodamage, 382
Imaging protein oligomerization, neurodegeneration
aggregates, misfolded proteins, 158
equipment, 161*t*
fundamentals, BiFC, 159–160, 160*f*
immunoblotting, 158–159
localization and composition, protein aggregation, 158, 159*t*
materials, 161*t*
PPIs, 160
protocol, 162–172
Immobilization, microbial cells
chemical attachment
cationic surface coatings, 10–11
covalent bonding, 11
gelatin matrix, 10
drying microbes, 7
isolated cell wall, 6–7
physical entrapment, live cell imaging
advantages, 8
agar gel, 10
Aspergillus fumigatus, 9–10, 9*f*
drawbacks, 8
polymer membranes, 7
porous membranes, 8–9
Immunohistochemistry (IHC)
EGFR pathway, 241
patient prognosis, 240
postimaging analysis, 57–58
Immunotyrosine activation motifs (ITAMs), 266, 267–268
Influenza A virus (strain X31) labeling, 72–73
Intratubular micropuncture, UPEC
blue and green fluorescent, use, 48
infusion, 47
injection, intravenous, 48
Leitz micromanipulator and mercury leveling bulb, use, 47
microperfusion, 48, 48*f*
time points, 48, 49*f*
In vivo bioluminescence imaging
chemical reaction, 125–126, 125*f*
ectopic/saturated signals, 131–132
impact, mouse fur color, 130–131
luciferin kinetic controls, 128
necrotic and hypoxic tissue, 132
preparation, luciferin solution
D-luciferin, 126
material, 126
neuroinflammation, 127
planning, 127
procedure, 126–127
protocol, 2D
D-luciferin, 128–129
ROI measurements, 128–129
protocol, 3D reconstruction
bioluminescent signals, 130, 131*f*
DLIT, 129–130
emissions, 130
luciferase detection, 129–130
structured light image, mouse, 130
TLR2 reporter mice, 128
In vivo models, parasite transmission
arthropod vector, 21–22
Leishmania species, 24
mice, 20–21
mosquito, 20–22, 22f
Plasmodium species, 23–24
sporozoites, 22–24
In vivo real-time visualization, zebrafish
HyPer, H_2O_2 levels determination, 148–153

leukocyte behavior, 137–143
macrophages, 136
multichannel fluorescence imaging
aberrations, 147
acquisition software, 147
acquisition speed, 143–145
automatic focus, 147–148
control, temperature, 145–146
data management, 147
gas and humidity, 146
microscopes and camera/detector types, 143, 144*t*, 145*t*
numerical aperture (NA) objectives, 146
stage movement, 147
'sticky water', 146
transgenic lines, 136–137, 138*t*
triggers, 136
ITAMs. *See* Immunotyrosine activation motifs

K

Kymograph, 27

L

Lenti-viral constructs, dual promoter reporter gene vector
imaging chondrogenic differentiation, 253–254
imaging osteogenic differentiation, 251–253
Leukocyte behavior visualization, zebrafish
anatomical features, 137
3 dpf embryos, 139
protocol, caudal fin amputation
method, 140*t*
required materials, 140*t*
tail fin transection, 139–140, 139*f*
time-lapse imaging, 137–139
wounded embryos
agarose mounting, 142*t*
fluorescent macrophages and neutrophils, 140–141, 141*f*
preparation, 141*t*
required materials, 141*t*
Leukocyte units (LUs)
and actual leukocyte numbers, 428–430
calculation, 428
imaging, 425–426
process, 426–428
Lymph node explants, live imaging, 442
Lymphocyte migration, two-photon microscopy
adoptive transfer and immunization
anti-CD62L antibody, 438–439
CFA, 439–440
HEL-OVA preparation, 440
polyclonal B cells, 440
pregerminal center B cell migration, 440
purification, B cells, 440
two-photon imaging, 441–442
visualization, B cells, 440, 441*f*
data analysis
B cell-T cell interactions, 450
cell tracking analysis, 445–449
pregerminal center, 449–450
dynamic tissue imaging, 436
imaging preparations and data
image data acquisition, 443–445
lymph node explants, 442
postimaging samples, 445
labeling, terget cells
fluorophores, 437–438
signals, 437
visualization, endogenous immune cells, 437–438
microscopes, 436–437

M

Malaria blood stages, imaging
ambient temperature
Bodipy-TR-C5-ceramide, 86
DAPI, 83, 84*f*
devices, 85
disposables, 85
hemozoin, 84
in vitro culture, 83
parasites, 85
preparation and viewing, 85–86
reagents, 85
red blood cells, 83, 84*f*
cryo-electron tomography, 82
fresh blood samples, 82
P. falciparum, 86–91
MAME models. *See* Mammary architecture and microenvironment engineering models
Mammary architecture and microenvironment engineering (MAME) models
cell-based 3d models, 186
in vitro 3D/4D model, 186–187
labeled collagens, 188
monolayer culture, 188–189
tripartite cocultures, 183*f*, 187–188
Matrix metalloproteinases (MMPs)
activity, FAP, 180–181
inhibitors (MMPIs), 177–178
Matrix metalloproteinases inhbitors (MMPIs)
breast cancer, 177–178
roles, proteases, 177–178
Mean square displacement (MSD), 282
Mesenchymal stem cells (MSCs)
in vivo imaging methods, 249
lenti-viral constructs
imaging chondrogenic differentiation, 253–254
imaging osteogenic differentiation, 251–253
marker genes identification, 250–251
molecular imaging, 249

Mesenchymal stem cells (MSCs) (*cont.*)
nonmurine cell, 250
osteoblasts, 248–249
promoters, 250
reporter genes, 249
tissue repair and regeneration, reporter system
chondrogene assays, *in vitro*, 258–260
in vitro osteogenic induction and col1α1 promoter activity, 254–256
in vivo osteogenesis assays, 256–258
tissue repair functions, 248
Microbial cells analysis
AFM (*see* Atomic force microscopy)
fluorescence techniques, 4
immobilization
imaging, living cells, 7–11
isolated cell wall, 6–7
wall functions, 4
MIRR. *See* Multichain immune recognition receptor
MMPIs. *See* Matrix metalloproteinases inhbitors
MMPs. *See* Matrix metalloproteinases
Moloney murine leukemia virus (M-MLV)
Monomeric red fluorescent protein (mRFP)
EGFP-PDK1, 238*f*
and mCherry, 229, 230–231
PKB conformation dynamics, 236*f*
PKB construct, 228
sensor, 234
Morphogenetic events, early embryo mouse
fusion tags, FPs
3D structure, histone, 366–368, 368*f*
transgenic cell populations, 368–369, 369*f*
multispectral FPs
fluorescent proteins, characteristics, 363–365, 364*t*
GFP, green light emission, 365–366
monomeric version, mKate, 366
mRFP and mCherry, 365–366
red region, Katushka, 366
photoactivatable fluorescent protein, 371, 371*f*
photomodulatable FPs, 369–371, 370*t*
single residue substitution, 371
transgenic mice, 371–372
Motile cells, cellular functions
alignment, 27, 28*f*
fluorescent approaches, 27
in vivo imaging, 28
MPLSM. *See* Multiphoton laser-scanning microscopy
MPM. *See* Multiphoton microscopy
mRFP. *See* Monomeric red fluorescent protein
MSD. *See* Mean square displacement
Multichain immune recognition receptor (MIRR), 268
Multiphoton laser-scanning microscopy (MPLSM), 202
Multiphoton microscopy (MPM), 37

N

ND-GFP. *See* Nestin promoter-driven GFP
Nestin promoter-driven GFP (ND-GFP), 201
Neutral phototoxicity, 301
NIH3T3 cells, 253

O

Olympus OV100 small-animal imaging system, 204, 212–213
Oviduct
digital videomicroscopic system, 415, 416*f*
in vitro cell cultures, 414
live cell imaging
qualitative analysis, 417–418
quantitative analysis, 418–420
pick-up and transport, oocyte, 414
preparation
abdomen, 415
bovine, 415, 417*f*
ciliary beating, 415–416
cow and pig, 415
Delta T-dish, 415, 417*f*
mesosalpinx, 415
mice and cows, 415–416
Sylgard gel, 415

P

Paraformaldehyde (PFA), 228–229
Parasite transmission
cellular functions, motile cells (*see* Motile cells, cellular functions)
dissect cell migration
in vitro and *in vivo* approaches, 27
motile parasites, 25–26
obstacle-to-obstacle distance, 26–27, 26*f*
sporozoites, 26
drug discovery
bioluminescent parasites, 30
short movies, 29–30, 30*f*
sporozoite motility, 29
staining, molecules, 29
in vivo models, 20–24
imaging, 20
microscope
confocal/ two-photon system, 24
two-photon microscopy, 25
wide bandwidth emission filter, 25
Parathyroid hormone (PTH), 256
Particle transport speed (PTS)
analysis, 419*f*
bovine ampulla, 419–420
measurement, 418–419
murine ampulla, 419
PBMCs. *See* Peripheral blood mononuclear cells
PBS. *See* Phosphate-buffered saline
PCR. *See* Polymerase chain reaction

PDGFR. *See* Platelet-derived growth factor receptor
Peripheral blood mononuclear cells (PBMCs), 268–269
PET. *See* Positron emission tomography
PFA. *See* Paraformaldehyde
1-Phenyl-2-thiourea (PTU), 425
Phosphate-buffered saline (PBS), 408
Phosphoinositide (PtdIns)
  phosphorylation, 234, 236–237
  substrate, 234, 236–237
3-Phosphoinositide-dependent protein kinase 1 (PDK1)
  EGFP construct, 227
  heterodimerisation (*see* Heterodimerisation)
Phosphoinositide 3-kinase (PI3-kinase), 234
Photomultiplier tube (PMT), 443
Phototoxicity
  delivery rate, light, 306–307
  vs. illumination modalities, 307
  indirect effects, 293
  light budget, 292–293
  live specimen-based metrology, 304–306
  measurement
    longer image exposure times, 302
    microscope-based imaging systems, 302–304
    microscopy system, 295–302
    quantitative, generic and convenient, 293–295
  multidimensional live cell imaging, 292–293
  photobleaching, 293
PI3-kinase. *See* Phosphoinositide 3-kinase
PKB. *See* Protein kinase B
Planar lipid bilayers (PLBs)
  and Ags
    glass items cleans, 270–271
    PLBs, 272–274
    SUVs preparation, 271–272
  making and tethering antigens, 272–274
Plasmodium falciparum, long-term imaging
  Dendra2 conversion
    materials, 91
    photoconvertible proteins, 90
    UV-laser, 91
  devices and materials, 87
  disposables, 88
  experiment and troubleshooting, evaluation, 90
  media and solutions, 88
  parasites, 87
  preparation
    concanavalin A, 88
    steps, 88–89
  time-lapse imaging, 89
Platelet-derived growth factor receptor (PDGFR), 234
PLBs. *See* Planar lipid bilayers
Polymerase chain reaction (PCR)
  amplified fragments, 120
  col1α1 gene expression, RT-PCR, 255–256
  detection, 122
  products, 252
Positron emission tomography (PET), 178
Pour Schneider's *Drosophila* medium, 408
PPIs. *See* Protein–protein interactions
Primary cultures and transient transfections
  coverslips and dishes preparation, 392
  E17 hippocampal neurons, 392–393
  mouse and rat hippocampal neuron, 392
  neurobasal medium, 392
  spine imaging experiment, 391
Primary neuron cultures
  mouse cortices, dissection
    dissociation medium, 335
    method, 335*t*
    neurobasal medium, 335
    OptiMEM/glucose, 335
    papain solution, 335
Probes, functional imaging
  ABPs, 180
  encounters, 179
  FAP, 180–181
  fluorescently tagged proteins
    collagenases, 180
    quenched fluorescent, 179
  substrates, 178–179
Protein kinase B (PKB)
  and Akt
    cell seeding and transfection, 234–235
    data analysis, 235–236
    stimulation, fixation and mounting, 235
  determination, 227
  EGFP-labeled, 232
  heterodimerisation (*see* Heterodimerisation)
  and mRFP, 228
Protein–protein interactions (PPIs)
  cell processes, 160
  N-terminal fragments, 160
  protein misfolding disorders, 158–159
Proteolysis
  3D/4D models
    cancer therapies, stroma, 185
    cocultures, 185
    fibril formation, 186
    MMPs, 184
    signaling pathways, 184
  functional imaging
    analysis, 181–183
    probes, 178–181
  live-cell imaging, MAME models, 186–189
  MMPs, 177–178
  protease activity
    interactions, 176
    "tumor degradome", 176
Protocol, imaging protein oligomerization
  BiFC
    creation, platform, 165–167

Protocol, imaging protein oligomerization (*cont.*)
optimization, 170–172
testing, 167–170
fluorescent reporter protein
color, limiting, 162–165
GFP protein, 165
temperatures, 162
PtdIns. *See* Phosphoinositide
PTH. *See* Parathyroid hormone
PTS. *See* Particle transport speed
PTU. *See* 1-phenyl-2-thiourea

R

Reactive oxygen species (ROS), 306–307
Real-time intravital imaging, bacterial infections *in vivo*
anesthesia and surgical creation
description, 44
incisions, 45–46
materials requirement, 44–45
presurgery preparation, 45
endothelial lining, integrity, 54
fluorescent probes changes, mucosal lining
exfoliation, 51
integrity, 51
kidney tissue, structure, 41*f*, 50
materials requirement, 50
neutrophil infiltration, 51
fluorophore-conjugated dextran, 51
homeostasis, tissue, 36, 50
host pathogen interactions, 37
infection initiation, UPEC
description, 46
intratubular micropuncture, 47–48
materials requirement, 47
positioning, rodent, 46, 46*f*
retrograde model, acute pyelonephritis, 49
local tissue oxygen tension
description, 54
materials requirement, 54–55
measurement, PO_2, 55–56
surgical preparation, animal, 55
molecular analysis
bacterial genetics, 59
host responses, 58–59
MPM, 37
multiphoton imaging platform
digital image analysis, 42
"homemade" system, 41–42
image capture, 42
inverted system, 42
pathogen and host
culture construction, 43–44
isogenic mutants, 43–44
materials requirement, 43
Sprague-Dawley and Munich-Wister, 44
perfused environment, imaging obstruction
colonization, 56, 57*f*
fluorophore signal, 56
intravital imaging model, 56
platelets and clots, anticoagulant therapy
description, 53–54
multiphoton analysis, UPEC, 53–54, 53*f*
postimaging analysis
description, 57
immunohistochemistry, 57–58
systemic dissemination, 56
tissues visualization
description, 39–40
dextran, 39–40
filtrate flow, 40
fluorescent probes, 39–40
materials requirement, 40
renal vasculature, 40–41, 41*f*
vascular flow estimation
description, 52–53
line-scan method, 52–53, 52*f*
visualization
description, 38
materials requirement, 38
UPEC strains, 39
Receptor tyrosine kinases (RTKs), 234
Red fluorescent protein (RFP)
Discosoma, 202
emission maximum, 202–203
and GFP gene transduction, cancer cells, 211, 215
and nucleus, 203
quantum yields, 202–203
retrovirus production, 210–211, 213–214
transgenic nude mouse, 205–206
Region of interest (ROI), 128–129, 130–132
ROI. *See* Region of interest
ROS. *See* Reactive oxygen species
RTKs. *See* Receptor tyrosine kinases

S

Screening, chemical libraries
automated quantitation
algorithm, incucyte device, 322–323
CellProfiler, 320
cell profiler analyst, 321
cell protusions and filopodia segmentation, 323
correctilluminationapply and calculate model, 325
culture media, 326–327
data analysis, R script, 322
equipment and software, 326
GFP reporter gene, 324–325
image extraction, 321
morphological parameters, 321
phase-contrast images, 320

pipeline, modules parameterization, 320–321
reshape package, 324
score image, 324
SQLite/MySQL database, 324
tracking, cells, 325
Xeon processors, CellProfiler, 323
glioblastoma derived neutral stem cells
astrocytomas, brain cancer, 313
GBM biopsies, 314
HCS, 314
human NS & GNS cells
addition, compounds, 318
cell culture, 317
CyBi-Selma, 319–320
doubling time, 96-well plates, 316–319
immunocytochemistry end-point assays, 318
incucyte imaging platform, 315–316
"N2B27" formulation, WM and CM, 319
preparation, 6×96-well plates, 317
in vivo and *in vitro*
EGF and FGF-2 mitogens, 313
GNS, 313
multipotent differentiation capacity, NS cell culture, 312
NS cell definition, 312
live cell imaging, CellProfiler software, 315, 316*f*
phenotypic information, 315
Screening, primary neurons
data analysis
image software, 353, 354*t*
pipeline image, 347–353
HCS and multivariate data
CPH model, R packages, 355–356
Z-factor, 356–357
HTS workflow, 333, 334*f*
htt levels and neuronal death, 333
in vitro assays measurement, 332–333
image acquisition
data storage and retrieval systems, 344–346
HTS microscopy platform, 343–344
plate management, 341–343
neurodegenerative diseases, 333
sample preparation
cellular and subcellular labeling, 337–340
dissection, mouse cortices, 335–337
outbred, inbred/hybrid mouse, 333–335
transfection, 340–341
structure–function relationships, 332
Small unilamellar vesicle (SUV), 271–272

T

T cell receptor (TCR), 268
Time correlated single photon counting (TCSPC), 231
TIRFM. *See* Total internal reflection fluorescence microscopy
Tissue repair and regeneration
chondrogenesis assays, *in vitro*, 258–260
in vitro osteogenic induction and col1α1 promoter activity
Col2.3 promoter expression, 255
extracellular matrix mineralization hMSCs, 256
hMSCs transduction, 254
RT-PCR, 255–256
in vivo osteogenesis assays
histology, 258
hMSCs encapsulation, 257
imaging, 257
segmental bone defect model, 257
TLR2/luc/gfp mouse model
activation, microglial cells, 119–120
PCR-amplified fragments, 120
plasmids and vectors, 120*t*
Total internal reflection fluorescence microscopy (TIRFM)
advantage, 267
application, 280
availability, 277
challenge, 278
design
ancillary equipment, 279–280
excitation light source, 274–276
filters, mirrors, and lenses, 277–279
illuminator port, 276–277
image acquisition hardware and software, 280–281
principles, 274
system design, 274, 275*f*
"turn-key" system, 274
imaging B cell activation, 284*t*
in-house system, 268
objective lenses, 279
single molecule, 283*f*
Transfection, primary neurons
calcium phosphate, 340
Lipofectamine method, 341
Transgenic nude mouse
CFP, 206
GFP, 205
RFP, 205–206
Tumor micro arrays (TMAs), protein activation
antibodies labels, Cy dyes, 241–242
EGFR measurement, 240
EGFR pathway, 241
frequency-domain FLIM and high-throughput process, 242
molecular targets, cancer therapy, 240–241
preparation, two-site FRET assay, 243–245
statistical analysis, 245
two-site FRET assay, EGFR phosphorylation, 242–243

Two-photon acquisition parameters
image bit number, 444
laser
intensity, 443
wavelength, 443
PMT sensitivity, 443
scan averaging, 443
time interval, 444
z-interval, 444
zooming factor, 444

U

UPEC. *See* Uropathogenic *Escherichia coli*
Uropathogenic *Escherichia coli* (UPEC)
infection initiation
description, 46
intratubular micropuncture, 47–48
materials requirement, 47
positioning, rodent, 46, 46*f*
strains construction, 39

V

Vascular endothelial growth factor (VEGF), 202
VEGF. *See* Vascular endothelial growth factor
Viral trajectory
diffusion coefficient, 77
parameters, 77, 78*f*
square displacement plot, 77
transition point, 78
Virus entry and cellular membrane dynamics
"biological noise", 66
data acquisition and analysis
cellular markers, 76–77
microscopy, 73–74
tracking virus particles, 74–76
viral trajectory, 77–78
description, 64
endocytic internalization mechanisms, 65
fluorescent labeling
approaches, 66
chemical labeling, 67–71
description, 67, 67*f*
experimental challenge, 66–67
lipid tracers, 71–73
motion quantification, 65, 66
single particle tracking, 64–65, 65*t*
visualization, 64

W

Wall-teichoic acids (WTAs), 14
Wounded embryos
agarose mounting, 142*t*
fluorescent macrophages and neutrophils, 140–141, 141*f*
preparation, 141*t*
required materials, 141*t*
WTAs. *See* Wall-teichoic acids

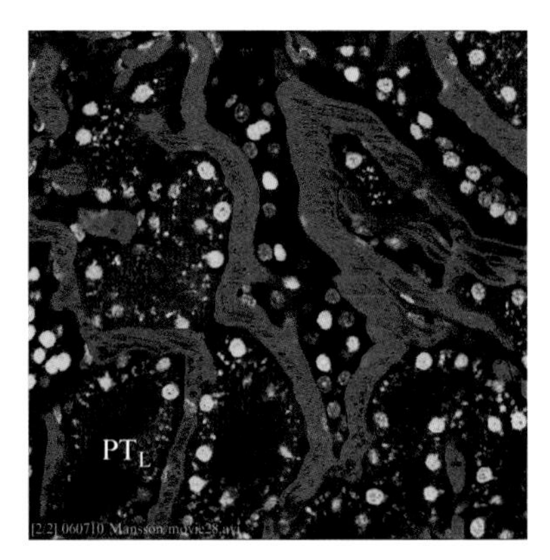

Ferdinand X. Choong *et al.*, Figure 3.1 Fluorescence image of live kidney tissue morphology under multiphoton. Blood plasma (red) is labeled by fluorophore-conjugated 500 kDa dextran, black streaks represent erythrocytes. Epithelial lining of the proximal tubules exhibits autofluorescence and appears dull green, PT_L indicates the lumen of the proximal tubule, and cell nuclei are stained by Hoechst 33342 (blue). Image adapted from Movie S1 in Månsson *et al.* (2007a,b).

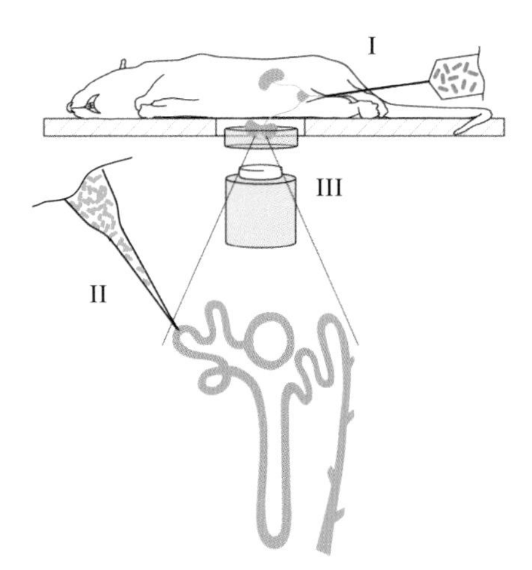

Ferdinand X. Choong *et al.*, Figure 3.2 Graphical depiction showing the positioning of a live rodent on an inverted multiphoton microscope. The left kidney is stabilized in a cell culture dish and immersed in isotonic saline. The microscope stage is heated, and the rat is wrapped in a heating pad (not shown). Green fluorescent bacteria can be introduced either via (I) retrograde infection where bacteria are infused into the bladder or (II) injected by micropuncture directly into the proximal tubule of one nephron (yellow) in an exposed kidney. Infection is imaged by multiphoton microscopy (III). Image adapted from Månsson *et al.* (2007a,b).

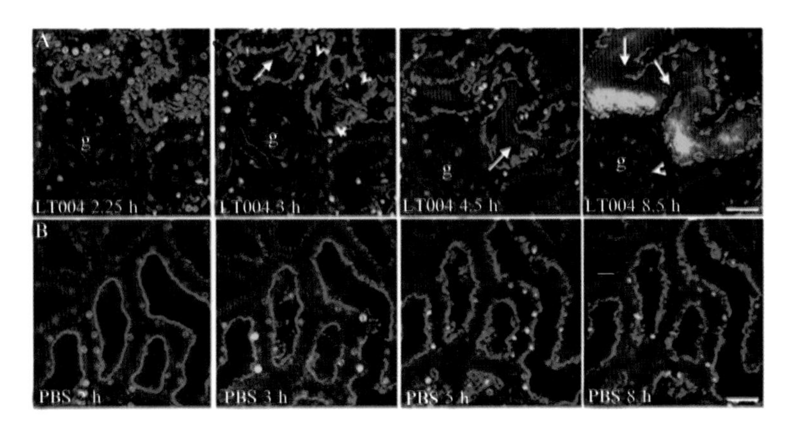

Ferdinand X. Choong *et al.*, Figure 3.4 Selected time points obtained from real-time multiphoton imaging of UPEC infected (A, top panel) and PBS control-injected (B, lower panel) proximal tubules. The injected tubules are outlined by endocytosed blue dextran while tubules in uninfected nephrons exhibit green autofluorescence. In (A), normal blood flow, visualized by infusion of large Mw fluorescently labeled dextran (red), is observed in peritubular and glomerular (g) capillaries 2.25 h after onset of infection. Erythrocytes are seen as black streaks within vessels. At 3 h, initiation of capillary collapse and altered blood flow (arrowheads) occur as a consequence of infection. At 4.5 h, bacteria fill the tubule lumen (arrow). At 8.5 h, with persistent multiplication of the pathogen, fluorescence signal of proximal tubule-specific labeling disappears indicative of epithelial linings disintegration. A single glomerular loop with slowed flow of erythrocytes is shown (arrowhead). In (B), a PBS sham-injected nephron (blue outline) shows no indication of abnormal function. Image adapted from Melican *et al.* (2008).

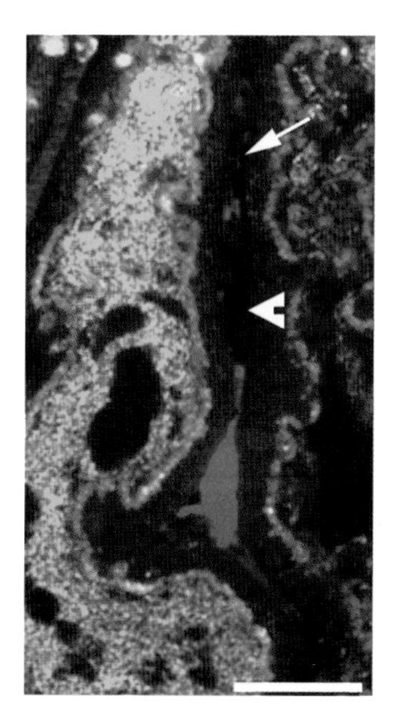

Ferdinand X. Choong *et al.*, Figure 3.6 Images obtained during live multiphoton analysis of an UPEC (strain LT004, green) infected proximal tubule (blue) and adjacent blood vessel (red). A lack of erythrocyte movement is seen in the area. Black silhouettes within vessels are indicative of platelets (arrow). Aggregation of black masses adhering to the vessel wall (arrowhead) suggests platelet aggregates. Image adapted from Melican *et al.* (2008).

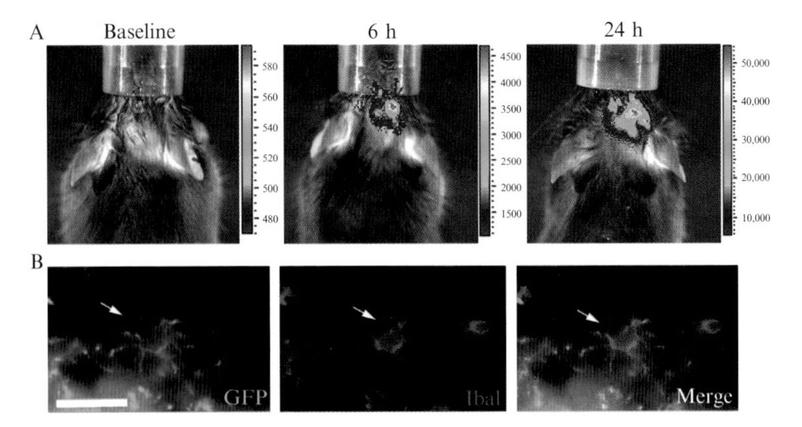

Pierre Cordeau and Jasna Kriz, Figure 7.2 The TLR2-luc/gfp mouse is a bicistronic reporter system in (A) the TLR2 induction after LPS injection in the brain of a male mouse at baseline, 6 and 24 h is measured using the luciferase reporter. While in (B) to achieve microscope resolution and to identify the induction of the transgene in microglial cells using the microglia specific Iba1 marker the GFP reporter was used. Note colocalization of the Iba1 with the TLR2 driven transgene GFP. Scale bar = 25 μm.

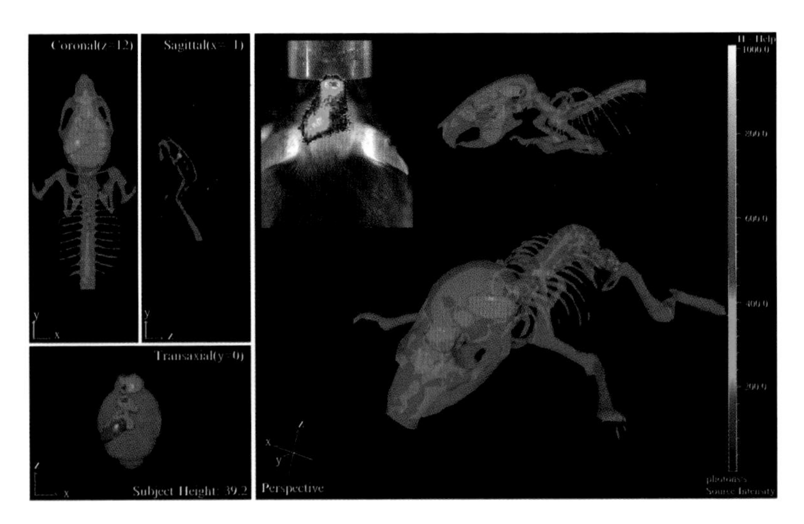

Pierre Cordeau and Jasna Kriz, Figure 7.4 Three-dimensional reconstruction of bioluminescent signals emitted from the brain of a 3-month-old TLR2-luc/gfp mouse 24 h after MCAO. Reconstruction is build using three different wavelengths (600, 620, and 640 nm) across the emission spectrum of the bioluminescent source (Living Image software, Xenogen). Green area, concentrated in the ischemic lesion, represents areas of the brain with the highest photons emission thus the biggest TLR2 activation. Scales bar on the right are the color maps for source intensity in photons per second. Insert image was taken with the A field of view with a 2 min exposition prior to the 3D sequence image.

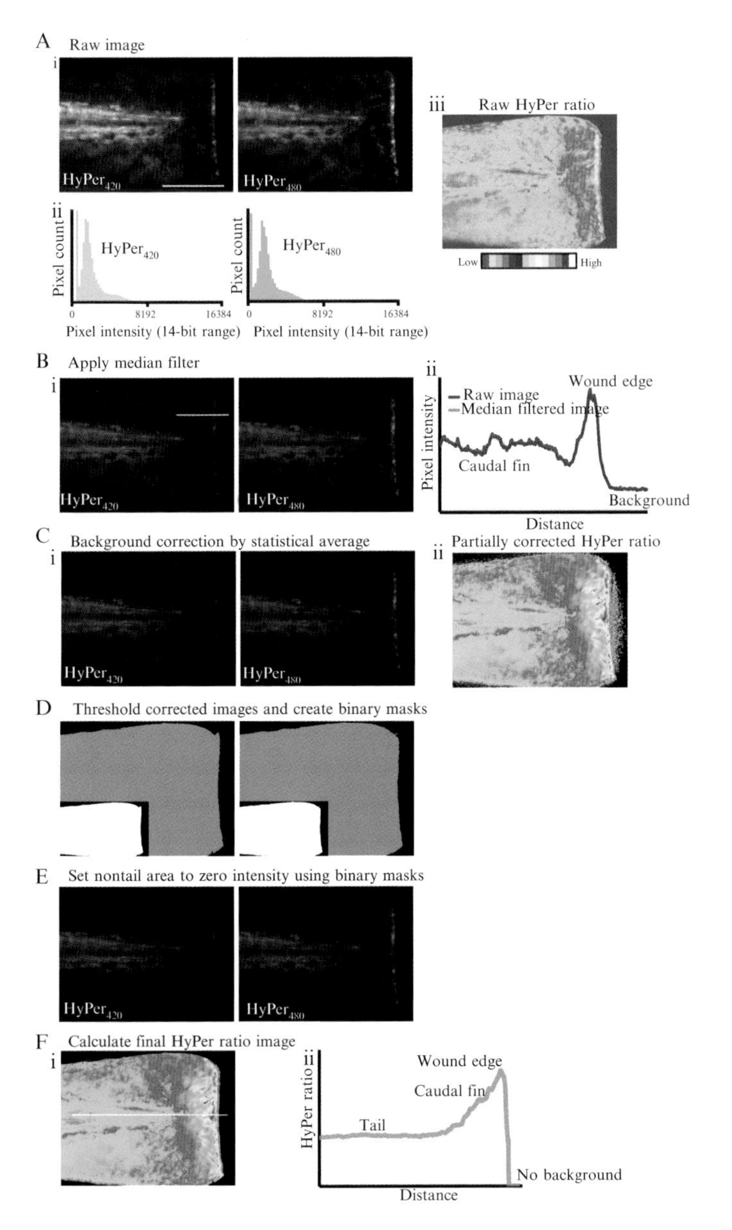

Luke Pase *et al.*, Figure 8.3 A method for processing images to generate specific HyPer ratios reflecting intracellular hydrogen peroxide concentrations. Still images are of a HyPer-expressing 3 days postfertilization zebrafish embryo, 20 min post-caudal fin injury. (A) (i) Acquire a raw HyPer fluorescent images. This example was acquired on a Nikon Ti-E inverted microscope using a (ex420/20, 505LP, em535/20) filter cube ($HyPer_{420}$) and a (ex480/15, 505LP, em535/20) filter cube ($HyPer_{480}$). (ii) Histogram illustrating the distribution of pixel intensities across the 14-bit range for both $HyPer_{420}$

and $HyPer_{480}$ images. Importantly, the settings result in a signal intensity from the embryo tail that provides distinction from the background without there being any saturated pixels. (iii) Raw HyPer ratio ($HyPer_{480}/HyPer_{420}$) represented by a rainbow color spectrum. High HyPer ratio values (i.e., high H_2O_2 concentrations) are represented by yellow and red colors, while low HyPer ratio values (low H_2O_2 levels) are represented by violet and blue colors. Note that background noise generates a ratio value that will be removed by the end of this process. (B) Apply a median filter (5,5,1) to reduce subtle fluctuations in the pixel values that may be amplified in the ratio image. (i) Median filtered $HyPer_{420}$ and $HyPer_{480}$ images. (ii) Graph of the pixel intensity along the white line drawn on the $HyPer_{420}$ image. Note that the small fluctuations between adjacent pixels in the raw image are smoothed out in the median filtered image without affecting the trend in pixel intensity. (C) Correct for background noise. (i) Background corrected $HyPer_{420}$ and $HyPer_{480}$ images, correction performed using the statistical average algorithm. (ii) HyPer ratio generated from background corrected images. Note that correcting for the background noise slightly increased the HyPer ratio values throughout the tail compared to panel (A, iii) (i.e., there are more red assigned pixels). However, there is still background noise in the HyPer ratio image. (D) Remove all background pixel values using a binary mask. (i) Images depicting how the whole tail could be selected using a simple pixel intensity threshold (orange) and the resulting binary images (insert). The binary mask provides a method to extract pixel values within a defined area (white area). (E) Assign zero intensity to background using the binary mask. $HyPer_{420}$ and $HyPer_{480}$ with zero intensity background. Note there are no changes to tail intensity values. (F) Final HyPer ratio image that is specific to the tissue expressing the HyPer fluorophore. (i) Corrected HyPer ratio image. (ii) A graph showing the average HyPer ratio 25 μm either side of the white line in panel (F, i). Note that the HyPer ratio demonstrates a H_2O_2 concentration gradient extending from the wound edge. All images and data were generated using the MetaMorph journal presented in Fig. 8.4, except of the panel (F, ii) which was generated using the MetaMorph line scan function to show average intensities along the user described line. Scale bar (in panel (A, i) applies to all panels): 200 μm.

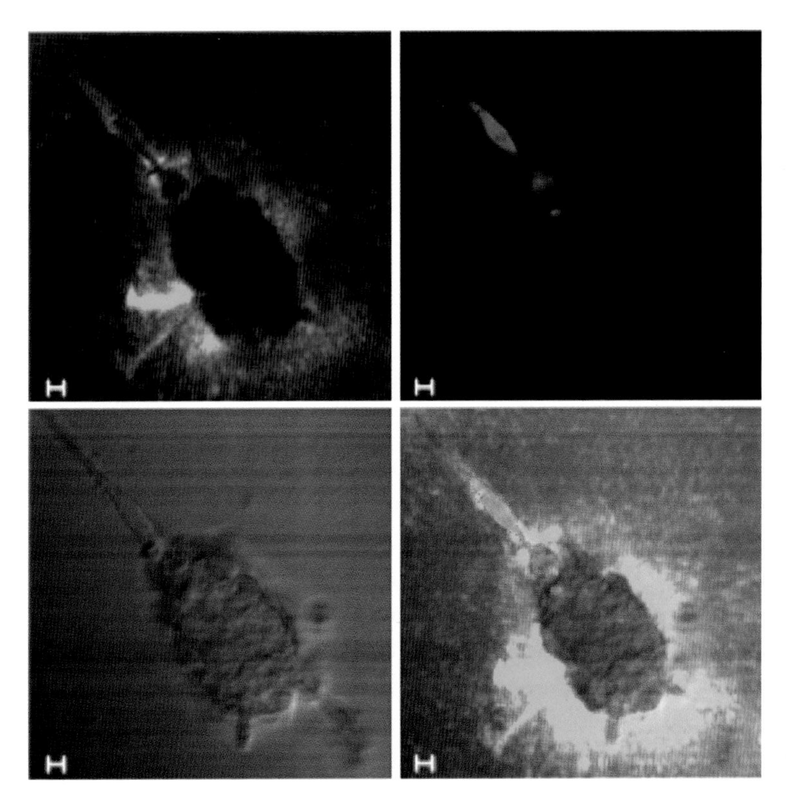

Kamiar Moin *et al.*, Figure 10.1 Still images of time-lapse series (90 min total) to follow degradation of DQ-collagen IV (green fluorescence) by carcinoma cells in coculture with fibroblasts. CCD-112CoN colon fibroblasts (red due to prestaining with CellTracker Orange) were cocultured with a spheroid of HCT 116 colon carcinoma cells. In this image taken after an overnight period of coculture, the fibroblasts can be seen to be moving toward and entering the HCT 116 spheroid. Note the pericellular fluorescent cleavage products due to degradation of DQ-collagen IV by the fibroblasts as they infiltrate into the HCT 116 spheroid, which is more readily apparent in video format. The four panels of this figure represent degradation products of DQ-collagen IV (green), fibroblasts (red), spheroid of HCT 116 cells (DIC), and a composite of the other three images. Bar, 10 μm.

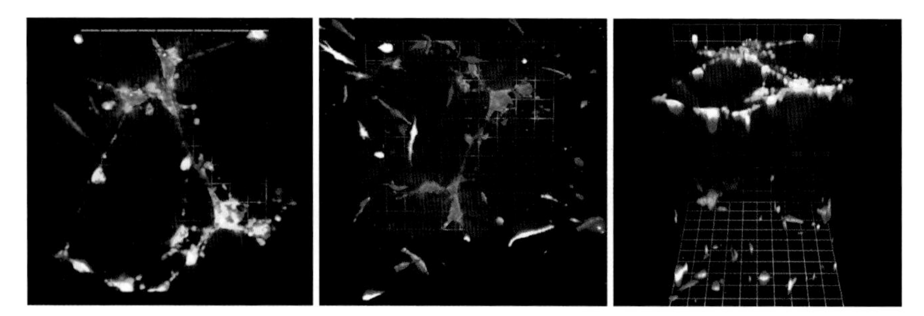

Kamiar Moin *et al.*, Figure 10.2 MAME tripartite cocultures of human SUM102 breast carcinoma cells (red; transduced with Lenti-RFP) with human umbilicial vein endothelial cells (HUVEC; blue; stained with CellTrace Far Red) in reconstituted basement membrane above a bottom layer of WS-12Ti human breast tumor-associated fibroblasts (magenta; transduced with Lenti-YFP) in interstitial collagen I (see also Fig. 10.3A). At 2 days of coculture as illustrated here, the SUM 102 cells cluster around the branching networks formed by the HUVECs. Degradation products of DQ-collagen IV (green) are apparent around the interacting cells. The fibroblasts, which are primarily in the layer of collagen I, are associated with pericellular degradation products of DQ-collagen I (white). A few fibroblasts, also associated with proteolysis, can be seen to have migrated into the upper layer. The three panels from left to right are 3D reconstructions of the MAME tripartite coculture from the top, the bottom and at a 45° angle. Magnification, 20×.

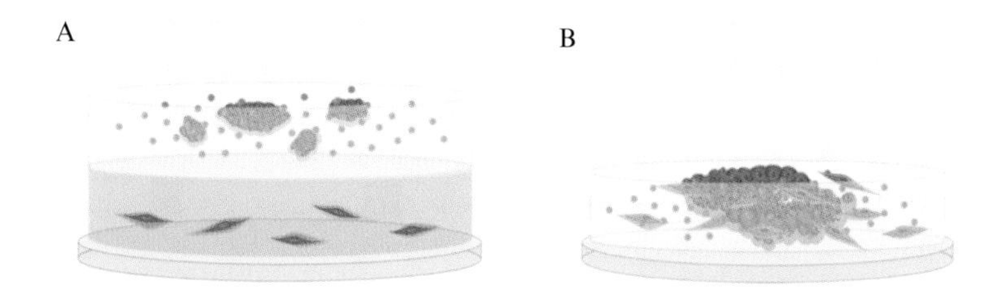

Kamiar Moin *et al.*, Figure 10.3 Schematic of MAME tripartite and mixed cocultures of tumor cells, fibroblasts, and macrophages. (A) Coverslips are coated with collagen I containing DQ-collagen I^{TM} and fibroblasts (elongated red cells). A second layer of rBM containing DQ-collagen IV^{TM} is added and tumor cells (round red cells) plated on top along with macrophages (blue). The cultures are then overlaid with a third layer of 2% rBM, which also is included in subsequent changes of media. (B) Coverslips are coated with rBM containing DQ-collagen IV^{TM} and a mixture of fibroblasts, tumor cells, and macrophages plated on top. Cocultures are imaged live to follow changes in morphogenesis and collagen degradation, depicted here in green.

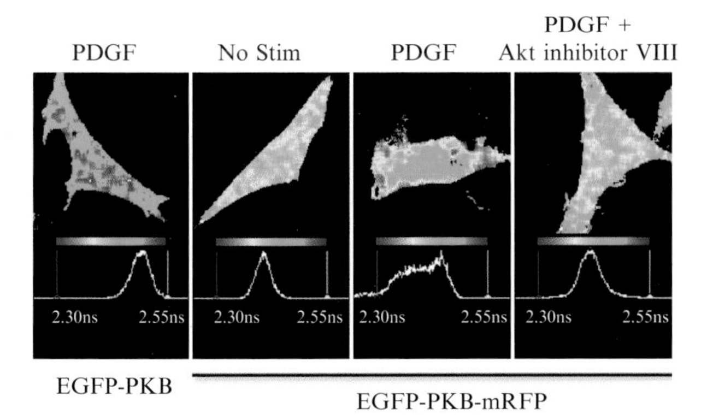

Véronique Calleja *et al.*, Figure 12.1 PKB conformation dynamics are prevented by an allosteric inhibitor Lifetime maps of fixed NIH3T3 cells expressing either EGFP-PKB or our EGFP-PKB-mRFP conformation sensor. The cells are treated as indicated. Akt inhibitor VIII prevents the change in conformation of the PKB sensor upon PDGF stimulation. The lifetime scale ranges from 2.30 (red) to 2.55 ns (blue).

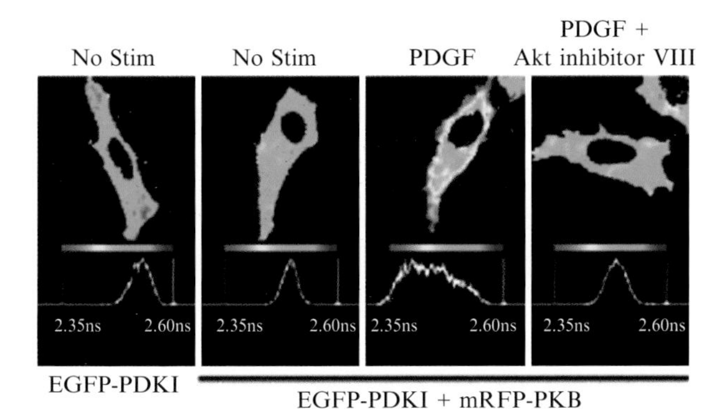

Véronique Calleja *et al.*, Figure 12.2 Inhibition of PKB and PDK1 interaction by an allosteric inhibitor. Lifetime maps of fixed NIH3T3 cells expressing EGFP-PDK1 alone or with mRFP-PKB. The cells are treated as indicated. Akt inhibitor VIII prevents the heterodimerisation of PKB and PDK1 at the plasma membrane upon PDGF stimulation. The basal interaction is not affected. The scale is 2.35 (red) to 2.60 ns (blue).

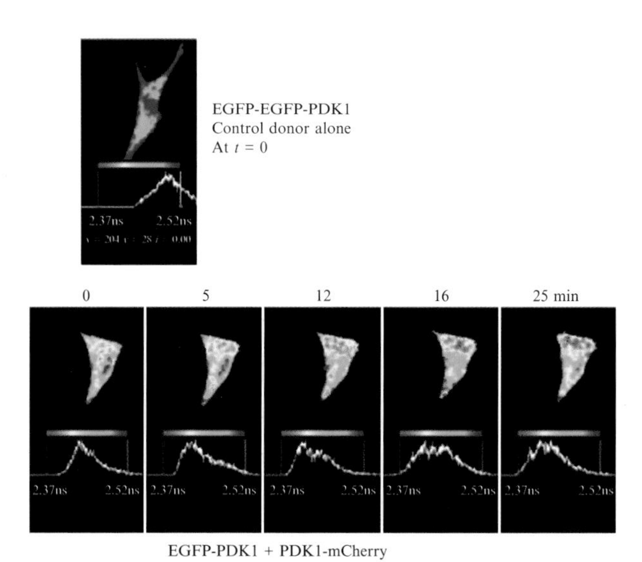

Véronique Calleja *et al.*, Figure 12.3 PDK1 homodimerisation monitored by FRET-two-photon FLIM lifetime maps of live NIH 3T3 cells expressing EGFP-PDK1 alone or with PDK1-mCherry. The time series shows the increase in PDK1 homodimerisation after stimulation with PDGF. The lifetime map of the donor alone is used as a control to show the lifetime distribution when FRET does not occur. The scale is 2.37 (red) to 2.52 ns (blue).

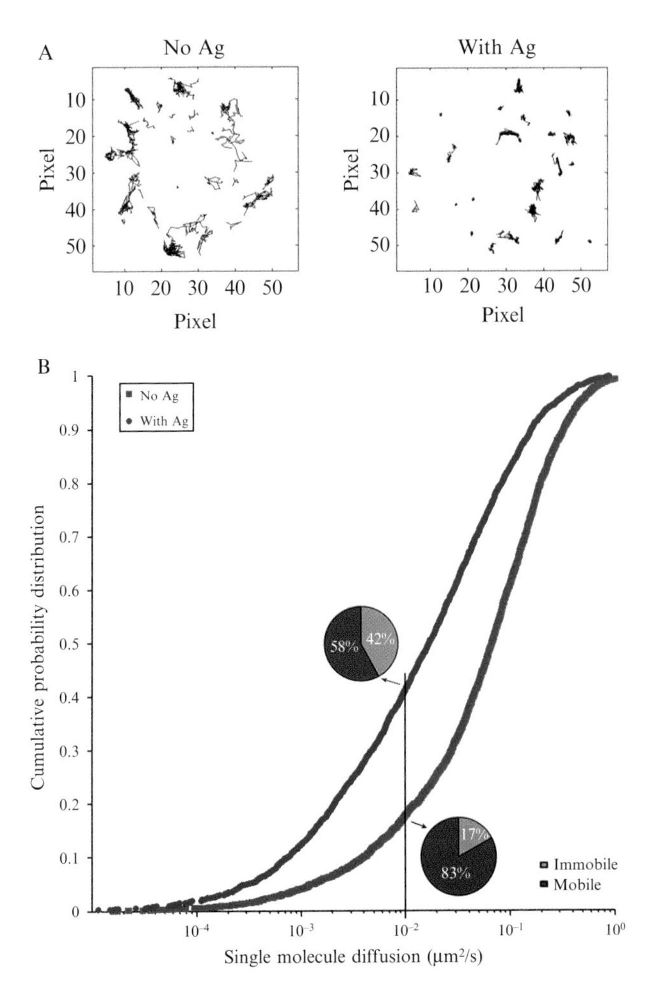

Angel M. Davey *et al.*, Figure 14.2 Imaging BCR oligomerization by single molecule TIRFM. (A) Trajectories of individual BCR molecules accumulated over the entire time course of Supplementary Movie 1. (B) Cumulative probability plots of the diffusion coefficients of individual BCR molecules were obtained from time-lapse TIRF movies (such as those shown in Supplementary Movie 1, http://www.elsevierdirect.com/companions/9780123918574) of human peripheral blood B cells labeled with DyLight 649-Fab anti-IgM and placed on bilayers with (blue curve) or without (red curve) goat anti-human IgM F(ab′)$_2$ Ag. The trajectories used to construct each probability curve were collected from two independent experiments ($n = 1871$ with Ag; $n = 3622$ without Ag). Also given are the percent of mobile and immobile BCRs for cells with and without Ag.

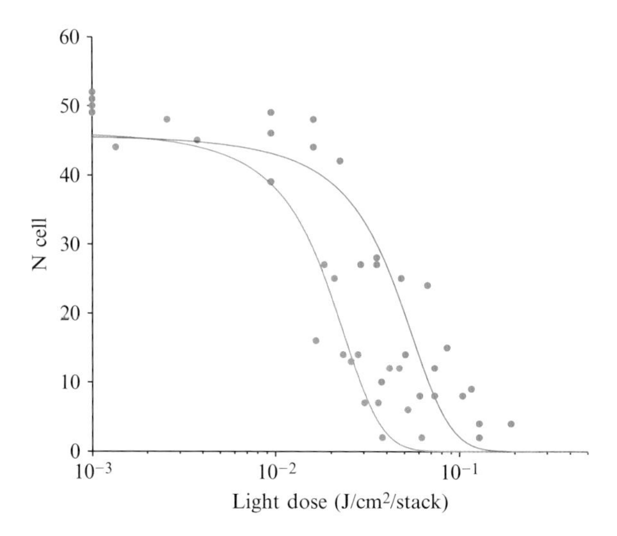

Jean-Yves Tinevez *et al.*, Figure 15.5 Two phototoxicity curves, acquired on the same wide-field system, differing only in the exposure time per slice. Orange: exposure time is 100 ms. Green: 500 ms. The phototoxicity thresholds are, respectively, 1.85×10^{-2} [1.57×10^{-2} to 2.13×10^{-2}] and 4.27×10^{-2} [3.65×10^{-2} to 4.88×10^{-2} J/cm^2/stack (value and (95% confidence interval)].

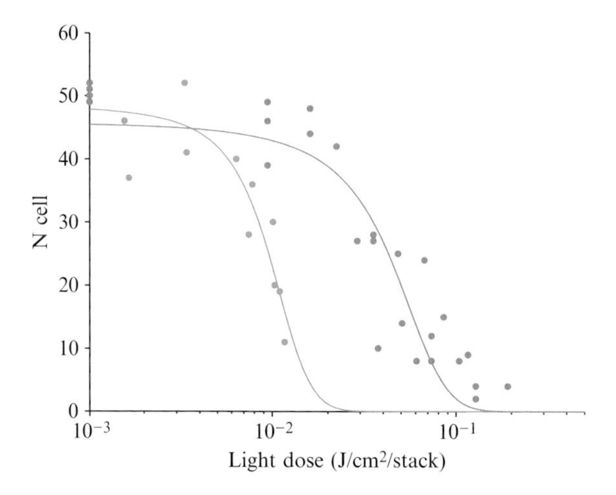

Jean-Yves Tinevez *et al.*, Figure 15.6 Two phototoxicity curves for two different systems using the same objective. Blue: a spinning-disk microscope using a Yokagawa CSU10 head. Green: a wide-field microscope. The phototoxicity thresholds are, respectively, 9.58×10^{-3} [8.58×10^{-3} to 1.06×10^{-2}] and 4.27×10^{-2} [3.65×10^{-2} to 4.88×10^{-2}] J/cm^2/stack.

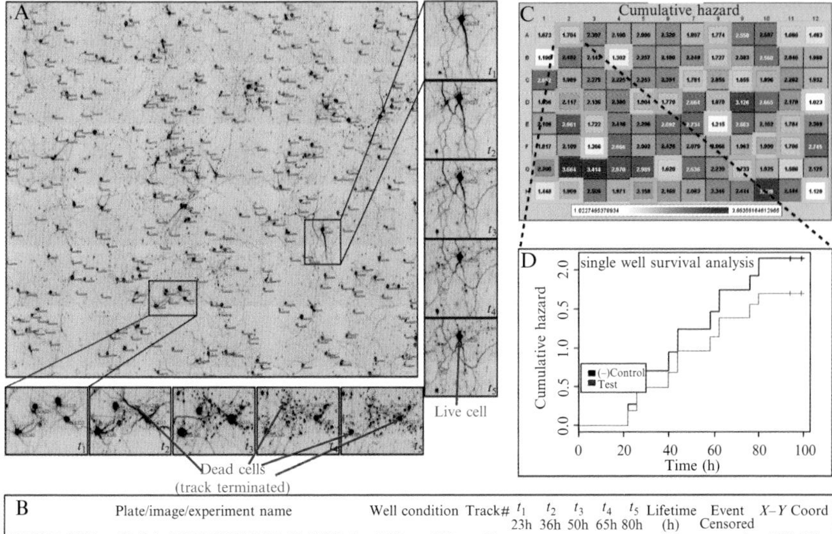

B Plate/image/experiment name	Well condition		Track#	t_1 23h	t_2 36h	t_3 50h	t_4 65h	t_5 80h	Lifetime (h)	Event Censored	X–Y Coord
PID111809_HTTSurv_T1_23_hr-A12_01_FP-RFP_TILE_012.tif.SEQ_aligned	A12	97Q	32	P	P	P	P	P	80	censored	(4270,440)
PID111809_HTTSurv_T1_23_hr-A12_01_FP-RFP_TILE_012.tif.SEQ_aligned	A12	97Q	33	P	P	P	P	P	80	censored	(5339,463)
PID111809_HTTSurv_T1_23_hr-A12_01_FP-RFP_TILE_012.tif.SEQ_aligned	A12	97Q	36	P	P	P	P	P	80	censored	(5552,479)
PID111809_HTTSurv_T1_23_hr-A12_01_FP-RFP_TILE_012.tif.SEQ_aligned	A12	97Q	40	P	P	P			50	uncensored	(3128,619)
PID111809_HTTSurv_T1_23_hr-A12_01_FP-RFP_TILE_012.tif.SEQ_aligned	A12	97Q	41	P	P	P	P	P	80	censored	(678,616)
PID111809_HTTSurv_T1_23_hr-A12_01_FP-RFP_TILE_012.tif.SEQ_aligned	A12	97Q	44	P	P	P	P	P	80	censored	(5725,692)
PID111809_HTTSurv_T1_23_hr-A12_01_FP-RFP_TILE_012.tif.SEQ_aligned	A12	97Q	45	P					23	uncensored	(5452,697)
PID111809_HTTSurv_T1_23_hr-A12_01_FP-RFP_TILE_012.tif.SEQ_aligned	A12	97Q	46	P	P				36	uncensored	(7703,732)
PID111809_HTTSurv_T1_23_hr-A12_01_FP-RFP_TILE_012.tif.SEQ_aligned	A12	97Q	47	P	P	P	P		65	uncensored	(3056,754)

Punita Sharma *et al.*, Figure 17.4 Image segmentation, cell tracking, feature extraction, and data reporting. (A) Shows a typical montaged image showing the first time point after image segmentation. Unique cell track numbers are shown in red next to each cell. Two regions within this image have been zoomed in to show cell lifetime detection from the entire time lapse. (B) A typical delimited text file showing the arrangement of data output and feature extraction by the analysis program. (C) Shows a representative heat map of a control plate where a known modifier (blue outlined wells) was spotted throughout the plate in the midst of positive control wells (green outlined wells). White shaded wells indicate longer survival and dark red shaded wells indicate reduced survival. The numbers represent the cumulative death in each well at the end of the experiment. (D) A survival curve from a modifier well showing decreased hazard (increased survival) compared to the negative control wells.

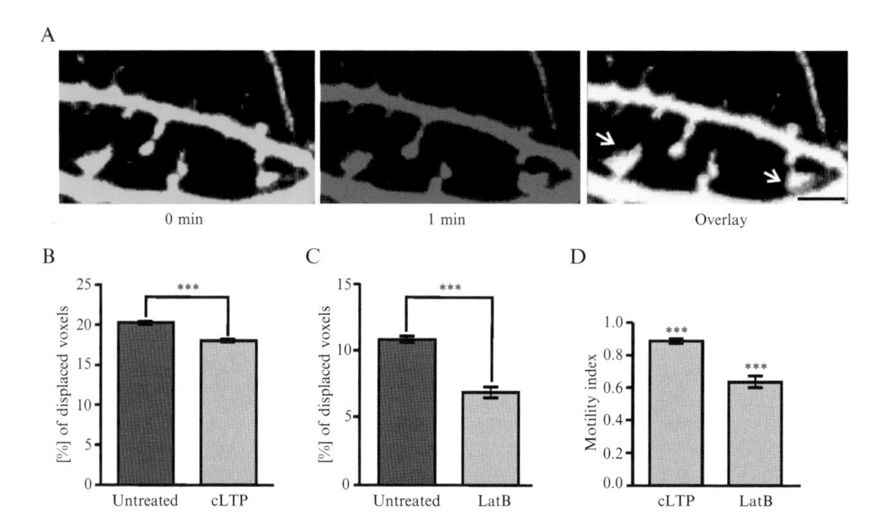

Enni Bertling *et al.*, Figure 19.4 Spine motility analysis. (A) A representative image of a dendritic branch with spines scanned at two time points with 1-min intervals. An overlay panel shows stable areas in yellow and motile areas in either red or green. Note spine heads morphing marked with arrows. Scale bar, 2 μm. (B) Quantification of cLTP experiment. (C) Quantification of latrunculin B experiment. (D) Spine motility index for cLTP and latrunculin B experiments. Bars are means ± SEM; $^{***}p < 0.001$ (B and C—Mann–Whitney test, D—one sample t-test).

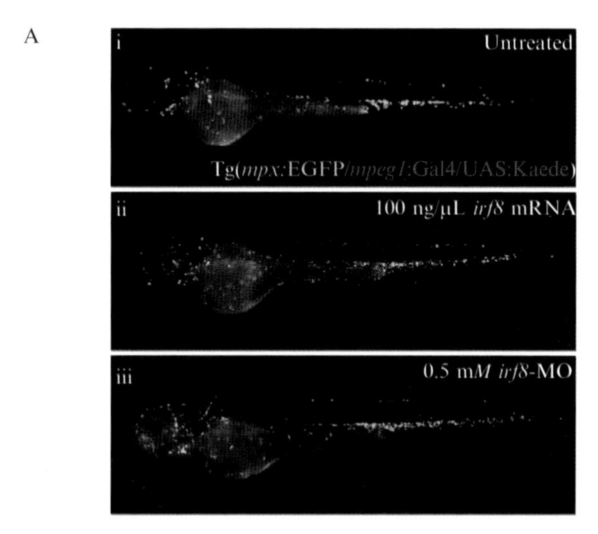

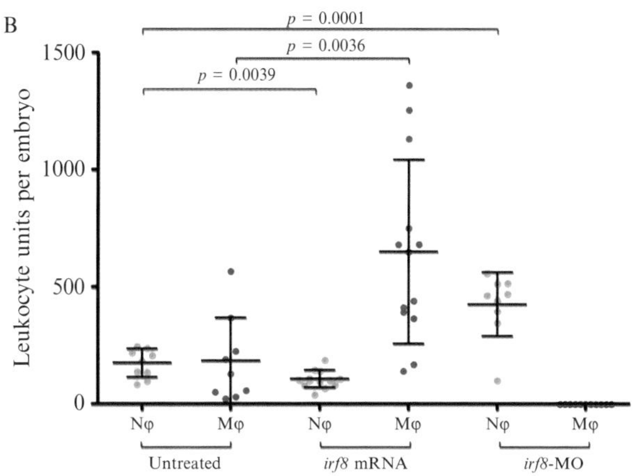

Felix Ellett and Graham J. Lieschke, Figure 22.2 Using leukocyte units to document changes in leukocyte populations resulting from *irf8* misexpression. (A) Images demonstrating the altered balance of macrophage/neutrophil specification following *irf8* overexpression and knockdown in Tg(*mpeg1*:Gal4/UAS:Kaede/*mpx*:EGFP) embryos. In this compound transgenic line, macrophages fluoresce red from photoconverted Kaede with its lineage specificity conferred by the *mpeg1* promoter (Ellett *et al.*, 2011) and neutrophils fluoresce green from an EGFP transgene driven by the *mpx* promoter (Renshaw *et al.*, 2006). This arrangement allows for comparison not only between treatment groups but also between cell types. Compared to untreated groups (i), knockdown of *irf8* results in loss of macrophage specification and increased granulopoiesis in *irf8*-MO treated embryos (ii), while overexpression by *irf8* mRNA treatment leads to increased macrophage specification and reduced granulopoiesis (iii).

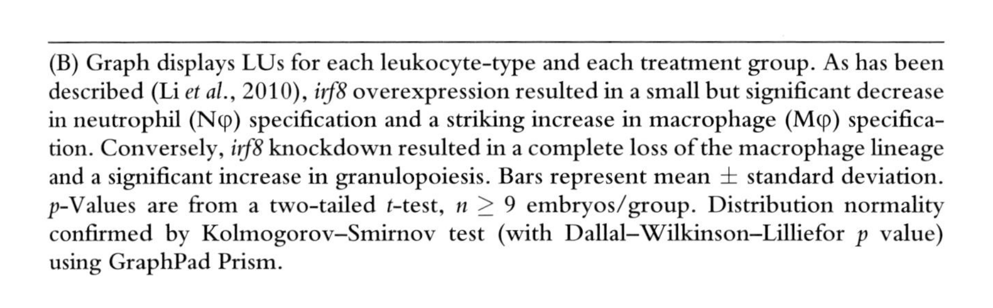

(B) Graph displays LUs for each leukocyte-type and each treatment group. As has been described (Li *et al.*, 2010), *irf8* overexpression resulted in a small but significant decrease in neutrophil (Nφ) specification and a striking increase in macrophage (Mφ) specification. Conversely, *irf8* knockdown resulted in a complete loss of the macrophage lineage and a significant increase in granulopoiesis. Bars represent mean ± standard deviation. p-Values are from a two-tailed t-test, $n \geq 9$ embryos/group. Distribution normality confirmed by Kolmogorov–Smirnov test (with Dallal–Wilkinson–Lilliefor p value) using GraphPad Prism.